MATHEMATICAL MODE
AND NUMERICAL METH
IN CHEMICAL PHYSI
AND MECHANICS

Innovations in Chemical Physics and Mesoscopy

MATHEMATICAL MODELING AND NUMERICAL METHODS IN CHEMICAL PHYSICS AND MECHANICS

Ali V. Aliev, DSc
Olga V. Mishchenkova, PhD

Alexey M. Lipanov, DSc, *Editor*

AAP | APPLE ACADEMIC PRESS

Apple Academic Press Inc.
3333 Mistwell Crescent
Oakville, ON L6L 0A2
Canada

Apple Academic Press Inc.
9 Spinnaker Way
Waretown, NJ 08758
USA

Exclusive worldwide distribution by CRC Press, a member of Taylor & Francis Group
No claim to original U.S. Government works
Printed in the United States of America on acid-free paper

International Standard Book Number-13: 978-1-77188-151-7 (Hardcover)

International Standard Book Number-13: 978-1-77188-290-3 (eBook)

Library and Archives Canada Cataloguing in Publication

Aliev, A. V. (Ali Ve˘isovich), author
Mathematical modeling and numerical methods in chemical physics and mechanics / Ali V. Aliev, DSc, Olga V. Mishchenkova, PhD ; Alexey M. Lipanov, DSc, editor.

(Innovations in chemical physics and mesoscopy)
Includes bibliographical references and index.
Issued in print and electronic formats.
ISBN 978-1-77188-151-7 (hardcover).--ISBN 978-1-77188-290-3 (pdf)
1. Engineering--Mathematical models. 2. Engineering mathematics--Data processing.
3. Numerical analysis--Data processing. I. Mishchenkova, Olga V., author II. Lipanov, A. M. (Alekse˘i Matveevich), editor III. Title. IV. Series: Innovations in chemical physics and mesoscopy

TA342.A45 2016 620.001'518 C2016-901753-2 C2016-901754-0

CIP data on file with US Library of Congress

Apple Academic Press also publishes its books in a variety of electronic formats. Some content that appears in print may not be available in electronic format. For information about Apple Academic Press products, visit our website at **www.appleacademicpress.com** and the CRC Press website at **www.crcpress.com**

About the Series INNOVATIONS IN CHEMICAL PHYSICS AND MESOSCOPY

The Innovations in Chemical Physics and Mesoscopybook series publishes books containing original papers and reviews as well as monographs. These books and monographs report on research developments in the following fields: nanochemistry, mesoscopic physics, computer modeling, and technical engineering, including chemical engineering. The books in this series will prove very useful for academic institutes and industrial sectors round the world interested in advanced research.

BOOKS IN THE SERIES

Multifunctional Materials and Modeling
Editors: Mikhail. A. Korepanov, DSc, and Alexey M. Lipanov, DSc
Reviewers and Advisory Board Members: Gennady E. Zaikov, DSc, and A. K. Haghi, PhD

Mathematical Modeling and Numerical Methods in Chemical Physics and Mechanics
Ali V. Aliev, DSc, and Olga V. Mishchenkova, PhD
Editor: Alexey M. Lipanov, DSc

Applied Mathematical Models and Experimental Approaches in Chemical Science
Editors: Vladimir I. Kodolov, DSc, and Mikhail. A. Korepanov, DSc

CONTENTS

CONTENTS

LIST OF ABBREVIATIONS

AU	actuating unit
CC	combustion chamber
CO	control object
DII	device of instruction input
EO	executive organ
ICD	intensive-converter device
LFD	local feedback device
MNFD	main negative feedback device
PLS	pressure leveling system
PS	pressure sensor
SE	sensing element
SFRE	solid fuel rocket engine
SIMPLE	Semi-Implicit Method for Pressure-Linked

LIST OF SYMBOLS

$(\mathbf{a} \times \mathbf{b})$	scalar product of vectors ***a*** and ***b***
[Q]	dimension of physical variable **Q**
$\mathbf{a} \times \mathbf{b}$	vector product of vectors ***a*** and ***b***
div ***a***	divergence of the vector ***a***
E	total energy (the sum of internal and kinetic energy) continuous medium
F	vector of bulk forces
grad f	gradient of scalar function f
H_1, H_2, H_3	Lame coefficients
k, c_p, c_v	adiabatic index, values of specific heat of gas (liquid)
$q_1, q_2, q_3,$	axes of arbitrary curvilinear coordinate system
t	current time of a process
v	velocity vector of continuous medium
$v_1, v_2, v_3; v_{q1}, v_{q2}, v_{q3}; u, v, w$	components of velocity vector projected on axes
x, y, z	axes of Cartesian coordinate system
ε	source power (by mass and energy returns)
Π, F, W	perimeter of side face of channel and area of its cross-section, volume
ρ, p, T	density, pressure and temperature of continuous medium (gas, liquid, solid)
ρ_m, u_m, H_m (ρ_T, u_T, H_T)	fuel density, its combustion rate and heat content
τ	stress tensor

PREFACE

The rapid development of computer science creates the confidence that a day is near when all existing on the Earth, animate and inanimate, could be represented as a digit combination processed by powerful computers. That is why the number of works associated with mathematical modeling has significantly increased recently. Suffice it to note that the use of mathematical modeling, for instance in engineering, allows us to reduce significantly material costs associated with design, production and operation of technical objects. This is primarily due to the reduction of full-scale tests for benefit of problem-solving using techniques of mathematical modeling.

As a technique of mathematical modeling we will understand the complex of operations and methods to ensure the solution of problems by mathematics. Defined mathematical model of studied object or process is the first stage of the technique. On the next stages one should select mathematical methods ensuring the solution of equations written in a mathematical model, then write an algorithm of problem solving and ensure its computer implementation. Finally, the received calculated results should be compared with any direct or indirect information about the studied object or process, etc.

It can be stated that the current development of mathematical models and their application for problem solving in various areas of science and technology is still a very creative process. There is a limited number of monographs and textbooks that methodically recite the issues of mathematical modeling and methods of mathematical modeling. Moreover, most of works recite just applied model, typical for particular fields without an attempt to make a generalization. The content of this book has been emerged from an experience of the authors in the field of power and rocket engineering. However, a lot of issues stated below (for instance, associated with mathematical modeling) can be useful for those involved in modeling of processes in mechanics and engineering.

One more feature of the book should be noted. A practical use of the technique of mathematical modeling involves computational approach and computers. At the present time, there are lots of worthy monographs, textbooks and tutorials on computational mathematics (they are in the list of recommended literature). However, from the point of view of engineers most of the available literature has one shortcoming. In such literature oriented mainly on students of classical universities studying mathematical and physical courses, all the presented methods and statements are proved strictly. Such description of computational mathematics teaches readers the mathematical culture undoubtedly. However, one can agree that engineers do not need the proof of theorems, they have no doubts about mathematicians' rightness. It is much more important for an engineer to use available computational approaches correctly. Moreover, if to take into account the level of modern computer technology development, an engineer should properly define physical and mathematical problem statement, choose the computational approach and solve the problem by proven reliable computational approach using computer and software applications during the solution of a particular problem. Such approach is quite justified and it is increasingly used in engineering.

During the study of various methods of mathematical modeling in the book, the basic information about mathematical methods of work with these models is given. In addition, the use of such methods applying computers is discussed. All the examples are based on two approaches. The first approach involves the use of relatively simple mathematical system (package) ***MathCad***. The second one involves solution of the tasks using ***Intel Visual Fortran*** compiler with ***IMSL*** library. It is expected that readers are familiar with programming and have minimum knowledge of the above-mentioned mathematical packages and compliers. It should be noted that the use of other software packages (e.g., ***Maple***, ***MathLab***, ***Matematica***, etc.) or compliers (***C***, ***C++***, ***Visual Basic***, etc.) to get the code is equally acceptable in solution of the tasks given in the book, at the same time, their use is comparable with the given examples in the work content.

ABOUT THE AUTHORS

Ali V. Aliev, DSc

Ali V. Aliev, DSc, is a professor at the M.T. Kalashnikov Izhevsk State Technical University in Izhevsk, Russia, where he teaches theory, calculating, and design of rocket engines as well as mathematical modeling in rocket engines. He has published over 300 articles and seven books and hold 25 patents. He has received several awards, including an Honorable Worker of High Education of the Russian Federation and Doctor Honoris Causa of Trenchin University (Slovakia). His research interests include intrachamber processes in solid propellant rocket engines and numerical simulation of processes in technical system.

Olga V. Mishchenkova, PhD

Olga V. Mishchenkova, PhD, is a senior lecturer at the M.T. Kalashnikov Izhevsk State Technical University in Izhevsk, Russia, where she teaches numerical methods and mathematical modeling. She has published over 50 articles and one book. Her research interests include numerical simulation of processes in technical system and mathematical modeling in technology.

ABOUT THE EDITOR AND REVIEWERS

Editor: A. M. Lipanov, DSc

A. M. Lipanov, DSc, is a Professor at the M. T. Kalashnikov Izhevsk State Technical Institute in Izhevsk, Russia, where he teaches on applied fluid mechanics, internal ballistics of rocket engines, and numerical simulation of processes in technical systems. He is senior editor of the journal *Chemical Physics and Mesoscopy* (Russia) and is on the editorial boards of several international journals. He has published over 500 articles and 12 books and holds 65 patents. He has received several awards, including Honorable Scientist of the Russian Federation and Honorable Worker of High Education of the Russian Federation. His research interests include hydrodynamics, turbulence, computational mathematics, intrachamber processes in solid propellant rocket engines, numerical simulation of processes in technical systems, and natural phenomena.

Reviewers:

Igor G. Assovskiy is a Doctor of Science and a Leading Scientist of the Semenov Institute of Chemical Physics, the Russian Academy of Sciences (Moscow City). He is the author of more than 200 scientific publications, including two books. His fields of scientific activity are chemical physics of combustion of energetic materials, interior ballistics of rocket engines, and artillery systems.

Victor I. Sarabiev is a Doctor of Science and a Professor and the Chief of Department of Applied Chemistry at the Scientific Research Institute (Sergiev Posad, Russia). He is the author of more than 150 scientific publications. Honorable Chemist of the Russian Federation, Dr. Sarabiev's fields of scientific activity are chemical physics, chemical kinetic and interior ballistics of rocket engines, and artillery systems.

INTRODUCTION

During research activity a scientist has to deal with the study of complex systems, which includes systems of different areas of social and economic life, science and technology, etc. The study of such systems requires the use of large material resources that are not available all the time. The wish to solve a problem requires its simplification. Such simplification can be achieved when it is possible to neglect certain properties of the system that are not significant at a particular stage of study of this system. It is clear that if some features in properties of complex systems are excluded, then it is a different system, its image. Properties of a simplified system do not fully reproduce properties of the original system. Moreover, in comparison with the original system in a simplified system, new properties, which can be regarded as "parasitic," can emerge. Indeed, there is such danger; however, there are no worthy alternatives to the methods of complex systems study associated with the original problem simplification, that is why, these methods are widely used in the work of a researcher.

The technology of complex system study related to its simplification and change of some properties is called modeling. The simplified system is called a model. Thus, modeling is the replacement process of an examined complex system with a similar simplified system and the simplified system research to obtain the information on the initial system. The model can be examined as "an object-substitute which under certain conditions can replace the object-original reproducing properties and characteristics of the original interesting for us, and also it has important advantages" [172].

Modeling is a tool used in any purposeful activity. For example, a study can be seen as a model of future work, sportsmen training – as a model of participation in the competition, training simulators – as models of real-life situations, etc. The quality of taken assumptions in complex systems simplifying and their quantity can be varied, and therefore, the quantity of the system models, in general, can be large. One can build a well-defined hierarchical structure of models; the main condition in this structure will be the requirement of the adequate description of properties of the examined original system by any model [157].

All the known models, used by a researcher, can be classified according to different criteria and categories. All the models in terms of their "value" in the practical activity of a person can be divided into two classes:

1. Cognitive models are the models, which are the form of organization and representation of knowledge, a tool to link the existing knowledge with the new one. Examples of these models can be models in astronomy, theory of nuclear matter, philosophical theories of the structure and development of the world, etc. The feature of these models is their constant development and wishes to "take up" the reality;
2. Pragmatic models are the models, which are the tool of management or organization of actions. In distinction from the previous type of models, in pragmatic models the reality is simplified. This type of models is basically used in engineering applications.

Both classes of models discussed above can be examined with the other positions. All the models can be divided into classes of physical (real and material) and abstract (ideal) models.

1. Physical models are the models that are produced from the set of material objects (systems). The coincidence of the nature of the original object and model is not necessary.

   If the nature of the original object and model are the same, then this is a direct similarity. Examples are scale models and models of real objects (airplanes, cars, and buildings), pictures and photographs of real objects, etc. At the same time, the nature of the original object and the model can be different, but the behavior patterns are similar. These models are analog or indirect ones. The analog simulation is widely used in the control theory, the theory of heat conduction and other sciences. The study of patterns in these objects is realized, for example, by means of electrical circuits. In particular, the use of electrical circuits can solve differential equations (Cauchy problem and boundary-value problem), systems of nonlinear algebraic equations, describe particle motion in a potential field, examine the kinetics of homogeneous chemical reactions, kinematics of crack gear, solve game problems, solve problems of beams simple bending, problems of linear programming, etc. [123, 219].

We can give other examples of analog simulation. For example, a clock is an analog of time. Experimental animals in medicine are the analogs of a human body. Commands of autopilot on board are analogs of the pilot's behavior, etc. Let us note that some examples, given above, are conditional models sometimes. These models include, for example, money (they reflect cost model), identity card (expresses an official model of owner), maps, etc.

2. Abstract (ideal) models are the models that are constructed by means of thinking and consciousness. Its components are the notions, signs, signals, etc. Linguistic and non-linguistic means, in particular, emotions, intuition, insight, creative thinking, subconscious and other means play a certain role in the construction of the abstract model.

Abstract models can be of several types [213]:

- Epistemological models. These models are concentrated on the study of objective laws of nature. Examples can be models of solar system, biosphere, oceans, models of catastrophic nature phenomena, etc.;
- Information models. These models describe the behavior of object-original, but do not copy it. The description can be, for example, verbal. Examples of information models include the descriptions of phenomena, objects of technology, personality, etc.;
- Sensual models. Examples of sensual models can be models of feelings, emotions, or models that impact on person feelings;
- Conceptual models. This group of models reveals cause-effect relations inherent to the studied object and significant within a given study. The same object can be represented by different conceptual models according to the chosen aim of study. Examples of such models can be pneumatic-hydraulic circuits in rocket systems that display the operation of the system over time, carry out analysis of failure effect on the system operability, etc.;
- Mathematical models. Abstract models, presented usually with the use of mathematical and (or) logical relations, are referred to mathematical models;
- Algorithmic (discrete) models. Models where the trajectory of the change of studied object is planned step-by-step are referred to the algorithmic models. Note that using computational methods, mathematical models can be moved to the algorithmic group;
- Computer models. Computer models are algorithmic models programmed for solutions on digital computers.

We note in addition the following important properties, which are true for the above-mentioned groups of models [172]. Any model tries to describe an infinite set of the existing relations by finite means that makes the use of simplifications necessary. Simplifications are achieved by assumptions and degree of the model approximation to the original object depends on the nature of the assumptions. Depending on the aim set in the beginning of the study, the number and nature of assumptions can be different, but, in any case, the model corresponds to the level of civilization development. Indeed, the construction of any model assumes some knowledge about the studied object, requires the use of specific material and human resources. These factors lead to the conclusion that models predict the behavior of a real object approximately, and as the historical development models are constantly developing and are subject to the certain dynamics. It should be noted that studying complex systems and object mathematical relations anyway are not able to fully reproduce the qualitative and quantitative behavior of the object.

Three last groups of models (mathematical, algorithmic and computer) are of interest for us in what follows. Note that the division of these three groups in some cases is relative. Moreover, all groups of models stated above (epistemological, information, sensual, conceptual) do not preclude the use of mathematical technologies and, in fact, can be classified as mathematical models.

Mathematical modeling of the technical object or processes of its operation involves the construction of a closed mathematical model, adequately within the specified requirements describing the real object or process. The construction of a mathematical model consists of several stages. Particularly, when constructing the mathematical model of a technical object, one should perform the following:

1. Description of an object and characteristics of its operation. The better description is to be performed, the easier mathematical model of the object will be formulated.
2. List of design factors that is relevant in the object analysis, their mathematical description. Errors at this stage of mathematical model formulating get more complicated the problem being solved (for example, when the excessive amount of significant parameters is listed), or make the diverse solution (if any significant parameter is overlooked in the description).

3. List of assumptions, which simplify the problem solution and are relevant to these assumptions restrictions on the basic design parameters. It should be noted that all assumptions should be explained. The explanation should be based either on the existing experimental or computational evidence, or follow the objectives of the problem to be solved.
4. Mathematical relations that characterize the studied technical object. Among the mathematical relations there are systems of partial differential equations, systems of ordinary differential equations, systems of linear and non-linear algebraic equations, algebraic and (or) logical relations, conditions written in the form of equalities and inequalities, etc. At the same time, some fragments can be included in the part of mathematical model; the description of these fragments can be presented in the form of simple table relations that bind together two or more variables.
5. List of additional conditions and restrictions to ensure the uniqueness of the problem solution. Particularly, when writing the differential equations (ordinary and partial derivatives), initial and boundary conditions should be given. As additional conditions, limits to invalid combinations of design parameters can be written, etc.

Mathematical modeling is a tool whose effectiveness is ensured by special technologies. The potential of mathematical modeling technologies continually increases after the growth of the power of applied computers and plays an important role in engineering in creation of new technical objects, forcing out the physical and (or) natural modeling. Increasingly, the results obtained with the mathematical modeling are more accurate and more acceptable in practice than the results obtained in experimental studies.

The construction (formulation) of the mathematical model is the first stage of mathematical modeling. The next stage involves the selection of mathematical methods providing the solution of problems in the mathematical model, and this stage is associated with certain difficulties. First of all, there is no single method to solve any mathematical problem. Another difficulty is that all the problems should be solved simultaneously (together) because all the equations of the mathematical model are interconnected. Taking into account the use of computers for solving problems of mathematical modeling, mathematical methods are based, primarily, on the system of computational mathematics. The final result of this stage is getting a computational algorithm.

Considering the fact that the complexity of mathematical models used in engineering is great and is constantly increasing (it is caused by the need to improve the prediction quality of the technical object actual behavior), then the computational engineering is the only tool for solving the problems set in the mathematical model.

Computational mathematics and computer technology are closely related. The development of computer technologies stimulates the development of computational mathematics, and the development of computational mathematics establishes new opportunities for the development of computer technology architecture. The authorship in the creation of the first computer and first computational algorithms belongs to Charles Babbage, Professor of Mathematics at Cambridge University [118, 209]. He created the computer system of electromechanical type, including the resolver and printing installation, allowing to automate the creation of tables necessary to ship navigators and determined the development of a "decisive" technology for more than a hundred years ahead. The most intense stage of the development of computational mathematics begins in the forties of twentieth century and is connected with the appearance of an electronic digital computer (EDC). The first EDC (computer ENIAC created in the United States of America) was primitive on the modern view; however, its role in the implementation of the U.S. military programs is difficult to overestimate.

The development of computer technologies stimulates the development of computational mathematics. Qualitatively, new computational algorithms and methods, whose use was impossible twenty-thirty years ago, are constantly created. The methods used today can be obsolete in the nearest future. At the same time, it is possible that the methods that are not effective today can suddenly become popular again.

Tendencies in the development of computational methods are related to several factors, some of which are discussed below.

1. Results of the solution of engineering problems obtained using computational methods contain the error, which should be estimated by an engineer. This is due to the fact that the creation of a new technical object or quality analysis of its operation are related to the performance of numerous calculations. Most of them are mathematical operations with approximate numbers. Indeed, an engineer in his

daily work never knows the exact values of quantities with which he has to work. He can estimate their level on the approximate value. The work with approximate numbers is fundamental due to the fact that for the solution of even simple engineering problems on a computer, millions elementary arithmetic operations can be performed. In this regard, the following aspects are important:

- computational algorithms should be constructed in the way to avoid the uncontrolled growth of computational errors;
- in the created algorithms of calculations whose solution is expected by means of computer, you should always tend to reduce the number of arithmetic operations, as with the increase in the number of arithmetic operations, as a rule, the calculation error increases proportionally;
- in the realization of the complex computational algorithms related to the mathematical problems solution, the calculation of values of unknown variables should be supplemented by the calculation of their errors.

2. Engineering calculations are performed with a limited number of significant digits. The number of significant digits is directly related to the accuracy of calculations. So, if the required accuracy of calculations is from 0.1% to 10%, then in calculations 3…5 significant digits should be left. If it is necessary to increase the calculation accuracy, the number of significant digits can be increased. Calculations on computers introduce an additional aspect related to the accuracy of calculations. This aspect is related to the digit capacity of EDC used. We can recall the small "revolution" occurred when replacing computers BESM-6 (machines with the element base of the second generation, made in Russia) by the new generation machines of a single system (ES EVM – Unified System of Electronic Computers). On the new computers many problems (e.g., optimization problems) stopped to provide the previous accuracy of calculations. The reason was simple. The digit capacity of real numbers in ES EVM in calculations with the accuracy by default was only 4 bites (32 binary numbers are used to represent real number in the form 0.000000e+00). At the same time, digit capacity of numbers was almost 2 times higher on BESM-6 computer. The way out was found by the use of real number of double accuracy (64 binary numbers were used to represent numbers). However, the use of real numbers

with double-precision led to the increase of problem solving time. In modern computers (including personal computers) digit capacity can be set by software tools, which are the part of operating system. Modern PCs are 32-bit machines by default (32 binary numbers are used for real numbers). Today 64-bit operating systems are used.

A limited number of significant numbers used in calculations on a computer is only the part of problem. Another problem is that the rounding rules on computers differ from the rounding rules used in approximate calculations. It is correct to say that a computer makes not number rounding but its cutting. The calculation result is always reproduced with a number of significant digits corresponding to the digit capacity of performed calculations. The digits belonging to the significant ones, by means of which to round the calculation results correctly could be possible, simply get lost.

In the literature containing the systematic description of computational algorithms for the different classes of mathematical problems, the listed above problems of non-increase of calculation errors (the problems of stability and convergence of the computational algorithms) are compulsory discussed.

3. Unique computer system ILLIAC – IV was created on the elemental basis of the second generation (transistor element base) in 1972 in the USA [118]. One of the characteristics of the system ILLIAC – IV was the multi-processor architecture (in ILLIAC – IV 256 processors were used). Several processors in the computer allow performing arithmetic or logical operations on each processor at the same time that theoretically improves the system performance directly proportional to the number of the processors used. After the system ILLIAC – IV, multi-processor computers have appeared in all countries of the world, including the USSR (computer systems "Elbrus," "ПС-2000," МВК-100, МВК-1000, etc.) It is hard to overestimate the possibilities offered by multi-processor computer application. Currently, the use of such computer systems provides the solution of the problems of the weather forecasting, nuclear physics, problems of rocket aircraft creation. Without the multi-processor computers the development of nanotechnology is difficult to imagine.

It should be noted that computational methods efficient for solving problems on monoprocessor computers can be useless using the multiprocessor computers. Aspects related to the possibility of computational methods application in the multiprocessor computers will be discussed below.

4. Effectiveness of computational methods application is determined by the applicable software [103]. As a part of operating systems there are some primitive computational means (e.g., program "Calculator" in the operating system Windows). More powerful computational tools are provided by Excel program, which is a part of Microsoft Office software package. Mathematical systems Maple, MathCad, MathLab, Matematica are widely used [64, 86]. However, works connected with serious calculations should be performed using special software. Compliers (or algorithmic programming languages) ALGOL, FORTRAN, PASCAL, PL-1 and others were widespread developed during calculations on the digital computers of the past century [35]. Some of these compliers are functioning successfully at the present time. For example, in complex engineering calculations algorithmic language FORTRAN is indispensable [159, 186]. An advantage of this complier can be the high efficiency of the program code ensuring the computational problem calculation. Usually, the speed work of the program written in FORTRAN language is the highest among all the used compliers, and RAM used in calculations is minimal. There are many compliers FORTRAN. Particularly, we should note compliers Compaq Visual Fortran and Intel Visual Fortran Compiler built-in the environment of development Microsoft Visual C++.net [33]. Another advantage of compliers of algorithmic language FORTRAN is the experience in solving mathematical problems accumulated for the long history of its operation (the first commercial version of the language was created in the middle of XX century). There are many libraries of computational methods, which can be used in a program that use algorithmic language FORTRAN. Particularly, these are software packages IMSL (software company is Visual Numerics), NAG (software company is The Numerical Algorithms Group Ltd, Oxford). The listed packages contain more than one thousand software components that allow solving the problems of computational mathematics and mathematical statistics, including for computers with parallel architecture. Package Visual Array enables the efficient graphic processing of calculation results obtained in FORTRAN environment.

At present, other compliers are being rapidly developed. For example, systems Visual Basic [40, 96], Visual C, C++ [33], Borland Delphi [17, 238], Borland C [73] are of interest. Losing as a resultant program code that solves a computational problem, these compliers provide the rich opportunities in processing the results of calculations that is important in solving engineering problems.

The final stage of works, related to the mathematical modeling implementation, is the analysis of calculation results, which allows us to estimate the correctness of the model creation and perform any necessary corrections. At this stage, the well-known analytical and numerical solutions should be applied to estimate the correctness and accuracy of the models. A special role at this stage is given to comparing the results of numerical analysis with the results of natural (physical) experiments.

CHAPTER 1

MATHEMATICAL MODELING AS A METHOD OF TECHNICAL OBJECTS RESEARCH

CONTENTS

1.1 MATHEMATICAL RELATIONS USED IN MODELS

Mathematical modeling of complex systems can be performed with a different degree of detailing and accuracy. This can be due to the set of requirements to develop models and methods for making these models. It was noted in the introduction that models may include mathematical relations of various complexity. The simplest ones are functional relations presented by tables or in the form of separate analytical relations. More complex functional relations are based on the fundamental laws of physics (mechanics) including the use of phenomenological approaches [9]. Below we discuss examples of various options of the functional relations used.

1. Table relations are the simplest relations used for constructing a mathematical model. Such relations are characterized by a finite set of elements, which can be placed by a hierarchical scheme (structure). A complete mathematical

model of an object or a process may contain the relations presented only in a table form. In this case, a mathematical model can be called a table model. Table models will be referred to mathematical models due to the reason that an effective work with tables can be performed using mathematical techniques. Particularly, it is an application of sorting algorithms, interpolation algorithms and extrapolation algorithms, etc. An arbitrary table element can be characterized by the set of quantitative and (or) logical (qualitative) parameters. And the table dimension is defined not necessarily by the number of parameters characterizing each element of the system.

Databases widely used in various areas of human activity can be referred to table models. Examples can be:

- schedule of passenger transport;
- catalog of books in a library;
- list of finished products of an enterprise;
- catalog of component parts for a technical object;
- library of prototypes of technical objects, etc.

Let us note that the work of data retrieval systems used in different areas of human activity (service, medicine, equipment, etc.) is based on a table representation of a complex object.

At the dawn of computer-aided design (1970–80 years) the consuming optimal design decisions of analytical and design services were tried to get within one working day. Based on the computer technologies available at that time (the best computer of that period was BESM-6 with a speed up to one million operations per second) mathematical models of complex objects and processes should be simplified. One way to simplify is the preliminary serial calculation of these objects and processes and subsequent processing of calculations with the construction of tables with digital information by the basic parameters used in the design. The results of experimental studies of objects or processes are presented in the form of tables.

In the cases discussed above and solving the other practical problems a need for further processing of tabular data may exist. In particular, these can be problems of object behavior forecasting under dynamic conditions. Such problems can be solved only by using interpolation procedures, numerical differentiation, table data integration, etc.

2. Analytical relations are ratios written in the form of algebraic equations, or as a set of algebraic and logical equations. The application of analytical relations as a part of mathematical models does not require the use of complex

calculations, but, at the same time, such relations are not versatile and can be used only in strictly limited ranges of relevant parameters. In particular, analytical relations can be obtained on processing the results of a scientific experiment (regression equation). In some cases, the analytical relations can be obtained with considerable properties simplifications of the studied system (object, phenomenon). For example, writing Newton second law (relations of force $\boldsymbol{F}$ with the material body mass $\boldsymbol{m}$ and its acceleration $\boldsymbol{a}$) in the form $F = ma$ suggests that the physical body can be represented as a material point of mass $\boldsymbol{m} = \boldsymbol{const}$, and body movement occurs along a rectilinear trajectory with acceleration $\boldsymbol{a} = \boldsymbol{const}$.

If an object or a process under study are simple, their mathematical model can contain only analytical relations. In this case, we can say that the model is analytical.

Analytical models are widely used in engineering, and this is due to their relative simplicity. However, it should be noted that the use of analytical models is limited to the narrow limits of changing the parameters in the formulas. These boundaries are due to the assumptions by which the formulas are made, or the conditions of the experiment and its processing.

3. The basis of phenomenological models is models, in which the mathematical description of the individual microscopic or small-scale processes or object sets of macroscopic equations are used.

Phenomenological models are more complex than tabular and analytical ones; however, they have a significant error. The problem solving with the use of such models can be difficult without computer technologies.

The examples of phenomenological models can be:

- representation of chemical reactions in a gas mixture of one brutto-reaction with the given values of kinetic parameters;
- representation of turbulent effects while liquid or gas motions by the models of small-scale and (or) large-scale turbulence;
- representation of the laws of behavior of gas molecules aggregate in the form of gas laws (Mendeleev-Clapeyron), etc.

4. Basis of the fundamental model is models where the system under study (object, phenomenon) is examined as much as possible setting of parts list, taking into account the fundamental laws of physics. The study of the properties of these models is the most time consuming, that is why, now it is possible to give only a small number of problems solved in this statement. Most of these problems are considered for separate, isolated processes.

The examples can be quantum-mechanical calculations of the potential energy surface for molecules, Navier-Stokes equations, turbulence development model, etc.

The above models are special cases. As a rule, modern mathematical models of complex objects and complex processes are combined models, where the mathematical relations in a table form, in the form of analytical relations, equations written using phenomenological approaches and fundamental laws of physics are included. The examples of such models can be found, for instance, in Refs. [195, 244].

The creation of new mathematical models of various objects or technologies and development of the existing mathematical models are closely related to the technology of problems solving with the use of these models. The trends in mathematical modeling are to move from relatively simple models (e.g., analytical) to more complex (phenomenological, fundamental). The stimulating effect in this case is the rapid development of computer technologies.

Indeed, the emergence of computer technologies has allowed people to solve interesting problems in a new way and contributed to the establishment and development of numerical modeling technology [25, 27, 169, 256]. On the basis of the developed physical and mathematical models, the algorithms and software oriented at computing are being created and improved.

The technology to solve engineering problems using numerical modeling is very fruitful and effective, however, in its present form, it has certain disadvantages. Particularly, the developed applications are focused on the solution of narrow class problems. Adapting these packages to meet new challenges, even close in content to the same problems, is associated with considerable difficulties. The performing of additional adaptation procedures may involve new labor and material resources. In addition (which may be even more important), these works can be extended in time. In this connection, it is possible that the development of the new software system morally obsoletes and ends in the completion of the newly designed technical object or process, and the need for developed software disappears. The numerical modeling technology is uncomfortable when studying the physical nature of the processes investigated.

Modern computer technologies and prospects for their further development permit now lead the development of mathematical modeling at a higher level than the numerical modeling. The theory of the new technology of mathematical modeling, called numerical or computational experiment, applied to the problems of fluid mechanics and gas is developed in the works of academics Samarsky [189, 194, 195], Belotserkovsky [25–28], Yanenko [113, 256], Marchuk [144–146] and their scientific schools. The main feature of the computational experiment that distinguishes it from the numerical simulation is the systematic approach to the problem solution. The systematic approach involves the interconnectedness of all stages of problem solving (including the stage of the mathematical model construction, algorithmization, programming and carrying out the calculations on a computer), structuredness and hierarchical construction of models, algorithms and programs, which subordinate the main problem solution. In accordance with the requirement, the computational experiments can be numerical modeling variants that satisfy such conditions in solving complex physical problems:

- calculations are performed for the complex processes in which several physical and (or) chemical phenomena are realized simultaneously;
- mathematical models that form the numerical solution of the problem checked by the set of test problems, comparisons with experimental results and with other known solutions;
- mathematical models algorithmically and software implemented by a set of different techniques, the use of which is equally allowable for solving the problems. The most valuable is the application of methods built on qualitatively different numerical methods and algorithms (for example, to solve the partial differential equations, finite difference and variational methods can be applied, etc.);
- methodical filling should provide the operation of the software package implementing an experiment in a wide range of variable parameters. Partial "overlap" of the applicability of various numerical techniques for the same initial data should be provided;
- structure of the software package to be "open," that is, to ensure the possibility of changes in the subprogram (removal of the old or addition of the new subprograms changed within the software modules);
- users' work with the software providing the computational experiment should be as simplified as possible (input information with the original data, the output results of the calculations, the dialogue with the computer when calculating, etc.);
- in the software package the possibility to flexibly change the physical setting of the individual blocks of the problem should be provided in order to clarify the nature of the physical processes.

The latter requirement is one of the most interesting and it significantly differs the computational experiment from the usual calculation of a complex problem on a computer. A good illustration of the application of this requirement in practice is the work [151], in which the mechanism of separated flows over the finite-dimensional and axisymmetric bodies by the flow of an ideal gas was analyzed. In the cited work, the analysis was carried out by two fundamentally different numerical methods – large particles [28] and discrete vortices [29]. An identity of the results confirms the assumption that the large-scale motion on the inertial range of turbulence viscosity of the gas is negligible. The computational experiment allowed to discover in magneto hydrodynamics solving the problem of plasma interaction with the magnetic field in railotron channel the formation of hot T-layers separated by zones of cold gas [194]. Later experimentalists actually detected T-layer [158]. However, this happened only five years after the discovery was made by computational means.

With the development of the computer power the role of computational experiment is constantly increasing, and this is due to the following factors:

- reducing commercial value of computer operation can significantly reduce the amount of expensive full-scale simulation by increasing the volume of numerical modeling;
- using computational experiment it is possible to obtain the results that are difficult or impossible to get at natural modeling;
- computational experiment provides an opportunity to establish the physical laws in the complex phenomena.

The implementation of computational experiment involves the following stages of work related to each other: the development of physical and mathematical models of the problem being solved; the development of the discrete model; the development of software complex; the operation of the finished complex; the analysis of the calculation results.

The development of the physical model greatly depends on the computer class supposed to be used. So, if in 1980s, using the third-generation computers it was difficult to find the solution to the problem of the release of solid fuel rocket engine (SFRE) to the mode using three-dimensional models of the dynamics of an ideal gas, two-dimensional (and, moreover, three-dimensional) models based on the solution of the Navier-Stokes equations [113, 134, 138, 163] within the reasonable time with the accuracy reasonable for practice, then using the modern computers that have the speed of more than a hundred billion operations per second, the level of "physicality" of computer simulation models is much higher.

Developing the physical model, the basic phenomena occurring in this process should be listed, and also taken simplifying assumptions should be formulated.

The mathematical model presents the record of physical model by a set of mathematical tools – algebraic or differential equations with initial and boundary conditions, etc.

The discrete model development involves the reduction of the mathematical problem formulation to a sequence of arithmetic and logical operations performed by a computer. The development of algorithms for solving mathematical problems is performed using numerical methods. The content of discrete models is largely dependent on the architecture of the computers used.

At the development stage of the software system (computer model) the computational schemes, text programs on algorithmic programming languages are developed, test cases, which will verify the correctness of the programs, are selected and software debug is performed. The software system development involves the use of advanced programming techniques.

For the purposes of the numerical experiment the stage of a finished complex operation should ensure the possibility of some restructuring, allowing the use of the software package for the solution of various problems.

At the stage of analysis of the calculation results the fact about the need to return to any of the previous stages in order to correct a certain fragments of documentation can be concluded.

The congruent set of stages of physical (natural) experiment can be set to the specified sequence of computational experiment steps. This, apparently, was the reason for putting into practice the term "computational experiment." The following important practical fact should further be noted. One of the tests for debugging the software system is to compare the numerical results with the known solutions and natural experiments. The computational experiment cannot replace the natural experiment. It can only reduce the amount of full-scale modeling in favor of increasing the amount of numerical simulation. Also, you should not try to completely match the numerical results with the results of physical experiments. Such outcome should be considered valid only when the mathematical models used will be most appropriate to the real process and will be built without any assumptions. Therefore, the most rational strategy of computational experiment implementation should be considered as the one in which the field of numerical results obtained by varying the initial data with regard to their uncertainty and varying options of physical models embedded in the functional content of the application package are accumulated. The field of numerical results formed can be considered as the forecast of change in the performance of the object during its operation. At the same time, the application of the experimental curves to this field allows to specify the nature of the processes, identify conditions that provide the greatest impact on workflow, reduce the level of uncertainty in a number of original data. This strategy of stages implementation of the finished complex operation and analysis of the calculation results will be called the predictive. It brings the computational experiment ideology to the ideology of simulation maximally [143].

1.2 ACCURACY OF A PROBLEM SOLVING IN THE TECHNOLOGY OF MATHEMATICAL MODELING

1.2.1 ERROR SOURCES IN ENGINEERING PROBLEMS

Carrying out of engineering calculations using mathematical modeling techniques should be accompanied with the evaluation of the accuracy of the results. Indeed, there should be no errors – any mathematical model is only the way of approaching the reality, and therefore, has a particular error. First of all, this is due to the fact that the source of faults is errors determined by the approximate nature of the mathematical model. The same object or process can be described using different models of various complexities. Specifically, for the analysis of processes in the rocket engine combustion chamber one can use a three-dimensional spatial model of the ideal gas flow. Neglecting viscous properties of the gas, in this case, is the source of additional errors. The use of relatively simple mathematical models, in which it is assumed that the parameters for the internal volume of the rocket engine combustion chamber can be averaged, will make the error due to the absence of any information about the speed of the combustion products. Such examples could be any number for technology objects.

Errors associated with the use of a particular mathematical model cannot be eliminated completely. However, it is possible to reduce their number, and it can be provided by reducing the number of assumptions made in the mathematical formulation of the problem being solved. Usually, the most accurate models are the mathematical ones based on the fundamental laws of physics and chemistry. However, this rule is not always true.

Other sources of errors in the calculations are the errors contained in the original information. As an example, the strength analysis of the crankshaft of internal combustion engine can be considered. In the calculations it is necessary to define the diameter of the shaft, the magnitude of the load acting on this shaft, the strength characteristics of the material from which the shaft is formed. None of the above mentioned parameters used in the calculation of the shaft strength can be defined with absolute precision. Thus, the geometric dimensions of the shaft can be defined by preliminary measurements. However, any measurement tool has errors specified in the passport. Strength characteristics of the material from which the shaft is made of are also approximate. This is due to the fact, for example, that the processes do not permit to produce two or more parts with exactly the same structure of material. Consequently, the material properties are variables with statistical character and are approximate numbers. It is not possible to correct the error in the source data, but the use of more accurate measuring instruments can reduce the error in the source data.

Inaccuracy of the original data is a part of the problem. In the calculations additional factors appear that increase the error. The next group of causes that affect the accuracy of calculations is formed by applied numerical methods and computers. The simplest example that confirms this statement can be the calculating of transcendental functions. Any tools used in the computation of logarithms, exponents, trigonometric functions, etc. produce the results with a certain number of significant digits. Even the computation of transcendental functions on computers does not allow getting an accurate result. This is due to the fact that a computer processor only performs the basic arithmetic operations (addition, subtraction, multiplication, division). In connection with this, the calculation of such trigonometric functions as $\sin x$ on a computer or scientific calculator is provided by a formula using the trigonometric function expansion in a Taylor series:

$$\sin x = \sum_{i=1}^{\infty} (-1)^{i+1} \cdot \frac{x^{2i-1}}{(2i-1)!} = x - \frac{x^3}{3!} + \frac{x^5}{5!} - \frac{x^7}{7!} + \ldots$$

Naturally, in the calculations due to the limited capacity of the computational tools used, a row consisting of an infinite number of terms is replaced by the finite number of terms. Such “voluntarism” leads to the increase of computational errors. The calculation error is the greater if the less terms are used in calculating the transcendent function.

Another example can be connected, for example, with the solution of systems of linear equations. Gauss or rotations method can be applied for their solution. In textbooks on computational mathematics it is shown that the method of rotation is substantially more accurate than Gauss method [79, 102]. In large systems of equations (thousands or more unknown) the usage of Gauss method cannot be generally valid because of the increase of calculation errors by decades. Such computational errors, which are associated with applied computational methods, are sometimes called the method errors and they can be reduced.

Let us note another source of errors associated with the fact that any engineering calculations are performed with a limited number of significant digits. Let us recall that a significant figure for the approximate number written in a decimal form is any number other than zero, or zero if it is recorded between the significant figures, or at the end of the number (otherwise – all the digits in the decimal number starting with the first non-zero at the left side). The number of significant digits is associated with the precision of calculations performing. Thus, in engineering at the early stages of design the computational accuracy is small and the error in the calculations can be from 3% up to 10%. Therefore, in the calculation at this stage it is sufficient to use from 3 to 5 significant digits. The replacement of exact number by its approximate value with a limited number of significant digits causes the accumulation of errors during the elementary calculations and it will be shown below.

Errors due to the restriction of the number of significant digits used in the calculations are referred to rounding errors. In engineering practice, while working with approximate numbers, the so-called symmetrical rounding rule is used:

- to round the approximate number to n significant digits all the numbers from the right side of nth significant digit are discarded, while if the discarded figures exceed 5 or 50, then the last n^{th} significant digit is left unchanged, and otherwise it is incremented by one; if the first discarded rounding digit equals 5 and all the other are zeros, then the rounding rule of even number (rounding the last n^{th} significant digit must be even) is used.

1.2.2 ERROR DEFINITION WHEN CALCULATING

Using the technology of mathematical models involves a lot of computation. The processed digital information has initially an error that will be changing (usually increase) during the calculations being performed. Let us consider the basic terms of the theory of approximate calculations, which allows the change in the error magnitude in the elementary mathematical operations [142, 251].

Let us introduce the definitions necessary to understand the material presented below.

Definition 1. The approximate number **a** is the number slightly different from the exact number **A** and replacing it in the calculations.

Engineers in their routine work almost never know the exact quantities of the calculated values. They can judge about their level, in particular, by an approximate value. It seems to be important to estimate errors of the exact magnitude of the calculated value from its known approximate magnitude. In calculations, the error of the approximate number, its absolute error and relative error of the approximate number are separated.

Definition 2. **Δa** error of approximate number **a** is the difference between the exact and approximate values **Δa=A—a.**

Definition 3. Absolute error Δ_a of the approximate number is the absolute value of the difference between its exact and approximate values $-\Delta_a = |\mathbf{a} - \mathbf{A}|$.

Definition 4. Relative error δ of the approximate number **a** is the ratio of the absolute error to the module of the corresponding exact number **A** – $\delta = \frac{\Delta_a}{|\mathbf{A}|}$.

It was noted above that the exact meaning of the number **A** is usually unknown, so it is more convenient to define the relative error from the approximate expression $\delta \approx \frac{\Delta_a}{|\mathbf{a}|}$.

The written definitions allow setting the values of the errors of the approximate numbers corresponding to the rounding procedure. The rounding rules stated above imply that the relative error of the approximate number satisfies the expression $\delta \le \frac{1}{\alpha_m} \cdot 10^{-(n-1)}$. Here a_m – the first significant digit of the approximate number **a**. It can be proved that if the number of significant digits exceeds 2 (***n*>2**), then the estimate for the limit of the relative error d_a of the form $\delta_a \le \frac{1}{2 \cdot \alpha_m} \cdot 10^{-(n-1)}$ is true. Knowing the relative error of the approximate number, it is possible to determine the absolute error and, accordingly, the number Δ of the correct significant digits $-\Delta \leq \frac{a \cdot \delta}{1-\delta}$, $(0 \leq \delta \leq 1)$.

The results of arithmetic operations with approximate numbers are also approximate numbers. It is interesting how the number error changes during the elementary computational procedures.

1. Error of adding (subtracting) the approximate number.

Assume that there are N approximate numbers x_n (lower index is from 1 to N) whose error is Δx_n, the absolute error $-\Delta_{xn}$ and relative error $-\delta_{xn}$. The task is to find the error Δ_u, the absolute error $-\delta u$ and relative error for the sum $-\delta_u$.

$$u = x_1 + x_2 + \ldots + x_n + \ldots + x_N = \sum_{n=1}^{N} x_n \tag{1.1}$$

The Eq. (1.1) taking into account the errors of all terms can be written down as

$$u \pm \Delta u = (x_1 \pm \Delta x_1) + (x_2 \pm \Delta x_2) + \ldots + (x_N \pm \Delta x_N) = \sum_{n=1}^{N} (x_n \pm \Delta x_n)$$

The latter ratio in view of Eq. (1.1) can be re-written as follows:

$$\pm \Delta u = \pm \Delta x_1 \pm \Delta x_2 \pm \ldots \pm \Delta x_N = \sum_{n=1}^{N} \pm \Delta x_n$$

Replacing the right-hand side of the equation errors Δ_{xn} by their absolute values Δ_{xn}, we obtain

$$|\Delta u| \leq |\Delta x_1| + |\Delta x_2| + \ldots + |\Delta x_N| = \sum_{n=1}^{N} |\Delta x_n|$$

or

$$\Delta_u = \sum_{n=1}^{N} \Delta_{xn} \tag{1.2}$$

The sum relative error in accordance with its definition can be set by

$$\delta_u = \frac{\Delta_u}{\sum_{n=1}^{N} x_n} = \frac{\sum_{n=1}^{N} x_n \cdot \delta_{xn}}{\sum_{n=1}^{N} x_n} \le \max(\delta_{x1}, \delta_{x2}, \ldots \delta_{xN}) \qquad (1.3)$$

The recorded ratios (1.2) and (1.3) allow us to formulate the rule for adding the approximate numbers that is used in practice for small-scale calculations [251]:

- result of the addition will have the accuracy of no less than the uncertainty of the most inaccurate number from terms, so you should select this number from terms and round the remaining numbers, leaving one – two bits more digits than in the preferred the least accurate term, after the addition the rounding by one character should be performed.

The Eqs. (1.2) and (1.3) allowing to evaluate the error in the addition, remain valid in the subtraction (indeed, some terms in the Eq. (1.1) could be negative). However, one should keep in mind the following considerations. If you are subtracting two numbers close in value, the denominator in the equation for the relative error becomes very small, and the relative error can grow to unacceptable values. As an example, the calculation of the following difference can be given [142]

$$u = \sqrt{a + 0{,}001} - \sqrt{a}$$

You can be sure that the calculation u with a given number of significant figures requires to write the number a with a larger number of significant digits. Nevertheless, one should look for (and find) techniques that can help to ease the calculation of such difference. For example, the transformation of the original recording to a new form

$$u = \sqrt{a+0{,}001} - \sqrt{a} = \left(\frac{\sqrt{a+0{,}001} - \sqrt{a}) \cdot (\sqrt{a+0{,}001} + \sqrt{a}}{\sqrt{a+0{,}001} + \sqrt{a}} \right)$$

$$= \frac{0{,}001}{\sqrt{a+0{,}001} + \sqrt{a}}$$

allows to replace a subtraction of close numbers by summation and division operations.

Such case can occur in solving the quadratic equation in which the value of the constant term is relatively small (example can be the equation of the form $x^2 + a \cdot x + 0{,}0001 = 0$). More similar examples could be given.

2. Error of multiplication (division) of approximate numbers.

Let there be given N approximate numbers x_n (lower index from 1 to N) whose error is Δx_n, the absolute error $-\Delta x_n$ and relative error – δ_{xn}. The task is to find the error Δ_u, absolute error δ_u and relative error δ_u for the multiplication product

$$u = x_1 x_2 \cdot \ldots \cdot x_n \cdot \ldots x_N = \prod_{n=1}^{N} x_n \tag{1.4}$$

After taking the logarithm of the Eq. (1.4), the original problem can be reduced to the problem of determining the approximate errors when adding numbers. Indeed, we have consistently

$$\ln u = \ln x_1 + \ln x_2 + \ldots + \ln x_N = \sum_{n=1}^{N} \ln x_n;$$

$$\frac{\Delta u}{u} = \sum_{n=1}^{N} \frac{\Delta x_n}{x_n};$$

$$\delta \le \delta_u = \sum_{n=1}^{N} \delta_{xn};$$

$$\Delta_u = |u| \cdot \delta_u$$

A small number of multiplications of approximate numbers can be guided by the rule [251]:

- set the number of significant figures in the least precise factor, round all the other factors leaving one or two characters more than in the less accurate factor; round the product of factors keeping the number of significant figures such as in the least precise factor (or one significant figure more).

The division of approximate numbers can be regarded as a special case of the multiplication of numbers. In particular, when dividing two close numbers (x_1, x_2), the following correlations are true:

$$u = \frac{x_1}{x_2};$$

$$\ln u = \ln x_1 - \ln x_2;$$

$$\frac{\Delta u}{u} = \frac{\Delta x_1}{x_1} - \frac{\Delta x_2}{x_2};$$

$$\left|\frac{\Delta u}{u}\right| \le \left|\frac{\Delta x_1}{x_1}\right| + \left|\frac{\Delta x_2}{x_2}\right|;$$

$$\delta \le \delta_u = \delta_{x1} + \delta_{x2}$$

Doing the division, the possibility of the loss of some significant digits should be taken into account as in the subtraction of approximate numbers.

Let us consider the special case of the division operation. Let the dividend and divisor have ***m*** correct significant digits. In accordance with the above-mentioned ratios, the value of the relative error of the quotient can be determined by $\delta_u = \frac{1}{2} \cdot (\frac{1}{\alpha} + \frac{1}{\beta}) \cdot 10^{-(m-1)}$. Here α, β – the first significant digit of the dividend and divisor. If the value is $\alpha \ge 2$, $\beta \ge 2$, then the quotient $(\boldsymbol{m} - \boldsymbol{1})$ has the significant digit, and if $\alpha = 1$, $\beta = 1 - (\boldsymbol{m} - \boldsymbol{2})$ significant digit. This conclusion shows that in the division of approximate numbers one can use the same rounding rule as in the multiplication of approximate numbers.

3. Error of the arbitrary algebraic function computation.

Let there be given a differentiable function $u = u(x_1, x_2, ..., x_N)$ and Δ_{xN} is the absolute errors of the function arguments (n = 1, N). The task is to find the absolute error Δ_u and relative error δ_u of the function u when calculating its argument values in the $(x_1^0, x_2^0, ..., x_N^0)$.

To calculate the errors Δ_u, δ_u, one can use the expansion of u in Taylor series limiting the first-order accuracy:

$$u + \Delta_u \approx u\left(x_1^0, x_2^0, ..., x_N^0\right) + \sum_{n=1}^{N} \frac{\partial u}{\partial x_n} \cdot \Delta_{xn}$$

The values of partial derivatives $\frac{\partial u}{\partial x_n}$ are calculated with the value of the arguments $(x_1^0, x_2^0, ..., x_N^0)$. To calculate the absolute error Δ_u and relative error δ_u of the function $u = u(x_1, x_2, ..., x_N)$ the following correlations are true

$$\Delta_u \approx \sum_{n=1}^{N} \frac{\partial u}{\partial x_n} \Delta_{xn};$$

$$\delta_u \approx \sum_{n=1}^{N} \frac{\partial \ln u}{\partial x_n} \Delta_{xn}$$

It should be noted that for the arbitrary function $u = u(x_1, x_2, ..., x_N)$ the values of the absolute and relative errors depend on the value of arguments $(x_1^0, x_2^0, ..., x_N^0)$ at which these errors are calculated.

It would seem that the development of computer technologies should eliminate the accumulation of errors in calculations. However, it is not true. A little humorous story comes to mind. An inspector comes into the store under the mask of a buyer and asks to weigh 50 grams of different sausages. The seller weighs and tells him the amount that has to be paid for these sausages. The buyer thoroughly recounts it on paper with a pencil and then gleefully shouts: “We will draw up a report, you must have holes in the weights, magnets are suspended, you cheated me 2 rubles.” The seller says: “Excuse me, what a protocol, what holes in the weights? You have calculated manually and I have done it on a calculator.” Thereafter, he re-calculates everything on the calculator in front of the buyer and shows that the amount, which he called initially is correct. The amazed buyer goes away puzzled and the seller mutters after him: “Just think – holes in the weights! Why should we make holes in the weights if we can make holes in the calculator?” Indeed, the seller is right – there are “holes” in computer technologies.

Operations with approximate numbers on computers have some features that have a direct impact on the amount of arithmetic calculations error. The first feature is due to the limited number of significant digits used in the presentation of approximate numbers in machine code. The amount of significant digits in the calculations depends on the class of computer technology and software used. As a rule, in modern personal computer the real number in computer memory is represented by four bytes (this corresponds to 32 binary bits), which in decimal code corresponds to the number ***±0.000000 e±00*** (code contains the sign before the number, the first significant digit before the comma, six significant digits after the decimal point, the exponent indication, the exponent sign, the degree value). The computer bit capacity can be changed by the software. So, FORTRAN compiler allows the use of double-precision real numbers (8 bytes of computer memory is used for the number representation). Moreover, in the latest standards of the algorithmic language FORTRAN it is allowed to use real numbers of arbitrary capacity. Such means of representation of numbers are found in other compilers and mathematical systems of computing.

Another feature of the computations performed by a computer is the aspects of rounding of the calculations results. In computer calculations digits located to the right of the last significant digit during the next arithmetical operation are simply cut off and not rounded upwards or downwards. This type of rounding differs from the symmetrical rounding discussed above. Indeed, if the number of significant digits used equals $\boldsymbol{n}$, the limiting relative error of machine rounding will equal $\delta_u = 10^{-(n-1)}$. The comparison with symmetrical rounding indicates that the machine rounding accuracy is worse than symmetrical. It is not difficult to provide the symmetric mathematical rounding in the computer software, but the need for such rounding is arguable. In particular, the accuracy of computer calculations can always be significantly improved by increasing the bit capacity of real numbers.

As a result, the following conclusion should be made. Modern problems solved using computers suggest millions and billions of simple arithmetic operations. In this regard, it is important to create such computational algorithms, the application of which on computers would not lead to the increase of errors presented in the source data of engineering problems.

KEYWORDS

- **accuracy of a problem solving**
- **algorithms**
- **complex physical problems**
- **computational experiment**
- **mathematical modeling**
- **physical model**

CHAPTER 2

TABULAR DEPENDENCIES AND TECHNIQUES OF WORK WITH TABLES

CONTENTS

2.1 FIELDS OF USE OF TABLES

The main task, which a mathematical model should provide, is an adequate description of objects or processes within the boundaries established for parameters that describe the object or process. In the simplest case, the solution of such problem can be provided with a tabular description of an object or process. A tabular description can be obtained as a result of natural experiment, processing a series of computer calculations, etc. It should be noted that tabular description in these cases is not the only way to present the experimental results and results of processing a series of calculations. The other methods that, for instance, come to the record of analytic dependence (equations or equation systems, algebraic or differential, linear or non-linear equations, etc.) will be discussed in the following chapters. Features of the experiments or calculations in tables will not be analyzed. They are presented in the available literature. Particularly, we can note [43, 74, 245, 252] and others.

Table functions can depend on one or more arguments. The difference between table functions and continuous functions is that their values are given in a discrete set of values of arguments. To store the table data in the operational or long-term computer memory, one should sort out a certain space whose sizes notably increase with the growing number of arguments, on which the function depends. Let the function depend on N arguments, each is defined in m points. In this case, to store the table function information it is necessary to provide, at least, $4\times N^m$ bytes of computer memory. With a large number of arguments ($N>3$) the work with tables becomes cumbersome and inconvenient.

In practice, table functions are relatively common, however, they are usually limited to functions depending on one (N = i) or two (N = i) arguments.

In Figure 2.1 the gas-flow controller used in gas industry and power plant engineering is shown. The controller consists of body *1*, saddle *2*, inlet *3* and outlet fittings *5*, damper *4*. The change in the gas flow G inflowing the outlet fitting *5* is ensured by damper *4* rotating around its axis at an arbitrary angle φ. The metering characteristic G(φ) depends on the section between saddle *2* and damper body *4*. The metering characteristic is considerably nonlinear and can be found as a result of physical experiment or calculation of gas-dynamic problem in three-dimensional setting (in [252] this task is solved experimentally). In both cases, the total dependence G(φ) can be presented as a table function for the gas flow, depending on one argument – the angle φ of damper rotation. The rotation angle varies from 0 to π/2. The number of intermediate values of damper rotation angle φ_i is selected depending on the required accuracy of metering characteristic G(φ) representation.

The above-discussed problem of the controller metering characteristics can be of interest in more complex problem solving – the problem of the flight of an aircraft with adjustable power plant. In this case, the tabular presentation of the metering characteristic $G(\varphi)$ instead of the numerical calculation of three-dimensional gas-dynamic problem allows significantly reducing the computer time required to calculate the problem of the aircraft flight.

Figure 2.2 shows the diagram of solid charge fuel used in rocket engines [136]. Basic geometrical dimensions of charge used in design are the length L, the outer D and inner d diameters. The change in charge surface is determined by the listed geometrical dimensions and also by coordinates of points 1–20 presented in Figure 2.2 (solid fuel burning takes place inside the normal to the solid fuel surface). For the practice charge forms, where the

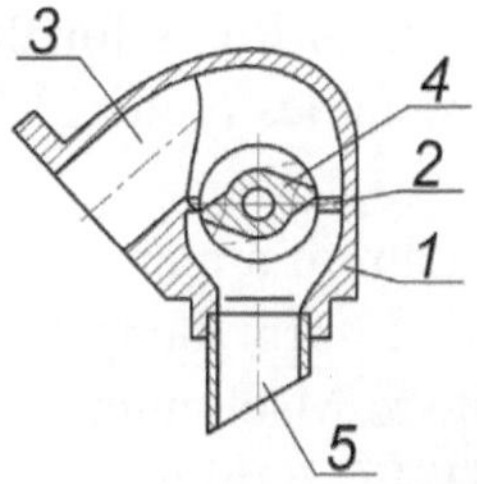

FIGURE 2.1 Gas-flow controller (*1* – body; *2* – saddle; 3 -inlet and 5 - outlet fittings; 4– damper).

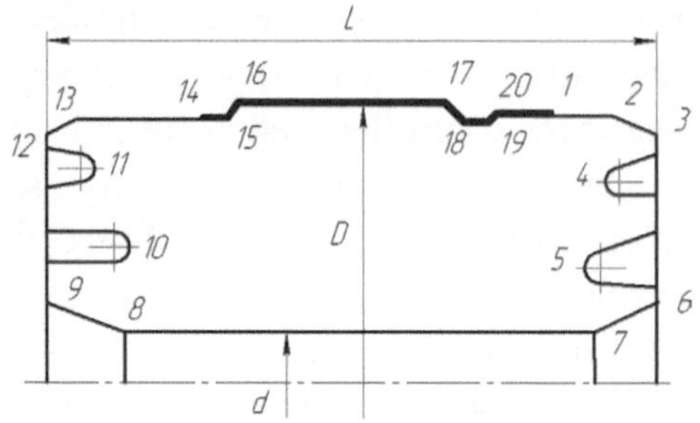

FIGURE 2.2 Diagram of the solid fuel solid-fuel charge (L, D, d, e_{max}, h_5 – geometrical dimensions; 1–20 – characteristic points in charge section).

value of surface of solid fuel burning remains constant (or similar to constant value) for the whole work period with the values of burned vault e from 0 to e_{max}, are of interest. The coordinates of points 1–20 ensuring such charge forms are found by consuming the geometric calculations and can be presented in a tabular form depending on dimensionless parameters $\bar{L} = \frac{L}{D}$ and $\bar{d} = \frac{d}{D}$. In this example, it is necessary to construct the table (for all geometrical dimensions of solid fuel charge) depending on two arguments L and d.

In the early 1980-s the problem of choice of the solid fuel charge was solved as a part of the problem of a multi-stage rocket system design. One option of the optimal fuel charge calculation is found after solving the problem for 30 minutes of computer time (the calculations are made with the computer M-220).

Such duration of the particular problem calculation was unacceptably long (not more than 6 hours of computer time were envisaged for the whole design stage). A preliminary calculation of a series of optimal fuel charge options followed by the tabular interpretation of the calculation results was the way out. The algorithms for the selection of the optimal fuel charge geometry from the tables required 1 second of computer time, at the most.

Figure 2.3 shows the diagram of a planar segmental body flowed (lines of the same values of Mach number) along the transonic gas flow with a rate in incoming flow suitable to Mach number M=0.9 [72]. The results of flow calculation process on HSEC-6 (high-speed electronic computer) (calculation of one option demanded ~ 2 hours of machine time) in Ref. [72]

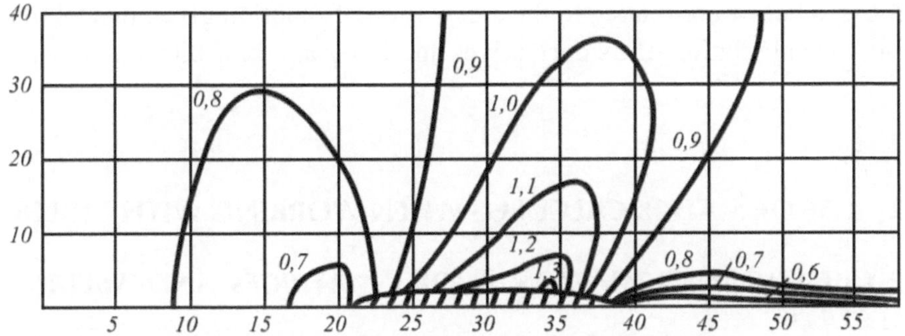

FIGURE 2.3 Lines of identical values of Mach numbers (M_∞=0.9; δ=12%).

are presented in a tabular form. The tabular functions for fields of gas density, pressure, Mach number, values of the longitudinal and transverse gas velocities are constructed. The tables are constructed as functions of four arguments – longitudinal x and transverse y in coordinates of calculation conditions, Mach number M_∞ in the incoming flow and relative thickness of segmental profile δ.

As in the previous case, the use of tables with the results of flow calculation of the segment profile significantly reduces the machine time necessary to complete the aircraft design.

We should mention another use of table models, which mathematicians and engineers are accustomed to. Finite-difference methods for problem-solving described by ordinary differential equations and partial differential equations are used in cases where it is impossible to find an analytical solution for the current task or analytical solution is unnecessarily cumbersome and one cannot use it. Using finite-difference methods, the initial problem on a set of continuous function f depending on one or more variables (arguments) x_n $(n = 1, N)$ is replaced by the discrete function $f_{(h)}$ given on a grid, where the nodes are the discrete values of the arguments $x_{n(h)}$. In the latter records, h is the parameter, which characterizes the discretion of the arguments x_n $(n = 1, N)$ change. In fact, with such approach the object or process behavior is defined by the table where the dimension is equal to the number of variables (arguments) that affect the function, and the number of tabular values used for each argument is determined by additional conditions (accuracy, stability, etc.).

Finite-difference methods for problem-solving described by differential equations will be discussed in the following chapters. The issues discussed below are related to the calculation of function given in a tabular form, the calculation of its derivatives and integrals for arbitrary values of arguments.

2.2 USE OF MATRIX CALCULUS WHEN WORKING WITH TABLES

2.2.1 BASIC INFORMATION ABOUT MATRICES AND MATRIX CALCULUS

In mathematics there is an object called a matrix and recorded in the form of a rectangular table. Certain rules are established for matrices, which are algebraic operations that allow using them efficiently in solving different mathematical problems. Matrices are very convenient objects that can be used to describe table functions and models. Rectangular (or two-dimensional) matrices can be used for functions depending on one or two arguments. However, with a more arguments of table functions, the multi-dimensional matrices (matrices of non-rectangular form) can be used.

Matrix calculus plays an important role in mathematical modeling. The use of matrices and matrix calculus significantly simplifies the problem definition and problem-solving that can appear in engineering. Particularly, these are the problems related to the solution of systems of linear equations, determination of eigen values and eigen vectors of a linear system, solution of linear programming problems, etc. The detailed description of matrix calculus is given in Refs. [39, 55]. The basic definitions and properties related to the theory of matrices and matrix calculus are examined below. This material will be supplemented whenever necessary.

2.2.1.1 Basic Definitions

Suppose we are given a table of numbers (real or complex) containing n rows and m columns. Let us designate each element in the table as $a_{i,j}$ (or a_{ij}) where the first index is the row number, and the second – the column number. The table determined this way will be named the matrix **A** (or {A}) and designated as follows

$$\mathbf{A} = \begin{pmatrix} a_{1,1} & a_{1,2} & \cdots & a_{1,i} & \cdots & a_{1,n-1} & a_{1,n} \\ a_{2,1} & a_{2,2} & \cdots & a_{2,i} & \cdots & a_{2,n-1} & a_{2,n} \\ \cdots & \cdots & \cdots & \cdots & \cdots & \cdots & \cdots \\ a_{i,1} & a_{i,2} & \cdots & a_{i,i} & \cdots & a_{i,n-1} & a_{i,n} \\ \cdots & \cdots & \cdots & \cdots & \cdots & \cdots & \cdots \\ a_{n-1,1} & a_{n-1,2} & \cdots & a_{n-1,i} & \cdots & a_{n-1,n-1} & a_{n-1,n} \\ a_{n,1} & a_{n,2} & \cdots & a_{n,i} & \cdots & a_{n,n-1} & a_{n,n} \end{pmatrix}$$

In the recorded matrix, the row number n cannot match the column number m, such matrix is a rectangular one. In a particular case, if m=n, the matrix is square, and number n is the matrix order. Furthermore, a matrix can contain only one row n=1 (matrix **B** recorded below) or only one column m=1 (matrix **C** recorded below).

$$\mathbf{B} = \begin{pmatrix} b_1 \; b_2 \; b_3 \ldots b_i \ldots b_n \end{pmatrix}; \quad \mathbf{C} = \begin{pmatrix} c_1 \\ c_2 \\ c_3 \\ \ldots \\ c_i \\ \ldots \\ c_n \end{pmatrix}$$

Let us examine other particular cases of matrix used in matrix calculus.

The square matrix **D** is a diagonal matrix. Its feature is that all the elements a_{ij}, which do not lie on the central diagonal ($i \neq j$), equal zero. If all diagonal elements of such matrix equal one ($a_{ij} = 1, (i = j)$), the matrix is a unit matrix and can be designated as $\mathbf{E}_n$ or **E**.

If the matrix elements located on several diagonals adjoining the central diagonal are non-zero, such matrix is banded. The banded matrix **N** recorded below, where besides the central diagonal another two diagonals (one is below the central diagonal, the second is above) are filled, is tridiagonal.

$$\mathbf{D}=\begin{pmatrix} a_{1,1} & 0 & \dots & 0 & \dots & 0 & 0 \\ 0 & a_{2,2} & \dots & 0 & \dots & 0 & 0 \\ \dots & \dots & \dots & \dots & \dots & \dots & \dots \\ 0 & 0 & \dots & a_{i,i} & \dots & 0 & 0 \\ \dots & \dots & \dots & \dots & \dots & \dots & \dots \\ 0 & 0 & \dots & 0 & \dots & a_{n-1,n-1} & 0 \\ 0 & 0 & \dots & 0 & \dots & 0 & a_{n,n} \end{pmatrix};$$

$$\mathbf{E}=\begin{pmatrix} 1 & 0 & \dots & 0 & \dots & 0 & 0 \\ 0 & 1 & \dots & 0 & \dots & 0 & 0 \\ \dots & \dots & \dots & \dots & \dots & \dots & \dots \\ 0 & 0 & \dots & 1 & \dots & 0 & 0 \\ \dots & \dots & \dots & \dots & \dots & \dots & \dots \\ 0 & 0 & \dots & 0 & \dots & 1 & 0 \\ 0 & 0 & \dots & 0 & \dots & 0 & 1 \end{pmatrix};$$

$$\mathbf{N}=\begin{pmatrix} a_{1,1} & a_{1,2} & 0 & \dots & 0 & \dots & 0 & 0 \\ a_{2,1} & a_{2,2} & a_{2,3} & \dots & 0 & \dots & 0 & 0 \\ 0 & a_{3,2} & a_{3,3} & \dots & 0 & \dots & 0 & 0 \\ \dots & \dots & \dots & \dots & \dots & \dots & \dots & \dots \\ 0 & 0 & 0 & \dots & a_{i,i} & \dots & 0 & 0 \\ \dots & \dots & \dots & \dots & \dots & \dots & \dots & \dots \\ 0 & 0 & 0 & \dots & 0 & \dots & a_{n-1,n-1} & a_{n-1,n} \\ 0 & 0 & 0 & \dots & 0 & \dots & a_{n,n-1} & a_{n,n} \end{pmatrix}$$

The square matrix **U** is called an upper matrix (or right), if all the elements of the matrix a_{ij} lying below the central diagonal ($i > j$) equal zero. The square matrix **L** is called a lower matrix (or left), if all the elements of the matrix a_{ij} lying above the central diagonal ($i < j$) equal zero.

$$\mathbf{U} = \begin{pmatrix} a_{1,1} & a_{1,2} & \dots & a_{1,i} & \dots & a_{1,n-1} & a_{1,n} \\ 0 & a_{2,2} & \dots & a_{2,i} & \dots & a_{2,n-1} & a_{2,n} \\ \dots & \dots & \dots & \dots & \dots & \dots & \dots \\ 0 & 0 & \dots & a_{i,i} & \dots & a_{i,n-1} & a_{i,n} \\ \dots & \dots & \dots & \dots & \dots & \dots & \dots \\ 0 & 0 & \dots & 0 & \dots & a_{n-1,n-1} & a_{n-1,n} \\ 0 & 0 & \dots & 0 & \dots & 0 & a_{n,n} \end{pmatrix};$$

$$\mathbf{L} = \begin{pmatrix} a_{1,1} & 0 & \dots & 0 & \dots & 0 & 0 \\ a_{2,1} & a_{2,2} & \dots & 0 & \dots & 0 & 0 \\ \dots & \dots & \dots & \dots & \dots & \dots & \dots \\ a_{i,1} & a_{i,2} & \dots & a_{i,i} & \dots & 0 & 0 \\ \dots & \dots & \dots & \dots & \dots & \dots & \dots \\ a_{n-1,1} & a_{n-1,2} & \dots & a_{n-1,i} & \dots & a_{n-1,n-1} & 0 \\ a_{n,1} & a_{n,2} & \dots & a_{n,i} & \dots & a_{n,n-1} & a_{n,n} \end{pmatrix}$$

The square matrix **H** is called upper (or right) Hessenberg matrix, if all the elements of the matrix a_{ij} lying below the diagonal with the number $i+1=j$ equal zero. The square matrix **F** is called lower (or left) Hessenberg matrix, if all the elements of the matrix a_{ij} lying above the diagonal with the number $i-1=j$ equal zero.

The square matrix **C** is called transposed against the matrix **A**, if the equality $c_{ij} = a_{ji}$ is true for the matrix components. The transposed matrix is denoted as $\mathbf{A}^T$. If the matrix elements are complex numbers, the transposed matrix is called conjugate (the elements c_{ij}, a_{ji} are complex conjugates).

The square matrix **A** is called symmetrical (this matrix is denoted as **S** below), if the equality $\mathbf{A}^T = \mathbf{A}$ is true (for all values i and j the equality $a_{ij} = a_{ji}$ is true). If the matrix elements are complex numbers, the symmetrical matrix is called Hermitian matrix (the elements a_{ij}, a_{ji} are complex conjugates). The types of square matrices discussed are given below:

$$\mathbf{H}=\begin{pmatrix} a_{1,1} & a_{1,2} & \dots & a_{1,i} & \dots & a_{1,n-1} & a_{1,n} \\ a_{2,1} & a_{2,2} & \dots & a_{2,i} & \dots & a_{2,n-1} & a_{2,n} \\ \dots & \dots & \dots & \dots & \dots & \dots & \dots \\ 0 & 0 & \dots & a_{i,i} & \dots & a_{i,n-1} & a_{i,n} \\ \dots & \dots & \dots & \dots & \dots & \dots & \dots \\ 0 & 0 & \dots & 0 & \dots & a_{n-1,n-1} & a_{n-1,n} \\ 0 & 0 & \dots & 0 & \dots & a_{n,n-1} & a_{n,n} \end{pmatrix};$$

$$\mathbf{F}=\begin{pmatrix} a_{1,1} & a_{1,2} & \dots & 0 & \dots & 0 & 0 \\ a_{2,1} & a_{2,2} & \dots & 0 & \dots & 0 & 0 \\ \dots & \dots & \dots & \dots & \dots & \dots & \dots \\ a_{i,1} & a_{i,2} & \dots & a_{i,i} & \dots & 0 & 0 \\ \dots & \dots & \dots & \dots & \dots & \dots & \dots \\ a_{n-1,1} & a_{n-1,2} & \dots & a_{n-1,i} & \dots & a_{n-1,n-1} & a_{n-1,n} \\ a_{n,1} & a_{n,2} & \dots & a_{n,i} & \dots & a_{n,n-1} & a_{n,n} \end{pmatrix};$$

$$\mathbf{A}^T=\begin{pmatrix} a_{1,1} & a_{2,1} & \dots & a_{i,1} & \dots & a_{n-1,1} & a_{n,1} \\ a_{1,2} & a_{2,2} & \dots & a_{i,2} & \dots & a_{n-1,2} & a_{n,2} \\ \dots & \dots & \dots & \dots & \dots & \dots & \dots \\ a_{1,i} & a_{2,i} & \dots & a_{i,i} & \dots & a_{n-1,i} & a_{n,i} \\ \dots & \dots & \dots & \dots & \dots & \dots & \dots \\ a_{1,n-1} & a_{2,n-1} & \dots & a_{i,n-1} & \dots & a_{n-1,n-1} & a_{n,n-1} \\ a_{1,n} & a_{2,n} & \dots & a_{i,n} & \dots & a_{n-1,n} & a_{n,n} \end{pmatrix};$$

$$\mathbf{S}=\begin{pmatrix} a_{1,1} & a_{1,2} & \dots & a_{1,i} & \dots & a_{1,n-1} & a_{1,n} \\ a_{1,2} & a_{2,2} & \dots & a_{2,i} & \dots & a_{2,n-1} & a_{2,n} \\ \dots & \dots & \dots & \dots & \dots & \dots & \dots \\ a_{1,i} & a_{2,i} & \dots & a_{i,i} & \dots & a_{i,n-1} & a_{i,n} \\ \dots & \dots & \dots & \dots & \dots & \dots & \dots \\ a_{1,n-1} & a_{2,n-1} & \dots & a_{i,n-1} & \dots & a_{n-1,n-1} & a_{n-1,n} \\ a_{1,n} & a_{2,n} & \dots & a_{i,n} & \dots & a_{n-1,n} & a_{n,n} \end{pmatrix}$$

2.2.1.2 Elements of Matrix Algebra

It was noted that the foregoing matrices are useful in solving the problems in calculus. Particularly, the use of matrices can significantly simplify the writing of systems of linear equations. Moreover, different transformations in linear algebra and calculus are easier to perform using the matrix notation. Advantages of matrix representation of linear systems are more attractive after introducing the algebraic operations with matrices. Two operations are introduced in matrix calculus – the addition of matrices and multiplication of matrices.

The sum of matrices **A** and **B** is the matrix **C=A+B**, where the elements are calculated by the ratio $c_{ij} = a_{ij} + b_{ij}$.

The sum of matrices makes sense only if the number of rows and columns in both matrices coincide. In the expanded form the addition of matrices **A** and **B** is presented in the following way:

$$\begin{pmatrix} a_{11} & a_{12} & \dots & a_{1i} & \dots & a_{1n} \\ a_{21} & a_{22} & \dots & a_{2i} & \dots & a_{2n} \\ \dots & \dots & \dots & \dots & \dots & \dots \\ a_{i1} & a_{i2} & \dots & a_{ii} & \dots & a_{in} \\ \dots & \dots & \dots & \dots & \dots & \dots \\ a_{n1} & a_{n2} & \dots & a_{ni} & \dots & a_{nn} \end{pmatrix} + \begin{pmatrix} b_{11} & b_{12} & \dots & b_{1i} & \dots & b_{1n} \\ b_{21} & b_{22} & \dots & b_{2i} & \dots & b_{2n} \\ \dots & \dots & \dots & \dots & \dots & \dots \\ b_{i1} & b_{i2} & \dots & b_{ii} & \dots & b_{in} \\ \dots & \dots & \dots & \dots & \dots & \dots \\ b_{n1} & b_{n2} & \dots & b_{ni} & \dots & b_{nn} \end{pmatrix}$$

$$= \begin{pmatrix} a_{11}+b_{11} & a_{12}+b_{12} & \dots & a_{1i}+b_{1i} & \dots & a_{1n}+b_{1n} \\ a_{21}+b_{21} & a_{22}+b_{22} & \dots & a_{2i}+b_{2i} & \dots & a_{2n}+b_{2n} \\ \dots & \dots & \dots & \dots & \dots & \dots \\ a_{i1}+b_{i1} & a_{i2}+b_{i2} & \dots & a_{ii}+b_{ii} & \dots & a_{in}+b_{in} \\ \dots & \dots & \dots & \dots & \dots & \dots \\ a_{n1}+b_{n1} & a_{n2}+b_{n2} & \dots & a_{ni}+b_{ni} & \dots & a_{nn}+b_{nn} \end{pmatrix}$$

The product of the matrices **A** and **B** is the matrix **C=AB**, where the elements are calculated by the ratio $c_{ij} = \sum_{k=1}^{n} a_{ik} \cdot b_{kj}$.

$$\begin{pmatrix} a_{11} & a_{12} & \cdots & a_{1i} & \cdots & a_{1n} \\ a_{21} & a_{22} & \cdots & a_{2i} & \cdots & a_{2n} \\ \cdots & \cdots & \cdots & \cdots & \cdots & \cdots \\ a_{i1} & a_{i2} & \cdots & a_{ii} & \cdots & a_{in} \\ \cdots & \cdots & \cdots & \cdots & \cdots & \cdots \\ a_{n1} & a_{n2} & \cdots & a_{ni} & \cdots & a_{nn} \end{pmatrix} \times \begin{pmatrix} b_{11} & b_{12} & \cdots & b_{1i} & \cdots & b_{1n} \\ b_{21} & b_{22} & \cdots & b_{2i} & \cdots & b_{2n} \\ \cdots & \cdots & \cdots & \cdots & \cdots & \cdots \\ b_{i1} & b_{i2} & \cdots & b_{ii} & \cdots & b_{in} \\ \cdots & \cdots & \cdots & \cdots & \cdots & \cdots \\ b_{n1} & b_{n2} & \cdots & b_{ni} & \cdots & b_{nn} \end{pmatrix}$$

$$= \begin{pmatrix} \sum_{k=1}^{n} a_{1k}b_{k1} & \sum_{k=1}^{n} a_{1k}b_{k2} & \cdots & \sum_{k=1}^{n} a_{1k}b_{ki} & \cdots & \sum_{k=1}^{n} a_{1k}b_{kn} \\ \sum_{k=1}^{n} a_{2k}b_{k1} & \sum_{k=1}^{n} a_{2k}b_{k2} & \cdots & \sum_{k=1}^{n} a_{2k}b_{ki} & \cdots & \sum_{k=1}^{n} a_{2k}b_{kn} \\ \cdots & \cdots & \cdots & \cdots & \cdots & \cdots \\ \sum_{k=1}^{n} a_{ik}b_{k1} & \sum_{k=1}^{n} a_{ik}b_{k2} & \cdots & \sum_{k=1}^{n} a_{ik}b_{ki} & \cdots & \sum_{k=1}^{n} a_{ik}b_{kn} \\ \cdots & \cdots & \cdots & \cdots & \cdots & \cdots \\ \sum_{k=1}^{n} a_{nk}b_{k1} & \sum_{k=1}^{n} a_{nk}b_{k2} & \cdots & \sum_{k=1}^{n} a_{nk}b_{ki} & \cdots & \sum_{k=1}^{n} a_{nk}b_{kn} \end{pmatrix}$$

According to the definition, the multiplication of matrices makes sense only if the number of rows and columns in the first matrix coincide with the number of rows and columns in the second matrix. The matrix **C** formed as a result of the product of the matrices **A** and **B** contains as many rows as the matrix **A**, and the number of columns in it coincides with the number of columns in the matrix **B**. The multiplication of two matrices makes sense if matrices are square and have the same size $n \times n$. An output matrix also has the size $n \times n$.

What is the purpose of introducing such a difficult definition of the multiplication product of two matrices? For example, it was possible to use the operation of the product similar to the addition operation of the matrices. The answer is simple – the matrix algebra formed in this case would have been useless in solving many problems of calculus. To evaluate striking opportunities of the matrix algebra, let us examine some examples of the matrix multiplication:

a) multiplication of the matrix **A** by the unit matrix **E**:

$$\begin{pmatrix} a_{11} & a_{12} & \dots & a_{1i} & \dots & a_{1n} \\ a_{21} & a_{22} & \dots & a_{2i} & \dots & a_{2n} \\ \dots & \dots & \dots & \dots & \dots & \dots \\ a_{i1} & a_{i2} & \dots & a_{ii} & \dots & a_{in} \\ \dots & \dots & \dots & \dots & \dots & \dots \\ a_{n1} & a_{n2} & \dots & a_{ni} & \dots & a_{nn} \end{pmatrix} \times \begin{pmatrix} 1 & 0 & \dots & 0 & \dots & 0 \\ 0 & 1 & \dots & 0 & \dots & 0 \\ \dots & \dots & \dots & \dots & \dots & \dots \\ 0 & 0 & \dots & 1 & \dots & 0 \\ \dots & \dots & \dots & \dots & \dots & \dots \\ 0 & 0 & \dots & 0 & \dots & 1 \end{pmatrix}$$

$$= \begin{pmatrix} a_{11} & a_{12} & \dots & a_{1i} & \dots & a_{1n} \\ a_{21} & a_{22} & \dots & a_{2i} & \dots & a_{2n} \\ \dots & \dots & \dots & \dots & \dots & \dots \\ a_{i1} & a_{i2} & \dots & a_{ii} & \dots & a_{in} \\ \dots & \dots & \dots & \dots & \dots & \dots \\ a_{n1} & a_{n2} & \dots & a_{ni} & \dots & a_{nn} \end{pmatrix}$$

These examples show that the multiplication of the square matrix by the unit matrix (in this multiplication the unit matrix can be placed to the left or right side) does not change the content of the original matrix (**AE=EA=A**);

b) multiplication of the matrix **A** by the permutation matrix $\mathbf{P}_{ij}$ (permutation matrix $\mathbf{P}_{ij}$ differs from the unit matrix **E** in permutation of ith and j^{th} rows):

$$\begin{pmatrix} a_{11} & \dots & a_{1i} & \dots & a_{1j} & \dots & a_{1n} \\ \dots & \dots & \dots & \dots & \dots & \dots & \dots \\ a_{i1} & \dots & a_{ii} & \dots & a_{ij} & \dots & a_{in} \\ \dots & \dots & \dots & \dots & \dots & \dots & \dots \\ a_{j1} & \dots & a_{ji} & \dots & a_{jj} & \dots & a_{jn} \\ \dots & \dots & \dots & \dots & \dots & \dots & \dots \\ a_{n1} & \dots & a_{ni} & \dots & a_{nj} & \dots & a_{nn} \end{pmatrix} \times \begin{pmatrix} 1 & \dots & 0 & \dots & 0 & \dots & 0 \\ \dots & \dots & \dots & \dots & \dots & \dots & \dots \\ 0 & \dots & 0 & \dots & 1 & \dots & 0 \\ \dots & \dots & \dots & \dots & \dots & \dots & \dots \\ 0 & \dots & 1 & \dots & 0 & \dots & 0 \\ \dots & \dots & \dots & \dots & \dots & \dots & \dots \\ 0 & \dots & 0 & \dots & 0 & \dots & 1 \end{pmatrix}$$

$$= \begin{pmatrix} a_{11} & \dots & a_{1j} & \dots & a_{1i} & \dots & a_{1n} \\ \dots & \dots & \dots & \dots & \dots & \dots & \dots \\ a_{i1} & \dots & a_{ij} & \dots & a_{ii} & \dots & a_{in} \\ \dots & \dots & \dots & \dots & \dots & \dots & \dots \\ a_{j1} & \dots & a_{jj} & \dots & a_{ji} & \dots & a_{jn} \\ \dots & \dots & \dots & \dots & \dots & \dots & \dots \\ a_{n1} & \dots & a_{nj} & \dots & a_{ni} & \dots & a_{nn} \end{pmatrix}$$

$$\begin{pmatrix} 1 & \dots & 0 & \dots & 0 & \dots & 0 \\ \dots & \dots & \dots & \dots & \dots & \dots & \dots \\ 0 & \dots & 0 & \dots & 1 & \dots & 0 \\ \dots & \dots & \dots & \dots & \dots & \dots & \dots \\ 0 & \dots & 1 & \dots & 0 & \dots & 0 \\ \dots & \dots & \dots & \dots & \dots & \dots & \dots \\ 0 & \dots & 0 & \dots & 0 & \dots & 1 \end{pmatrix} \times \begin{pmatrix} a_{11} & \dots & a_{1i} & \dots & a_{1j} & \dots & a_{1n} \\ \dots & \dots & \dots & \dots & \dots & \dots & \dots \\ a_{i1} & \dots & a_{ii} & \dots & a_{ij} & \dots & a_{in} \\ \dots & \dots & \dots & \dots & \dots & \dots & \dots \\ a_{j1} & \dots & a_{ji} & \dots & a_{jj} & \dots & a_{jn} \\ \dots & \dots & \dots & \dots & \dots & \dots & \dots \\ a_{n1} & \dots & a_{ni} & \dots & a_{nj} & \dots & a_{nn} \end{pmatrix}$$

$$= \begin{pmatrix} a_{11} & \dots & a_{1i} & \dots & a_{1j} & \dots & a_{1n} \\ \dots & \dots & \dots & \dots & \dots & \dots & \dots \\ a_{j1} & \dots & a_{ji} & \dots & a_{jj} & \dots & a_{jn} \\ \dots & \dots & \dots & \dots & \dots & \dots & \dots \\ a_{i1} & \dots & a_{ij} & \dots & a_{ii} & \dots & a_{in} \\ \dots & \dots & \dots & \dots & \dots & \dots & \dots \\ a_{n1} & \dots & a_{ni} & \dots & a_{nj} & \dots & a_{nn} \end{pmatrix}$$

The above examples demonstrate that multiplying the matrix **A** by the matrix $\mathbf{P}_{ij}$ ($\mathbf{AP}_{ij}$), the matrix elements placed in i^{th} and j^{th} columns are interchanged. The multiplication of the matrix $\mathbf{P}_{ij}$ by the matrix **A** ($\mathbf{P}_{ij}\mathbf{A}$) leads to the change in the elements location placed in i^{th} and j^{th} rows;

c) multiplication of the matrix **A** by the modified unit diagonal matrix $\mathbf{E}_i(\delta)$ (the modified matrix $\mathbf{E}_i(\delta)$ differs from the unit matrix E only in the element – $e_{ii} = \delta$)

$$\begin{pmatrix} a_{11} & a_{12} & \dots & a_{1i} & \dots & a_{1n} \\ a_{21} & a_{22} & \dots & a_{2i} & \dots & a_{2n} \\ \dots & \dots & \dots & \dots & \dots & \dots \\ a_{i1} & a_{i2} & \dots & a_{ii} & \dots & a_{in} \\ \dots & \dots & \dots & \dots & \dots & \dots \\ a_{n1} & a_{n2} & \dots & a_{ni} & \dots & a_{nn} \end{pmatrix} \times \begin{pmatrix} 1 & 0 & \dots & 0 & \dots & 0 \\ 0 & 1 & \dots & 0 & \dots & 0 \\ \dots & \dots & \dots & \dots & \dots & \dots \\ 0 & 0 & \dots & \delta & \dots & 0 \\ \dots & \dots & \dots & \dots & \dots & \dots \\ 0 & 0 & \dots & 0 & \dots & 1 \end{pmatrix}$$

$$= \begin{pmatrix} a_{11} & a_{12} & \dots & \delta \cdot a_{1i} & \dots & a_{1n} \\ a_{21} & a_{22} & \dots & \delta \cdot a_{2i} & \dots & a_{2n} \\ \dots & \dots & \dots & \dots & \dots & \dots \\ a_{i1} & a_{i2} & \dots & \delta \cdot a_{ii} & \dots & a_{in} \\ \dots & \dots & \dots & \dots & \dots & \dots \\ a_{n1} & a_{n2} & \dots & \delta \cdot a_{ni} & \dots & a_{nn} \end{pmatrix}$$

$$\begin{pmatrix} 1 & 0 & \dots & 0 & \dots & 0 \\ 0 & 1 & \dots & 0 & \dots & 0 \\ \dots & \dots & \dots & \dots & \dots & \dots \\ 0 & 0 & \dots & \delta & \dots & 0 \\ \dots & \dots & \dots & \dots & \dots & \dots \\ 0 & 0 & \dots & 0 & \dots & 1 \end{pmatrix} \times \begin{pmatrix} a_{11} & a_{12} & \dots & a_{1i} & \dots & a_{1n} \\ a_{21} & a_{22} & \dots & a_{2i} & \dots & a_{2n} \\ \dots & \dots & \dots & \dots & \dots & \dots \\ a_{i1} & a_{i2} & \dots & a_{ii} & \dots & a_{in} \\ \dots & \dots & \dots & \dots & \dots & \dots \\ a_{n1} & a_{n2} & \dots & a_{ni} & \dots & a_{nn} \end{pmatrix}$$

$$= \begin{pmatrix} a_{11} & a_{12} & \dots & a_{1i} & \dots & a_{1n} \\ a_{21} & a_{22} & \dots & a_{2i} & \dots & a_{2n} \\ \dots & \dots & \dots & \dots & \dots & \dots \\ \delta \cdot a_{i1} & \delta \cdot a_{i2} & \dots & \delta \cdot a_{ii} & \dots & \delta \cdot a_{in} \\ \dots & \dots & \dots & \dots & \dots & \dots \\ a_{n1} & a_{n2} & \dots & a_{ni} & \dots & a_{nn} \end{pmatrix}$$

As a result of the matrix multiplication $\mathbf{AE}_i(\delta)$, the matrix **A** elements placed in the ith column are increasing in **δ** times. As a result of the matrix multiplication $\mathbf{E}_i(\delta)\mathbf{A}$, the matrix **A** elements placed in the ith row are increasing in **δ** times;

d) multiplication of the matrix **A** by the modified unit diagonal matrix $\mathbf{M}_{ij}(\delta)$ (matrix $\mathbf{M}_{ij}(\delta)$ differs from the unit matrix E only in the element – $e_{ij}=\delta$)

$$\begin{pmatrix} a_{11} & \dots & a_{1i} & \dots & a_{1j} & \dots & a_{1n} \\ \dots & \dots & \dots & \dots & \dots & \dots & \dots \\ a_{i1} & \dots & a_{ii} & \dots & a_{ij} & \dots & a_{in} \\ \dots & \dots & \dots & \dots & \dots & \dots & \dots \\ a_{j1} & \dots & a_{ji} & \dots & a_{jj} & \dots & a_{jn} \\ \dots & \dots & \dots & \dots & \dots & \dots & \dots \\ a_{n1} & \dots & a_{ni} & \dots & a_{nj} & \dots & a_{nn} \end{pmatrix} \times \begin{pmatrix} 1 & \dots & 0 & \dots & 0 & \dots & 0 \\ \dots & \dots & \dots & \dots & \dots & \dots & \dots \\ 0 & \dots & 1 & \dots & 0 & \dots & 0 \\ \dots & \dots & \dots & \dots & \dots & \dots & \dots \\ 0 & \dots & \delta & \dots & 1 & \dots & 0 \\ \dots & \dots & \dots & \dots & \dots & \dots & \dots \\ 0 & \dots & 0 & \dots & 0 & \dots & 1 \end{pmatrix}$$

$$= \begin{pmatrix} a_{11} & \dots & a_{1i} + \delta \cdot a_{1j} & \dots & a_{1j} & \dots & a_{1n} \\ \dots & \dots & \dots & \dots & \dots & \dots & \dots \\ a_{i1} & \dots & a_{ii} + \delta \cdot a_{ij} & \dots & a_{ij} & \dots & a_{in} \\ \dots & \dots & \dots & \dots & \dots & \dots & \dots \\ a_{j1} & \dots & a_{ji} + \delta \cdot a_{jj} & \dots & a_{jj} & \dots & a_{jn} \\ \dots & \dots & \dots & \dots & \dots & \dots & \dots \\ a_{n1} & \dots & a_{ni} + \delta \cdot a_{nj} & \dots & a_{nj} & \dots & a_{nn} \end{pmatrix}$$

$$\begin{pmatrix} 1 & \dots & 0 & \dots & 0 & \dots & 0 \\ \dots & \dots & \dots & \dots & \dots & \dots & \dots \\ 0 & \dots & 1 & \dots & 0 & \dots & 0 \\ \dots & \dots & \dots & \dots & \dots & \dots & \dots \\ 0 & \dots & \delta & \dots & 1 & \dots & 0 \\ \dots & \dots & \dots & \dots & \dots & \dots & \dots \\ 0 & \dots & 0 & \dots & 0 & \dots & 1 \end{pmatrix} \times \begin{pmatrix} a_{11} & \dots & a_{1i} & \dots & a_{1j} & \dots & a_{1n} \\ \dots & \dots & \dots & \dots & \dots & \dots & \dots \\ a_{i1} & \dots & a_{ii} & \dots & a_{ij} & \dots & a_{in} \\ \dots & \dots & \dots & \dots & \dots & \dots & \dots \\ a_{j1} & \dots & a_{ji} & \dots & a_{jj} & \dots & a_{jn} \\ \dots & \dots & \dots & \dots & \dots & \dots & \dots \\ a_{n1} & \dots & a_{ni} & \dots & a_{nj} & \dots & a_{nn} \end{pmatrix}$$

$$= \begin{pmatrix} a_{11} & \dots & a_{1i} & \dots & a_{1j} & \dots & a_{1n} \\ \dots & \dots & \dots & \dots & \dots & \dots & \dots \\ a_{i1} & \dots & a_{ii} & \dots & a_{ij} & \dots & a_{in} \\ \dots & \dots & \dots & \dots & \dots & \dots & \dots \\ a_{j1}+\delta\cdot a_{i1} & \dots & a_{ji}+\delta\cdot a_{ii} & \dots & a_{jj}+\delta\cdot a_{ij} & \dots & a_{jn}+\delta\cdot a_{in} \\ \dots & \dots & \dots & \dots & \dots & \dots & \dots \\ a_{n1} & \dots & a_{ni} & \dots & a_{nj} & \dots & a_{nn} \end{pmatrix}$$

As a result of the matrix multiplication $\mathbf{AM}_{ij}(\delta)$, the matrix **A** elements placed in the ith column are added to the elements of the matrix **A** placed in the jth column increased in **δ** times. As a result of the matrix multiplication $\mathbf{M}_{ij}(\delta)$A, the matrix **A** elements placed in the jth row are added to the elements of the matrix **A** placed in the ith row increased in **δ** times;

e) multiplication of the square matrix **A** of $n \times n$ size by the matrix **X** containing n rows and one column (column vector):

$$\begin{pmatrix} a_{11}\ a_{12}\ a_{13}\ \dots a_{1n} \\ a_{21}\ a_{22}\ a_{23}\ \dots a_{2n} \\ \dots\dots\dots\dots \\ a_{n1}\ a_{n2}\ a_{n3}\ \dots a_{nn} \end{pmatrix} \times \begin{pmatrix} x_1 \\ x_2 \\ \dots \\ x_n \end{pmatrix} = \begin{pmatrix} a_{11}\cdot x_1 + a_{12}\cdot x_2 + a_{13}\cdot x_3 + \dots + a_{1n}\cdot x_n \\ a_{21}\cdot x_1 + a_{22}\cdot x_2 + a_{23}\cdot x_3 + \dots + a_{2n}\cdot x_n \\ \dots\dots\dots\dots\dots\dots \\ a_{n1}\cdot x_1 + a_{n2}\cdot x_2 + a_{n3}\cdot x_3 + \dots + a_{nn}\cdot x_n \end{pmatrix} \tag{2.1}$$

The given example (2.1) is a matter of principle. Later this example will be used to write the systems of linear algebraic equations;

f) multiplication of matrices where the number of rows or columns equals one:

$$\left(b_1\ b_2\ b_3\ \ldots b_i\ \ldots b_n\right)\times\begin{pmatrix}c_1\\c_2\\c_3\\\ldots\\c_i\\\ldots\\c_n\end{pmatrix}=\left(b_1\cdot c_1+b_2\cdot c_2+b_3\cdot c_3+\ldots+b_i\cdot c_i+\ldots+b_n\cdot c_n\right);$$

$$\begin{pmatrix}c_1\\c_2\\c_3\\\ldots\\c_i\\\ldots\\c_n\end{pmatrix}\times\left(b_1\ b_2\ b_3\ \ldots b_i\ \ldots b_n\right)=\begin{pmatrix}c_1b_1 & c_1b_2 & c_1b_3 & \ldots & c_1b_i & \ldots & c_1b_n\\c_2b_1 & c_2b_2 & c_2b_3 & \ldots & c_2b_i & \ldots & c_2b_n\\c_3b_1 & c_3b_2 & c_3b_3 & \ldots & c_3b_i & \ldots & c_3b_n\\\ldots & \ldots & \ldots & \ldots & \ldots & \ldots & \ldots\\c_ib_1 & c_ib_2 & c_ib_3 & \ldots & c_ib_i & \ldots & c_ib_n\\\ldots & \ldots & \ldots & \ldots & \ldots & \ldots & \ldots\\c_nb_1 & c_nb_2 & c_nb_3 & \ldots & c_nb_i & \ldots & c_nb_n\end{pmatrix}$$

The latter two examples highlight the importance of the multipliers location in the matrix multiplication. In the first case, the multiplication result is a number (or a matrix consisting of one element). In the second case, the original matrices (row vector and column vector) form the square matrix in the multiplication result.

It can be assumed that the unit square matrix can be the multiplication result of two certain matrices. Indeed, such matrices exist and are used in the matrix calculation. Let the matrix product **AC=E** be true. In this case, the matrix **C** is inversed to the matrix **A**. The inverse matrix **C** is designated as $\mathbf{A}^{-1}$. In the matrix calculus it is proved that if the matrix **A**determinant (definition of the determinant is given below in the same section) is not zero, then the inverse matrix $\mathbf{A}^{-1}$ exists and its value is the only one. The inverse matrix $\mathbf{A}^{-1}$ can be not similar in appearance to the matrix **A**. As an example, we can give the following matrix multiplication of a matrix and inverse matrix:

$$\mathbf{A}\mathbf{A}^{-1}=\begin{pmatrix}1&4&1\\4&2&1\\2&3&1\end{pmatrix}\times\begin{pmatrix}1&1&-2\\2&1&-3\\-8&-5&14\end{pmatrix}=\begin{pmatrix}1&0&0\\0&1&0\\0&0&1\end{pmatrix}=\mathbf{E}$$

The inverse matrix calculation is the time-consuming computational procedure. Its optimization is still an urgent problem.

The following matrix product $\mathbf{A} \bullet \mathbf{A}^T = \mathbf{E}$ can also be true. These unique real matrices where the inverse matrix $\mathbf{A}^{-1}$ coincides with the transposed matrix $\mathbf{A}^T$ are called orthogonal. If the matrix **A** elements are complex numbers and satisfy the condition $\mathbf{A} \bullet \mathbf{A}^* = \mathbf{E}$, the matrix is called unitary. The value of orthogonal matrices in calculus is difficult to overestimate. A lot of new high-performance computational procedures are based on these matrices. Some of them will be discussed later.

The definitions of matrix, sum and multiplication allow creating the matrix algebra. The basic properties of matrix calculus are listed below. Each of them should be seen as a theorem whose proof can be found in books on linear and (or) matrix algebra (e.g., [39, 55]).

1. $\mathbf{A} + \mathbf{B} = \mathbf{B} + \mathbf{A}$ – the addition of two matrices possesses a commutative property;
   $\mathbf{AB} \neq \mathbf{BA}$ – the multiplication of two matrices does not possess a commutative property;
2. $(\mathbf{A} + \mathbf{B}) + \mathbf{C} = \mathbf{A} + (\mathbf{B} + \mathbf{C})$ – the addition of matrices possesses an associativity property;
   $(\mathbf{AB})\mathbf{C} = \mathbf{A}(\mathbf{BC})$ – the multiplication of matrices possesses an associativity property;
3. $(\mathbf{A} + \mathbf{B})\mathbf{C} = \mathbf{AC} + \mathbf{BC}$;
   $\mathbf{C}(\mathbf{A} + \mathbf{B}) = \mathbf{CA} + \mathbf{CB}$ – the addition of matrices is linked to the multiplication by the distributive law;
4. $\alpha(\mathbf{AB}) = (\alpha\mathbf{A})\mathbf{B} = \mathbf{A}(\alpha\mathbf{B})$ – the multiplication of the product of two matrices by a number is equivalent to the prior multiplication of any of the matrices by that number;
5. $\mathbf{AE} = \mathbf{EA} = \mathbf{A}$ – the multiplication of any matrix by the unit (right or left side) matrix does not change the matrix content;
6. $(\mathbf{A}^T)^T = \mathbf{A}$ – the matrix **A** is the transposition result of the transposed matrix $\mathbf{A}^T$;
7. $(\mathbf{A}^{-1})^{-1} = \mathbf{A}$ – the matrix **A** is the inverse matrix of the inverse matrix $\mathbf{A}^{-1}$;
8. $(\mathbf{AB})^T = \mathbf{B}^T\mathbf{A}^T$ – the transposition of the product of two matrices is equivalent to the product of two transposed matrices. In this case, the multiplied matrices change their location;
9. $(\mathbf{AB})^{-1} = \mathbf{B}^{-1}\mathbf{A}^{-1}$) – the matrix inversed to the product of any two matrices can be calculated as the product of two matrices

inversed to the initial matrices; the multiplied matrices change their location.

2.2.1.3 Additional Properties of Matrices

Some additional properties can be set for the determined above square matrix **A** of $n \times n$ order that is used later in the creation of computational algorithms. In particular, the list of such qualities can be determinant (or determiner) and minors of matrices, proper and singular numbers of matrices, and matrix norm.

The matrix **A** determinant of $n \times n$ order (det**A**) is the sum $\det \mathbf{A} = \sum_{j=1}^{n} (-1)^{j+1} a_{1j} \det \mathbf{A}_{1j}$, where $\mathbf{A}_{1j}$ is the matrix obtained from the matrix **A** by deleting the first row and the jth column.

The definition implies that the determinant is the number that is obtained by the sum of all possible products of n matrix elements, which are in different rows and columns. The total number of possible products is $n!$, and the product sign can be positive or negative. For many applications, including technical ones, the numerical value of the determinant is essential. In particular, if the matrix A discriminant is zero (this matrix is called a degenerate matrix), then there is no inverse matrix $\mathbf{A}^{-1}$ for it.

The above formula to calculate the determinant could be re-written in other forms. In particular, in summation it would be possible to use any row with i-number as the base. In this case, the formula for determiner calculation can be re-written as

$$\det \mathbf{A} = \sum_{j=1}^{n} (-1)^{j+i} a_{ij} \det \mathbf{A}_{ij} \tag{2.2}$$

Moreover, the summation can be carried out by the matrix rows and not by columns

$$\det \mathbf{A} = \sum_{i=1}^{n} (-1)^{j+i} a_{ij} \det \mathbf{A}_{ij} \tag{2.3}$$

Other options of the matrix determinant calculation can be proposed, but the calculation result will be the same, in any case.

To record the matrix **A** determinant, another designation frequently applied in practice is used

$$\det \mathbf{A} = \begin{vmatrix} a_{11} & a_{12} & \dots & a_{1i} & \dots & a_{1m} \\ a_{21} & a_{22} & \dots & a_{2i} & \dots & a_{2m} \\ \dots & \dots & \dots & \dots & \dots & \dots \\ a_{i1} & a_{i2} & \dots & a_{ii} & \dots & a_{im} \\ \dots & \dots & \dots & \dots & \dots & \dots \\ a_{n1} & a_{n2} & \dots & a_{ni} & \dots & a_{nm} \end{vmatrix} \tag{2.4}$$

The determinant definitions imply that its computation involves the calculation of n determinants of $(n\text{-}1)\times(n\text{-}1)$ size obtained from the original matrix **A**. The computation of the determinants, in turn, comprises the calculations of $(n\text{-}1)$ determinants of $(n\text{-}2)\times(n\text{-}2)$ size, and so on, until the computational process comes to the matrix determinants consisting of one element.

The determinants of the matrices of lower orders, which can be obtained from the matrix **A** by deleting certain rows and columns, are called minors. Obviously, minors can be determined both for square and rectangular matrices. For square matrices of $n\times n$ size, the submatrices built along its diagonal are of interest.

$$\mathbf{A}_1 = (a_{11}); \quad \mathbf{A}_2 = \begin{pmatrix} a_{11} & a_{12} \\ a_{21} & a_{22} \end{pmatrix}; \quad \mathbf{A}_3 = \begin{pmatrix} a_{11} & a_{12} & a_{13} \\ a_{21} & a_{22} & a_{23} \\ a_{31} & a_{32} & a_{33} \end{pmatrix};$$

$$\mathbf{A}_4 = \begin{pmatrix} a_{11} & a_{12} & a_{13} & a_{14} \\ a_{21} & a_{22} & a_{23} & a_{24} \\ a_{31} & a_{32} & a_{33} & a_{34} \\ a_{41} & a_{42} & a_{43} & a_{44} \end{pmatrix} \text{ etc.}$$

Minors det$\mathbf{A}_1$, det$\mathbf{A}_2$, det$\mathbf{A}_3$, det$\mathbf{A}_4$ and others are called principal diagonal minors, or simply, principal minors.

Another important characteristic of a matrix is its rank. It can be used both for square and rectangular matrices. We define the rank of the rectangular matrix **A** of $m\times n$ size. Let us first consider all the minors of the original matrix with the first order (matrix size is 1×1). If there is one non-zero minor among them, then we consider all minors of the original matrix with the second order (matrix size is 2×2). Continuing this procedure many times, let us consider all minors of the original matrix with the $r\times r$ size. Let among all minors of $r\times r$ order be at least one non-zero minor, but all minors of $(r+1)\times(r+1)$ order are zero. In this case, the r number is the rank of the original rectangular matrix **A**. It is obvious that the inequality $r \leq \min(m,n)$, stating that the matrix rank cannot be greater than the number of rows or columns in the original matrix, is true.

We note some useful properties of square matrices simplifying the calculation of their determinants:

- if in the matrix **A** two rows or two columns coincide, then the determinant of this matrix is zero;
- if any two rows or two columns in the matrix are interchanged, then the determinant changes its sign to the opposite;
- if all the elements of any row or column in the matrix **A** are multiplied by $k\neq0$ number, then the determinant value increases in k times;
- the value of the matrix determinant is not changed if the elements of any row (any column) are added to the elements of another row (column) multiplied by an arbitrary number $k\neq0$;
- if all the elements of the matrix placed above the central diagonal equal zero (lower or left matrix – **L**), or all the elements of the matrix placed below the central diagonal (upper or right matrix – **U**), then the determinant value is calculated as the product of the matrix diagonal terms.

After defining the determinant of the square matrix **A,** we can introduce the definition of eigen numbers (eigen values) of this matrix.

The eigen numbers λ_i of the matrix A are roots of λ_i equation, resulting in the determinant expansion

$$\begin{vmatrix} a_{11}-\lambda & a_{12} & \dots & a_{1n} \\ a_{21} & a_{22}-\lambda & \dots & a_{2n} \\ \dots & \dots & \dots & \dots \\ a_{n1} & a_{n2} & \dots & a_{nn}-\lambda \end{vmatrix} = 0 \qquad (2.5)$$

The Eq. (2.5) is often called as characteristic or age-old. The determinant expansion leads to the polynomial of n degree, and its solution can present difficulties. Nevertheless, among the above-listed matrices there are such ones, which do not have problems with determining the eigen numbers (eigen values). These are the diagonal matrix **D**, the upper matrix **U**, and the lower matrix **L**. For these matrices their eigen numbers coincide with the values of diagonal elements $\boldsymbol{a}_{ii}$. If the elements $\boldsymbol{a}_{ii}$ of the matrix are real numbers, then the eigen numbers of the original matrix are real numbers. Let us note that in symmetric matrices **S** all eigen numbers are also real.

Such matrices will be used in the following sections. Thereby we note that two matrices are called similar if all eigen numbers of two matrices coincide.

In practical applications the eigen numbers λ_i of the matrix **C** which is obtained as the product of two matrices – **A** and transposed one $\mathbf{A}^{\mathrm{T}}$ ($\mathbf{C} = \mathbf{A}\mathbf{A}^{\mathrm{T}}$) are of interest. It is known [39, 55] that for any real matrix **A** the eigen numbers λ_i of the matrix **C** are positive, and for these matrices so-called singular values s_i calculated as $s_i = \sqrt{\lambda_i}$ are introduced.

To analyze the stability of computational procedures working with matrices (e.g., solving the system of linear equations), the notion of matrix norm is introduced. For the matrix A the norm is designated as $\|\mathbf{A}\|$. The norm can be calculated in different ways, each of them is used in practice:

1. Sums of matrix elements taken by the absolute value and arranged in one row are calculated. Among n sums calculated for each row, the maximum one is selected. This sum (the number) is accepted as the matrix $\|\mathbf{A}\|_1$ norm. The formula for norm calculation is written down as:

$$\|\mathbf{A}\|_1 = \max \sum_{i=1}^{m} |a_{ij}|, \quad 1 \le j \le n; \tag{2.6}$$

2. The first singular value of the matrix **A** is taken as the norm $\|\mathbf{A}\|_2$:

$$\|\mathbf{A}\|_2 = s_1(\mathbf{A}); \tag{2.7}$$

3. The sum of squares of all the matrix elements is calculated. The value of the square root of the sum is taken as the norm $\|\mathbf{A}\|_E$ of the matrix **A**:

$$\|\mathbf{A}\|_E = (\sum_{i=1}^{m} \sum_{j=1}^{n} a_{ij}^2)^{0,5}; \tag{2.8}$$

The introduced norm (2.8) is called Frobenius norm or Euclidean norm.

4. Sums of matrix elements taken by the absolute value and arranged in one column are calculated. Among m sums calculated for each column, the maximum one is selected. This sum (the number) is accepted as the matrix norm and is designated as $\|\mathbf{A}\|_\infty$. The formula for norm calculation is written down as:

$$\|\mathbf{A}\|_\infty = \max \sum_{j=1}^{n} |a_{ij}|, \quad 1 \le i \le m \tag{2.9}$$

For all the above norms the following correlations are true:

$$\|\mathbf{AB}\| \le \|\mathbf{A}\| \|\mathbf{B}\|; \tag{2.10}$$

$$\|\mathbf{A}\|_2 \le \|\mathbf{A}\|_E \le \min(\sqrt{m}, \sqrt{n}) \|\mathbf{A}\|_2; \tag{2.11}$$

$$\operatorname{cond} \mathbf{A} = \|\mathbf{A}\| \|\mathbf{A}^{-1}\| \ge 1 \tag{2.12}$$

The correlations (2.10)–(2.12) should be understood as theorems, the proof of which can be found in additional literature (e.g., [39, 55]). The number cond **A** written down in (2.12) is called the matrix condition number. The matrix condition number is a useful criterion for assessing the stability of matrix calculations. As a rule, the closer is the value of the condition number to one, the slower errors are accumulated when performing the matrix transformations.

Let us note one important property of symmetric matrices that will be mentioned below:

The symmetric matrix A with the size $n \times n$ is called positive-definite, if for any random vector **X** (matrix with the size $n \times 1$) the condition $\mathbf{X}^T \mathbf{A} \mathbf{X} > \mathbf{0}$ is true.

2.2.2 APPLICATION OF MATRIX CALCULUS TO SOLVE THE SYSTEMS OF LINEAR ALGEBRAIC EQUATIONS

In problems related to the mathematical treatment of table functions, there is a need to solve systems of linear algebraic equations. As a system of linear equations we will assign systems of the following form

$$a_{11}x_1 + a_{12}x_2 + ... + a_{1i}x_i + ... + a_{1n}x_n = b_1$$

$$a_{21}x_1 + a_{22}x_2 + ... + a_{2i}x_i + ... + a_{2n}x_n = b_2$$

$$...$$

$$a_{i1}x_1 + a_{i2}x_2 + ... + a_{ii}x_i + ... + a_{in}x_n = b_i$$

$$...$$

$$a_{n1}x_1 + a_{n2}x_2 + ... + a_{ni}x_i + ... + a_{nn}x_n = b_n \qquad (2.13)$$

In the written equation system x_i, $(i = 1,...,n)$ are unknown quantities and coefficients a_{ij}, b_i $(i, j = 1,...,n)$ – any real or complex numbers. The equation system (2.13) can be re-written in matrix form. We introduce matrices

$$\mathbf{A} = \begin{pmatrix} a_{11} & a_{12} & ... & a_{1i} & ... & a_{1n} \\ a_{21} & a_{22} & ... & a_{2i} & ... & a_{2n} \\ ... & ... & ... & ... & ... & ... \\ a_{i1} & a_{i2} & ... & a_{ii} & ... & a_{in} \\ ... & ... & ... & ... & ... & ... \\ a_{n1} & a_{n2} & ... & a_{ni} & ... & a_{nn} \end{pmatrix}; \quad \mathbf{X} = \begin{pmatrix} x_1 \\ x_2 \\ ... \\ x_i \\ ... \\ x_n \end{pmatrix}; \quad \mathbf{B} = \begin{pmatrix} b_1 \\ b_2 \\ ... \\ b_i \\ ... \\ b_n \end{pmatrix} \qquad (2.14)$$

In accordance with the rules of matrix calculus, the original system of Eqs. (2.13) can be re-written in matrix form as

$$\mathbf{AX} = \mathbf{B} \qquad (2.15)$$

Now it is possible to evaluate the merits of multiplication accepted in matrix calculus. Writing down the system of linear algebraic equations in the matrix form (2.15) is much simpler than (2.13). But the most important thing is that to solve the equation system (2.15) a very well-developed apparatus of matrix calculus can be used. Moreover, in many mathematical systems and mathematical libraries for compliers the solving of systems of linear algebraic equations can be performed [22] directly in a matrix form. We note that when solving systems of linear algebraic equations, the extended matrix $\mathbf{A}_\mathbf{p}$ can be used instead of the matrix **A** and vector **B**; the dimension of such matrix is $n \times (n+1)$ and the last column is the vector of the right sides of the equation system:

$$\mathbf{A}_p = \begin{pmatrix} a_{11} & a_{12} & \dots & a_{1i} & \dots & a_{1n} & b_1 \\ a_{21} & a_{22} & \dots & a_{2i} & \dots & a_{2n} & b_2 \\ \dots & \dots & \dots & \dots & \dots & \dots & \dots \\ a_{i1} & a_{i2} & \dots & a_{ii} & \dots & a_{in} & b_i \\ \dots & \dots & \dots & \dots & \dots & \dots & \dots \\ a_{n1} & a_{n2} & \dots & a_{ni} & \dots & a_{nn} & b_n \end{pmatrix}$$

Solving the systems of linear algebraic equations is one of the most common procedures in the solution of engineering problems. The classical linear algebra studies a class of systems of linear algebraic equations where the number of equations and number of unknowns are equal. If the matrix **A** determinant is not zero (det $\mathbf{A} \neq 0$), then the equation system (2.15) has the unique solution. However, it should be noted that in engineering, generally speaking, the number of equations in the linear system may not coincide with the number of unknowns. This situation is normal because an approximate solution of the equation system is enough for engineering applications. Systems, where the number of equations does not equal the number of unknowns, are called underdetermined (number of equations is less than the number of unknowns) or overdetermined (number of equations is larger than the number of unknowns). The solution of these equation systems will be examined in Chapter 4. This chapter discusses elementary methods of solving systems like (2.13) (or (2.15)), for which the condition det $\mathbf{A} \neq 0$ is true. More complex, but more effective methods will be discussed in the following chapters.

2.2.2.1 Cramer's Rule (Determinants)

It should be noted that Cramer's rule is the simplest computational method used for solving systems of linear algebraic equations; however, its use is reasonable with a small number of solvable equations (no more than 5–7). The solution of the equation system comes to the calculation of $(n+1)$ determinant – Δ, Δ_1, …, Δ_i, …, Δ_n:

$$\Delta = \det \mathbf{A} = \begin{vmatrix} a_{11} & a_{12} & \dots & a_{1i} & \dots & a_{1n} \\ a_{21} & a_{22} & \dots & a_{2i} & \dots & a_{2n} \\ \dots & \dots & \dots & \dots & \dots & \dots \\ a_{i1} & a_{i2} & \dots & a_{ii} & \dots & a_{in} \\ \dots & \dots & \dots & \dots & \dots & \dots \\ a_{n1} & a_{n2} & \dots & a_{ni} & \dots & a_{nn} \end{vmatrix};$$

$$\Delta_i = \det \mathbf{A}_i = \begin{vmatrix} a_{11} & a_{12} & \dots & b_1 & \dots & a_{1n} \\ a_{21} & a_{22} & \dots & b_2 & \dots & a_{2n} \\ \dots & \dots & \dots & \dots & \dots & \dots \\ a_{i1} & a_{i2} & \dots & b_i & \dots & a_{in} \\ \dots & \dots & \dots & \dots & \dots & \dots \\ a_{n1} & a_{n2} & \dots & b_n & \dots & a_{nn} \end{vmatrix}, \; i=1, n$$

The final solution of the problem (2.13) by Cramer's rule is written as

$$x_i = \frac{\Delta_i}{\Delta}, \quad i = 1, n \tag{2.16}$$

Sometimes the solution of the equation system (2.13) by Cramer's rule is written in another form:

$$x_i = \frac{\sum_{k=1}^{n} (b_k \cdot \det \mathbf{A}_{ki})}{\det \mathbf{A}}$$

$\mathbf{A}_{ki}$ – minors of the matrix **A** (their number equals the number of equations n).

2.2.2.2 Inverse Matrix Method

We perform the multiplication of the matrix equation $\mathbf{A}\mathbf{X} = \mathbf{B}$ on the left by the matrix $\mathbf{A}^{-1}$ inversed to the matrix **A**

$$\mathbf{A}^{-1}\mathbf{A}\mathbf{X} = \mathbf{A}^{-1}\mathbf{B}$$

Taking into consideration the equality $\mathbf{A}^{-1}\mathbf{A} = \mathbf{E}$, this equation can be re-written

$$\mathbf{X} = \mathbf{A}^{-1} \cdot \mathbf{B} \tag{2.17}$$

Thus, the equation system solution comes to the calculation of the inverse matrix $\mathbf{A}^{-1}$. If the inverse matrix calculation is easy, the method can be the most effective known method. Let us take, for example, the orthogonal matrices. If the matrix **A** is orthogonal, then the inverse matrix $\mathbf{A}^{-1}$ coincides with the transposed one $\mathbf{A}^T = \mathbf{A}^{-1}$, and the equation system solution (2.13) can be

$$\mathbf{X} = \mathbf{A}^T\mathbf{B} \tag{2.18}$$

The assessments show that the use of the inverse matrix method, as well as the use of Cramer's rule are time-consuming and require performing at least $(n+1)!$ elementary arithmetic operations (additions and multiplications). With a large value of n the application of the inverse matrix method is cumbersome and cannot be recommended. Despite the fact that at the present there are new effective algorithms to calculate inverse matrices, the method (2.17) is still unpopular, because the complexity of inverse matrices calculation is comparable with the complexity of the equation system solution (2.13).

2.2.2.3 Gauss Method (Jordan-Gauss)

Let us discuss the special case of the equation system (2.13), where the matrix **U** is the upper triangular matrix

$$\begin{pmatrix} u_{11} & u_{12} & \ldots & u_{1i} & \ldots & u_{1n} \\ 0 & u_{22} & \ldots & u_{2i} & \ldots & u_{2n} \\ \ldots & \ldots & \ldots & \ldots & \ldots & \ldots \\ 0 & 0 & \ldots & u_{ii} & \ldots & u_{in} \\ \ldots & \ldots & \ldots & \ldots & \ldots & \ldots \\ 0 & 0 & \ldots & 0 & \ldots & u_{nn} \end{pmatrix} \cdot \begin{pmatrix} x_1 \\ x_2 \\ \ldots \\ x_i \\ \ldots \\ x_n \end{pmatrix} = \begin{pmatrix} b_1^* \\ b_2^* \\ \ldots \\ b_i^* \\ \ldots \\ b_n^* \end{pmatrix} \tag{2.19}$$

The solution of this system is not difficult if all the elements u_{ii} of the matrix **U** lying on the main diagonal are not zero. Indeed, from the latter equation of the linear system the value of the unknown x_n is calculated. On the basis of the known x_n the value x_{n-1} is found from the last but one system equation. Gradually, all the values x_i for arbitrary $i = 1, \ldots, n-1$ are established in the bottom-up way. The main calculation relations used in the algorithm are of the form

$$x_n = \frac{b_n^*}{u_{nn}},$$

$$x_{n-1} = \frac{b_{n-1}^* - u_{n-1,n}x_n}{u_{n-1,n-1}},$$

$$\ldots$$

$$x_i = \frac{b_i^* - \sum_{k=i+1}^{n} u_{ik} x_k}{u_{ii}},$$

$$\ldots$$

$$x_1 = \frac{b_1^* - \sum_{k=2}^{n} u_{1k} x_k}{u_{11}} \tag{2.20}$$

The main idea of Gauss method is to transform the original matrix **A** having an arbitrary form, until it takes the form of the upper triangular matrix **U**. The matrix transformation can be performed step-by-step excluding variables in the equations, or using the matrix transformation with the use of the foregoing modified unit matrix $\mathbf{M}_{ij}(\delta)$.

The algorithm of the matrix $\mathbf{M}_{ij}(\delta)$ application for the transformation of the original matrix **A** to the upper triangular form is as follows:

a) let us calculate the changed unit matrix $\mathbf{M}_1$ determined as the product of the matrices $\mathbf{M}_{i1}(\delta)$

$$\mathbf{M}_1 = \mathbf{M}_{21}(m_{21})\mathbf{M}_{31}(m_{31})\cdot\ldots\cdot\mathbf{M}_{i1}(m_{i1})\cdot\ldots\cdot\mathbf{M}_{n1}(m_{n1}) \tag{2.21}$$

The coefficients m_{i1} in the Eq. (2.21) are determined by the formula

$$m_{i1} = -\frac{a_{i1}}{a_{11}}$$

b) let us calculate the matrix $\mathbf{A}_1$ elements determined as the product $\mathbf{A}_1 = \mathbf{M}_1\mathbf{A}$

$$\begin{pmatrix} 1 & 0 & 0 & \ldots & 0 & \ldots & 0 \\ m_{21} & 1 & 0 & \ldots & 0 & \ldots & 0 \\ m_{31} & 0 & 1 & \ldots & 0 & \ldots & 0 \\ \ldots & \ldots & \ldots & \ldots & \ldots & \ldots & \ldots \\ m_{i1} & 0 & 0 & \ldots & 1 & \ldots & 0 \\ \ldots & \ldots & \ldots & \ldots & \ldots & \ldots & \ldots \\ m_{n1} & 0 & 0 & \ldots & 0 & \ldots & 1 \end{pmatrix} \times \begin{pmatrix} a_{11} & a_{12} & a_{13} & \ldots & a_{1i} & \ldots & a_{1n} \\ a_{21} & a_{22} & a_{23} & \ldots & a_{2i} & \ldots & a_{2n} \\ a_{31} & a_{32} & a_{33} & \ldots & a_{3i} & \ldots & a_{3n} \\ \ldots & \ldots & \ldots & \ldots & \ldots & \ldots & \ldots \\ a_{i1} & a_{i2} & a_{i3} & \ldots & a_{ii} & \ldots & a_{in} \\ \ldots & \ldots & \ldots & \ldots & \ldots & \ldots & \ldots \\ a_{n1} & a_{n2} & a_{n3} & \ldots & a_{ni} & \ldots & a_{nn} \end{pmatrix}$$

$$= \begin{pmatrix} a_{11} & a_{12} & a_{13} & \dots & a_{1i} & \dots & a_{1n} \\ 0 & a_{22}^{(1)} & a_{23}^{(1)} & \dots & a_{2i}^{(1)} & \dots & a_{2n}^{(1)} \\ 0 & a_{32}^{(1)} & a_{33}^{(1)} & \dots & a_{3i}^{(1)} & \dots & a_{3n}^{(1)} \\ \dots & \dots & \dots & \dots & \dots & \dots & \dots \\ 0 & a_{i2}^{(1)} & a_{i3}^{(1)} & \dots & a_{ii}^{(1)} & \dots & a_{in}^{(1)} \\ \dots & \dots & \dots & \dots & \dots & \dots & \dots \\ 0 & a_{n2}^{(1)} & a_{n3}^{(1)} & \dots & a_{ni}^{(1)} & \dots & a_{nn}^{(1)} \end{pmatrix}$$

The new matrix $\mathbf{A}_1$ has the following properties:

- the first row of the matrix $\mathbf{A}_1$ coincides with the first row of the matrix $\mathbf{A}$;
- elements a_{i1} $(i > 1)$ of the matrix $\mathbf{A}_1$ are zero – $a_{i1} = 0$.
- elements of the following i^{th} row change their values according to the correlation $a_{ij}^{(1)} = a_{ij} - \dfrac{a_{i1}}{a_{11}} \cdot a_{1j}$;

c) let us consider the matrix $\mathbf{A}_1$ and construct the modified unit matrix $\mathbf{M}_2$ for it determined as the product of the matrices $\mathbf{M}_{j2}(\delta)$

$$\mathbf{M}_2 = \mathbf{M}_{32}(m_{32})\mathbf{M}_{42}(m_{42}) \cdot \ldots \cdot \mathbf{M}_{i2}(m_{i2}) \cdot \ldots \cdot \mathbf{M}_{n2}(m_{n2}) \qquad (2.22)$$

The coefficients m_{i2} in the Eq. (2.22) are determined by the formula

$$m_{i2} = -\frac{a_{i2}^{(1)}}{a_{22}^{(1)}};$$

d) let us calculate the matrix $\mathbf{A}_2$ elements determined as the product $\mathbf{A}_2 = \mathbf{M}_2\mathbf{A}_1$

$$\begin{pmatrix} 1 & 0 & 0 & \dots & 0 & \dots & 0 \\ 0 & 1 & 0 & \dots & 0 & \dots & 0 \\ 0 & m_{32} & 1 & \dots & 0 & \dots & 0 \\ \dots & \dots & \dots & \dots & \dots & \dots & \dots \\ 0 & m_{i3} & 0 & \dots & 1 & \dots & 0 \\ \dots & \dots & \dots & \dots & \dots & \dots & \dots \\ 0 & m_{n3} & 0 & \dots & 0 & \dots & 1 \end{pmatrix} \times \begin{pmatrix} a_{11} & a_{12} & a_{13} & \dots & a_{1i} & \dots & a_{1n} \\ 0 & a_{22}^{(1)} & a_{23}^{(1)} & \dots & a_{2i}^{(1)} & \dots & a_{2n}^{(1)} \\ 0 & a_{32}^{(1)} & a_{33}^{(1)} & \dots & a_{3i}^{(1)} & \dots & a_{3n}^{(1)} \\ \dots & \dots & \dots & \dots & \dots & \dots & \dots \\ 0 & a_{i2}^{(1)} & a_{i3}^{(1)} & \dots & a_{ii}^{(1)} & \dots & a_{in}^{(1)} \\ \dots & \dots & \dots & \dots & \dots & \dots & \dots \\ 0 & a_{n2}^{(1)} & a_{n3}^{(1)} & \dots & a_{ni}^{(1)} & \dots & a_{nn}^{(1)} \end{pmatrix}$$

$$= \begin{pmatrix} a_{11} & a_{12} & a_{13} & \dots & a_{1i} & \dots & a_{1n} \\ 0 & a_{22}^{(1)} & a_{23}^{(1)} & \dots & a_{2i}^{(1)} & \dots & a_{2n}^{(1)} \\ 0 & 0 & a_{33}^{(2)} & \dots & a_{3i}^{(2)} & \dots & a_{3n}^{(2)} \\ \dots & \dots & \dots & \dots & \dots & \dots & \dots \\ 0 & 0 & a_{i3}^{(2)} & \dots & a_{ii}^{(2)} & \dots & a_{in}^{(2)} \\ \dots & \dots & \dots & \dots & \dots & \dots & \dots \\ 0 & 0 & a_{n3}^{(2)} & \dots & a_{ni}^{(2)} & \dots & a_{nn}^{(2)} \end{pmatrix}$$

The new matrix $\mathbf{A}_2$ has the following properties:

- the first row of the matrix $\mathbf{A}_2$ coincides with the first row of the matrix $\mathbf{A}$;
- the second row of the matrix $\mathbf{A}_2$ coincides with the second row of the matrix $\mathbf{A}_1$;
- elements $a_{i1}\ (i > 1), a_{i2}\ (i > 2)$ of the matrix $\mathbf{A}_2$ are zero;
- elements of any following i^{th} row change their values according to the relation $a_{ij}^{(2)} = a_{ij}^{(1)} - \dfrac{a_{i2}^{(1)}}{a_{22}^{(1)}} \cdot a_{2j}^{(1)}$;

e) similarly, we find the matrices $\mathbf{M}_3, \mathbf{M}_4, \dots, \mathbf{M}_i, \dots, \mathbf{M}_{n-1}$. The matrix $\mathbf{A}_{n-1}$ obtained in the result of the matrix multiplication

$$\mathbf{A}_{n-1} = \mathbf{M}_{n-1}\, \mathbf{M}_{n-2} \cdot \dots \cdot \mathbf{M}_i \cdot \dots \cdot \mathbf{M}_2 \mathbf{M}_1 \mathbf{A}$$

has the upper triangular form:

$$\mathbf{A}_{n-1} = \begin{pmatrix} a_{11} & a_{12} & a_{13} & \dots & a_{1i} & \dots & a_{1n} \\ 0 & a_{22}^{(1)} & a_{23}^{(1)} & \dots & a_{2i}^{(1)} & \dots & a_{2n}^{(1)} \\ 0 & 0 & a_{33}^{(2)} & \dots & a_{3i}^{(2)} & \dots & a_{3n}^{(2)} \\ \dots & \dots & \dots & \dots & \dots & \dots & \dots \\ 0 & 0 & 0 & \dots & a_{ii}^{(i-1)} & \dots & a_{in}^{(i-1)} \\ \dots & \dots & \dots & \dots & \dots & \dots & \dots \\ 0 & 0 & 0 & \dots & 0 & \dots & a_{nn}^{(n-1)} \end{pmatrix}$$

Let us point out the important properties of the matrix multiplication $\mathbf{M}_i$, the proof of which is not given because of its simplicity. These properties will be used hereinafter:

- the first property – the matrix **M** determined as the product of the matrices $\mathbf{M}_i$ is the lower triangular matrix

$$\mathbf{M} = \mathbf{M}_{n-1}\,\mathbf{M}_{n-2}\cdot\ldots\cdot\mathbf{M}_i\cdot\ldots\cdot\mathbf{M}_2\mathbf{M}_1 = \begin{pmatrix} 1 & 0 & 0 & \ldots & 0 & \ldots & 0 \\ m_{21} & 1 & 0 & \ldots & 0 & \ldots & 0 \\ m_{31} & m_{32} & 1 & \ldots & 0 & \ldots & 0 \\ \ldots & \ldots & \ldots & \ldots & \ldots & \ldots & \ldots \\ m_{i1} & m_{i2} & m_{i3} & \ldots & 1 & \ldots & 0 \\ \ldots & \ldots & \ldots & \ldots & \ldots & \ldots & \ldots \\ m_{n1} & m_{n2} & m_{n3} & \ldots & m_{n,i} & \ldots & 1 \end{pmatrix} \quad (2.23)$$

- the second property – the matrix **L**, the matrix **M** inverse, is the next lower triangular matrix

$$\mathbf{L} = \mathbf{M}^{-1} = \mathbf{M}_1^{-1}\,\mathbf{M}_2^{-1}\cdot\ldots\cdot\mathbf{M}_i^{-1}\cdot\ldots\cdot\mathbf{M}_{n-1}^{-1}$$

$$= \begin{pmatrix} 1 & 0 & 0 & \ldots & 0 & \ldots & 0 \\ -m_{21} & 1 & 0 & \ldots & 0 & \ldots & 0 \\ -m_{31} & -m_{32} & 1 & \ldots & 0 & \ldots & 0 \\ \ldots & \ldots & \ldots & \ldots & \ldots & \ldots & \ldots \\ -m_{i1} & -m_{i2} & -m_{i3} & \ldots & 1 & \ldots & 0 \\ \ldots & \ldots & \ldots & \ldots & \ldots & \ldots & \ldots \\ -m_{n1} & -m_{n2} & -m_{n3} & \ldots & -m_{n,i} & \ldots & 1 \end{pmatrix} \quad (2.24)$$

Taking into account (2.23), the system of linear Eq. (2.15) can be re-written as

$$\mathbf{U}\mathbf{X} = \mathbf{B}^* \quad (2.25)$$

In Eq. (2.25) **U=MA** – is the upper triangular matrix, **B*=MB**.

The subsequent solution of the equation system (2.25), which is the essence of the Gauss method (Jordan-Gauss), is described above (Eqs. (2.19) and (2.20)).

The implementation of Gauss method requires performing $\sim n^3$ arithmetic operations approximately (n – number of equations in the system). For the large values of n (for $n>5$), Gauss method is more efficient than Cramer's rule.

It should be noted that the calculation of the coefficients m_{ij} for the matrix $\mathbf{M}_i$ involves dividing the elements of the original matrix $\mathbf{A}$ and intermediate matrix $\mathbf{A}_i$ by their diagonal elements. It was noted in Chapter 1 that if in division the denominator value is close to zero, then the loss of accuracy takes place. In this regard, along with the examined method its modification called Gauss method with pivoting is used in practice. The essence of modifications is that at each stage of nulling the matrix $\mathbf{A}_{i-1}$ elements placed in the ith column ($i>1$), the permutation of rows is performed. As a result of permutation, the elements of the row $j \geq i$ should be placed in the ith row of the matrix $\mathbf{A}_{i-1}$, where the value of element $a_{ij}^{(i)}$ located in the ith column is maximal. Thus, unlike the standard Gauss method described above, in the finally solved Eq. (2.25) the resulting matrix $\mathbf{M}$ used in the calculation of the matrix $\mathbf{U}$ and vector $\mathbf{B}^*$ using Gauss method with pivoting has the form

$$\mathbf{M} = (\mathbf{M}_{n-1}\mathbf{P}_{n-1})\,(\mathbf{M}_{n-2}\mathbf{P}_{n-2})\cdot\ldots\cdot(\mathbf{M}_i\mathbf{P}_i)\cdot\ldots\cdot(\mathbf{M}_2\mathbf{P}_2)(\mathbf{M}_1\mathbf{P}_1) \quad (2.26)$$

In the written matrix product (2.26) the matrices $\mathbf{P}_i$ are permutation matrices. The multiplication by these matrices is performed before the multiplication by the matrix $\mathbf{M}_i$ nulling the elements placed in the ith column.

In comparison with the standard Gauss method, the modified version requires many arithmetic operations, but the accuracy of problem solution has significantly increased. It should be noted that Gauss method (including a modified variant with the rows permutation) is conditionally stable [46, 75, 102, 228]. Its stability to the accumulation of calculation errors can be additionally increased using the combination of both rows of intermediate matrices and their columns permutation.

The described methods for solving the systems of linear algebraic equations are enough for the problems solved in this chapter. We note that at present there are more powerful and reliable methods for solving such equations, and they will be discussed further in Chapter 8.

2.3 MATHEMATICAL PROCEDURES WHEN WORKING WITH TABLES

2.3.1 APPROXIMATION OF FUNCTIONS, DERIVATIVES AND INTEGRALS USING POLYNOMIALS AND NEWTON FORMULAS

In engineering, problems associated with interpolation and approximation of functions are frequent. The need to use the interpolation procedure appears in the work with tabular data, including tabular data obtained when processing the results of the experiments or complex calculations. With the use of interpolation algorithms it is possible to restore the unknown function values for all possible values of the argument. As a consequence, the solution of interpolation problem allows solving the related problems of function approximation – numerical differentiation and numerical integration. In engineering calculation, these problems are important, especially when it is difficult or even impossible to write down the functional relations in analytical form. The issues related to the interpolation, numerical differentiation and integration of functions given in tables will be examined below.

2.3.1.1 Approximation of Functions Depending on One Argument

Let us discuss the problem of function interpolation f depending only on one argument x. The problem of interpolation in this case is formulated as follows:

- let us assume that some function $f(x)$ whose values f_i are known for discrete values x_i (i=1, I; $x_1 < x_2 < ... < x_I$). We calculate the value of the function $f(x)$ for any $x \in (x_1, x_I)$.

Further, the discrete values $(x_i, f(x_i))$ corresponding to the tabular ones are called nodal points. From the problem definition it follows that the desired function $f(x)$ is not established in the explicit form, and it creates certain freedom of choice. Indeed, we can assume that the function $f(x)$ is linear, polynomial, logarithmic, trigonometric or any combination thereof. The only condition that is set by the definition of interpolation problem is the equivalence of the values of function f_i to the values of the argument x_i.

Let us consider the possible way of solving the interpolation problem. Linear interpolation is the simplest method. With the linear interpolation the restored function $f(x)$ is the set of line segments crossing in the points corresponding to the table values (x_i, $f(x_i)$, $i=1, I$):

$$f(x) = f(x_i)\frac{x_{i+1} - x}{x_{i+1} - x_i} + f(x_{i+1})\frac{x - x_i}{x_{i+1} - x_i}, \quad (x_i < x < x_{i+1}) \qquad (2.27)$$

The linear interpolation can be the acceptable approximation of function when there is no need to calculate the derivatives $\frac{d^{(n)} f(x)}{dx^n}$. With the linear interpolation the first derivatives have constant values within any range ($x_i < x < x_{i+1}$, $i=1, I$) and are discontinued in nodal points. In any case, the differentiation procedures with the use of linear interpolation have the intolerable error.

The interpolation error can be reduced by using the parabola passing through the three neighboring nodal points for restoring the function $f(x)$. With the number of nodal points more than three, the function can be interpolated as piecewise-parabolic

$$f(x) = f(x_{i-1})\frac{(x-x_i)(x-x_{i+1})}{(x_{i-1}-x_i)(x_{i-1}-x_{i+1})} + f(x_i)\frac{(x-x_{i-1})(x-x_{i+1})}{(x_i-x_{i-1})(x_i-x_{i+1})}$$
$$+ f(x_{i+1})\frac{(x-x_{i-1})(x-x_i)}{(x_{i+1}-x_{i-1})(x_{i+1}-x_i)}; \quad x_{i-1} \le x \le x_{i+1}, \quad i = 2, I+1 \tag{2.28}$$

However, the increase in the accuracy of the function reproduction $f(x)$ does not lead to the increase in the accuracy of the derivatives reproduction $\frac{d^{(n)} f(x)}{dx^n}$. In the nodal points, as in linear interpolation, the first and second derivatives $\frac{df(x)}{dx}, \frac{d^2 f(x)}{dx^2}$ will have gaps, and the values of higher derivatives (third and others) equal zero – $\frac{d^3 f(x)}{dx^3} = 0, \frac{d^4 f(x)}{dx^4} = 0, \ldots$

According to the table values of the function $(x_i, f(x_i))$, $i=1$, I, it is possible to construct a polynomial of Ith degree, which is called Lagrangian polynomial [23, 75]

$$f(x) = \sum_{i=1}^{I} f(x_i)\frac{(x-x_1)(x-x_2)\cdot\ldots\cdot(x-x_{i-1})(x-x_{i+1})\cdot\ldots\cdot(x-x_I)}{(x_i-x_1)(x_i-x_2)\cdot\ldots\cdot(x_i-x_{i-1})(x_i-x_{i+1})\cdot\ldots\cdot(x_i-x_I)} \tag{2.29}$$

The interpolation formulas can be constructed using the divided differences φ_i that are based on the function expansion in Taylor series. Such formulas are called Newton interpolation formulas [75, 117], and their use is reasonable if you need to increase the accuracy of the function definition $f(x)$ close to the table value of the argument x_i. Particularly, if more accuracy to calculate function $f(x)$at the point $x = x_1$ is required, then it is advisable to carry out the interpolation by the formula [117]

$$f(x) = f(x_1) + (x - x_1)\varphi_1(x_1, x_2) + (x - x_1)(x - x_2)\varphi_2(x_1, x_2, x_3) + \ldots$$
$$+ (x - x_1)(x - x_2) \cdot \ldots \cdot (x - x_I)\varphi_I(x_1, x_2, \ldots, x_I) \tag{2.30}$$

The functions φ_i here are established by the correlations

$$\varphi_1(x_1, x_2) = \frac{f(x_1)}{x_1 - x_2} + \frac{f(x_2)}{x_2 - x_1} = \frac{f(x_2) - f(x_1)}{x_2 - x_1}$$

$$\varphi_2(x_1, x_2, x_3) = \frac{f(x_1)}{(x_3 - x_1)(x_1 - x_3)} + \frac{f(x_2)}{(x_2 - x_1)(x_2 - x_3)} + \frac{f(x_3)}{(x_3 - x_1)(x_3 - x_2)}$$

$$\varphi_{I-1}(x_1, x_2, \ldots, x_I) = \sum_{i=1}^{L} \frac{f(x_i)}{(x_i - x_1)(x_i - x_2) \cdot \ldots \cdot (x_i - x_{i-1})(x_i - x_{i+1}) \cdot \ldots \cdot (x_i - x_I)}$$

It should be noted that the use of the Eq. (2.30) with the argument values $x > x_2$ results in the unacceptably great errors.

If it is necessary to interpolate the function with high accuracy on the right border of the table (at the points $x_{I-1} > x > x_I$), then the formula similar to the previous one can be used

$$f(x) = f(x_I) + (x - x_I)\varphi_1(x_I, x_{I-1}) + (x - x_I)(x - x_{I-1})\varphi_2(x_I, x_{I-1}, x_{I-2}) + \ldots$$
$$+ (x - x_I)(x - x_{I-1}) \cdot \ldots \cdot (x - x_1)\varphi_{I-1}(x_I, x_{I-1}, \ldots, x_1) \tag{2.31}$$

The functions φ_i here are established by the correlations

$$\varphi_1(x_I, x_{I-1}) = \frac{f(x_I)}{x_I - x_{I-1}} + \frac{f(x_{I-1})}{x_{I-1} - x_I},$$

$$\varphi_2(x_I, x_{I-1}, x_{I-2}) = \frac{f(x_I)}{(x_I - x_{I-1})(x_I - x_{I-2})} + \frac{f(x_{I-1})}{(x_{I-1} - x_I)(x_{I-1} - x_{I-2})} + \frac{f(x_{I-2})}{(x_{I-2} - x_I)(x_{I-2} - x_{I-1})},$$

$$\varphi_{I-1}(x_I, x_{I-1}, ..., x_1)$$
$$= \sum_{i=1}^{I} \frac{f(x_i)}{(x_i - x_I)(x_i - x_{I-1}) \cdot ... \cdot (x_i - x_{i+1})(x_i - x_{i-1}) \cdot ... \cdot (x_i - x_1)}$$

As in the previous case, the Eq. (2.31) provides good interpolation only in the points $x_{I-1} > x > x_I$. In [117] other variants of Newton interpolation formulas are given.

2.3.1.2 Approximation of Functions Depending on Two or More Arguments

The problem of interpolating the function *f* depending on two or more arguments can be solved by the methods for the function $f(x)$ of one variable discussed above. However, it should be noted that this approach is rarely used because of its bulkiness, especially with more than two arguments. It is relatively easy to solve the problem of linear interpolation in the function dependence on two arguments. The interpolation problem in this case is formulated as:

- let there be some function $f(x, y)$ whose values f_{ij} are known for the discrete values of x_i (i=1, I; $x_1 < x_2 < ... < x_I$) and y_j (j=1, J; $x_1 < x_2 < ... < x_J$). Let us define the values of the function $f(x)$ for any values of $x \in (x_1, x_I)$ and $y \in (y_1, y_J)$.
  In the studied case, the discrete values $(x_i, y_j, f(x_i, y_j))$ will be called nodal points. The simplest algorithm for solving the problem of linear interpolation for the table function $f(x, y)$ can be:
- by the given values of the arguments *x* and *y* we set the integer number of the node (*i, j*) which obeys the conditions

$$x_i > x \geq x_{i+1}; \quad y_j > y \geq y_{j+1};$$

- using the formulas of linear interpolation true for functions of one variable, we calculate the values of $f_1(y_j)$, $f_2(y_{j+1})$

$$f_1(x, y_j) = f(x_i, y_j)\frac{x_{i+1} - x}{x_{i+1} - x_i} + f(x_{i+1}, y_j)\frac{x - x_i}{x_{i+1} - x_i}$$

$$f_2(x, y_{j+1}) = f(x_i, y_{j+1})\frac{x_{i+1} - x}{x_{i+1} - x_i} + f(x_{i+1}, y_{j+1})\frac{x - x_i}{x_{i+1} - x_i};$$

- using the formulas of linear interpolation correct for functions of one variable, we calculate the final value $f(x, y)$

$$f(x,y) = f_1(x,y_j)\frac{y_{j+1}-y}{y_{j+1}-y_j} + f_2(x,y_{j+1})\frac{y-y_j}{y_{j+1}-y_j}$$

Similarly, we can construct the formulas of linear interpolation for the function of a large number of arguments, and also get the formulas carrying out a parabolic interpolation.

2.3.1.3 Approximate Differentiation and Integration of a Table Function

The use of Lagrangian and Newton formulas allows relatively easily calculating the derivatives and integrals of the table function $f(x)$ on the range of its existence $x \in (x_1, x_I)$. To do this, one should use the procedures of differentiation and integration for approximations written, for example, in the form of (2.27)–(2.31).

The value of derivative $\frac{df(x)}{dx}$ calculated from Lagrangian formula (2.29), can be written as

$$\frac{df}{dx} = \sum_{i=1}^{I}\left(\sum_{\substack{j=1 \\ j\neq i}}^{I} \frac{f(x_i)}{(x-x_j)} \cdot \frac{(x-x_1)(x-x_2)\cdot\ldots\cdot(x-x_{i-1})(x-x_{i+1})\cdot\ldots\cdot(x-x_I)}{(x_i-x_1)(x_i-x_2)\cdot\ldots\cdot(x_i-x_{i-1})(x_i-x_{i+1})\cdot\ldots\cdot(x_i-x_I)}\right) \tag{2.32}$$

The formula (2.32) remains correct when applied for the simpler cases of interpolation presented by the formulas (2.27), (2.28).

Similarly, the integration of the restored table function $f(x)$ can be performed. However, in practice, to calculate the function integral given in a tabular form (or discrete), Lagrangian formula based on a relatively small number of points x_i $(i \leq 5)$ is used. In this case, we can obtain the antiderivative $F(x)$ in analytical form satisfying the condition $\frac{dF(x)}{dx} = f(x)$, and the value of the definite integral $I = \int_{x_i}^{x_{i+1}} f(x)\,dx$ can be calculated by Newton-Leibniz formula $I = \int_{x_i}^{x_{i+1}} f(x)\,dx = F(x_{i+1}) - F(x_i)$.

In the simplest case, the integral can be calculated according to the mean value theorem [202]

$$I = \int_{x_i}^{x_{i+1}} f(x)dx = (x_{i+1} - x_i) f(\xi), \quad \xi \in (x_i, x_{i+1})$$

If we assume that the point ξ is in the middle of the range (x_i, x_{i+1}) – $\xi = \frac{x_i + x_{i+1}}{2}$, then the approximate quadrature formula can be written down as

$$I = \int_{x_i}^{x_{i+1}} f(x)dx \approx (x_{i+1} - x_i) f(\frac{x_i + x_{i+1}}{2}) \tag{2.33}$$

The latter formula is called the rectangular formula and its graphical interpretation is shown in Figure 2.4. When integrating the table function $f(x)$ in the range $x \in (x_i, x_{i+n})$, the formula (2.33) can be re-written as

$$I = \int_{x_i}^{x_{i+n}} f(x)dx \approx \sum_{t=i}^{i+n} (x_{t+1} - x_t) f(\frac{x_t + x_{t+1}}{2}) \tag{2.34}$$

The formula (2.34) can be simplified for the case if the table step (on argument) is the same: $x_{i+1} - x_i = h \; \forall i$. In [250] it is shown that in this case the error of integral calculations is $\sim R(f) = \max\left|\frac{df(x)}{dx}\right| \cdot \frac{h^2}{4}$.

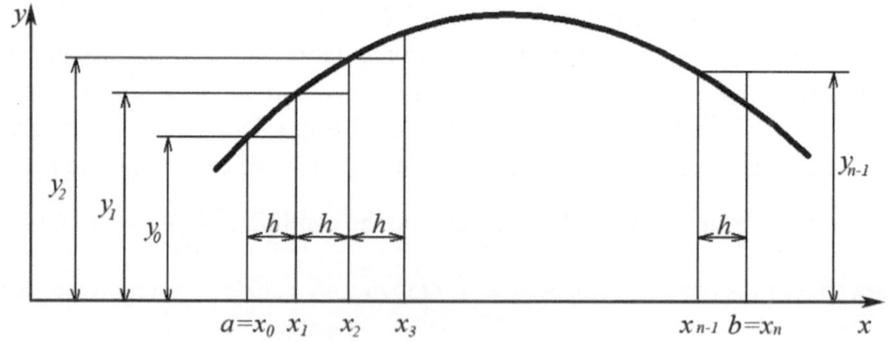

FIGURE 2.4 Method of rectangles in function integration.

More accurate calculation of the integral is provided when using piecewise linear formulas of interpolation to restore the function $f(x)$. On the range $x \in (x_i, x_{i+1})$ of the linear approximation the function $f(x)$ is written as $f(x) \approx \frac{(x_{i+1} - x) f(x_i) + (x - x_i) f(x_{i+1})}{x_{i+1} - x_i}$. The integration of the function $f(x)$ on the range from x_i to x_{i+n} allows you to obtain the following quadrature formula

$$I = \int_{x_i}^{x_{i+n}} f(x)\,dx \approx \sum_{t=i}^{i+n-1} \frac{(x_t + x_{t+1})(f(x_t) + f(x_{t+1}))}{2} \qquad (2.35)$$

which is called a trapezoid formula. The graphical interpretation of the trapezoid formula is given in Figure 2.5. The formula (2.35) can be simplified for the case if the table step (on argument) is the same: $x_{i+1} - x_i = h \ \forall i$. In this case, the integral calculation error is $\sim R(f) = \max|f''(x)| \cdot \frac{h^3}{12}$ [250].

We can replace the integrand $f(x)$ by the piecewise parabolic function of the form (2.28). If the parabola is constructed in the range $x \in (x_i, x_{i+1})$ on three values of the function – $f(x_i)$, $f(\frac{x_i + x_{i+1}}{2})$, $f(x_{i+1})$, then the integration of piecewise parabolic function in the range x_i to x_{i+n} allows to obtain the formula

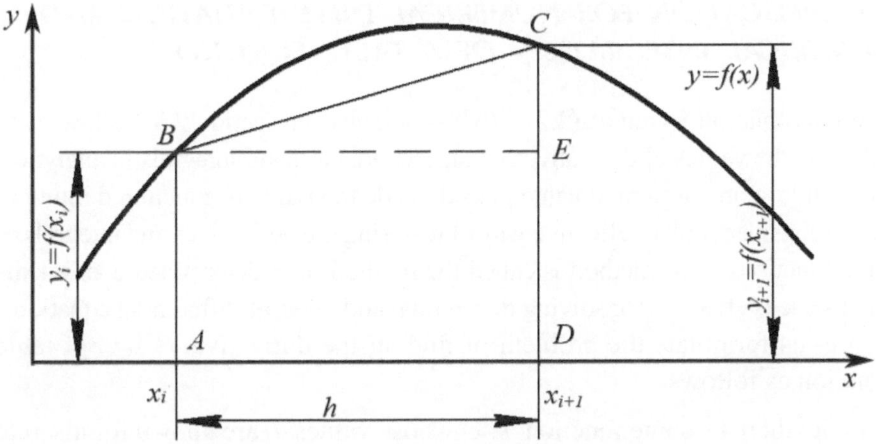

FIGURE 2.5 Trapezoidal method in function integration.

$$I = \int_{x_i}^{x_{i+n}} f(x)\,dx \approx \sum_{t=i}^{i+n-1} \left(\frac{x_{t+1} - x_t}{6} \cdot (f(x_t) + 4f(\frac{x_t + x_{t+1}}{2}) + f(x_{t+1})\right) \tag{2.36}$$

called Simpson formula. The use of Simpson formula for each segment split into n equal parts of length h, provides a high accuracy of the integral calculation. The remainder (error) in the integration by Simpson rule is calculated by the formula $R(f) = \max\left|f^{(4)}(x)\right| \cdot \frac{h^5}{2880}$.

The integrand replacement by any simple function can be continued. Particularly, it is possible to use the polynomial $L_n(x) = \sum_{i=0}^{n} a_i x^i$ of the arbitrary degree n requiring that the integrand $f(x)$ coincides with the polynomial in n points (sections), may not be equidistant to each other in the given range. Instead of the polynomial $L_n(x)$, we can use many other functions $\varphi_i(x)$ presenting the integrand as the combination $f(x) \approx \sum_{i=0}^{n} a_i \varphi_i(x)$. For example, the aggregate of orthogonal functions (Chebyshev polynomial, Legendre polynomial, Fourier functions, etc. [23, 75, 227, 230]) can be used as the functions $\varphi_i(x)$. The approximation $f(x) \approx \sum_{i=0}^{n} a_i\, p(x)\varphi_i(x)$ is convenient for obtaining quadrature formulas. $p(x)$ here is the weight function, which, in particular case, can be $p(x) \equiv 1$.

2.3.2 METHOD OF UNDETERMINED COEFFICIENTS AND ITS APPLICATION FOR NUMERICAL DIFFERENTIATION AND NUMERICAL INTEGRATION OF A TABLE FUNCTION

The interpolation formulas (2.27)–(2.29) obtained in Section 2.3.1 allow us to calculate the values of derivatives (first, second, etc.) and integrals in analytical form and for any argument value placed inside the range of function definition. Let us consider another effective way of restoring the derivatives and integrals of table functions. This method is called the method of undetermined coefficients and its use is effective for solving differential and integral-differential equations.

Let us formulate the problem of finding the derivative $f^{(k)}(x)$ of table function as follows

- let there be some function $f(x)$ whose values f_i are known for discrete values x_i, $(i = t, t+n;\ x_i < x_{i+1} < \ldots < x_{i+n})$). We determine the values of the derivatives $f^{(k)}(x)$ with any $x \in (x_t, x_{t+n})$.

We will seek the value of the derivate $f^{(k)}(x)$ in the form of

$$f^{(k)}(x) \approx \sum_{i=t}^{t+n} c_{i-t} f(x_i) \tag{2.37}$$

In the formula (2.37), the coefficients c_{i-t} are calculated from the condition that the formula is correct for the polynomial $f(x) = \sum_{j=0}^{M} a_j x^j$ in the highest possible degree M ($M<n$). The expression for $f(x)$ in the form of the polynomial can be differentiated. Then $f^{(k)}(x)$ takes the form

$$f^{(k)}(x) = \sum_{j=0}^{M} a_j (x^j)^{(k)} = \sum_{j=0}^{M} a_j \cdot (j(j-1)...(j-k-1)x^{j-k})$$

and the Eq. (2.37) can be re-written as

$$\sum_{j=0}^{M} a_j \cdot (j(j-1)...(j-k-1)x^{j-k}) = \sum_{i=t}^{t+n} c_{i-t} f(x_i) = \sum_{i=t}^{t+n} c_{i-t} (\sum_{j=0}^{M} a_j x_i^j)$$

In practice, the problem stated above is simplified. Indeed, the functions $f_0(x) = x^0 = 1;\ f_1(x) = x^1 = x;\ f_2(x) = x^2; \ldots f_j(x) = x^j; \ldots f_M(x) = x^M$ can be used as the polynomial of the arbitrary degree $j \le M$. The use of such functions does not reduce the approach generality, but simplifies the calculation of the coefficients c_{i-t}, in the Eq. (2.37). The coefficients c_{i-t} are defined by the solution of the system $(n+1)$ linear algebraic equations of the form

$$c_0 f_j(x_i) + c_1 f_j(x_{i+1}) + ... + c_{n-1} f_j(x_{i+n-1}) + c_n f_j(x_{i+n}) = \frac{d^k}{dx^k} f_j(x), \quad j = 0, n \tag{2.38}$$

In the Eq. (2.38) – $f_j(x) = x^j$, j=0, n. It should be remembered that the solutions obtained for the coefficients c_{i-t} are the function of the argument x values. In other words, with different values of x the solutions for the derivative $f^{(k)}(x)$ obtained from the system of linear algebraic Eq. (2.38) will be different. Let us examine the examples of the equation system construction of the form (2.38) for calculating the derivatives of table function.

Example 1. To calculate the value of the derivative $\frac{df(x_t)}{dx}$ of the known values $x_t, f(x_t); x_{t+1}, f(x_{t+1})$.

Only two points for calculating the derivative $\frac{df(x_t)}{dx}$ are given, so the Eq. (2.38) is in the form $c_0 f(x_t) + c_1 f(x_{t+1}) = \frac{d}{dx} f(x_t)$ (here $j = 0, 1$), and we use only two functions – $f_0(x) = x^0 = 1$ (for the first equation) and $f_1(x) = x^1 = x$ (for the second equation) as the basic polynomials;

- equation for calculating the coefficient c_0, c_1 will take the form:

$$c_0 \cdot 1 + c_1 \cdot 1 = 0$$

$$c_0 \cdot x_t + c_1 \cdot x_{t+1} = 1;$$

- matrix form of equations for the coefficients c_0, c_1

$$\begin{pmatrix} 1 & 1 \\ x_t & x_{t+1} \end{pmatrix} \cdot \begin{pmatrix} c_0 \\ c_1 \end{pmatrix} = \begin{pmatrix} 0 \\ 1 \end{pmatrix};$$

- solution of the equation system written above for the values of the coefficients c_0, c_1 in analytical form can be written down as

$$c_0 = -\frac{1}{x_{t+1} - x_t}, \quad c_1 = \frac{1}{x_{t+1} - x_t};$$

- final formula for the derivative values

$$\frac{df(x_t)}{dx} = -\frac{1}{x_{t+1} - x_t} f(x_t) + \frac{1}{x_{t+1} - x_t} f(x_{t+1}) = \frac{f(x_{t+1}) - f(x_t)}{x_{t+1} - x_t}$$

Example 2. To calculate the value of the derivative $\frac{df(x_t)}{dx}$ of the known values $x_t, f(x_t); x_{t+1}, f(x_{t+1}); x_{t+2}, f(x_{t+2})$.

Three points for calculating the derivative $\frac{df(x_t)}{dx}$ are given, so the Eq. (2.38) is in the form $c_0 f(x_t) + c_1 f(x_{t+1}) + c_2 f(x_{t+2}) = \frac{d}{dx} f(x_t)$ (here $j = 0, 1, 2$), and we use only two functions – $f_0(x) = x^0 = 1$ (for the first equation) and

$f_1(x)=x^1=x$ (for the second equation) and $f_2(x)=x^2$ (for the third equation) as the basic polynomials;

- equation for calculating the coefficient c_0, c_1, c_2 will take the form:

$$c_0\cdot 1+c_1\cdot 1+c_2\cdot 1=0$$

$$c_0\cdot x_t+c_1\cdot x_{t+1}+c_2\cdot x_{t+2}=1$$

$$c_0\cdot x_t^2+c_1\cdot x_{t+1}^2+c_2\cdot x_{t+2}^2=2x_t;$$

- matrix form of equations for the coefficients c_0, c_1, c_2

$$\begin{pmatrix}1 & 1 & 1\\ x_t & x_{t+1} & x_{t+2}\\ x_t^2 & x_{t+1}^2 & x_{t+2}^2\end{pmatrix}\cdot\begin{pmatrix}c_0\\ c_1\\ c_2\end{pmatrix}=\begin{pmatrix}0\\ 1\\ 2x_t\end{pmatrix};$$

- solution of the equation system written above for the values of the coefficients c_0, c_1, c_2 in analytical form can be written as

$$c_0=\frac{1-(x_{t+1}-x_{t+2})\cdot(x_{t+1}^2(x_t-x_{t+2})-x_t^2(x_{t+1}-x_{t+2})-x_{t+2}^2(x_t-x_{t+1}))}{x_t-x_{t+2}},$$

$$c_1=x_{t+1}^2(x_t-x_{t+2})-x_t^2(x_{t+1}-x_{t+2})-x_{t+2}^2(x_t-x_{t+1}),$$

$$c_2=\frac{1+(x_t-x_{t+1})\cdot(x_{t+1}^2(x_t-x_{t+2})-x_t^2(x_{t+1}-x_{t+2})-x_{t+2}^2(x_t-x_{t+1}))}{x_{t+2}-x_t};$$

- final expression for the derivative value

$$\frac{df(x_t)}{dx}=c_0f(x_t)+c_1f(x_{t+1})+c_2f(x_{t+2)}$$

The examples given above allow us without additional clarifications to write the matrix equation which lets to define the values of the coefficients c_{i-t} in the Eq. (2.38) and approximate the derivative $\frac{d^k f(x)}{dx^k}$ with the coordinate x, corresponding to $x=x_t$

$$\begin{pmatrix} 1 & 1 & 1 & \dots & 1 & 1 \\ x_i & x_{i+1} & x_{i+2} & \dots & x_{i+n-1} & x_{i+n} \\ x_i^2 & x_{i+1}^2 & x_{i+2}^2 & \dots & x_{i+n-1}^2 & x_{i+n}^2 \\ \dots & \dots & \dots & \dots & \dots & \dots \\ x_i^{n-1} & x_{i+1}^{n-1} & x_{i+2}^{n-1} & \dots & x_{i+n-1}^{n-1} & x_{i+n}^{n-1} \\ x_i^n & x_{i+1}^n & x_{i+2}^n & \dots & x_{i+n-1}^n & x_{i+n}^n \end{pmatrix} \cdot \begin{pmatrix} c_0 \\ c_1 \\ c_2 \\ \dots \\ c_{n-1} \\ c_n \end{pmatrix} = \begin{pmatrix} \left(d^k x^0 / dx^k\right)_{x=x_t} \\ \left(d^k x^1 / dx^k\right)_{x=x_t} \\ \left(d^k x^2 / dx^k\right)_{x=x_t} \\ \dots \\ \left(d^k x^{n-1} / dx^k\right)_{x=x_t} \\ \left(d^k x^n / dx^k\right)_{x=x_t} \end{pmatrix} \tag{2.39}$$

The approximations of derivatives obtained for several values of table functions will be used when writing down the ordinary differential equations and partial differential equations in a finite-difference way. We present without derivation some special cases of writing the values of derivatives $f^{(k)}(x_t)$ obtained for the case when the nodal values of the argument are equidistant ($x_{t+1} - x_t = x_t - x_{t-1} = h$). In this case, the values of derivatives $f^{(1)}(x_t)$, $f^{(2)}(x_t)$, $f^{(3)}(x_t)$ and their approximation errors $\delta_i^{(k)}$ can be written as

$$f^{(k)}(x_t) \approx \sum c_i f(x_i) + \delta_i^{(k)}$$

In the Tables 2.1–2.3 there are examples of approximation used in practice [250].

The problem of table function integration by the method of undetermined coefficients can be formulated as follows:

- let there is some function $f(x)$, which values f_i are known for discrete values x_i, ($i = t, t+n$; $x_t < x_{t+1} < \dots < x_{t+n}$)). Determine the values of integral $\int_a^b f(x)dx$ for the case, when $x_t \le a < b \le x_{t+n}$.

We will seek the value of integral in the form

$$\int_a^b f(x)dx \approx \sum_{i=t}^{t+n} c_{i-t} f(x_i) \tag{2.40}$$

TABLE 2.1 Approximation of Derivative $f^{(1)}(x_t)$ for Table Functions

Approximate expression to define the derivative	Error
$\frac{f(x_{t+1})-f(x_t)}{h}$	$-\frac{h}{2}f^{(2)}(x_t)$
$\frac{f(x_t)-f(x_{t-1})}{h}$	$\frac{h}{2}f^{(2)}(x_t)$
$\frac{f(x_{t+1})-f(x_{t-1})}{2h}$	$-\frac{h^2}{6}f^{(3)}(x_t)$
$\frac{-f(x_{t+2})+4f(x_{t+1})-3f(x_t)}{2h}$	$\frac{h^2}{3}f^{(3)}(x_t)$
$\frac{3f(x_t)-4f(x_{t-1})+f(x_{t-2})}{2h}$	$-\frac{h^2}{3}f^{(3)}(x_t)$
$\frac{2f(x_{t+3})-9f(x_{t+2})+18f(x_{t+1})-11f(x_t)}{6h}$	$-\frac{h^3}{4}f^{(4)}(x_t)$
$\frac{11f(x_t)-18f(x_{t-1})+9f(x_{t-2})-2f(x_{t-3})}{6h}$	$\frac{h^3}{4}f^{(4)}(x_t)$
$\frac{-3f(x_{t+4})+16f(x_{t+3})-36f(x_{t+2})+48f(x_{t+1})-25f(x_t)}{12h}$	$\frac{h^4}{5}f^{(5)}(x_t)$
$\frac{25f(x_t)-48f(x_{t-1})+36f(x_{t-2})-16f(x_{t-3})+3f(x_{t-4})}{12h}$	$-\frac{h^4}{5}f^{(5)}(x_t)$
$\frac{-f(x_{t+2})+8f(x_{t+1})-8f(x_{t-1})+f(x_{t-2})}{12h}$	$\frac{h^4}{30}f^{(5)}(x_t)$

In the formula (2.40) the coefficients c_{i-t} are determined with the condition that formula is correct for the polynomial $f(x)=\sum_{j=0}^{M} a_j x^j$ in the highest possible degree M ($M<n$). As a polynomial, like in derivative calculations, the functions $f_0(x)=x^0=1;\ f_1(x)=x^1=x;\ f_2(x)=x^2;\dots f_j(x)=x^j;\dots f_M(x)=x^M$ can be used. In this case, the coefficients c_{i-t} are determined by the solution of system ($n+1$) linear algebraic equations of the form

$$c_0 f_j(x_i)+c_1 f_j(x_{i+1})+\dots+c_{n-1}f_j(x_{i+n-1})+c_n f_j(x_{i+n})=\int_a^b f_j(x)dx,\quad j=0,n \tag{2.41}$$

It should be noted that solutions, obtained for the coefficients c_{i-t}, are function of integration limits values $a\le x\le b$. Values of coefficients c_{i-t} can be calculated by matrix equation

TABLE 2.2 Approximation of Derivative $f^{(2)}(x_t)$ for Table Functions

Approximate expression to define the derivative	Error
$\frac{f(x_{t+2})-2f(x_{t+1})+f(x_t)}{h^2}$	$-hf^{(3)}(x_t)$
$\frac{f(x_t)-2f(x_{t-1})+f(x_{t-2})}{h^2}$	$hf^{(3)}(x_t)$
$\frac{f(x_{t+1})-2f(x_t)+f(x_{t-1})}{h^2}$	$-\frac{h^2}{12}f^{(4)}(x_t)$
$\frac{-f(x_{t+3})+4f(x_{t+2})-5f(x_{t+1})+2f(x_t)}{h^2}$	$\frac{11h^2}{12}f^{(4)}(x_t)$
$\frac{2f(x_t)-5f(x_{t-1})+4f(x_{t-2})-f(x_{t-2})}{h^2}$	$-\frac{11h^2}{12}f^{(4)}(x_t)$
$\frac{11f(x_{t+3})-56f(x_{t+2})+114f(x_{t+2})-104f(x_{t+1})+35f(x_t)}{12h^2}$	$\frac{5h^3}{6}f^{(5)}(x_t)$
$\frac{35f(x_t)-104f(x_{t-1})+114f(x_{t-2})-56f(x_{t-3})+11f(x_{t-4})}{12h^2}$	$-\frac{5h^3}{6}f^{(5)}(x_t)$
$\frac{-f(x_{t+2})+16f(x_{t+1})-30f(x_t)+16f(x_{t-1})-f(x_{t-2})}{12h^2}$	$-\frac{h^4}{90}f^{(6)}(x_t)$

TABLE 2.3 Approximation of Derivative $f^{(3)}(x_t)$ for Table Functions

Approximate expression to define the derivative	Error
$\frac{f(x_{t+3})-3f(x_{t+2})+3f(x_{t+1})-f(x_t)}{h^3}$	$-\frac{3h}{2}f^{(4)}(x_t)$
$\frac{f(x_t)-3f(x_{t-1})+3f(x_{t-2})-f(x_{t-3})}{h^3}$	$\frac{3h}{2}f^{(4)}(x_t)$
$\frac{-3f(x_{t+4})+14f(x_{t+3})-24f(x_{t+2})+18f(x_{t+1})-5f(x_t)}{2h^3}$	$\frac{21h^2}{12}f^{(5)}(x_t)$
$\frac{5f(x_t)-18f(x_{t-1})+24f(x_{t-2})-14f(x_{t-3})+3f(x_{t-4})}{2h^3}$	$-\frac{21h^2}{12}f^{(5)}(x_t)$
$\frac{f(x_{t+2})-2f(x_{t+1})+2f(x_{t-1})-f(x_{t-2})}{2h^3}$	$-\frac{h^2}{4}f^{(5)}(x_t)$

$$\begin{pmatrix} 1 & 1 & 1 & \dots & 1 & 1 \\ x_i & x_{i+1} & x_{i+2} & \dots & x_{i+n-1} & x_{i+n} \\ x_i^2 & x_{i+1}^2 & x_{i+2}^2 & \dots & x_{i+n-1}^2 & x_{i+n}^2 \\ \dots & \dots & \dots & \dots & \dots & \dots \\ x_i^{n-1} & x_{i+1}^{n-1} & x_{i+2}^{n-1} & \dots & x_{i+n-1}^{n-1} & x_{i+n}^{n-1} \\ x_i^n & x_{i+1}^n & x_{i+2}^n & \dots & x_{i+n-1}^n & x_{i+n}^n \end{pmatrix} \cdot \begin{pmatrix} c_0 \\ c_1 \\ c_2 \\ \dots \\ c_{n-1} \\ c_n \end{pmatrix} = \begin{pmatrix} \int_a^b x^0 dx \\ \int_a^b x^1 dx \\ \int_a^b x^2 dx \\ \dots \\ \int_a^b x^{n-1} dx \\ \int_a^b x^n dx \end{pmatrix} \tag{2.42}$$

Highly effective methods of Gauss and Chebyshev, based on the use of the system of orthogonal functions, can be used for the numerical integration of table functions [75, 76]. It should be noted that the use of these methods is limited to strict requirements to the numerical values of an argument x_t that may not always be achieved working with table functions.

2.3.3 SPLINE FUNCTIONS WHEN WORKING WITH TABLES

Interpolation of arbitrary functions $f(x)$ by the Lagrange polynomial of n degree, constructed in accordance to the condition that $f(x_i) = f_i$ with all $x_i, (i = 1, n)$, is cumbersome and, in general, does not provide the same accuracy of interpolation for the entire range $x_1 > x > x_n$ of function $f(x)$. At the present time interpolation, based on the spline functions, became widespread [44, 95, 208, 250]. Method, based on the use of spline function, can be regarded as the development of interpolation method with the use of piecewise polynomial approximation, but free from its drawbacks. Particularly, for large values of n interpolation is performed by the polynomials of small degrees (second or third one). Coefficient of polynomials is recalculated for each value of an argument x_i in accordance with known table values of pairs $(x_i, f(x_i))$, calculated in a neighborhood of x, near to x_i, and regularly recalculated with increasing i.

One of the most effective methods of this group is interpolations with the use of cubic splines. The Figure 2.6 explain the reasoning of splines. Let some table function is defined on a set of points from 1 to n (i=1, n). We construct for each range between points i and i+1 the cubic polynomial of the form $q_i(x) = k_{1i} + k_{2i}x + k_{3i}x^2 + k_{4i}x^3$, $(i = 1, n-1)$. Coefficients $k_{1i}, k_{2i}, k_{3i}, k_{4i}$ in this polynomial are unknown, and the total number of unknown coefficients will be 4(n-1). For unambiguous definition of coefficients we formulate the following additional requirements:

- the first spline $q_1(x)$ lets calculate the values of function in the point with coordinate x_1 ($f(x_1) = q_1(x_1)$), and spline $q_{n-1}(x)$ allows us to calculate the value of function in the point with coordinate x_n ($f(x_n) = q_{n-1}(x_n)$);
- to determine the value of function in the nodal point with the coordinate x_i $(i = 2, n-1)$ splines $q_{i-1}(x)$ or $q_i(x)$ can be used, at the same time the values of both splines are matched ($q_{i-1}(x_i) = q_i(x_i) = f(x_i)$);
- in the nodal points with coordinate x_i $(i = 2, n-1)$ the first and the second spline derivatives $q_{i-1}(x)$ and $q_i(x)$ coincide ($q'_{i-1}(x_i) = q'_i(x_i)$, $q''_{i-1}(x_i) = q''_i(x_i)$);
- behavior of table function for the values $x < x_1$ and $x > x_n$, generally speaking, is not defined. Therefore, on the bounds of these splines different conditions can be used. Conditions of spline second derivatives $q_1(x)$ and $q_{n-1}(x)$ to be equal-zero on bounds ($q''_1(x_1) = 0$, $q''_{n-1}(x_n) = 0$) are seemed the most natural. However, other conditions can be used.

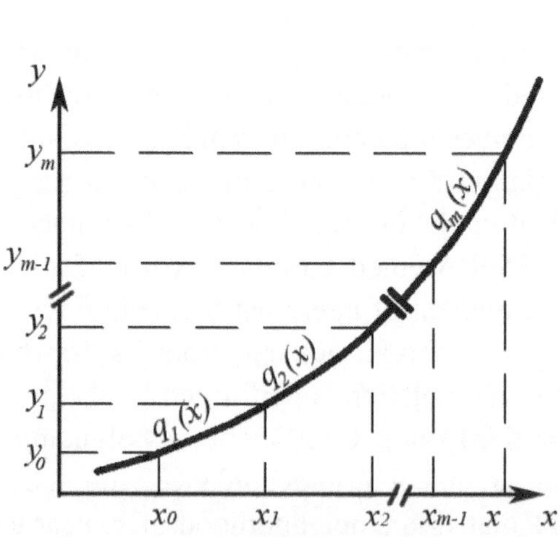

FIGURE 2.6 Method of spline function for the interpolation of functions.

For example, conditions of first derivatives to be equal-zero, periodicity conditions, etc.

On the bases of the requirements to splines, we write the system of equations, for calculation the values of coefficients in these splines:

- spline $q_i(x_i)$ should ensure the correct calculation of function with values of an argument $x = x_i$ and $x = x_{i+1}$ $(i = 1, n-1)$

$$k_{1i} + k_{2i}x_i + k_{3i}x_i^2 + k_{4i}x_i^3 = f(x_i), \quad i = 1, n-1;$$

$$k_{1i} + k_{2i}x_{i+1} + k_{3i}x_{i+1}^2 + k_{4i}x_{i+1}^3 = f(x_{i+1}), \quad i = 1, n-1;$$

- equality of the first derivatives for the neighboring splines $(q'_{i-1}(x_i) = q'_i(x_i))$

$$k_{2(i-1)} + 2k_{3(i-1)}x_i + 3k_{4(i-1)}x_i^2 - k_{2i} - 2k_{3i}x_i - 3k_{4i}x_i^2 = 0, \quad i = 2, n-1;$$

- equality of the second derivatives for the neighboring splines $(q''_{i-1}(x_i) = q''_i(x_i))$

$$2k_{3(i-1)} + 6k_{4(i-1)}x_i - 2k_{3i} - 6k_{4i}x_i = 0, \quad i = 2, n-1;$$

- conditions at the bounds of the table ($q''_1(x_1) = 0$, $q''_{n-1}(x_n) = 0$)

$$2k_{31} + 6k_{41}x_1 = 0;$$

$$2k_{3(n-1)} + 6k_{4(n-1)}x_n = 0$$

The system of equations written above $(4n-4)$ for $(4n-4)$ unknown coefficients $k_{1i}, k_{2i}, k_{3i}, k_{4i}$ after simplifications can be written in a matrix form

$$\mathbf{AX} = \mathbf{B} \tag{2.43}$$

where

$$\mathbf{A}=\begin{pmatrix}
0 & 0 & 1 & 3x_1 & 0 & 0 & 0 & 0 & \dots & 0 & 0 & 0 & 0 & 0 & 0 & 0\\
1 & x_1 & x_1^2 & x_1^3 & 0 & 0 & 0 & 0 & \dots & 0 & 0 & 0 & 0 & 0 & 0 & 0\\
1 & x_2 & x_2^2 & x_2^3 & 0 & 0 & 0 & 0 & \dots & 0 & 0 & 0 & 0 & 0 & 0 & 0\\
0 & 1 & 2x_1 & 3x_1^2 & 0 & -1 & -2x_1 & -3x_1^2 & \dots & 0 & 0 & 0 & 0 & 0 & 0 & 0\\
0 & 0 & 1 & 3x_1 & 0 & 0 & -1 & -3x_1 & \dots & 0 & 0 & 0 & 0 & 0 & 0 & 0\\
0 & 0 & 0 & 0 & 1 & x_2 & x_2^2 & x_2^3 & \dots & 0 & 0 & 0 & 0 & 0 & 0 & 0\\
0 & 0 & 0 & 0 & 1 & x_3 & x_3^2 & x_3^3 & \dots & 0 & 0 & 0 & 0 & 0 & 0 & 0\\
0 & 0 & 0 & 0 & 0 & 1 & 2x_2 & 3x_2^2 & \dots & 0 & 0 & 0 & 0 & 0 & 0 & 0\\
0 & 0 & 0 & 0 & 0 & 0 & 1 & 3x_2 & \dots & 0 & 0 & 0 & 0 & 0 & 0 & 0\\
\dots & \dots & \dots & \dots & \dots & \dots & \dots & \dots & \dots & \dots & \dots & \dots & \dots & \dots & \dots & \dots\\
0 & 0 & 0 & 0 & 0 & 0 & 0 & 0 & \dots & x_{n-1} & x_{n-1}^2 & x_{n-1}^3 & 0 & 0 & 0 & 0\\
0 & 0 & 0 & 0 & 0 & 0 & 0 & 0 & \dots & 0 & 1 & 3x_1 & 0 & 0 & -1 & -3x_1\\
0 & 0 & 0 & 0 & 0 & 0 & 0 & 0 & \dots & 1 & 2x_n & 3x_n^2 & 0 & -1 & -2x_n & -3x_n^2\\
0 & 0 & 0 & 0 & 0 & 0 & 0 & 0 & \dots & 0 & 0 & 0 & 1 & x_{n-1} & x_{n-1}^2 & x_{n-1}^3\\
0 & 0 & 0 & 0 & 0 & 0 & 0 & 0 & \dots & 0 & 0 & 0 & 1 & x_n & x_n^2 & x_n^3\\
0 & 0 & 0 & 0 & 0 & 0 & 0 & 0 & \dots & 0 & 0 & 0 & 0 & 0 & 1 & 3x_n
\end{pmatrix}$$

$$\mathbf{X}=\begin{pmatrix} k_{11}\\ k_{21}\\ k_{31}\\ k_{41}\\ k_{12}\\ k_{22}\\ k_{32}\\ k_{42}\\ \dots\\ k_{2(n-2)}\\ k_{3(n-2)}\\ k_{4(n-2)}\\ k_{1(n-1)}\\ k_{2(n-1)}\\ k_{3(n-1)}\\ k_{4(n-1)} \end{pmatrix};\qquad
\mathbf{B}=\begin{pmatrix} 0\\ f(x_1)\\ f(x_2)\\ 0\\ 0\\ f(x_2)\\ f(x_3)\\ 0\\ \dots\\ f(x_{n-2})\\ f(x_{n-1})\\ 0\\ 0\\ f(x_{n-1})\\ f(x_n)\\ 0 \end{pmatrix}$$

The system of linear algebraic equations for $4(n–1)$ coefficients of cubic splines can be solved by any of methods, discussed above in Section 2.2.

Cubic splines are widely used working with the table data. Their use for calculation of derivatives and integrals of the table function is not difficult. It should be noted that polynomial splines (quadratic, cubic) are convenient to use, but also there are used other types of splines. Particularly, Hermitian splines, constructed with the use of Hermitian interpolation polynomial are of interest [208]. The main advantage of the use of Hermitian splines is the possibility of more accurate calculation of derivatives of table function in comparison with polynomial splines. In addition, splines (both polynomial and Hermitian) can be used for the tables of greater dimension (two- and three-dimensional). Examples of software using different types of splines are given, for example, in [22].

2.4 GRAPHICAL METHODS OF TABLE INFORMATION PROCESSING

Graphics of functional dependence, including in a table form, is natural, and in some cases obligatory, stage of research. At the present time there is a variety of software that provides graphics of function. For example, these tools are included in used widely in practice software package *Microsoft Excel* [56, 160]. Graphic tools are included in the system of computer mathematics *MathCad* [86, 141]. Specialized software, such as *Visual Studio* [33, 40, 96], *Array Visualizer* [186], *Borland Delphi* [73], *Open GL* [20] and others, contain graphic tools, that allow you to create highly professional graphic processing of digital information. *GnuPlot* program is the most attractive of free graphics software programs [186]. There are program versions to work in different operating systems. Work with the program is similar to the work with the simplest programming language. The number of *GnuPlot* users is constantly increasing, there are rich libraries with the worked out graphic software.

Below we consider only the simplest opportunities of software application of general use to construct graphics.

2.4.1 CONSTRUCTION OF GRAPHICS BY THE MEANS OF EXCEL

The first step, related to the graph construction, suggests the record of table data into file with the extension **.xls*. If table data is placed in a text file (extension **.txt*), then module "Text Wizard" allows to convert them into *Excel* file. Further steps, necessary to construct the charts, are provided by step-by-step operations with the module "Chart Wizard." Module connecting happens after selecting the menu item "Insert" → "Picture" → "Chart" (Figure 2.7).

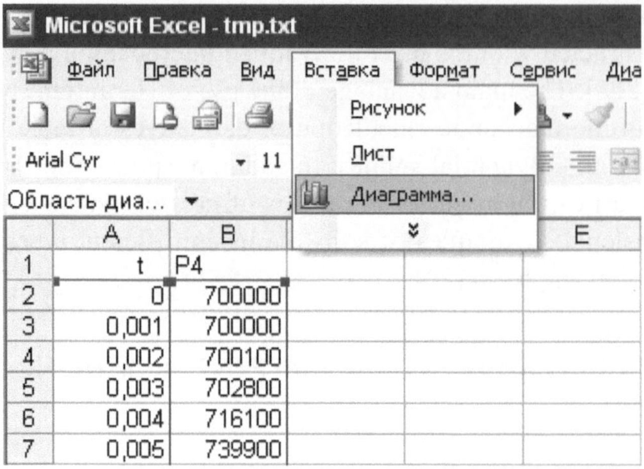

FIGURE 2.7 Construction of chart in Excel – selection from the chart menu.

The first dialog box of "Chart Wizard" allows you to choose the type of chart (Figure 2.8).

The second dialog box (Figure 2.9), opened by "Chart Wizard," lets to the range of data, on the basis of which chart should be constructed (or define it, if it is not done yet). In addition, you can specify how data series are presented in this range. After display of this dialog box, cell range, selected in the table before activation of chart wizard, will be surrounded by a moving dotted line and represented as a formula (with the absolute coordinates of the cells) in the text filed Range.

The third dialog box, opened by module "Chart Wizard" (Figure 2.10), allows you to set a number of parameters that determine the presence of titles, axes, gridlines, legends, data labels, data tables in the chart.

The work of "Chart Wizard" is completed with the identification of location where should be newly constructed chart (Figure 2.11). The results of the work with *Excel* program are presented in the Figure 2.12.

2.4.2 CONSTRUCTION OF GRAPHICS BY THE MEANS OF MATHCAD

Let examine the case, when the table with calculation results is in the file, whose location is determined by the command file *G:\tmp.txt*. In the table the first column is an argument (process time), and the other columns

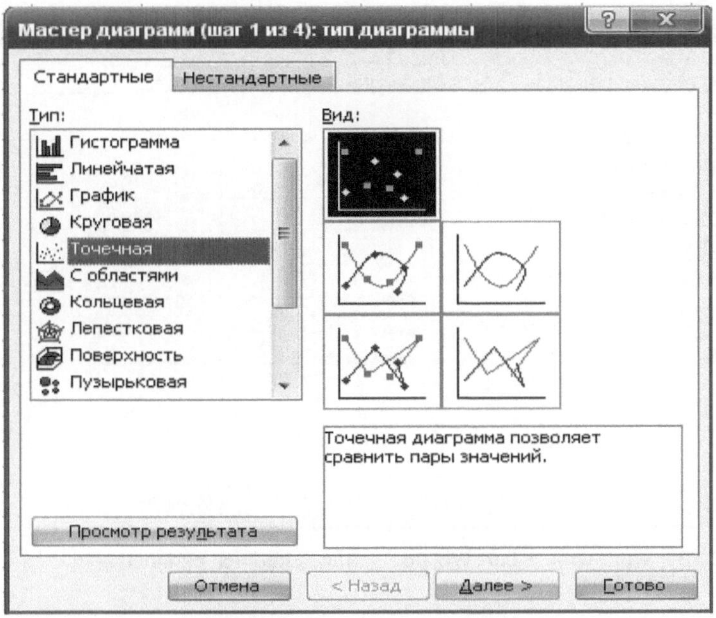

FIGURE 2.8 Work with "Chart Wizard," selection of chart type.

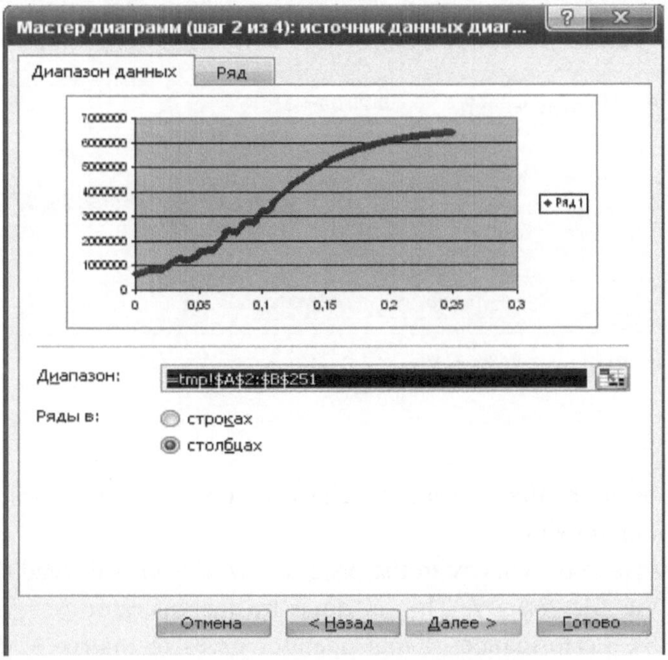

FIGURE 2.9 Work with "Chart Wizard," selection of data range.

FIGURE 2.10 Work with "Chart Wizard," setting additional parameters.

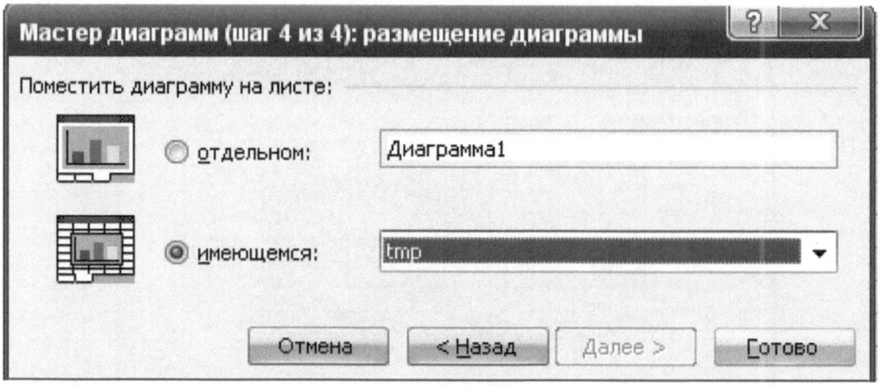

FIGURE 2.11 Work with "Chart Wizard," final stage.

(they may be more than one) are values of function (the pressure in the combustion chamber).

The first stage of work with the program *MathCad* is to read the source data from the file *tmp.txt*. This is done by the operator *N:=READPRN* (Figure 2.13). Performance of this operator gives to matrix *N* the values from the file *tmp.txt*. Next operators give to vector *x* the numerical values,

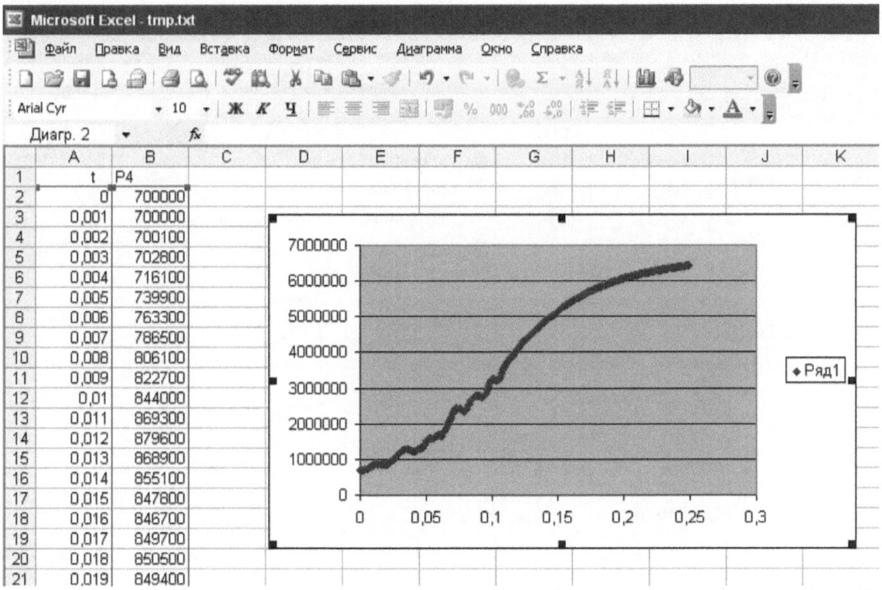

FIGURE 2.12 Result of chart creating by means of *Excel.*

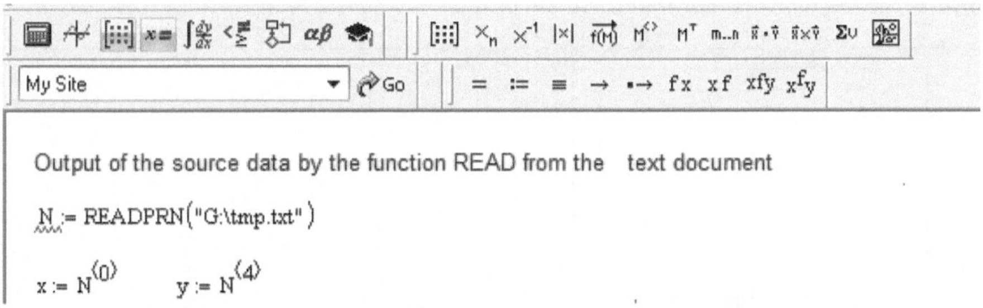

FIGURE 2.13 Reading the source data from the file.

placed in the zero (first) column of matrix, and to vector y – the numerical values, placed in the forth column of matrix N.

In the second stage the type of planned version of the graph is selected. For this reason in the command line we choose the command Insert → Graph → Type of graph or in the panel of mathematical symbols the button with a graph image is chosen. Clicking on this button leads to the appearance of graph palette on the screen (Figure 2.14).

The final part of the graph construction is presented in Figure 2.15, 2.16. Graph template is shown on the screen (Figure 2.15). In the place of template

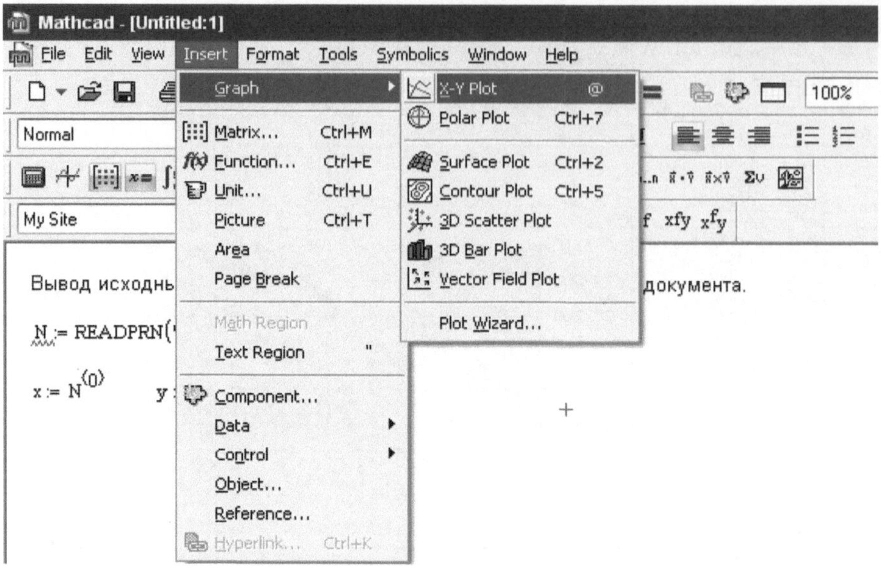

FIGURE 2.14 Selection the chart type.

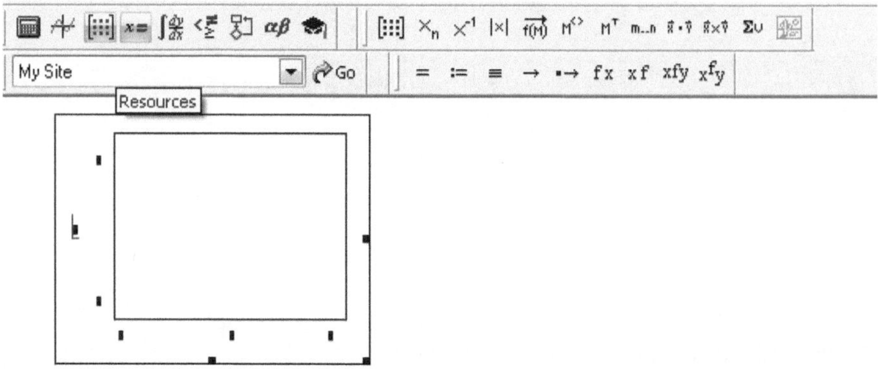

FIGURE 2.15 Work with the template.

input at X axis the name of independent variable x is included, and in the place of template input at Y axis the name of dependent variable is $y(x)$. At the end of templates input the final graph of function is drawn (Figure 2.16).

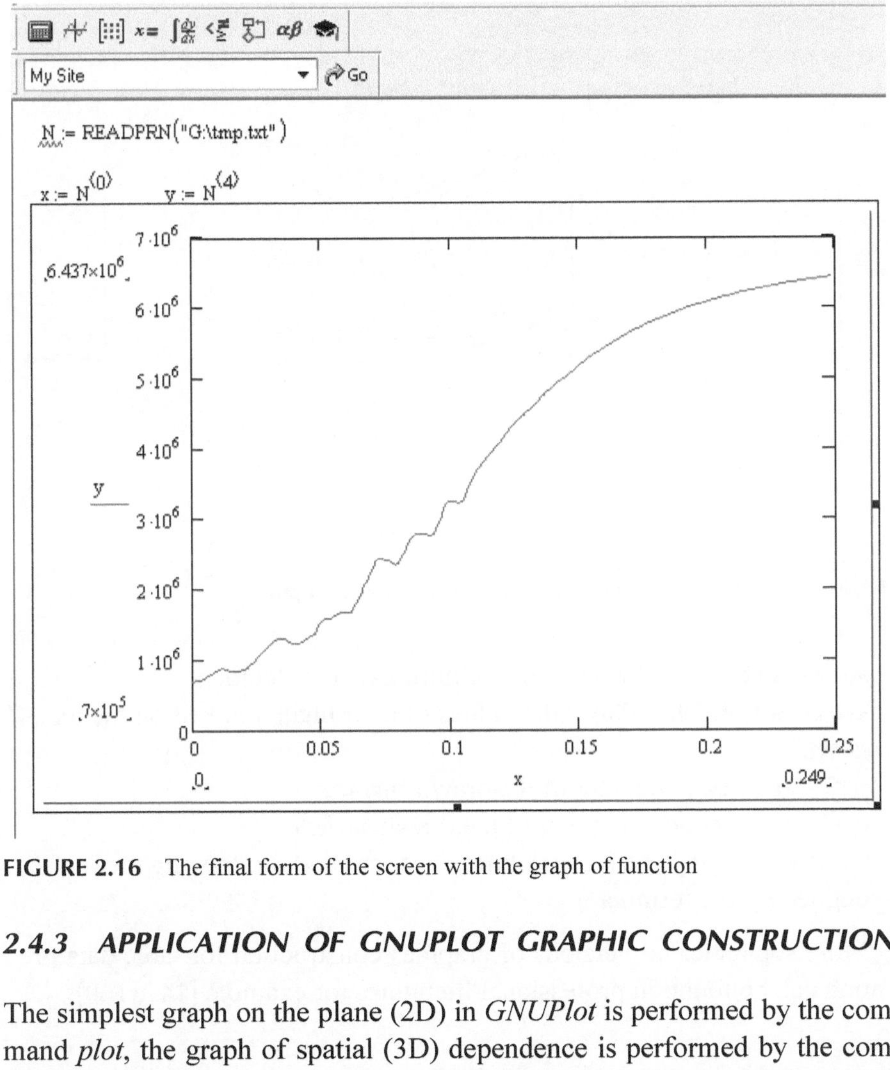

FIGURE 2.16 The final form of the screen with the graph of function

2.4.3 *APPLICATION OF GNUPLOT GRAPHIC CONSTRUCTION*

The simplest graph on the plane (2D) in *GNUPlot* is performed by the command *plot*, the graph of spatial (3D) dependence is performed by the command *splot*. Command for graph construction contains the parameters, which the user should set up. Some of the parameters can be taken by the program on default. Command can be written in one line, containing a list of all parameters, or can be written in several lines. Particularly, the Figure 2.17, given below, reproduced after performing the following commands, processed by the *GNUPlot* program:

set title "Paraboloid" – the name of graph is entered;
set x-label "X" – the name of the first axis is entered;
set y-label "Y" – the name of the second axis is entered;

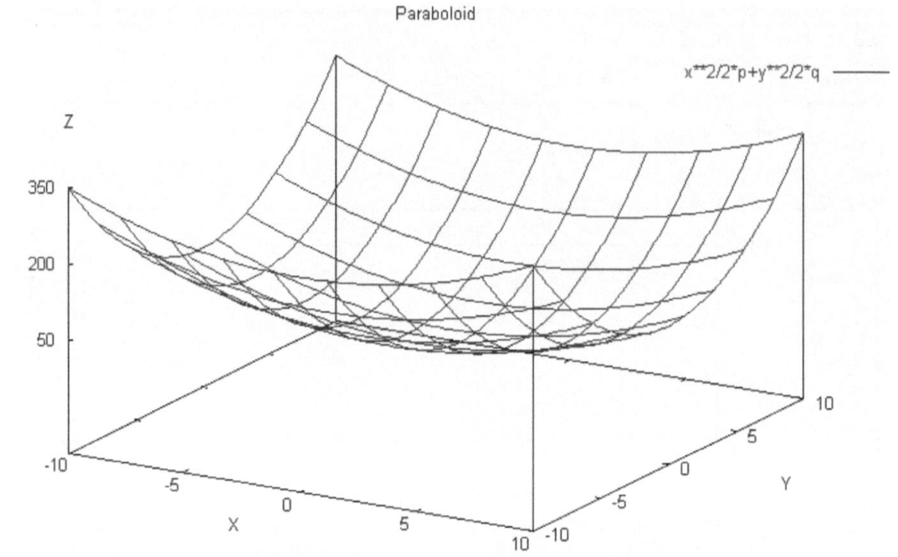

FIGURE 2.17 The paraboloid, constructed by the *GNUPlot* program.

set z-label "Z" – the name of the third axis is entered;
set ztics (50, 200, 350) – the values of coordinate marked on the axis ***Z*** is set;
p=2 – the first parameter of the drawn surface;
q=4 – the second parameter of the drawn surface;
splot x2/2p+y**2/2q with lines 5** – command for drawing the surface, defined by the formula.

More sophisticated versions of graphics construction for table data presenting can be found in professional literature, for example [141, 160].

2.5 SOFTWARE APPLICATION TO WORK WITH TABLE FUNCTIONS

Modern mathematical software packages and compliers contain a large number of computational procedures that significantly reduces the time to solve a particular problem. These computational procedures take the shape of standard function (e.g., in the *MathCad* package) or standard subprogram (e.g., in *IMSL* for complier *Compaq Visual Fortran*), and their number is increasing. Here and in the following chapters, information on existing standard functions and subprograms is presented briefly and does not attempt to list all available computational tools in the mathematical packages. For the more detailed acquaintance with these software, one should refer to the professional literature (e.g., [20–22, 40, 86, 141, 198]).

In the Table 2.4 there are given some functions that are part of mathematical system *MathCad.* Table 2.5 provides some subprograms and functions

TABLE 2.4 Examples of *MathCad* Functions to Work with Tables and Matrices

№	Function name	Purpose of function	Description of function arguments
1	***linterp*** (*VX, VY, x*)	Calculation of function values, given in a table form (linear interpolation is used)	*VX, VY* – tables (vectors) with the values of an argument and a function; *x* – the value of argument, for which the function is calculated
2	***predict*** (*DATA, k, N*)	Calculation of *N* points, located on the right side after the last table value of function (extrapolation involves the construction of polynomial of *k* degree)	*DATA* – vector defined the values of function (argument is integer values of natural numbers); *k* – polynomial degree; *N* – number of calculated values of function
3a	***cspline*** *(VX, VY);* ***pspline*** *(VX, VY);* ***lspline*** *(VX, VY);*	Calculation of spline-functions coefficients (cubic, parabolic or linear), which helps to calculate the value of function on the second step	*VX, VY* – tables (vectors) with the values of an argument and a function
3b	***interp*** *(VS, VX, VY, x)*	Calculation of function value, corresponding to the value of argument *x*	*VS* – table (vector) with the values of coefficients of spline function; *x* – the value of an argument for which the function is calculated
4	***rank***(*A*)	Calculation of matrix rank	**A** – matrix, which size should be defined in the *MathCad* program
5	***norm1***(*A*), ***norm2***(*A*), ***norme***(*A*)	Calculation of matrix norms $\|\mathbf{A}\|_1$, $\|\mathbf{A}\|_2$, $\|\mathbf{A}\|_E$	**A** – matrix, which size should be defined in the *MathCad* program
6	***cond1****(A),* ***cond2****(A),* ***conde****(A)*	Calculation the condition number of matrix in the norms $\|\mathbf{A}\|_1$, $\|\mathbf{A}\|_2$, $\|\mathbf{A}\|_E$	**A** – matrix, which size should be defined in the *MathCad* program

TABLE 2.4 Continued

№	Function name	Purpose of function	Description of function arguments
7	***eigenvals***(*A*)	Calculation of vector, containing the eigenvalues of the matrix	**A** – matrix, which size should be defined in the *MathCad* program
8	***eigenvec***(*A*, *z*)	Calculation of eigenvector, corresponding to the eigenvalue *z* matrix	**A** – matrix, which size should be defined in the *MathCad* program
9	***svds***(*A*)	Calculation of vector, containing the singular values of matrix	**A** – matrix, which size should be defined in the *MathCad* program
10	***lsolve***(*A*, *B*)	Calculation of the vector **X**, which is the solution of the system of linear equations **AX** = **B**	**A** – matrix of coefficients in the system of linear equations; **B** – vector, defining the right sides of equations

TABLE 2.5 Examples of *IMSL* Subprogram to Work with Tables and Matrices

№	Call to subprogram or function	Purpose of subprogram or function	Description of function arguments or subprogram arguments
1	*call* ***csdec*** (*ndata, xdata, fdata, ileft, dleft, iright, dright, break, cscoef);* ***csval*** (*x, ndata-1, break, cscoef);*	Calculation of coefficients of interpolation spline with the given values of derivatives on the bounds of the range; Calculated on spline value of table function, corresponding to the argument *x*	*ndata* – number of table values; *xdata* – vector, containing the values of an argument of table function; *fdata* – values of table function; *ileft, iright* – a sign, indicating the type of boundary condition on the left and right bounds of the table (condition is selected on default or the values of the first or the second derivatives are set); *dltft, droght* – values of the derivatives on the bounds of the table; *break, cscoef* – vectors, containing the spline information calculated in a subprogram
2	*call* ***nr1rr*** (*nra, nca, a, Lda, anorm);* *call* ***nr2rr*** (*nra, nca, a, Lda, anorm);*	Calculation of matrix norms $\|\mathbf{A}\|_1$, $\|\mathbf{A}\|_E$	*nra, nca* – number of rows and columns of matrix **A**; *a* – array, containing the elements of matrix; *Lda* – leading size of array **A**; *anorm* – the value of norm

TABLE 2.5 Continued

№	Call to subprogram or function	Purpose of subprogram or function	Description of function arguments or subprogram arguments
3	*call* **lslrg** (*n, a, Lda, b, ipath, x*) *call* **lslxg** (*n, nz, a, irow, jcol, b, ipath, iparam, rparam, x*)	Solution of the system of linear equations **AX** = **B** using ***LU***–decomposition Solutions of the system of linear equations **AX** = **B** by Gauss method	*n* – order of matrix **A**; *a* – vector of *nz* size, containing the non-zero values of matrix **A**; *Lda* – leading size of array **A**; *b* – vector of n size, containing the right part of equations; ipath=1 – solving the ordinary system of equations; *x* – vector – the solution of the system of equations; *nz* – number of non-zero elements of matrix **A**; *irow, jcol* – vectors of *nz* size, containing the numbers of rows and columns of non-zero elements of matrix **A**; *iparam, rparam* – working vectors

that are part of *IMSL* of complier *Compaq Visual Fortran*. It should be noted that the actual number of existing subprograms in the *IMSL* exceeds significantly the number of functions in the *MathCad*.

In conclusion, in Figure 2.18–2.20 the solution of interpolation problem by means of *MathCad* is given. In Figure 2.18 the stage of reading the table from the file *tmp.txt*, located in the folder *L*, is presented. Reading the table is performed by the *ReadPrn* operator. For the further work columns numbered "0" and "4" are allocated.

In the second stage (Figure 2.19) for three values of an argument (zero column of the table function), – *t1, t, t2*, – appropriate values of function (forth column of the table function) are calculated by the linear interpolation (function *linterp* is used).

At the third stage (Figure 2.20) for three values of argument the values of function are calculated with the use of linear, parabolic and cubic spline.

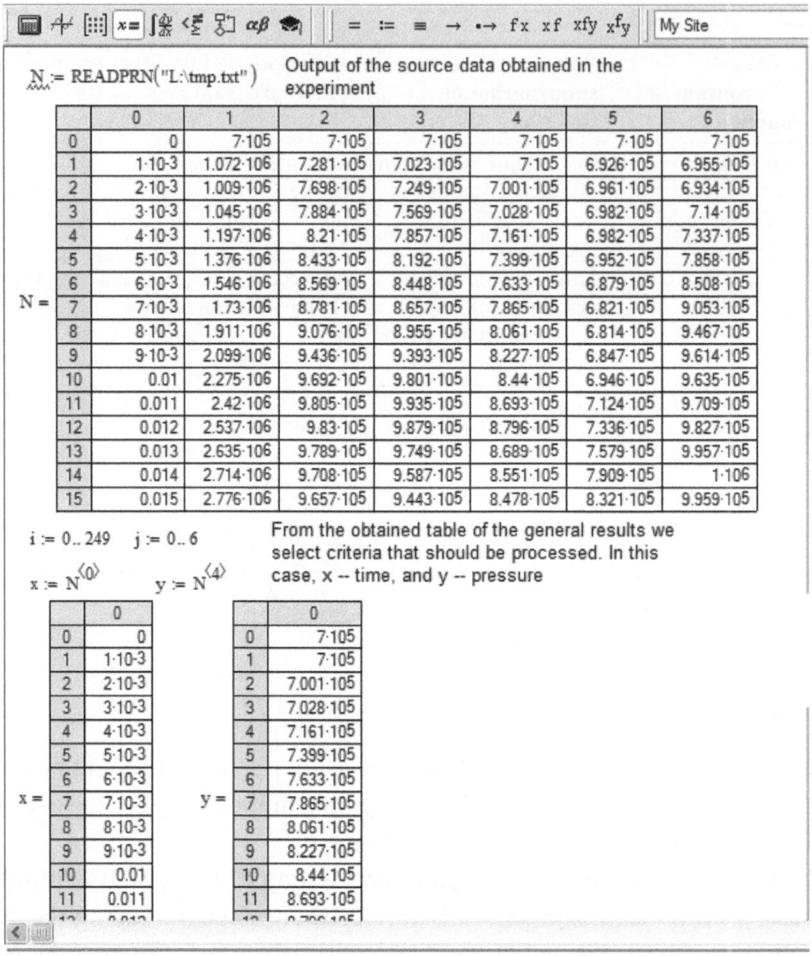

N := READPRN("L:\tmp.txt")

Output of the source data obtained in the experiment

N =

	0	1	2	3	4	5	6
0	0	7·105	7·105	7·105	7·105	7·105	7·105
1	1·10-3	1.072·106	7.281·105	7.023·105	7·105	6.926·105	6.955·105
2	2·10-3	1.009·106	7.698·105	7.249·105	7.001·105	6.961·105	6.934·105
3	3·10-3	1.045·106	7.884·105	7.569·105	7.028·105	6.982·105	7.14·105
4	4·10-3	1.197·106	8.21·105	7.857·105	7.161·105	6.982·105	7.337·105
5	5·10-3	1.376·106	8.433·105	8.192·105	7.399·105	6.952·105	7.858·105
6	6·10-3	1.546·106	8.569·105	8.448·105	7.633·105	6.879·105	8.508·105
7	7·10-3	1.73·106	8.781·105	8.657·105	7.865·105	6.821·105	9.053·105
8	8·10-3	1.911·106	9.076·105	8.955·105	8.061·105	6.814·105	9.467·105
9	9·10-3	2.099·106	9.436·105	9.393·105	8.227·105	6.847·105	9.614·105
10	0.01	2.275·106	9.692·105	9.801·105	8.44·105	6.946·105	9.635·105
11	0.011	2.42·106	9.805·105	9.935·105	8.693·105	7.124·105	9.709·105
12	0.012	2.537·106	9.83·105	9.879·105	8.796·105	7.336·105	9.827·105
13	0.013	2.635·106	9.789·105	9.749·105	8.689·105	7.579·105	9.957·105
14	0.014	2.714·106	9.708·105	9.587·105	8.551·105	7.909·105	1·106
15	0.015	2.776·106	9.657·105	9.443·105	8.478·105	8.321·105	9.959·105

i := 0..249 j := 0..6

$x := N^{\langle 0\rangle}$ $y := N^{\langle 4\rangle}$

From the obtained table of the general results we select criteria that should be processed. In this case, x -- time, and y -- pressure

x =

	0
0	0
1	1·10-3
2	2·10-3
3	3·10-3
4	4·10-3
5	5·10-3
6	6·10-3
7	7·10-3
8	8·10-3
9	9·10-3
10	0.01
11	0.011

y =

	0
0	7·105
1	7·105
2	7.001·105
3	7.028·105
4	7.161·105
5	7.399·105
6	7.633·105
7	7.865·105
8	8.061·105
9	8.227·105
10	8.44·105
11	8.693·105

FIGURE 2.18 Reading the table data, obtained experimentally.

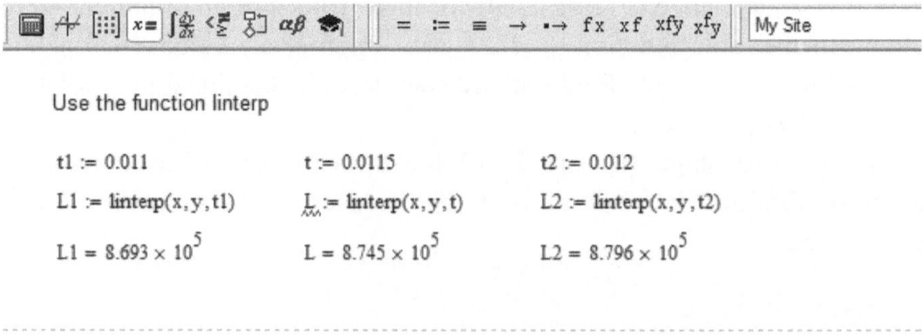

Use the function linterp

t1 := 0.011 t := 0.0115 t2 := 0.012

L1 := linterp(x,y,t1) L := linterp(x,y,t) L2 := linterp(x,y,t2)

$L1 = 8.693\times 10^5$ $L = 8.745\times 10^5$ $L2 = 8.796\times 10^5$

FIGURE 2.19 The procedure of linear interpolation.

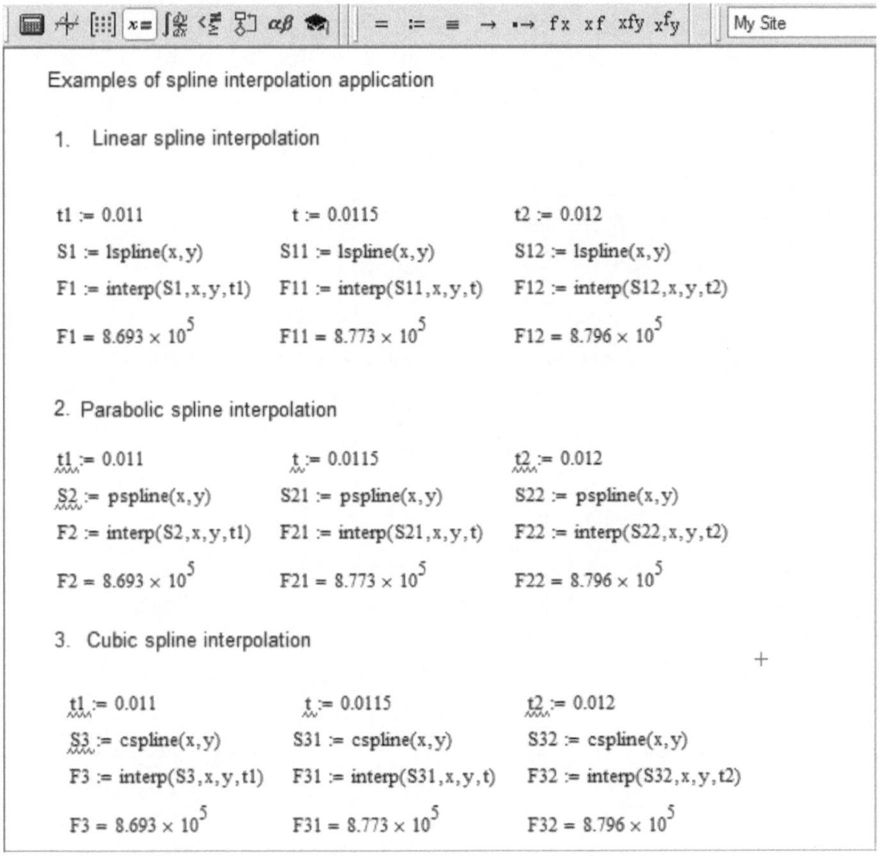

FIGURE 2.20 An example of spline interpolation.

KEYWORDS

- **Gauss method**
- **linear interpolation**
- **Mach numbers**
- **matrix calculus**
- **parabolic interpolation**
- **spline interpolation**

CHAPTER 3

ANALYTICAL DEPENDENCIES AND METHODS OF THEIR OBTAINMENT

CONTENTS

3.1 ANALYTICAL RELATIONS IN MATHEMATICAL MODELS

The ability to apply the analytical relations to predict the behavior of a technical object, as well as to manage this object, is a valuable fact from which no one refuses in practice. The examples of problems in which the analytical relations can be used to close the mathematical models are the following:

- to determine the discharge coefficient of gas from the rectangular cross section volume;
- to determine the aerodynamic drag coefficient of the aircraft moving with a subsonic speed in dense atmosphere;
- to determine the amount of convective heat current inflowing to the solid from the heated liquid, etc.

To obtain the analytical relations describing the technical object behavior, different approaches can be applied.

The first approach involves the application of the theory of dimensions and similarity which allows obtaining a relatively simple algebraic relations and relatively complex automodel solutions.

Another approach is to conduct and process the series of specially designed experiments. According to the experimental results, the empirical relations of linear and non-linear types can be built. Semi-empirical models can be built with the help of experimental approaches. In models of this class, the dependencies type describing the technical object behavior is set out of physical considerations. These can be algebraic equations and equation systems, differential equations or systems of differential equations, etc. Experimental studies in semi-empirical models allow you to determine the values of the missing coefficients, which cannot be determined theoretically.

To obtain analytical correlations, another approach is used which involves getting these relations from mathematical models of fundamental nature. The derivation of the relations in this case involves taking a significant number of assumptions that, at first stage, allow moving from differential equations of partial derivatives to ordinary differential equations or systems of algebraic equations, and at the next step – to implement the decision of the simplified equations in analytic form.

In this chapter, all of these approaches will be examined in one or another way.

3.2 APPLICATION OF THE THEORY OF SIMILARITY AND DIMENSIONS TO CONSTRUCT ANALYTICAL MODELS

3.2.1 UNITS AND THEIR RELATIONS TO EACH OTHER

The analysis of any physical phenomenon requires the researchers to take various measurements. As the measurement, we understand the procedure in which the measured value Q is compared with the reference value q, which has the same physical nature as Q. Each physical value can have its measurement unit and own definition. The physical value definition can be true within a single country or on the world level. Basic (absolute) units and derivatives are distinguished. Basic units are taken as the basis to designate dimensions, and all derived units are expressed in terms of these basic ones uniquely. Generally speaking, the basic units of measurement can be as much as there is a variety of physical phenomena. The problem in physics and philosophy is a question of the minimal necessary number of major units. For example, in mechanics, we cannot use three units – length, mass, time, but any two of them. The relation of the third quantity with the selected two basic ones can be found from the known laws of mechanics (law of gravitation, law of quantum energy emitted in the transition of atom element from one orbit to another, etc.). However, the use of only two units in mechanics would be very inconvenient. It would be necessary to introduce the system of additional coefficients. Moreover, for the convenience of physical laws and dimensions of quantities in these laws, in some cases it may be appropriate to take the excess unit as the basic unit.

The most widespread in most countries is the system of units SI (System International). In the system of SI units studying the mechanical phenomena as basic units, the linear size (measured in meters), mass of body (measured in kilograms), and process time (measured in seconds) are used. Dimensionalities of basic measurement units are designated as follows – length L, mass M, time T. The dimensionality of the physical unit derivative can be presented based on any physical formula consisting of units of interest and principal physical variables. The dimensionality of the physical unit can be determined based on its definition.

Dimensionalities of the values relating to various branches of physics can be taken from reference-books (e.g., [129, 224]). Nevertheless, we give some examples where the dimensionality $[Q]$ of any physical unit Q is determined from the dimensions of basic measurement units:

- force F dimensionality is determined using Newton second law binding the mass of body m with its acceleration a

$$F = ma$$

$$[m] = M; \qquad [a] = LT^{-2}$$

$$[F] = [m] \cdot [a] = MLT^{-2} \tag{3.1}$$

- power N dimensionality is determined using the definition that power is work (product of force and distance) performed in a unit of time

$$[N] = \frac{MLT^{-2} \cdot L}{T} = L^2 MT^{-3}; \tag{3.2}$$

- work (energy) A dimensionality is determined using the definition that work is determined by the product of power and time of the impact power

$$[A] = [N] \cdot [t] = L^2 MT^{-2} \tag{3.3}$$

For practical purposes, the following problem may be of interest. Suppose that the values *L, M, T* are selected as basic measurement units of mechanical values. Can we choose any of the other three units U_1, U_2, U_3 as the main? To simplify, let us assume that the new values are also related to mechanics.

The dimension theory [6, 229] argues that such transition can be achieved only if the following conditions are satisfied:

- dimensionalities of $[U_1]$, $[U_2]$, $[U_3]$ are independent functions $[L]$, $[M]$, $[T]$ for any α and β (i.e., $[U_1] \neq [U_3]^{\alpha} \cdot [U_2]^{\beta}$);
- reverse transformation can be done where $[L]$, $[M]$, $[T]$ are uniquely expressed by variables $[U_1]$, $[U_2]$, $[U_3]$.

We define the mathematical conditions under which the following recorded above requirements are true. Let the dimensionalities of $[U_1]$, $[U_2]$, $[U_3]$ be as follows:

$$\left.\begin{aligned} [U_1] &= [M]^{\mu_1}[L]^{\lambda_1}[T]^{\tau_1} \\ [U_2] &= [M]^{\mu_2}[L]^{\lambda_2}[T]^{\tau_2} \\ [U_3] &= [M]^{\mu_3}[L]^{\lambda_3}[T]^{\tau_3} \end{aligned}\right\} \tag{3.4}$$

We find the logarithm of these expressions:

$$\lg[U_1] = \mu_1 \lg[M] + \lambda_1 \lg[L] + \tau_1 \lg[T]$$

$$\lg[U_2] = \mu_2 \lg[M] + \lambda_2 \lg[L] + \tau_2 \lg[T]$$

$$\lg[U_3] = \mu_3 \lg[M] + \lambda_3 \lg[L] + \tau_3 \lg[T] \tag{3.5}$$

The equation system (3.5) has the unique solution for the variables $[U_1]$, $[U_2]$, $[U_3]$ only if the determinant Δ of the system

$$\Delta = \begin{vmatrix} \mu_1 & \lambda_1 & \tau_1 \\ \mu_2 & \lambda_2 & \tau_2 \\ \mu_3 & \lambda_3 & \tau_3 \end{vmatrix} \neq 0$$

is different from zero. The non-zero determinant will be the condition when the correct transition from one system of units to another is possible. This conclusion remains valid when considering the cases with a large number of independent (main) units (e.g., when solving problems in thermodynamics and electrodynamics, etc.).

3.2.2 DEVELOPMENT OF ALGEBRAIC TYPE MODELS USING DIMENSIONALITY THEORY

The dimensionality method can be used in practice to determine the conversion factors in the transition from one system of units to another, to summarize the information obtained from experiments on models and patterns, and as a tool for researchers providing obtaining the maximum of useful information in the experiment preparation. Besides, the dimensionality method can be used to obtain partial solutions of problems that cannot be solved analytically and require the use of complex mathematical apparatus.

The dimensionality method allows you to establish some formal relation for any physical phenomenon, along with advantages it has certain disadvantages. Suppose that we need to determine the functional dependence of the physical phenomenon of the variable A from the set of variables U_i $(i = 1,2,3, \ldots, n)$ – $A = f(U_1,U_2,\ldots,U_i,\ldots,U_n)$. Let us search this dependence in the form

$$A = C \cdot U_1^{\alpha_1} U_2^{\alpha_2} \cdot \ldots \cdot U_n^{\alpha_n} \tag{3.6}$$

In the Eq. (3.6) – C is some factor that has no dimension. The essence of dimensionality method consists in selecting the exponents $\alpha_1, \alpha_2, \ldots, \alpha_n$ to ensure the equality of outcome exponents for dimensions of basic physical units in both sides of the Eq. (3.6). The advantage of the dimensionality method is the fact that it is possible to select such values as $\alpha_1, \alpha_2, \ldots, \alpha_n$. The disadvantage of this method is that it is impossible to determine the value of the dimensionless coefficient C. In addition, it is easy to see that in many practical cases, the selection of values $\alpha_1, \alpha_2, \ldots, \alpha_n$ satisfying (3.6) can be ambiguous.

Let us consider an example. Suppose that you want to set the dependence of viscous fluid volume flowing through a pipe of circular cross section per time unit (Figure 3.1).

To solve this problem, at the first stage it is necessary to formulate the estimated functional dependence binding the volumetric flow rate W' with other physical units. The second stage is to set the dimensions of all physical quantities in the functional dependence. At the final stage, it is necessary to set up the equations for the unknown exponents in the determined functional correlation.

From physical considerations it is clear that the fluid flow in the pipe is supported by the pressure differential $(p_1 - p_2)$ $(p_1 - p_2)$. Here p_1 – pressure

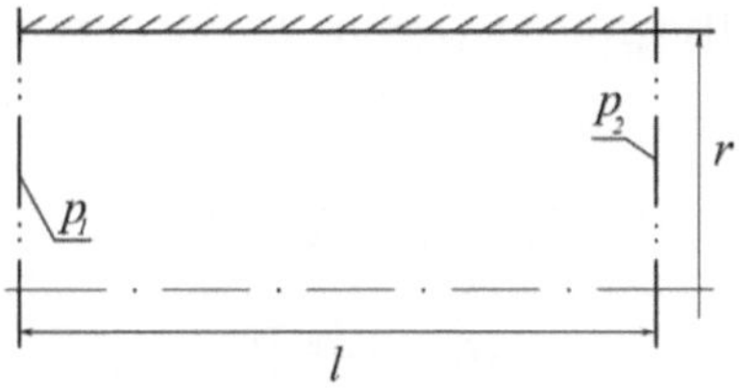

FIGURE 3.1 Diagram of the fluid flow in a pipe.

at the pipe inlet, and p_2 –pressure at the pipe outlet. The pressure gradient p_1 will affect the magnitude of the fluid flow through the pipe on the considered portion of the pipe from section *1* to section *2* – $P_l = \left(\frac{p_1 - p_2}{l}\right)$ (here l – distance between sections *1* and *2*). The tube radius r (tube radius increase leads to the increase in the fluid flow) and the value of the dynamic viscosity μ (viscosity increase decreases the fluid consumption flowing through the tube) will also affect the flow magnitude.

In accordance with the given problem analysis, the volumetric fluid discharge through the pipe can be written as the function $W' = f(P_l, r, \mu)$, and dependence for the rate of volumetric fluid flow will be sought as

$$W' = C \cdot P_l^{\alpha_1} r^{\alpha_2} \mu^{\alpha_3} \tag{3.7}$$

Let us define the dimensionalities of all values in the latter equation:

- dimensionality of the volumetric flow rate – $[W'] = L^3T^{-1}$,
- dimensionality of the pressure gradient – $[P_l] = L^{-2}MT^{-2}$,
- dimensionality of the pipe radius – $[r] = L$,
- in accordance with the definition, viscosity is the force acting on a unit area parallel to the flow surface and related to the velocity gradient observed in the direction perpendicular to the surface, and therefore its dimensionality is set by the ratio

$$[\mu] = \frac{force}{area} \Big/ \frac{velocity}{distance} = L^{-1}MT^{-1}$$

The Eq. (3.7) rewritten in the form for dimensionalities will be as follows

$$L^3T^{-1} = (L^{-2}MT^{-2})^{\alpha_1} L^{\alpha_2} (L^{-1}MT^{-1})^{\alpha_3} \tag{3.8}$$

The values of the coefficients $\alpha_1, \alpha_2, \alpha_3$ are determined by the solution of the equation system

$$2\alpha_1 + \alpha_2 - \alpha_3 = 3$$

$$\alpha_1 + \alpha_3 = 0$$

$$-2\alpha_1 - \alpha_3 = -1$$

Here, the first equation is written from the condition of degrees equality on both sides of the Eq. (3.8) with the dimensionality *L*, the second equation – from the condition of degrees equality with the dimensionality *M*, and the third equation – with the dimensionality *T*.

In the matrix form, the latter equation system is written down as follows

$$\begin{pmatrix} 2 & 1 & -1 \\ 1 & 0 & 1 \\ -2 & 0 & -1 \end{pmatrix} \cdot \begin{pmatrix} \alpha_1 \\ \alpha_2 \\ \alpha_3 \end{pmatrix} = \begin{pmatrix} 3 \\ 0 \\ -1 \end{pmatrix}$$

It is easy to verify that the solution of the equation system is the values $\alpha_1 = 1$, $\alpha_2 = 4$, $\alpha_3 = -1$, and the final record of the Eq. (3.7) is represented in the form

$$W' = C\frac{P_l r^4}{\mu} \tag{3.9}$$

Obtained by the problem solving the Eq. (3.9) is a well-known in physics Poiseuille equation which is correct when calculating flows in tubes with the large elongation ($l / r >> 1$) on the plot of stabilized flow when the boundary layer merges in the center of the pipe. Note that physicists long ago defined the aspect ratio C in the deduced above law $\left(C = \frac{\pi}{8}\right)$ [129].

The application of the dimensionality method in the form discussed above is not always successful. Indeed, we can give completely different physical units having the same dimensionality in the adopted system of measurement units (e.g., heat conductivity and viscosity). We can give examples of physical units without dimensionality (e.g., angle – flat or solid). At the same time, there are examples where the same physical units in the functional dependence can be interpreted ambiguously.

In Ref. [229] an interesting approach developing the classic method of dimensionalities and significantly enhancing its capabilities is considered. Here are the main features of the approach proposed in Ref. [229].

1. Mass in the analysis of any physical activity can be viewed from two completely different perspectives. The mass can be considered as the amount of substance (e.g., in determining the volume of a solid, when weighing, in the analysis of the laws of mass conservation, etc.). At the same time, the mass can be considered as a measure of the solid inertial motion (e.g., in determining the acceleration of a solid or system of solids, in solving problems related to the application of the law of conservation of momentum and moment of momentum, when there is an impact of various forces, etc.). These arguments make it appropriate to introduce two dimensionalities of mass – as the amount of substance $[M_m]$ and as a measure of inertia $[M_i]$.
2. Analyzing the effect of the linear size on the desired function we assume that the linear dimension is the vector value, and the dimensionality of the length L will be split by the unit vectors of the coordinate system (e.g., L_x, L_y, L_z – for Cartesian coordinate system). If is necessary in some cases, it is advisable to mark the orientation of the linear vector pointing in indices the positive L_x and (or) negative L_{-x} direction of the relevant linear dimension. Note some useful correlations which can be obtained by the offered method:
   - plane angle – $L_xL_y^{-1}$,
   - solid angle – $L_xL_yL_z^{-2}$,
   - angular velocity – $L_xL_y^{-1}T^{-1}$,
   - linear dimension has plane symmetry (e.g., the axes X, Y) – $L^2 \rightarrow L_xL_y$, $L \rightarrow \sqrt{L_xL_y}$,
   - linear dimension has symmetry in three spatial coordinates – $L^3 \rightarrow L_xL_yL_z$, $L \rightarrow \sqrt[3]{L_xL_yL_z}$.

Dimensional analysis technique after the introduction of two types of mass and vector of linear dimensions remains the same with the only difference, that now the degrees equality in both sides of the dimensionality equations for more variables should be considered. In mechanics – these are variables $L_x, L_y, L_z, M_m, M_i, T$.

Let us consider some applications of the above given approach.

Problem 1. Determine the mass flow rate *m*' of viscous fluid flowing through the pipe of circular cross section (Figure 3.1).

We seek a solution in the form $m' = C \cdot P_l^{\alpha_1} \rho^{\alpha_2} r^{\alpha_3} \mu^{\alpha_4}$. Let us write the dimensions of variables in the equation:

- m' – mass flow rate – $\left(M_m T^{-1} \right)$,
- P_l – pressure difference – $\left(L^{-2} M_i T^{-2} \right)$,
- ρ – fluid density – $\left(L^{-3} M_m \right)$,
- r – pipe radius – (L),
- μ – viscosity coefficient $\left(L^{-1} M_i T^{-1} \right)$.

Dimensional equation will be written down as,

$$M_m T^{-1} = \left(L^{-2} M_i T^{-2}\right)^{\alpha_1} \cdot \left(L^{-3} M_m\right)^{\alpha_2} \cdot L^{\alpha_3} \cdot \left(L^{-1} M_i T^{-1}\right)^{\alpha_4}$$

The corresponding equation system for determining the coefficients $\alpha_1, \alpha_2, \alpha_3, \alpha_4$ written in matrix form is as follows

$$\begin{pmatrix} -2 & -3 & 1 & -1 \\ 0 & 1 & 0 & 0 \\ 1 & 0 & 0 & 1 \\ -2 & 0 & 0 & -1 \end{pmatrix} \cdot \begin{pmatrix} \alpha_1 \\ \alpha_2 \\ \alpha_3 \\ \alpha_4 \end{pmatrix} = \begin{pmatrix} 0 \\ 1 \\ 0 \\ -1 \end{pmatrix}$$

In the written equation system the first equation establishes the degrees equality with the dimension L, the second equation – with M_m, the third equation – with M_i and the fourth equation – with T.

The solution of the equation system,

$$\alpha_1 = 1, \quad \alpha_2 = 1, \quad \alpha_3 = 4, \quad \alpha_4 = -1$$

and the desired formula for the mass flow rate has the form

$$m = C \cdot p \frac{\rho r^4}{\mu}$$

Problem 2. Find the velocity of the ball falling under gravity in viscous medium (Figure 3.2).

We assume that the ball is small, the rate of fall is also taken as small and uniform. In this case, we can presume that the hydrodynamics of the process of ball falling obeys the laminar law of a viscous fluid flow. Let us accept

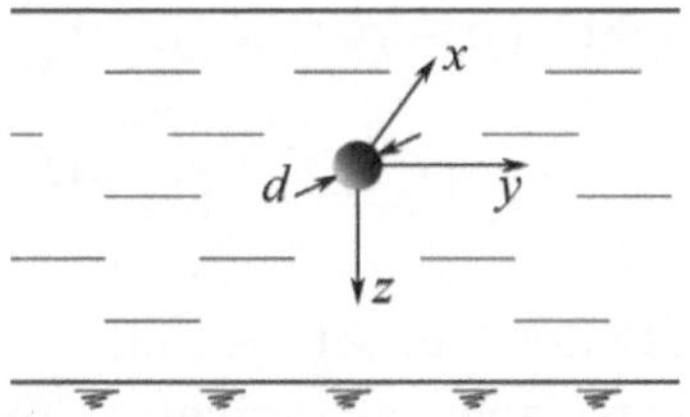

FIGURE 3.2 To the problem on the ball falling in the fluid.

that the *Z*-axis corresponds to the direction of gravity. Furthermore, the system symmetry about the axis *Z* should be taken into account. It can be assumed that the resistance due to the fluid viscosity depends on the cross-section of the ball (ball diameter *d*) perpendicular to the axis *Z*, for example, determined by the linear dimensions in the plane *XY*. Hence, the diameter dimensionality will further be written down as $L_x^{0,5}\ L_y^{0,5}$.

The viscosity dimensionality should be determined based on the frictional forces on the ball surface in the plane *XZ* (component μ_x)and in the plane *YZ* (component $\mu_{ó}$). The dimensionality [μ] taking the symmetry into account can be defined by the relation $[\mu]=\sqrt{[\mu_x][\mu_y]}$.

In accordance with the viscosity definition (see above), its dimension can be set as follows

$$[\mu_x]=\frac{L_z MT^{-2}}{L_y L_z}:\frac{L_z T^{-1}}{L_x}=L_x L_y^{-1} L_z^{-1} MT^{-1}$$

$$[\mu_y]=\frac{L_z MT^{-2}}{L_x L_z}:\frac{L_z T^{-1}}{L_y}=L_x^{-1} L_y L_z^{-1} MT^{-1}$$

$$[\mu]=\sqrt{[\mu_x][\mu_y]}=L_z^{-1} MT^{-1}$$

In addition to the frictional force, the gravity forces affect the ball falling proportional to the gravity acceleration g, and the buoyancy (Archimedean) force proportional to the density difference of the ball material ρ_1 and the fluid density ρ_2 $(\rho_1 > \rho_2)$.

Thus, the required relation to the velocity *v* of the ball fall and the corresponding equation for the dimensionalities can be written as:

$$v = C \cdot (\rho_1 - \rho_2)^{\alpha_1} d^{\alpha_2} \mu^{\alpha_3} g^{\alpha_4}$$

$$L_z T^{-1} = (L_x^{-1} L_y^{-1} L_z^{-1} M)^{\alpha_1} (L_x^{0,5} L_y^{0,5})^{\alpha_2} (L_z^{-1} M T^{-1})^{\alpha_3} (L_z T^{-2})^{\alpha_4}$$

The equation system that allows to set the values of the exponents α_1, α_2, α_3, α_4, is written for the exponents, respectively with L_x, L_y, L_z, M, T

$$\begin{pmatrix} -1 & 0,5 & 0 & 0 \\ -1 & 0,5 & 0 & 0 \\ -1 & 0 & -1 & 1 \\ 1 & 0 & 1 & 0 \\ 0 & 0 & -1 & -2 \end{pmatrix} \cdot \begin{pmatrix} \alpha_1 \\ \alpha_2 \\ \alpha_3 \\ \alpha_4 \end{pmatrix} = \begin{pmatrix} 0 \\ 0 \\ 1 \\ 0 \\ -1 \end{pmatrix}$$

The system comprises five equations for four unknowns, however, the first two equations are identical, so that one of them can be simply discarded. The solution for the coefficients $\alpha_1, \alpha_2, \alpha_3, \alpha_4$ exists and it is unique

$$\alpha_1 = 1, \quad \alpha_2 = 2, \quad \alpha_3 = -1, \quad \alpha_4 = 1$$

After substituting the coefficients obtained in the original equation, we have

$$v = C \cdot (\rho_1 - \rho_2) \frac{d^2 g}{\mu}$$

Note that in physics the final formula for the velocity of ball falling in viscous liquid is determined. It is defined [229] that the aspect ratio C value in the latter formula is $C = \frac{8}{9}$. In Ref. [229] you can find other interesting applications of the above-discussed approach.

The material presented below shows how the dimensionality method allows us to extrapolate the data of model experiments on full-scale objects. Modeling becomes necessary in those cases where the physical unit depends on a large number of factors. Formulas, composed in this case using the clauses of the dimensionality theory, are reasonably useful.

The examples of problems on mechanics were discussed so far. Naturally, the considered methods are acceptable in the solution of heat problems, problems of electrodynamics and other problems. Thus, when solving the heat problems along with the discussed basic measurement units (length, mass, time) we should introduce one more unit. In the unit system SI the temperature is introduced as the main heat unit. In the technical system of units (currently the system of units is not recommended to use) the quantity of heat is used as the basic unit instead of temperature.

3.2.3 AUTOMODEL DEVELOPMENT OF PROCESSES

In physics, such variants of process development can be referred to an automodel process, when the dependencies of some function $f_i(q_1, q_2, q_3, t)$ (density, energy, system, etc.) on spatial coordinates q_1, q_2, q_3 and time t can be presented as the function of only three dimensionless parameters ξ_1, ξ_2, ξ_3 – $f_i(\xi_1, \xi_2, \xi_3)$. New variables ξ_1, ξ_2, ξ_3 are functions of the original variables q_1, q_2, q_3, t and can be, for example, of the kind [200, 201] [200, 201] $\xi_1 = {q_1}/{bt^{\delta}}$, $\xi_2 = {q_2}/{bt^{\delta}}$, $\xi_3 = {q_3}/{bt^{\delta}}$. Here $[b] = LT^{-\delta}$ – parameter b dimension. In the particular case, in many automodel problems of gas dynamics the parameter b is the velocity of gas (or solid in gas) – b=v. In this case $\delta = 1$.

The selection of variables ξ_1, ξ_2, ξ_3 is symmetric relative to the original variables q_1, q_2, q_3, t. Therefore, the new variables to find automodel solutions can be constructed in other ways. When forming the self-similarity properties, we should talk about points, lines, planes, for which the self-similarity is provided. These points, lines, surfaces are called similitude or self-similarity centers.

The existence of automodel solutions is a special case of invariant solutions identified by the algorithms of the theory of group properties of differential equations [52, 163]. In some cases, constructing the automodel solutions it is possible to establish the validity of certain algebraic relations (invariants), which are the consequence of the conservation laws (mass, energy, entropy, momentum, etc.), their combinations. In some cases, to find automodel solutions the methods of dimension theory can be used. In Ref. [201] it is shown that the existence of self-similarity is sufficient to the system of the dimensional determinant parameters specified by initial and boundary conditions contain no more than two permanent parameters with independent dimensionalities. Permanent dimensionalities should be different from the length and (or) time. In particular, all the parameters that define the system changes are functionally determined by the values a, b; x, y, z, t; α_1, α_2,..., α_N. The dimensions of the values a, b here are defined by the correlations $[a] = M L^k T^s$, $[b] = L T^{-\delta}$. The parameters α_i are arbitrary combination of dimensional constants.

In Ref. [201] it is shown that the problem listed below can be attributed to the class of problems with automodel modes:

- diving of the cone in fluid or gas with constant velocity v and angle of attack α;
- plunger movement in a long cylindrical tube closed at one end (the speed of plunger movement $u = C t^n$, where C =const, $n \neq 0$);
- problem of the intense strong explosion localized in any point of space;
- problem of decay of the arbitrary discontinuity of parameters in the inert gas and combustible mixture, etc.

The latter problem in the list is particularly interesting because it is the basis of the famous numerical method for solving the problems of gas dynamics – method by S. K. Godunov [183, 241]. The solutions of this automodel problem and basic ideas of the numerical method by S. K. Godunov will be considered in Chapter 8.

Other interesting methods that allow applying the theory of similarity to the problems associated with the aerodynamics of aircraft are available, for example, in Ref. [80].

3.2.4 APPLICATION OF DIMENSIONLESS SYSTEMS TO CONSTRUCT MATHEMATICAL MODELS

From the dimensional physical variables we can form dimensionless systems of variables, further work with which can provide a lot of conveniences. In particular, in physics and engineering there is a well-established large set of dimensionless systems of physical variables called similarity criteria. The similarity criteria are widely used in the construction of dependency analysis, for example, when processing the experimental results. Indeed, if some dimensionless system of physical units that determine the development of the studied process coincide for full-scale product and its model used in the experiment, then there are bases for generalizing patterns obtained in the study of a model, on full-scale product, and vice versa. The dimensionless system (further we will call it a similarity criterion) in the theory of dimensionalities and similarity [6, 229] is usually denoted with the Greek letter π. The relations that exist in the studied complex system or technical object can be represented as a functional relation between the dimensional parameters. Likewise, the similar relation can be written between the dimensionless complexes. With the correct selection of dimensionless systems, the results obtained by the solution of both functional equations will coincide.

One of the main issues that can arise in selecting dimensionless systems is the question about the number of independent criteria, which can be constructed from N dimensional quantities. The independence criterion π_i (i=1, N) of each criterion π_j (j=1, N; $j \neq i$) means the validity of the following expression

$$\pi_i \neq f(\pi_1, \pi_2, \ldots \pi_{i-1}, \pi_{i+1}, \ldots, \pi_N) \tag{3.10}$$

Let us consider N physical quantities P_i (i=1, N) whose dimensionalities can be represented in the heat-mechanical system of units ([L] – linear dimension, meters, [M] – mass, kilograms, [T] – time, seconds, [θ]– temperature, Kelvin degrees). Then the dimensionality of any physical quantity P_i can be written as the ratio

$$[P_i] = [L]^{\lambda_i} [M]^{\mu_i} [T]^{\tau_i} [\theta]^{j_i}, \ i=1,\ n \tag{3.11}$$

and any similarity criterion can be expressed by the dependence

$$\pi = P_1^{\alpha_1} P_2^{\alpha_2} \ldots P_n^{\alpha_n} \tag{3.12}$$

The Eq. (3.12), taking into account the dimensionalities of the values P_i (Eq. (3.11)) can be rewritten as

$$[\pi] = [P_1]^{\alpha_1} [P_2]^{\alpha_2} \ldots [P]^{\alpha_n} = [L^{\lambda_1} M^{\mu_1} T^{\tau_1} \theta^{\kappa_1}]^{\alpha_1}$$
$$[L^{\lambda_2} M^{\mu_2} T^{\tau_2} \theta^{\kappa_2}]^{\alpha_2} \cdot \ldots [L^{\lambda_n} M^{\mu_n} T^{\tau_n} \theta^{\kappa_n}]^{\alpha_n}$$

or

$$[\pi] = [L]^{\sum_i^n \lambda_i \alpha_i} [M]^{\sum_i^n \mu_i \alpha_i} [T]^{\sum_i^n \eta_i \alpha_i} [\theta]^{\sum_i^n \kappa_i \alpha_i} \tag{3.13}$$

From this formulated problem comes that the similarity criterion – values of zero dimensionalities. In this case, the exponents in the Eq. (3.13) should equal zero which corresponds to the following system of homogeneous linear algebraic equations

$$\begin{pmatrix} \lambda_1 & \lambda_2 & \dots & \lambda_i & \dots & \lambda_{n-1} & \lambda_n \\ \mu_1 & \mu_2 & \dots & \mu_i & \dots & \mu_{n-1} & \mu_n \\ \tau_1 & \tau_2 & \dots & \tau_i & \dots & \tau_{n-1} & \tau_n \\ \kappa_1 & \kappa_2 & \dots & \kappa_i & \dots & \kappa_{n-1} & \kappa_n \end{pmatrix} \cdot \begin{pmatrix} \alpha_1 \\ \alpha_2 \\ \dots \\ \alpha_i \\ \dots \\ \alpha_{n-1} \\ \alpha_n \end{pmatrix} = \begin{pmatrix} 0 \\ 0 \\ 0 \\ 0 \end{pmatrix} \tag{3.14}$$

Let r – rank of the matrix of coefficients in the system of Eq. (3.14). It is known [39, 55] that in this case the system has $(n-r)$ independent solutions for the coefficients α_i. In the theory of dimensions, it is proved that every solution for α_i allows obtaining one similarity criterion. Consequently, it will be $(n-r)$ similarity criteria, independent of each other. The independent criteria are called fundamental. Using them, it is possible to obtain other similarity criteria, such as

$$\pi' = \prod_{i=1}^{n} \pi_i^{K_i}$$

New similarity criteria π'_j will be independent if the exponents K_i satisfy the condition $|K_i| \neq 0$.For the problem (3.14) the rank of the matrix $r = 4$, therefore, solving this problem we can obtain ($n-4$) independent (fundamental) similarity criteria. If to increase the number of basic units of measurement, but to ensure that emerging dimensional coefficients are not included in the mathematical model of the system, we can reduce the number of fundamental similarity criteria.

We should note an important fact. Among the parameters that determine the behavior of the studied system, we should include all the conditions that establish the uniqueness of its behavior. In particular, if the mathematical model of the system contains partial differential equations, it must contain the initial and boundary conditions, etc.

The construction of fundamental similarity criteria with the use of basic units of heat-mechanical measurement system was shown above. However, we can deduce the fundamental similarity criteria accepting arbitrary physical variables as basic. Naturally, the condition of independence of the physical variables from each other should be formulated.

Similarity criteria can be obtained in several ways. To illustrate these ways, let us consider the following problem:

- to determine the variation with time t of the flight height h of an aircraft (aircraft mass m) in dense atmosphere, if it is known that the driving force of the aircraft R, the gravity force G, aerodynamic resistance force X impact on the aircraft (Figure 3.3). For simplicity, we assume that the aircraft moves in the direction perpendicular to the Earth surface. In this case, the gravity force is defined by the equation $G = -mg$ (g – acceleration of gravity). We assume that the aerodynamic resistance force is directly proportional to the square of the velocity and is defined by the relation $X = -k v^2 = -k(\frac{dh}{dt})^2$.

The first method to obtain the similarity criteria for the formulated problem is to solve the system of homogeneous equations of the form (3.14), in which each value included in the functional equation is represented through the main units. From the problem condition we can assume that the aircraft height is established as the dependence of the aircraft mass m, thrust of the propulsion system R, gravity forces (from the gravitational acceleration g), friction forces (from the resistance coefficient k) and flight time t

$$h = f(m, R, g, k, t) \tag{3.15}$$

The dimensionless combination of variables in the Eq. (3.15) is set as

$$\pi = C \cdot h^{\alpha_1} m^{\alpha_2} R^{\alpha_3} g^{\alpha_4} k^{\alpha_5} t^{\alpha_6} \tag{3.16}$$

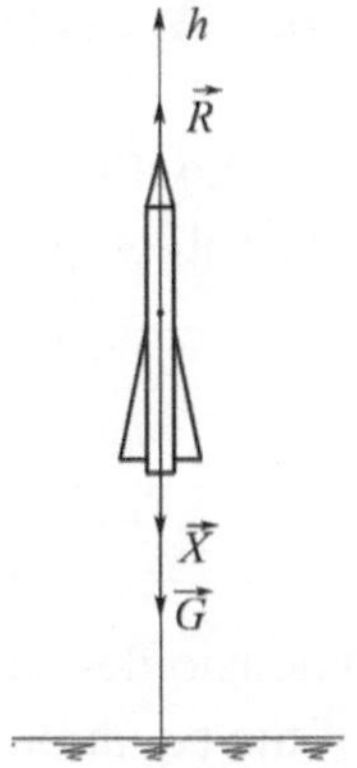

FIGURE 3.3 On the problem of aircraft movement.

The latter equation in the form for dimensionalities will be written as

$$[\pi]=(L^1)^{\alpha_1}(M^1)^{\alpha_2}(L^1M^1T^{-2})^{\alpha_3}(L^1T^{-2})^{\alpha_4}(L^{-1}M^1)^{\alpha_5}(T^1)^{\alpha_6}$$

Let us re-write the matrix Eq. (3.14) concerning the problem being solved

$$\begin{pmatrix} 1 & 0 & 1 & 1 & -1 & 0 \\ 0 & 1 & 1 & 0 & 1 & 0 \\ 0 & 0 & -2 & -2 & 0 & 1 \end{pmatrix} \cdot \begin{pmatrix} \alpha_1 \\ \alpha_2 \\ \alpha_3 \\ \alpha_4 \\ \alpha_5 \\ \alpha_6 \end{pmatrix} = \begin{pmatrix} 0 \\ 0 \\ 0 \end{pmatrix} \tag{3.17}$$

The equation system (3.17) for the values of the exponents α_i contains six unknowns, and the rank of the matrix coefficients of the unknowns is three. Consequently, the system has three linearly independent solutions. These three values α_i can take arbitrary values, and the rest should be determined from the Eqs. (3.17). Note that a certain arbitrariness in the selection of α_i in this case does not matter, because the main problem is the selection of an independent (fundamental) set of dimensionless complexes π_i.

Three possible solutions of the system (3.17) are written below. In this case, the values $\alpha_1, \alpha_2, \alpha_3$ are taken arbitrarily, and the values $\alpha_4, \alpha_5, \alpha_6$ are set by the system solution (3.17):

$$\alpha_1 = 1, \quad \alpha_2 = 1, \quad \alpha_3 = 0, \quad \alpha_4 = -2, \quad \alpha_5 = -1, \quad \alpha_6 = -4;$$

$$\alpha_1 = 0, \quad \alpha_2 = 0, \quad \alpha_3 = 1, \quad \alpha_4 = -2, \quad \alpha_5 = -1, \quad \alpha_6 = -2;$$

$$\alpha_1 = 0, \quad \alpha_2 = 1, \quad \alpha_3 = 0, \quad \alpha_4 = -1, \quad \alpha_5 = -1, \quad \alpha_6 = -2$$

Using the solutions obtained to the Eq. (3.16), we have the following three dimensionless combinations of variables (similarity criteria) contained in the dependence of (3.15)

$$\pi_1 = \frac{h \cdot m}{g^2 k t^4}; \quad \pi_2 = \frac{R}{g^2 k t^2}; \quad \pi_3 = \frac{m}{g k t^2} \tag{3.18}$$

The obtained combination of dimensionless complexes is linearly independent. Any additional solutions of the equation system (3.17) will be the linear combination of already written dimensionless complexes (3.18).

The second method of constructing the similarity criteria is also based on the method of dimensionalities. In this method, in the studied functional dependence (dependence (3.15)) any three variables are selected that are linearly independent from each other. Note that the number of selected independent variables coincide with the rank of the matrix of coefficients used in the first method of constructing dimensionless complexes. Let in the problem (3.15) be quantities *g*, *k*, *t*. The original functional relation (3.15), in this case, can be rewritten as

$$\frac{h}{[g]^{\alpha_h}[k]^{\beta_h}[t]^{\gamma_h}} = f\left(\frac{m}{[g]^{\alpha_m}[k]^{\beta_m}[t]^{\gamma_m}}, \frac{R}{[g]^{\alpha_R}[k]^{\beta_R}[t]^{\gamma_R}}\right) \tag{3.19}$$

Complexes $\pi_1' = \frac{h}{[g]^{\alpha_h}[k]^{\beta_h}[t]^{\gamma_h}}$, $\pi_2' = \frac{m}{[g]^{\alpha_m}[k]^{\beta_m}[t]^{\gamma_m}}$, $\pi_3' = \frac{R}{[g]^{\alpha_R}[k]^{\beta_R}[t]^{\gamma_R}}$ with proper selection of exponents $\alpha_h, \beta_h, \gamma_h$, $\alpha_m, \beta_m, \gamma_m$, $\alpha_R, \beta_R, \gamma_R$ can be dimensionless and it is easily provided. Taking into account the dimensions of the quantities g, k, t

$$[g] = M^0L^1T^{-2}, \quad [k] = M^1L^{-1}T^0, \quad [t] = M^0L^0T^1$$

the expressions for π_1', π_2', π_3' can be re-written as:

$$\frac{h}{[g]^{\alpha_h}[k]^{\beta_h}[t]^{\gamma_h}} = \frac{L^1M^0T^0}{(L^1T^{-2})^{\alpha_h}(M^1L^{-1})^{\beta_h}(T^1)^{\gamma_h}}$$

$$\frac{m}{[g]^{\alpha_m}[k]^{\beta_m}[t]^{\gamma_m}} = \frac{L^0M^1T^0}{(L^1T^{-2})^{\alpha_m}(M^1L^{-1})^{\beta_m}(T^1)^{\gamma_m}}$$

$$\frac{R}{[g]^{\alpha_R}[k]^{\beta_R}[t]^{\gamma_R}} = \frac{L^1M^1T^{-2}}{(L^1T^{-2})^{\alpha_R}(M^1L^{-1})^{\beta_R}(T^1)^{\gamma_R}}$$

The coefficients $\alpha_h, \beta_h, \gamma_h$ are found by solving the system of equations

$$\begin{pmatrix} 1 & -1 & 0 \\ 0 & 1 & 0 \\ -2 & 0 & 1 \end{pmatrix} \cdot \begin{pmatrix} \alpha_h \\ \beta_h \\ \gamma_h \end{pmatrix} = \begin{pmatrix} 1 \\ 0 \\ 0 \end{pmatrix}$$

and the appropriate solution for the coefficients

$$\alpha_h = 1, \quad \beta_h = 0, \quad \gamma_h = 2$$

The coefficients $\alpha_m, \beta_m, \gamma_m$ are found by solving the equation system

$$\begin{pmatrix} 1 & -1 & 0 \\ 0 & 1 & 0 \\ -2 & 0 & 1 \end{pmatrix} \cdot \begin{pmatrix} \alpha_m \\ \beta_m \\ \gamma_m \end{pmatrix} = \begin{pmatrix} 0 \\ 1 \\ 0 \end{pmatrix}$$

and the appropriate solution for the coefficients

$$\alpha_m = 1, \quad \beta_m = 1, \quad \gamma_m = 2$$

The coefficients $\alpha_R, \beta_R, \gamma_R$ are found by solving the equation system

$$\begin{pmatrix} 1 & -1 & 0 \\ 0 & 1 & 0 \\ -2 & 0 & 1 \end{pmatrix} \cdot \begin{pmatrix} \alpha_R \\ \beta_R \\ \gamma_R \end{pmatrix} = \begin{pmatrix} 1 \\ 1 \\ -2 \end{pmatrix}$$

and the appropriate solution for the coefficients

$$\alpha_R = 2, \quad \beta_R = 1, \quad \gamma_R = 2$$

Finally, we have

$$\pi_1' = \frac{h}{gt^2}; \quad \pi_2' = \frac{m}{gkt^2}; \quad \pi_3' = \frac{R}{g^2kt^2} \tag{3.20}$$

It is easy to find the connection between dimensionless complexes obtained by the second method (Eq. (3.20)), and with dimensionless complexes obtained by the first method (Eq. (3.18))

$$\pi_1' = \pi_1\pi_3^{-1}, \quad \pi_2' = \pi_3, \quad \pi_3' = \pi_2$$

In Ref. [6], we consider other ways of constructing dimensionless complexes from physical quantities.

Here are some of the dimensionless complexes (similarity criteria) composed of the physical quantities that are widely used in modern physics and engineering [71, 106, 123, 126, 127, 140, 199].

1. Heat Conductivity Processes

Biot number – $Bi = \frac{\alpha l}{\lambda}$ – characterizes the ratio of conductive and convective thermal resistance at the interface of two media (solid with fluid or gas). Here α – coefficient of heat emission from the heated to cold medium, λ – coefficient of heat conductivity, l – linear dimensions (length, diameter, thickness, etc.).

Fourier number – $Fo = \frac{at}{l^2}$ – homochronicity criterion, or "dimensionless" process time. Here a – thermal diffusivity, t – process time.

Criterion of the internal source – $\overline{q}_v = \frac{q_v l^2}{\lambda \Delta T}$ – characterizes the ratio of the quantity of heat q_v generated by internal sources to the quantity of heat transmitted by conductivity. Here ΔT – temperature difference that determines the heat conduction process.

2. Processes of Thermodynamics, Hydrodynamics and Gas Dynamics

Isentropy n – dimensionless complex whose value is determined by the exponent binding the change of pressure and density of gas at its compression and expansion occurring at constant entropy. If we denote two arbitrary values of pressure and density, respectively p_1, p_2 and ρ_1, ρ_2, the value of the isentropy can be set from the equation $\frac{p_1}{p_2} = (\frac{\rho_1}{\rho_2})^n$. The expression defining isentropy is written down in the form

$$n = \frac{\ln({}^{p_1}\!/\!{}_{p_2})}{\ln({}^{\rho_1}\!/\!{}_{\rho_2})}$$

Adiabat index – $k = \frac{c_p}{c_v}$ – is defined as the ratio of specific heats of the gas at constant pressure c_p and constant volume c_v.

Mach number – $M = \frac{v}{c}$ – is defined as the ratio of the gas velocity v to the speed of sound c in the gas (gas velocity and sound speed are calculated in the same point of space). Physically, Mach number characterizes the degree of the continuous medium compression. So, if we accept that M $\approx$ 0, then we can assert that the continuous medium is incompressible. Under normal conditions (moderate temperature and pressure quantities, for example, not exceeding the atmospheric conditions in more than ten times) most fluids satisfy the incompressibility condition. At the same time, under extreme conditions of pressure and temperature metals can be considered as compressible medium (e.g., when solving the problem on high-speed interaction of a metal rod with solid barrier). For an ideal gas, the sound speed can be set by one of the following equations – $c=\sqrt{\frac{kp}{\rho}}$, or $c=\sqrt{kRT}$, or $c=\sqrt{k(k-1)(E-\frac{v^2}{2})}$. Here R – gas constant, P – pressure, ρ – gas density, T – its temperature, E – gas total energy.

Strouhal number (homochronicity criterion Ho) – $Sh=\frac{vt}{l}$ – characterizes the degree of non-stationarity of the studied process.

Euler number (similarity criterion of pressure fields) – $Eu=\frac{p}{\rho v^2}$ – is defined as the ratio of pressure forces (or pressure differential Δp) to the inertial forces.

Weber number – $We=\frac{\sigma}{\rho v^2 l}$ – is defined as the ratio of surface tension forces in the fluid (σ-surface tension coefficient) to the inertial forces. Studying the flow of fluid particles (drops, films) in the gas stream, this criterion is used to analyze the stability of their form.

Froude number – $Fr=\frac{v^2}{gl}$ – is defined as the ratio of kinetic energy of the fluid (gas) to the potential energy due to the weight of the column of fluid (gas) in height l. Froude number can be determined also as the ratio of inertial forces to the gravity forces (g – gravitational constant).

Reynolds number – $Re=\frac{\rho vl}{\mu}$ – is defined as the ratio of inertial forces to viscous forces. Here μ – coefficient of dynamic viscosity. In equation for Reynolds number the linear dimension may correspond to the body length, its diameter, or to some conditional parameter (e.g., thickness of the boundary layer, displacement thickness in the boundary layer, etc.).

Lagrangian number – $La = Eu \cdot Re = \frac{pl}{\mu v}$ – is defined as the ratio of pressure forces to viscous forces.

Galilean number – $Ga = \frac{Re^2}{Fr}$ – is used as the similarity criterion of free flow fields.

Archimedes number – $Ar = \frac{\Delta\rho\rho^2 g l^3}{\mu^2}$ – characterizes the interaction of buoyancy force due to the difference in density of the medium $\Delta\rho$, and forces due to viscosity.

3. Processes of Heat Transfer and Heat Exchange

Prandtl number – $Pr = \frac{\mu c_p}{\lambda}$ – is defined as the ratio of the quantity of heat transferred by the fluid (gas) due to the work of the viscous forces to the heat transmitted through the conductivity mechanism. The important fact is that for most of gases and fluids at moderate quantities of pressure and temperature the values of Prandtl number coincide with the condition $0,7 < Pr < 1,0$.

Along with Prandtl number Pr the diffusion Prandtl number is used Pr_D, that is defined by the ratio of the kinematic viscosity v to the diffusion coefficient D – $Pr_D = \frac{v}{D}$.

Peclet number – $Pe = Re \cdot Pr$ – is defined as the ratio of the quantity of heat transferred by the inertia forces work to the heat transmitted by the conductivity mechanism.

Nusselt number – $Nu = \frac{\alpha l}{\lambda}$ – is defined as the ratio of the quantity of heat transferred from one medium to another by the mechanism of heat transfer to the quantity of heat transmitted by conductivity. Here the symbol α denotes the heat emission coefficient.

Stanton number – $St = \frac{\alpha}{c_p \rho v}$ – is defined as the ratio of the quantity of heat transferred from one medium to another by the heat transfer mechanism to the quantity of heat transferred by convection.

Grashof number – $Gr = \frac{g\rho^2 \beta_t \Delta T l^3}{\mu^2}$ – is the characteristic of heat-gravitational convection. Here β_t – the coefficient of thermal expansion of the medium, ΔT – the temperature difference.

Rayleigh number – $Ra = Gr \cdot Pr$ – is the characteristic of heat-gravitational convection in environments with Prandtl numbers $Pr \geq 1$.

4. Mechanics of Gas Mixtures (Fluid Mixtures)

Schmidt number for the ith component of the mixture $Sm_i = \frac{\rho D_i}{\mu}$ is defined as the ratio of mass transfer determined by diffusion and viscous mechanisms.

Lewis number for the ith component of the mixture $Le_i = Sm_i Pr$ is defined as the ratio of heat transfer due to diffusion and heat conductivity.

Damköhler number I – $Dam_i^{(1)} = \frac{l}{t_i v}$ in multicomponent chemically reacting media allows us to estimate the ratio of the characteristic time of mechanical processes to the characteristic time of ith chemical reaction t_i.

Damköhler number II – $Dam_i^{(2)} = \frac{q_i}{c_p T_0}$ in multicomponent chemically reacting media allows us to estimate the ratio of the heat q_i emitted (absorbed) in the flow of the ith chemical reaction to the characteristic thermodynamic enthalpy of medium (T_0 – characteristic temperature of environment).

The application of dimensionless complexes, generally speaking, increases the generality of the obtained solutions. Indeed, any combination of the physical parameters ρ, v, l, μ that satisfy the equation $Re = \frac{\rho v l}{\mu}$ can correspond to the same Reynolds number Re (or any other dimensionless complex). In Refs. [138, 249, 199] one can find examples of equations of gas motion in the partial derivatives written with the use of dimensionless complexes. However, dimensionless complexes are frequently used in the variant of ordinary algebraic equations obtained in the analysis of experimental results. For example, the dependence of the type $Nu = C Re^a Pr^b$ is widely used to estimate the heat transfer between gas (fluid) medium and solid in various technical fields (construction, power and mechanical engineering, instrumentation engineering, etc.). However, the value of the coefficients C, a, b in this formula is limited to the narrow limits of change of Reynolds number, Prandtl number and Nusselt number. The specific application of this and other formulas written in the form of criteria can be found, for example, in Refs. [106, 140].

It was shown above that the dimensionless complexes of physical quantities can be built arbitrarily many times. However, only a small part of them is used in physics and engineering. Only those criterial relations, by numerical values of which it is possible to recover these or other laws of the studied physical process "survive." For example, the adiabat *k* value of gas medium carries information about the number of atoms in the gas molecules, the presence of condensed particles in gas, dust, etc. Knowledge of Prandtl number *Pr* value allows relating the continuous medium to gas, fluid, metal melt. According to the known Mach number *M* it is possible to represent a qualitative picture of solid flow by gas flow. The information about Reynolds number *Re* indicates the nature of heat transfer between gas and solid, etc.

Physical science has a long history, and there can be the impression that all criterial relations were formulated long ago. Nevertheless, it is not the case. As an example, here are two dimensionless criteria used in the analysis of internal ballistics processes in solid fuel rocket engines (SFRE).

The first criterion – Pobedonostsev parameter κ [136] defined as the ratio of the burning solid fuel surface to the area of flow section of charge through which the combustion products emit (Figure 3.4) – $\kappa = \frac{S_m}{F_k}$. Pobedonostsev parameter is offered to assess the display of the effects of erosive burning of solid rocket fuel [136]. As a rule, the erosive burning fuel in a rocket projectile is absent with the values $\kappa < 70...100$.

Pobedonostsev parameter is convenient for rocket systems designers. However, it does not disclose the physics of processes running during the solid fuel combustion. In particular, using the parameter κ it is impossible to predict the negative effects of erosion of the solid rocket fuel. More versatile is the second criterion – parameter of V. N. Vilyunov *Vi* [41] defined by the formula

$$Vi = \frac{\rho_e v_e}{\rho_m u_m}\sqrt{\zeta_0}$$

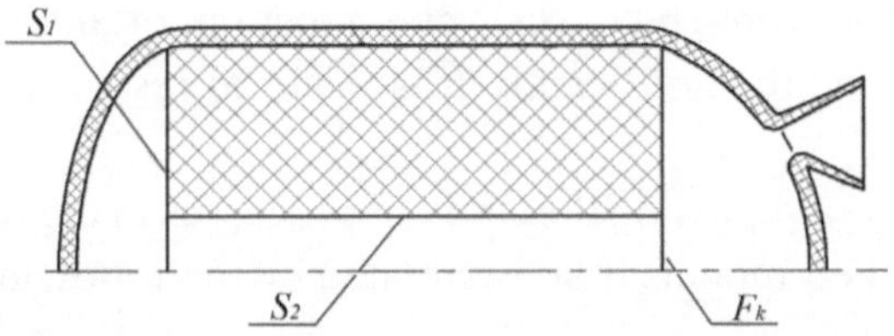

FIGURE 3.4 Scheme of the charge of solid fuel rocket engine to determine the Pobedonostsev parameter.

In the latter formula and below it is indicated:

- ρ_e, v_e – density and velocity of the combustion products at the outer edge of the boundary layer;
- ρ_m, u_m – fuel density and combustion rate;
- T_e, T_s – temperature of the combustion products at the outer edge of the boundary layer and temperature on the fuel surface;
- ζ_0 – coefficient of resistance whose value can be set according to the formula $\zeta_0 = 4c_{f0}$;
- c_{f0} – coefficient of resistance which can be calculated by the formula $c_{f0} = \frac{0,664}{\sqrt{\mathrm{Re}_\delta}}$; in this formula Re_δ – Reynolds number calculated on the thickness of the boundary layer δ.

The critical value Vi_* of V. N. Vilyunov parameter, excess of which leads to the positive erosion, is set by the formula

$$Vi_* = \frac{2T_s}{T_e} \cdot \frac{1}{\sqrt{\zeta_0}}$$

In Ref. [41] it is noted that the critical value Vi_* with sufficient accuracy for practical purposes can be accepted as $Vi_* \approx 8$. When $1,6 < Vi < 5,6$ there may be the effects of negative erosion during the combustion of solid fuels.

3.2.5 ANALYSIS OF THE SIMILARITY OF INTRABALLISTICS PROCESSES IN FULL-SCALE AND MODEL SOLID FUEL ROCKET ENGINES

Dimensionless systems of physical quantities are widely used in various fields of technology, and it is associated with the ability to study various processes not on the full-scale objects but on their models. Models can be performed in a more convenient scale and, most important, the studies using the models can require significantly less material resources than full-scale research.

The possibility of replacing the study of processes in the full-scale objects by investigating their models is substantiated by the following theorems [6]:

Theorem 1. The necessary and sufficient condition for the similarity of two systems is the equality of the corresponding similarity criteria of these systems composed of generalized coordinates and parameters of the systems ($\pi_k = idem$);

Theorem 2 (π-theorem). The functional relation between the quantities characterizing the process can be represented as the dependence between the similarity criteria composed of them.

The formulated theorems allow making the conclusions important for practice [6]:

- complex systems are similar if the corresponding subsystems are similar and similarity criteria are composed of units not included in any of the subsystems are equal;
- similarity conditions valid for the systems with constant parameters are extended to the systems with variable parameters on condition that relative characteristics of variables parameters coincide;
- similarity conditions valid for isotropic systems can be extended to anisotropic systems if the anisotropy in the compared systems are relatively the same.

In practice, solving the problems of object modeling using the theory of similarity, the following problems may occur:

- not all the defining parameters of the phenomenon are known;
- among the defining parameters we can mark such ones which significantly affect the process and those who have little influence;
- it is impossible to practically select the nature parameters in the way that the defining parameters of the nature and model are equal;
- it is impossible to satisfy the additional conditions of similarity if there are variable parameters or anisotropy.

These difficulties can be overcome by using an approximate similarity of the full-scale object and model. The approximate similarity suggests:

- elimination of the parameters whose influence is negligible;
- use only the known parameters;
- neglect the necessity of some criteria equality;
- use average values of variables.

Note that the use of the approximate similarity can lead to the increase of errors when processing the experiment. However, even in the case of large errors the modeling effect will not disappear, as it allows you to determine the right direction of search and order of the expected result.

As an example, let us consider the problem arising in the practice of designing ignition systems for large solid fuel rocket engines [3, 204, 205]. An important step of design is autonomous training of the igniter which is performed with the use of simulators of the engine internal volume. Such technology of training the ignition system can ensure the performance of geometric similarity with the experiment. In fact, the simulator can be made in the scale to 1:1. Nevertheless, the use of simulators of solid fuel rocket engines does not allow ensuring the similarity of the intrachamber processes because that prevents to realize the mode of co-burning of igniting composition and solid fuel. In this regard, the use of model SFREs for testing the initial part of work of the full-scale engine remains relevant.

Let us justify the similarity of processes in solid fuel rocket engine (Figure 3.5) and its model. We assume that the longitudinal and transverse dimensions of the full-scale engine and the model differ. The propulsion system shown in Figure 3.5 includes the tubular charge of the solid fuel *1* whose ignition is provided by the igniting device *2*. The combustion products of igniter and solid fuel expire through the nozzle block *3*.

In the further analysis we will denote the arbitrary parameter P established for the full-scale object and model P_n and P_m, respectively. The conversion factors from the model parameters to the parameters of full-scale object will be denoted as K_p. We assume that the conversion factors are constant. In this case, the conversion of the parameters of the full-scale object

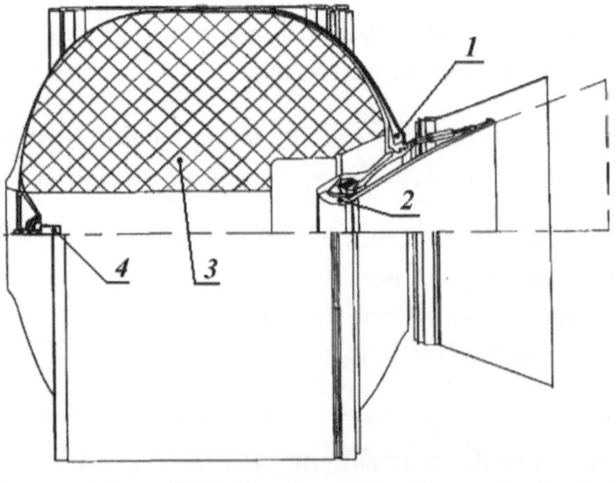

FIGURE 3.5 Scheme of the solid fuel rocket engine (1 – engine body; 2 – nozzle block; 3 – fuel charge; 4 – igniter).

on known values of the parameters set for the model will be performed on the relations of the form $P_n = K_P \cdot P_m$.

In the studied class of problems the following factors, processes and phenomena, the similarity of which should be ensured, can be sorted out:

- geometrical dimensions;
- parameters defining the dynamics of gas in the full-scale and model engines;
- chemical reactions occurring in the internal volume of objects;
- processes of heat exchange between the combustion products and fuel charge surface;
- process of heating the surface layer of the fuel and its ignition.

We assume that the change in the values of the free volume of intrachamber space due to the body deformation is not happening. This assumption is valid for the engine with a single-channel charge (face cavities are not taken into account). In addition, as shown by numerous calculations, the adoption of this assumption has practically no effect on the results of modeling the development of internal ballistics processes in the engine central channel.

In accordance with the theory of similarity and dimensionalities in determining the coefficients of conversion from full-scale engine to model parameters only four independent coefficients can be used. All the other coefficients can be re-calculated using the equations describing the relevant processes. The dimensionalities by spatial coordinates will be considered as the vector [229]. This allows setting the conversion factors in the case where the scale coefficients of the model and full-scale engine do not coincide in the longitudinal K_X and transverse K_R dimensions. Also, the number of independent coefficients increases by one.

The gas-dynamic processes are defining in the period of motor output to the mode. Therefore, the list of independent conversion factors is advisable to set from the analysis of these processes. The equations of gas dynamics contain the following set of basic physical unities – process time t, longitudinal x and transverse r coordinates, velocity v, density ρ, internal energy E, temperature T and pressure p of the combustion products.

It seems appropriate to make the selection of the minimal (critical) cross-section of the nozzle block so that the levels of quasi-stationary pressure values in the full-scale and model engines coincide. This will exclude from consideration the non-linearity in the law for the combustion rate of the fuel as the function of pressure of the combustion products. The same pressures can be provided when the ratio of the surface area of fuel burning S_m to the minimal section of the nozzle $F_{\min}$ in the full-scale and model engines are the same – $\frac{S_m}{F_{\min}} = idem$.

If we assume that in the full-scale and model engines the charge with the internal cylindrical channel is used, the latter relation can be re-written as

$$\frac{K_X}{K_R} = idem \quad (3.21)$$

The validity of relation (3.21), as a consequence, leads to the fact that the values of axial component (longitudinal) velocity of combustion products at the outlet of the channel charge in the full-scale and model engine coincide. Note the importance of the latter relation. For the conditions of the solid fuel rocket engine it means that in full-scale and model engines the values of Pobedonostsev parameter κ are close to each other. Pobedonostsev parameter is a dimensionless criterion which allows us to establish whether the solid fuel is burning erosionally. The appearance of the erosion effects corresponds to the values $\kappa > 80...120$ and is accompanied by a non-linear increase in the pressure of the engine combustion chamber. If the velocities of the combustion products at the outlet of the charge channel in the model and full-scale engines are the same and the values of Pobedonostsev parameter are below the written above critical value, then the value of the conversion factor for the rate of the combustion products has the specific value $K_V = 1$.

The acceptance of conditions that pressure levels in the full-scale and model engines are the same leads to the conclusion that the conversion factors in the density K_ρ and internal energy K_E also equal one (the composition of the combustion products and, consequently, the thermal-physical characteristics of the combustion products are the same) – $K_\rho = 1, \quad K_E = 1$.

Summarizing the analysis of dynamic processes, we can say that if the fuel used in the model and full-scale engine is the same, then the necessary and sufficient condition for the similarity of the processes is to perform the geometric condition (3.21), thus, K_X, K_R can take arbitrary values.

Let us now determine the dependent coefficients that are of interest in analyzing the experimental results. Among the parameters of non-stationary part of the engine performance except those studied above, we should highlight the following set of values:

- process time,
- mass of the ignition composition weigh,
- outlay of the combustion products from the igniter body and engine chamber.

Let us define the conversion factors for the weigh mass and size of the ignition composition tablets. We will use the requirements of the theory of dimension and relations for the coefficients K_V, K_ρ, K_E written above. Defining the conversion factor for the process time, let us use the known values K_X and K_V. Defining the mass of the ignition composition, we will consider that the conversion of arches used in the weigh of tablets should be performed with the coefficient K_X, and the current surface of the weigh of tablets burning – with the coefficient $K_R{}^2$. The same applies to defining the rate of the mass flow of combustion products from the igniter body and the engine body. Taking into account the above-given comments, we receive the following values of conversion factors:

- for the similar moments of processes time in a full-scale and model engines K_t,
- for the mass of ignition composition K_M,
- for the rate of mass consumption of the combustion products from the igniter and engine bodies K_G

$$K_t = \frac{K_X}{K_V},$$

$$K_M = K_\rho K_X K_R^2, \tag{3.22}$$

$$K_G = K_\rho K_V K_R^2$$

Analyzing the effect of chemical reactions on intrachamber processes, we will proceed from the fact that the rate of chemical reactions (catalytic and interaction of the combustion products of ignition composition and fuel) is high, and the characteristic time of the chemical interaction is much shorter than the time typical for all other processes that accompany the engine output to the mode of quasi-stationary work. If we consider that the compositions of combustion products in a full-scale engine and model engine are the same, then there is no need to enter the additional conversion factors caused by chemical reactions.

Let us perform the analysis of the processes of heat exchange between the combustion products and surface layer of fuel. We will proceed from the validity of the law of heat exchange of Dyunze-Zhimolokhin [89, 244] which is written in the general form

$$Nu_d = A \cdot (Re_d \Pr)^m (1 + \frac{x}{d})^{-n} \tag{3.23}$$

Here Nu_d, Re_d, Pr – similarity criteria of Nusselt, Reynolds and Prandtl. d – characteristic size (channel charge diameter), on the value of Nusselt and Reynolds numbers are calculated. The coefficients A, m, n – constants determined from the physical experiment.

From the Eq. (3.23) we can obtain the following conversion factor for the quantity of the heat transfer coefficient α

$$K_\alpha = K_R^{-1} (K_V K_R)^m (1 + \frac{K_X}{K_R})^{-n} \tag{3.24}$$

The conversion factor value for the heat flow K_q will coincide with the conversion factor for the heat transfer coefficient K_α, since the conversion factor value for the temperature $K_T = 1$.

The conversion factor value for the ignition time of the fuel surface layer (induction time) can be determined using the results obtained in Ref. [47] or from the heat conduction equation for the fuel surface layer written on the assumption that the temperature profile in the heated layer has an exponential form and the heat effect of the solid-state reactions in the fuel can be neglected [204, 205]:

$$\frac{d(T_S - T_H)^2}{dt} = \frac{2}{C_T \rho_T \lambda_T} |q_\Sigma| q_\Sigma$$

$$q_\Sigma = \alpha (T_g - T_S)$$

Here T_g, T_S, T_H – temperatures of the combustion products, fuel charge surface and the charge initial temperature, respectively.

If we additionally assume that the heat transfer coefficient α in the period prior to the fuel charge ignition is approximately constant (this corresponds to the constant mass flow rate of the combustion products of the ignition composition), the latter equation can be integrated, and for the time t_*, corresponding to the ignition of the fuel, the expression [136] will be correct

$$t_* \approx \frac{c_m \rho_m \lambda_m}{\alpha^2}\left(\frac{T_S - T_H}{T_g - T_S} + \ln\frac{T_g - T_S}{T_g - T_H}\right) \tag{3.25}$$

From the Eq. (3.25) the value of the conversion factor K_{t*} for the ignition time of the fuel charge t_* is defined

$$K_{t*} = K_R^2 (K_V \cdot K_R)^{-2m} (1 + \frac{K_X}{K_R})^{2n} \tag{3.26}$$

The correctness of the above similarity relations was tested using the software package developed on the basis of mathematical models of processes in SFREs [10, 244]. The calculations are performed for the propulsion system with the channel-type charge. The solid fuel charge has the length of 0.6 m. The cross-section area of the charge channel – 0.11 m^2, and the perimeter of the side surface (the solid fuel surface) – 0.19 m. In the calculations it was assumed that the model set has geometrical dimensions reduced tenfold in transverse direction. When calculating the full-scale engine, the mass of ignition composition varied in the range from 0.1 kg to 1.0 kg. The weigh mass in the model engines varied in the range from 0.0015 kg to 0.10 kg.

The main problem of analyzing the similarity of processes in full-scale and model engines is to assess the possibility of transferring the results of experiments conducted on model systems onto full-scale engines without additional experiments. Analyzing the processes occurring in the initial part of engines work, the most fundamental in the comparison are the rate of pressure rise in the engine chamber, time interval during which the fuel surface ignites and also the flame spread velocity.

Figures 3.6 and 3.7 demonstrate the dependence of changes in the operating pressure in engine chamber on the process time for the full-scale engine and model installation, respectively.

The regularities of the pressure level increase in the chamber for all considered propulsion systems are the same. At the first stage, there is an

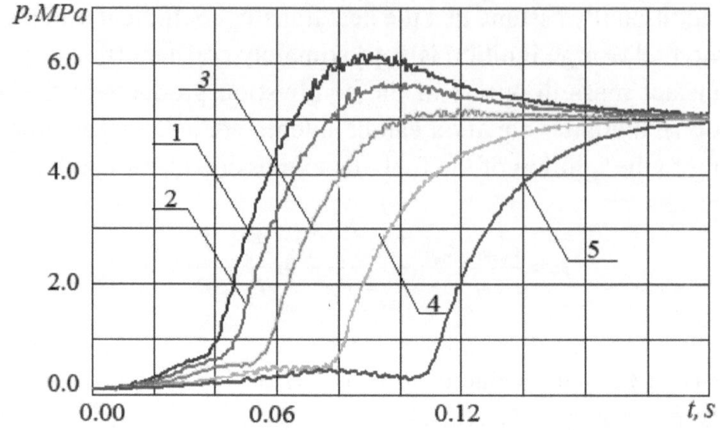

FIGURE 3.6 Change of the pressure in the front volume of SFRE with the process time (full-scale engine) (Weigh mass of the ignition composition; 1–1.00 kg, 2–0.75 kg, 3–0.50 kg, 4–0.25 kg, 5–0.15 kg).

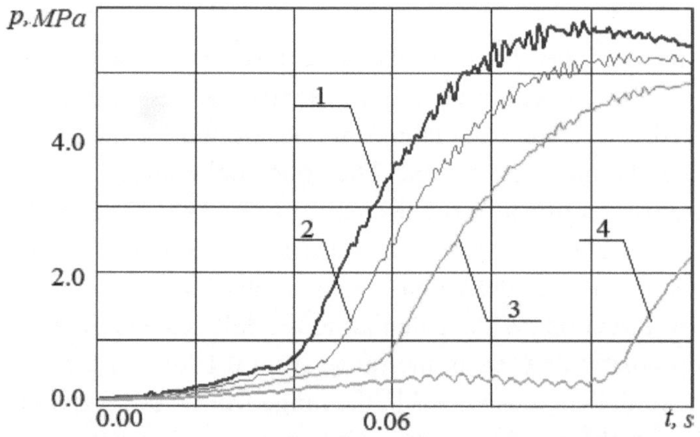

FIGURE 3.7 Change of the pressure in the front volume of SFRE with the process time (model engine) (Weigh mass of the ignition composition; 1–0.10 kg, 2–0.075 kg, 3–0.050 kg, 4–0.025 kg).

increase of pressure with relatively low values of the derivative $\frac{dp}{dt}$. At this stage, the intrachamber volume is filled only with the igniter weigh combustion products, and the value of the derivative $\frac{dp}{dt}$ at this point is completely determined by the incoming mass from the igniter. Reducing the weigh mass and, consequently, the mass income from the ignition system, we reduce the pressure rise rate in the intrachamber volume. To achieve the same values

$\frac{dp}{dt}$ at this stage of the development process is possible only after the geometric similarity of the intra-chamber volumes of full-scale engine and model with the mass flow ratio from the igniter body coincide.

At the second stage, the value of the pressure rise rate $\frac{dp}{dt}$ is substantially higher due to the connection to the solid fuel combustion. The value $\frac{dp}{dt}$ in this period of time is determined to a lesser extent by the igniter. The crucial value for the speed of the pressure rise during this period has the values of flame spread velocity on the fuel charge surface and the ratio of the fuel burning surface to the area of the minimal (critical) section of the engine. This conclusion is confirmed by the fact that the values $\frac{dp}{dt}$ for all the curves shown in the figures are almost identical. The time point, when the pressure rise velocity in the engine chamber changes, corresponds to the ignition moment of the more heat stressed part of fuel surface (usually near the front bottom of the engine). The calculations show that the rates of flame spread in the full-scale and model engines have similar values – ~ 80 m/s.

The analysis of the research shows that the results of mathematical modeling conform to conversion factors written above. Indeed, our calculations assume that $K_X = 0{,}1$; $K_R = 1$; $K_V = 1$. In this case, when $n = 0{,}59$ we have $K_{t*} \approx 1{,}1^{1{,}18} \approx 1{,}12$. It should also be noted that for a more accurate conversion of parameters of the full-scale engine and the model we should take the arch size of tablet with the ignition composition corresponding to coefficient K_X in the calculations. This would provide the similarity of mass income from the ignition composition weigh in the full-scale and model engines.

In general, based on the results of the calculations, we can conclude that the use of the model, in which the lateral dimension, surface combustion area, area of the minimum cross-section of the nozzle block, and values of internal volumes satisfy the geometric similarity, is quite acceptable. In this the basic laws of the initial part of engine work are correctly represented. The quantitative results are in close concordance if the ignition composition weigh masses relate to each other about the same proportion as the geometric dimensions of engines. However, it should be noted that the mathematical problem of the engine output to the quasi-stationary mode is essentially non-linear, and therefore, the application of the conversion factors of the model and full-scale parameters of engine is advantageously carried out in conjunction with the direct mathematical modeling of the problem of the engine output onto mode for both model and full-scale engines.

3.3 APPLICATION OF PROBABILISTIC APPROACHES TO CONSTRUCT THE ANALYTICAL MODELS

3.3.1 BASIC DEFINITIONS OF PROBABILITY THEORY AND MATHEMATICAL STATISTICS

The construction of analytical relations with experimental data is almost always associated with statistical data processing. Writing functional relations on the test results is the final stage of the experimental study. However, pre-planning the form of the empirical dependence, the order and sequence of the various tests can significantly reduce the volume and complexity of experiments.

Further analysis will consider two options to construct empirical relations:

- form of the functional dependence is defined, however, in the experiments, the values of the unknown coefficients available in these curves are defined;
- number of physical parameters (factors) that affect the object under study is relatively large, and the functional dependence type before the experiment is not determined.

The use of both variants of the empirical relationships (models) involves the use of elements of probability theory and mathematical statistics. Some definitions, terms and important theorem [4, 66, 68, 88, 178, 245, 251] to be further needed are given below.

Variable is any variable physical unit.

A variable is independent if its variation happens independently from other variables. If the change of any value results in the change of the variable, it is called the dependent. The variable is called stochastic (random) quantity if its value depends on random factors.

The complete set of all possible values which the stochastic variable takes, is called the population.

The value of the measured quantity differs from the real value by an amount called the absolute error. The ratio of the absolute error to the value of the measured quantity is called the relative error. The relative error can be expressed in percent. The absolute and relative errors are random numbers.

The probability theory defines two functions that characterize the deviation of the random variables from the general population from the real value – the cumulative distribution function and density function of the random distribution.

Cumulative distribution function is defined as the probability P of event $x \le x'$ where x' is some number

$$F(x') = P(x \le x') \tag{3.27}$$

We assume that the cumulative distribution function is inseparable and differentiable. From the above definition of the cumulative distribution function its properties follow:

$$0 \le F(x') \le 1; \quad F(-\infty) = 0; \quad F(\infty) = 1 \tag{3.28}$$

$$P(a < x \le b) = F(b) - F(a)$$

Density function f of the random variable x is defined as

$$f(x) = \frac{dF}{dx} \tag{3.29}$$

Using the Eqs. (3.27) and (3.28) for the density of probability $f(x)$ we can write down the following useful relations:

$$\int_{-\infty}^{\infty} f(x)dx = 1$$

$$P(a < x \le b) = \int_a^b f(x)dx$$

$$P(-\infty < x < \infty) = \int_{-\infty}^{\infty} f(x)\, dx = 1 \tag{3.30}$$

Mathematical expectation of the continuous random variable x defined in the interval $a < x \le b$ and having the probability density function $f(x)$ is called the quantity $\overline{x}$ defined by the expression:

$$\overline{x} = \int_a^b x f(x)dx \tag{3.31}$$

Initial time of the order S of the continuous random variable x defined in the interval $a < x \le b$ and having the probability density function *f(x)* is called the number v_S determined by the expression

$$v_S = \int_a^b f(x) x^S dx \tag{3.32}$$

Central moment of the order S of the random variable x defined in the interval $a < x \le b$ and having the probability density function *f(x)* is called number μ_s determined by the expression

$$\mu_S = \int_a^b (x - \bar{x})^S f(x) dx \tag{3.33}$$

From the definition of the initial and central moments we can write the following special cases:

$$v_o = 1$$

$$v_1 = \bar{x}$$

$$\mu_0 = 1$$

$$\mu_1 = \int_a^b x \, f(x) dx - \bar{x} \int_a^b f(x) dx = 0$$

$$\mu_2 = \sigma_x^2 \tag{3.34}$$

In the latter equation of the system (3.34) for the second-order central moment μ_2 the value σ_x (can also be indicated σ) called the mean-square deviation of the random variable is included. The central moment $\mu_2 = \sigma_x^2$ is called the dispersion of the random variable.

The above listed definitions for the random variable x can be generalized for the vector of random variables $\mathbf{x} \equiv (x_1, x_2, ..., x_N)$. In this case, if we denote the density distribution of the vector $\mathbf{x}$ of random variables – $f(\mathbf{x}) \equiv f(x_1, x_2, ..., x_N)$, then the basic definitions and laws are re-written as:

- equation for the probability value

$$P(a_1 < x_1 \le b_1, a_2 < x_2 \le b_2, \dots, a_N < x_N \le b_N)$$

$$= \int_{a_1}^{b_1}\int_{a_2}^{b_2}\dots\int_{a_N}^{b_N} f(x_1, x_2, \dots, x_N)\, dx_1\, dx_2 \cdot \dots \cdot dx_N$$

$$P\left(-\infty < x_1 \le \infty, -\infty < x_2 \le \infty, \dots, -\infty < x_N \le \infty\right)$$

$$= \int_{-\infty}^{\infty}\int_{-\infty}^{\infty}\dots\int_{-\infty}^{\infty} f(x_1, x_2, \dots, x_N)\, dx_1\, dx_2 \cdot \dots \cdot dx_N = 1$$

- equation for the mathematical expectation of the vector element x_i

$$E(x_i) = \bar{x}_i = \int_{-\infty}^{\infty}\int_{-\infty}^{\infty}\dots\int_{-\infty}^{\infty} x_i f(x_1, x_2, \dots, x_N)\, dx_1\, dx_2 \cdot \dots \cdot dx_N$$

$$= \int_{-\infty}^{\infty}\int_{-\infty}^{\infty}\dots\int_{-\infty}^{\infty} x_i\, f(\mathbf{x})\, dx_1\, dx_2 \cdot \dots \cdot dx_N$$

- initial moment v of the order $(m_1, m_2, \dots, m_n)$ (the analog of the Eq. (3.32)) can be written

$$v_{m_1, m_2, \dots, m_N} = \int_{-\infty}^{\infty}\int_{-\infty}^{\infty}\dots\int_{-\infty}^{\infty} (x_1^{m_1} x_2^{m_2} \cdot \dots \cdot x_N^{m_N})\, f(\mathbf{x})\, dx_1\, dx_2 \cdot \dots \cdot dx_N;$$

- central moment μ of the order $(m_1, m_2, \dots, m_N)$ (the analog of Eq. (3.33)) can be written

$$\mu_{m_1, m_2, \dots, m_N} = \int_{-\infty}^{\infty}\int_{-\infty}^{\infty}\dots\int_{-\infty}^{\infty} \begin{matrix} (x_1 - \bar{x}_1)^{m_1} (x_2 - \bar{x}_2)^{m_2} \cdot \dots \cdot (x_N - \bar{x}_N)^{m_N} \\ f(\mathbf{x})\, dx_1\, dx_2 \cdot \dots \cdot dx_N \end{matrix}$$

We have considered the basic definitions and equations valid when considering the continuous random variables. However, the experimental research data processing is carried out at the known (finite) number of tests, and every test corresponds to the measurement results which are also random numbers. Suppose that I experiments on the measurement of the random variable x are conducted. Let us denote x_i measured in i^{th} experience value of the random variable x. Let us denote P_i – the probability of the event $x = x_i$ in the ith experience. In this case, the following equations are correct for the discrete random variable:

$$\sum_{i=1}^{I} P_i = 1$$

$$P(x < x_I) = \sum_{i=1}^{I} P_i$$

$$\bar{x} = \sum_{i=1}^{I} x_i P_i$$

$$\sigma_x^2 = \sum_{i=1}^{I} (x_i - \bar{x})^2 P_i \qquad (3.35)$$

During the experiments it is often assumed that each of these results is equally probable. In this case, we can assume that $P_i \approx \frac{1}{I}$, and the latter two equations of the system (3.35) take the simpler form:

$$\bar{x} = \frac{\sum_{i=1}^{I} x_i}{I}; \quad \sigma_x = \sqrt{\frac{\sum_{i=1}^{I} (x_i - \bar{x})^2}{I}} \qquad (3.36)$$

In probability theory it is established that the deviations of random numbers from their mathematical expectations are subject to certain laws. In particular, these laws are manifested through the cumulative distribution functions of random numbers. In real processes associated with random factors, the limited number of options for integrated distribution functions is implemented. In engineering the most common are normally distributed and binomially distributed integral functions (distributions).

Random variables have the deviation from the real value, normally distributed, if they satisfy the conditions of the theorem by Chebyshev [251]:

if the random variables x_i, the number of which can be arbitrarily large, are mutually independent, if they have certain mathematical expectations $\bar{x}_1, \bar{x}_2, \ldots \bar{x}_I$ and certain uniformly bounded dispersions, then with probability arbitrarily close to one, we can expect that the absolute value of the difference between the arithmetic mean of these values and arithmetic average of their mathematical expectations is arbitrarily small in absolute value when the number of components is large enough.

Mathematically the recording of the theorem conditions is as follows

$$P\left\{\left|\frac{x_1+x_2+\ldots+x_I}{I}-\frac{\overline{x}_1+\overline{x}_2+\ldots+\overline{x}_I}{I}\right|<\varepsilon\right\}>1-\delta \tag{3.37}$$

Here ε, δ – arbitrarily small positive numbers $(\delta=\delta(\varepsilon))$, and the number of components (number of experiments or tests) I in this expression for any assumed values ε, δ satisfies the condition $I>I(\varepsilon,\delta)$.

For normally distributed random variable x the probability density function is described by the relation:

$$f(x)=\frac{1}{\sigma_x\sqrt{2\pi}}\exp\left(-\frac{(x-\overline{x})^2}{2\sigma_x^2}\right) \tag{3.38}$$

The probability of the event that the measured value x is in the range from a to b $(a<x\leq b)$ can be written

$$P(a<x\leq b)=\int_a^b\frac{1}{\sigma_x\sqrt{2\pi}}\exp\left(\frac{-(x-\overline{x})^2}{2\sigma_x^2}\right)dx \tag{3.39}$$

It is useful to note that for a normally distributed random variable the relations are correct:

$$P(\overline{x}\pm\sigma_x)=0{,}683;\quad P(\overline{x}\pm2\sigma_x)=0{,}955;\quad P(\overline{x}\pm3\sigma_x)=0{,}997 \tag{3.40}$$

Also, note that if the random variable is a vector and vector coordinates are mutually independent, then with the correctness of the assumptions of normal distribution for the probability density the following relation can be written:

$$f(\mathbf{x})=f(x_1,x_2,\ldots,x_I)=f_1(x_1)f_2(x_2)\cdot\ldots\cdot f_I(x_I)$$
$$=\frac{1}{(\sqrt{2\pi})^I}\cdot\frac{1}{\sigma_{x_1}\sigma_{x_2}\cdot\ldots\cdot\sigma_{x_I}}\exp\left(-\sum_{i=1}^{I}\frac{(x_i-\overline{x}_i)^2}{2\sigma_{xi}^2}\right)$$

Binomial distribution is correct for random quantities of discrete type, the result of which can take two values – *0* or *1* (otherwise – "yes" or "no"). For the binomial distribution formula for the probability of event in the experiment with the number i (I experiments in total) is determined by the relation:

$$P_n(i) = \frac{I!}{i!(I-i)!} p^i (1-p)^{I-i} \tag{3.41}$$

For the values observed for the time t and with $I \to \infty$, the Poisson ratio is correct at the average intensity of the events ν

$$P(i) = \frac{(\nu t)^i}{i!} e^{-\nu t} \tag{3.42}$$

The dispersion in Poisson statistics is given by the expression:

$$D(i) = \sum_{i=0}^{\infty} (i - \bar{i})^2 P(i) = \nu t = \bar{i} \tag{3.43}$$

If $\nu t >> 1$, then Poisson distribution becomes practically continuous and close to normal distribution.

The initial and central moments of the distribution allow determining to what extent the random quantity satisfies the normal distribution law. So, if the normal distribution law is correct, the following relations are performed

$$\mu_0 = 1$$

$$\mu_2 = \sigma_x^2$$

$$\mu_{2K} = (2K-1)(2K-3) \ldots \cdot 3 \cdot 1 \cdot (\sigma_x^2)^K$$

$$\mu_{2K-1} = 0; \quad A = \frac{\mu_3}{\sigma_x^3} = 0; \quad E = \frac{\mu_4}{\sigma_x^4} - 3 = 0 \tag{3.44}$$

The recorded values in the equations A and E, respectively, are called asymmetry and excess. If they are different from zero, the distribution of the random value, generally speaking, does not correspond to the normal law [245, 251]. Figure 3.8 shows how the law of density distribution of the random value f can look like for different values of the asymmetry A. Figure 3.9 shows how the law of density distribution of the random quantity f can look like for different values of the excess E. At non-zero values of the asymmetry

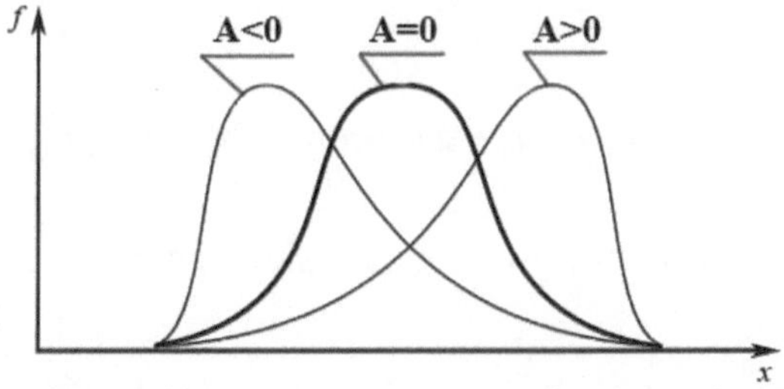

FIGURE 3.8 Density of *f* distribution of the random value with different values of the asymmetry A.

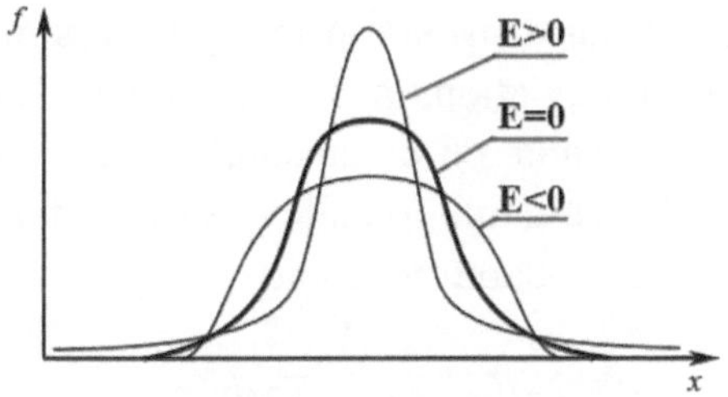

FIGURE 3.9 Density of *f* distribution of the random value with different values of the excess E.

and (or) the excess, however, you can apply the mathematical apparatus correct for laws of normally distributed random values.

It is possible to use, for example, the following considerations. Let the density of the random value not satisfy the density corresponding to the normal distribution (Eq. (3.38)). We assume that the density can be represented as

$$\varphi(x) = f(x)\Pi(x)$$

Here $\Pi(x)$ denotes the perturbation (disturbing) polynomial, the specific form of which to be selected. If we denote $u = x - \overline{x}$, the latter expression can be re-written as

$$\varphi(u) = f(u)\Pi(u) \tag{3.45}$$

$\Pi(u)$ can be written as the polynomial of the 4th order (Charlier curve for the law $\varphi(u)$ [251]) with the following values of the coefficients

$$a_0 = 1 + \frac{1}{8}E;\ a_1 = \frac{1}{2}\frac{A}{\sigma_x};\ a_2 = -\frac{1}{4}\frac{E}{\sigma_x^2};\ a_3 = \frac{1}{6}\frac{A}{\sigma_x^3};\ a_4 = \frac{1}{24}\frac{E}{\sigma_x^4} \tag{3.46}$$

In this case

$$\varphi(u) = f(u)(a_0 + a_1u + a_2u^2 + a_3u^3 + a_4u^4) \quad (3.47)$$

and we can show that the following expression is correct:

$$\varphi(u) = f(u) - \frac{1}{6}A\sigma_x^3 f'''(u) + \frac{1}{24}E\sigma_x^4 f^{IV}(u) \quad (3.48)$$

The derivation of the equation for the distribution density of the random value satisfying the normal distribution law involves making the assumption that the elementary deviations ε of the random value from its mathematical expectation, the same on the absolute value, are equally probable ($P(\varepsilon) = P(-\varepsilon) = 0,5$). This assumption allows us to obtain the following differential equation for the probability density

$$\frac{df}{f} = -\frac{(x - \overline{x})dx}{\sigma_x^2}$$

Pearson curves are obtained if in presuppositions, under which the normal distribution law is deduced, we refuse from the assumption that $P(\varepsilon) = P(-\varepsilon) = 0,5$ and instead of the latter equation for the distribution density the equation of the form [251] is solved

$$\frac{df}{f} = \frac{(x - a)dx}{b_0 + b_1x + b_2x^2}$$

Here a, b_0, b_1, b_2 – parameters to be defined. There can be a lot of Pearson distribution functions. Their use is particularly attractive, for example, if processing the experiments where there are two peaks on the curve of the probability density distribution (note that one way of solving this problem can be its partition into two, each of which corresponds to the normal distribution). As an example, we give the dependence of one variant of Pearson curves (Pearson curves of the 3rd type):

$$f = f_0 \exp(-\gamma\, x)\left(1 + \frac{x}{l}\right)^m$$

Here $\gamma = \frac{2\mu_2}{\mu_3}$; $l = \frac{2\mu_2^2}{\mu_3}$; $m = \frac{4\mu_2^3}{\mu_3^2} - 1$; $f_0 = \frac{\gamma^{m+1} \cdot l^m \cdot \exp(-l\gamma)}{\Gamma(m+1)}$;

$\Gamma(m+1) = \int_0^\infty x^m \exp(-x)\,dx$.

Maxwell distribution can be used to determine the distribution of the molecules velocity modulus, particles obtained on the assumption that the velocity components along the axes of Cartesian coordinates have the same dispersions. Maxwell distribution is essentially very close to the normal distribution. For the density of Maxwell distribution the following dependence is correct:

$$f(\vartheta) = \frac{2}{\sqrt{2\pi}} \frac{\vartheta^2}{\sigma_0^3} \exp\left(-\frac{\vartheta^2}{2\sigma_0^2}\right) \tag{3.49}$$

Here ϑ – velocity modulus, σ_0 – mean-square deviation for each of the coordinates.

Student distribution is widely used in the theory of errors in solving the problems with a small number of observations *I*. (It is assumed that at a large number of observations the random value is subject to the normal law). The event probability in Student distribution is defined by the expression

$$P(t) = C(I)\left(1 + \frac{t^2}{I-1}\right)^{-\frac{I}{2}} \tag{3.50}$$

where I – a number of observations, $C(I)$ is determined from the tables $t = \frac{\bar{x} - x_0}{\sigma_x}$, $\bar{x} = \frac{1}{I}\sum_{i=1}^{I} x_i$, $\sigma_x = \frac{\bar{\sigma}}{\sqrt{I}}$, $\bar{\sigma} = \sqrt{\frac{\sum_{i=1}^{I}(x_i - \bar{x})^2}{I-1}}$.

Student distribution is useful for the number of observations $I < 20$. For a large number of tests, Student distribution does not differ from the normal distribution law.

In engineering, other laws of the distribution of random numbers are implemented. However, the analysis demonstrates that if we take in the object analysis the assumption that the distribution of the random value deviations from its real value obeys the normal distribution law, then the final error will not be large. Therefore, further we will use exactly this kind of distribution. Moreover, we note the following fact. If in the analysis of the random values their probability distributions change over the time, or the random quantities depend on each other, then, in accordance with the definition [66], we can talk about the analysis of random processes. Random processes can be stationary, and this definition of random processes is correct for the case when the probability distribution for all random quantities does not vary with the argument (time). In practice, random processes are of interest, in which the transition from the state corresponding to any point of time t_n to the state corresponding to the point of time t_{n+1} happens in accordance with known transition probability $P(t_n \rightarrow t_{n+1})$. Such processes are called Markovian processes. They are relevant for solving problems of sociology, economics, demography, medicine, etc. In what follows, Gaussian random processes where all the probability distributions are normal for all the random variables determining the random process will to be of interest.

3.3.2 CONSTRUCTION OF STOCHASTIC MODELS WITH THE KNOWN FUNCTIONAL DEPENDENCE

The following assumption is fundamental in the theory of probability and mathematical statistics

- measurements (counting) of the random quantity $x_1,...,x_I$ in I experiments collectively deviate from the real value x_c so that the total probability of occurrence of these deviations is maximized.

  This assumption leads to equity ratio

$$\Phi(x_c) = (x_c - x_1)^2 + (x_c - x_2)^2 + ... + (x_c - x_I)^2 \rightarrow \min \tag{3.51}$$

We can say that the latter expression is correct when

$$x_c = \frac{x_1 + x_2 + ... + x_I}{I} = \overline{x} \tag{3.52}$$

Taking into account the latter equations, it is advisable to represent the model of the measured value by its mathematical expectation $\overline{x}$ (this follows from the Eq. (3.52)) and dispersion σ_x^2 (it comes from the Eq. (3.51)). In this case, the assumption that the absolute error of the measured value does not exceed $3\sigma_x$ $\left(\Delta x = |x - \overline{x}| \leq 3\sigma_x\right)$ has the probability not less than 99.7%.

Naturally, the Eqs. (3.51) and (3.52), as well as the conclusions formulated above, remain correct if the random values are vectors.

In carrying out the experimental research and processing the experimental results, the following problem occurs:

- let the searched value Φ be the function of several variables $x_i \quad (i = 1, I)$. The functional dependence binding the desired value Φ with the variables x_i is known and has the form $\Phi = \Phi(x_1, x_2, ..., x_I)$. Let us define the mathematical expectation $\bar{\Phi}$ of the value Φ and its dispersion σ_Φ^2, if the mathematical expectation of the variables $\bar{x}_i$ and their dispersion $\sigma_{x_i}^2$ are known.

The problem solution will start with a simple example. Suppose you want to estimate the mathematical expectation and dispersion for the mass flow rate of the combustion products G from the rocket engine chamber, if the values of mathematical expectations and dispersion of the area of minimal section of the nozzle $F_{\min}$ and pressure in the chamber p_κ. The dependence connecting the combustion products flow, the pressure in the engine chamber and area of minimal section of the nozzle is of the form:

$$G = A_c p_\kappa F_{\min} \tag{3.53}$$

Here A_c – efflux coefficient. We assume that its value is known. We assume that any of the values f in a particular experiment with the number i can be represented by the sum of the mean value f_c and deviation from it $r_f - f_i = f_c + r_{fi}$. Applying the above formula to the Eq. (3.53), we can write for the experiment with the number i:

$$G_c + r_{Gi} = A_c (p_{kc} + r_{pi})(F_{\min c} + r_{Fi})$$

Let us sum up the latter equation over all the experiments from $i = 1$ to $i = I$:

$$\sum_{i=1}^{I} (G_c + r_{Gi}) = A_c \sum_{i=1}^{I} (p_{kc} + r_{pi})(F_{\min c} + r_{Fi}) \tag{3.54}$$

Assuming that $G_c = A_c p_{kc} F_{\min c}$, as well as the positive and negative deviations r_{pi}, r_{Fi} in each experiment are equally probable (in this case $\sum_{i=1}^{I} r_{pi} r_{Fi} \approx 0$), the Eq. (3.54) can be sequentially converted

$$\sum_{i=1}^{I} r_{Gi} \approx A_c \sum_{i=1}^{I} (p_{kc} r_{Fi} + F_{\min c} r_{pi})$$

$$\sum_{i=1}^{I} (r_{Gi})^2 \approx A_c^{\,2} \sum_{i=1}^{I} (p_{kc} r_{Fi} + F_{\min c} r_{pi})^2$$

$$\sum_{i=1}^{I} \frac{(r_{Gi})^2}{G_c^2} \approx \sum_{i=1}^{I} \frac{(p_{kc} r_{Fi} + F_{\min c} r_{pi})^2}{p_{kc}^2 F_{\min c}^2}$$

$$\sum_{i=1}^{I} \frac{(r_{Gi})^2}{G_c^2} \approx \sum_{i=1}^{I} \left(\frac{r_{Fi}^2}{F_{\min c}^2} + \frac{r_{pi}^2}{p_{kc}^2}\right)$$

In accordance with the formulas (3.36) and (3.37), the latter equation finally takes the form

$$\frac{\sigma_G^2}{G_c^2} \approx \frac{\sigma_F^2}{F_{\min c}^2} + \frac{\sigma_p^2}{p_{kc}^2} \tag{3.55}$$

The Eqs. (3.53) and (3.55) allow us to consistently find the values $\bar{G}, \sigma_G^2$ on the basis of the known values $\bar{p}_k, \sigma_p^2;\ \bar{F}_{\min}, \sigma_F^2$.

The problem (3.53) can be extended to any kind of a more complicated case

$$\Phi = K \frac{x_1 x_2 \cdot \ldots \cdot x_n}{x_{n+1} x_{n+2} \cdot \ldots \cdot x_{n+m}} \tag{3.56}$$

Transformations, similar to the above performed with the Eq. (3.53), allow obtaining the following relation for the dispersions

$$\frac{\sigma_\Phi^2}{\Phi^2} = \sum_{i=1}^{n+m} \frac{\sigma_{x_i}^2}{x_i^2} \tag{3.57}$$

In the general case, if the functional dependence of several variables has the arbitrary form

$$\Phi = \Phi(x_1, x_2, \ldots, x_I) \tag{3.58}$$

then the total error and dispersion for the function *f* can be determined by expanding the function Φ in Taylor series and retaining the two or more components in the series. The dispersion equation for the Eq. (3.58) in this case will be written down

$$\sigma_f^2 \approx \sum_{i=1}^{I} \left(\frac{\partial \Phi}{\partial x_i} \right)^2 \sigma_{x_i}^2 \tag{3.59}$$

The Eqs. (3.59) is the most common, and we can prove that the Eqs. (3.55) and (3.57) are the special cases of (3.59).

Denote ω – the range of uncertainty for the measured quantity, binding it with a mean-square deviation by ratio $\omega = K\sigma$. Here is K=0.67, if in the range $\pm\omega$ a half of all its results of performed experiments is located; K=1, if in the range $\pm\omega$ 68.3% of all performed experiments is located; K=2–95.5%, and K=3–99.7% of all the experiments. Then, instead of the expression of total dispersion, we can write a similar expression of the total uncertainty interval

$$\omega_f^2 \approx \sum_{i=1}^{I} \left(\frac{\partial \Phi}{\partial x_i} \right)^2 \omega_{x_i}^2 \tag{3.60}$$

3.3.3 CONSTRUCTION OF STOCHASTIC MODELS AN THE UNCERTAIN FUNCTIONAL DEPENDENCE

The previous section dealt with issues related to the estimation of mathematical expectation and dispersion for the function Φ dependent on several random variables x_i $(i = 1, I)$. It was assumed that the form of the functional dependence $\Phi = \Phi(x_1, x_2, ..., x_I)$ is known. Processing the experimental results, the problem can be formulated in another way.

- let us define the dependence of the function Φ on several independent variables x_i $(i = 1, I)$. As a result of experiments, the values of mathematical expectations of the variables $\bar{\Phi}$, $\bar{x}_1$, $\bar{x}_2, ..., \bar{x}_I$ and their dispersion $\sigma_\Phi^2, \sigma_{x_1}^2, \sigma_{x_2}^2, ..., \sigma_{x_I}^2$ are known. Let us define the constructive form of the functional dependence $\Phi = \Phi(x_1, x_2, ..., x_I, a_1, a_2, ..., a_m)$ binding the function Φ with independent variables x_i $(i = 1, I)$ and set the values of the coefficients $a_1, a_2, ..., a_m$.

The problem solution consists of three stages:

1. Selection of the type of dependence $\Phi = \Phi(x_1, x_2, ..., x_I)$ binding the function Φ with the independent variables x_i $(i = 1, I)$;

2. Selection of the method of processing the experimental results;
3. Determination of the unknown coefficients (parameters) included in the selected dependence $\Phi = \Phi(x_1, x_2, ..., x_I)$.

In addition, it should be noted that processing the experimental results the following situations affecting the selection of mathematical methods of processing can occur:

- number of the available experimental material is redundant and has no restrictions on the selection of the experimental dependence;
- number of the experimental material is limited, but allows you to search the experimental dependence in the class of functions of defined type;
- search for the experimental dependence is carried out in the class of functions of the defined type, but the number of experimental results is limited and pre-planned.

3.3.4 SELECTION OF FUNCTIONAL DEPENDENCE

Selecting the functional dependence binding the function Φ with independent variables x_i $(i = \overline{1, I})$, relatively simple linear equations of the following form can be used

$$\Phi(x_1, x_2, \dots, x_I) = a_0 + \sum_{i=1}^{I} a_i x_i \tag{3.61}$$

Such models are useful in the processing of the experimental results, as it allows relatively easy to calculate the values of the coefficients a_i, $(i=\overline{0, I})$. If the accuracy of processing the experimental results using the models (3.61) turns out to be unsatisfactory, the quadratic model of the following form can be applied

$$\Phi(x_1, x_2, \dots, x_I) = a_0 + \sum_{i=1}^{I} a_n x_n + \sum_{n=1}^{N} \sum_{j=1}^{J} a_{nj} x_n x_j \tag{3.62}$$

In the most general case, the experiment processing can be carried out suggesting that the functional dependence $\Phi = \Phi(x_1, x_2, ..., x_I)$ has the form

$$\Phi(x_1, x_2, \dots, x_I) = \varphi_0(x_1, x_2, \dots, x_I) + \sum_{m=1}^{M} a_m \varphi_m(x_1, x_2, \dots, x_I) \tag{3.63}$$

or in the vector form

$$\Phi(\mathbf{x}) = \varphi_0(\mathbf{x}) + \sum_{m=1}^{M} a_m \varphi_m(\mathbf{x}) \tag{3.64}$$

Here $\varphi_m(\mathbf{x}) = \varphi_m(x_1, x_2, ..., x_I)$ – any algebraic function. The set of functions $\varphi_m(\mathbf{x})$ *(m=0, M,* used in the Eq. (3.64) is called the basic system of approximating functions.

Further, the so-called orthogonal functions that can be used as the base will be of interest. The set of functions $\varphi_m(\mathbf{x})$ *(m=0, M)*, the scalar products of which satisfy the following conditions, refers to orthogonal ones:

$$(\varphi_k(\mathbf{x}) \cdot \varphi_j(\mathbf{x})) = 0 \text{, if } k \neq j \tag{3.65}$$

If we talk about the orthogonal functions of one variable (*N=1*), the scalar product (3.65) is defined as follows [23]

- if H – space of complex-valued functions defined in the interval $[a, b]$ with the limited integral $\int_a^b |f(x)|^2 p(x)\,dx$;

  in this case, the scalar product is given by

$$(f(x) \cdot g(x)) = \int_a^b f(x)\overline{g}(x)\, p(x) dx$$

Here $\overline{g}(x)$ – function complexly conjugated with $g(x)$, $p > 0$ almost everywhere on $[a, b]$. The value *p(x)* is called the weight. In the particular case it can be *p(x)≡1.* When carrying out a limited number of experiments (*i=1, I*), the selection of the system of orthogonal functions is simplified, since the Eq. (3.65) is re-written for this case as:

$$\sum_{i=1}^{I} \varphi_k(x_i)\varphi_j(x_i) = 0, \ at\ k \neq j$$

Examples of orthogonal functions can be Chebyshev polynomials, Legendre polynomials, Fourier functions and others [23,76].

For the case of an integer variable *t* determined at *t=0, 1, 2, ..., N,* Chebyshev polynomials can be written as

$$P_{k,N}(t) = \sum_{s=0}^{N} (-1)^s C_k^s C_{k+s}^s \frac{t^{[s]}}{n^{[s]}} \tag{3.66}$$

Here $k = 0, 1, ..., M$, $M < N$, $t^{[s]} = t(t-1) \cdot ... \cdot (t-s+1)$, $n^{[s]} = n(n-1) \cdot ... \cdot (n-s+1)$, $C_k^s = \frac{k!}{s!\ (k-s)!}$, $C_{k+s}^s = \frac{(k+s)!}{s!\ k!}$.

If a physical experiment is carried out for the system $(I+1)$ of equally spaced points $x = \{x_0, x_1, ..., x_I\}$, then we can use Chebyshev polynomials using the conversion $t = \frac{x - x_0}{h}$. Here $h = x_1 - x_0 = x_2 - x_1 = ... = x_{i+1} - x_i = ... = x_I - x_{I-1}$.

It is comfortable to use polynomials whose values in this interval are limited by the absolute value (e.g., vary in the range from –1 to +1). Orthogonal polynomials can be reduced to this kind by the normalization procedure. Normalized orthogonal polynomials are called orthonormal. Orthonormal Chebyshev polynomials $\tilde{P}_{k,N}(t)$ are linked with the orthogonal polynomials $P_{k,N}(t)$ by introducing the normalizing factor $\|P_{k,N}(t)\|$:

$$\tilde{P}_{k,N}(t) = \frac{P_{k,N}(t)}{\|P_{k,N}(t)\|} \tag{3.67}$$

The value of the norm $\|P_{k,N}(t)\|$ is established by the expression

$$\|P_{k,N}(t)\|^2 = \sum_{i=0}^{N} P_{k,N}^2(i) = \frac{(N+k+1)^{[k+1]}}{(2 \cdot k+1) \cdot N^{[k]}} \tag{3.68}$$

In the applications two variants of Chebyshev polynomials orthogonal in the interval $-1 \le x \le 1$ are used:

- Chebyshev polynomials of the first kind (orthogonal with the mass $p(x) = \frac{1}{\sqrt{1-x^2}}$) are defined by the ratios

$$T_0(x) = 1$$

$$T_1(x) = x$$

$$T_2(x) = 2x^2 - 1$$

....

$$T_N(x) = 2xT_{N-1}(x) - T_{N-2}(x)$$

Note that changing the values of the basic polynomials in the interval $-1 \le x \le 1$ is restricted by the interval $-1 \le T_i(x) \le 1$;

- Chebyshev polynomials of the second kind (orthogonal with the weight $p(x) = \sqrt{1-x^2}$) are defined by the ratio

$$U_n(x) = \frac{T'_{n+1}(x)}{n+1}, \quad n = 0,1,\ldots.$$

Legendre polynomials are also defined in the interval $-1 \le x \le 1$ (weighing function $p(x) \equiv 1$), and the values of the basic polynomials are limited by the interval $-1 \le P_i(x) \le 1$:

$$P_0(x) = 1$$

$$P_1(x) = x$$

$$P_2(x) = \frac{1}{2}(3x^2 - 1)$$

....

$$P_N(x) = \frac{2N-1}{N} x P_{N-1}(x) - \frac{N-1}{N} P_{N-2}(x)$$

Fourier functions are defined in the interval $-\pi \le x \le \pi$, the values of basic functions are limited by the interval $-1 \le u_i(x) \le 1$:

$$u_0(x) = 1$$

$$u_1(x) = \sin x$$

$$u_2(x) = \cos x$$

....

$$u_{2N-1}(x) = \sin(Nx)$$

$$u_{2N}(x) = \cos(Nx)$$

3.3.5 APPLICATION OF THE MEAN-VALUE METHOD AND METHOD OF LEAST SQUARES

If there is no deficit in the amount of experimental results, the selection of the functional dependence is not restricted. Therefore, the problem of determining the unknown coefficients (parameters) included in the selected functional dependence can be in the following formulation:

- let the dependence $\Phi = \Phi(\mathbf{x}, a_1, a_2, \dots, a_M)$ be used to process the experimental results given in the tabular set $M_i(\mathbf{x}_i, \Phi_i)$ assigning the measured value of the function Φ_i to the independent measurable parameter $\mathbf{x}_i$. Here i – number of experiment, and the total number of performed experiments is $I>>M$. Let us set the values of the coefficients a_m, $m = \overline{1, M}$ ensuring the best coincidence of the selected empirical dependence $\Phi = \Phi(\mathbf{x}, a_1, a_2, \dots, a_M)$ with the results of the experiments $M_i(\mathbf{x}_i, \Phi_i)$.

One of the simplest methods to construct empirical formulas in the excessive amount of the experimental results is the mean-value method [251]. Its essence is as follows. Let us group all the experimental points dividing them into M groups ($m = \overline{1, M}$). The number of the experimental points J_m within each group can be different. Let us require that within each of the M groups the following condition is performed

$$\sum_{i=j_m}^{J_m+j_m-1} (\Phi(\mathbf{x}_i, a_1, a_2, \dots, a_M) - \Phi_i) = 0 \tag{3.69}$$

The system of M Eqs. (3.69) allows setting the values of M unknown coefficients $a_1, a_2, \dots, a_M$.

Obviously, the above-given method of determining the functional dependence $\Phi = \Phi(\mathbf{x}, a_1, a_2, \dots, a_M)$ provides the coincidence of mathematical expectations of the initial experimental sample $M_i(\mathbf{x}_i, f_i)$ and defined dependence $\Phi = \Phi(\mathbf{x}, a_1, a_2, \dots, a_M)$. However, the nature of experimental research to the points is probabilistic in nature and refers to mathematical statistics. Therefore, the requirement of coincidence of mathematical sample expectations $M_i(\mathbf{x}_i, f_i)$ and dependence $\Phi = \Phi(\mathbf{x}, a_1, a_2, \dots, a_M)$ is inadequate because it does not consider the dispersion of experimental results deviations.

The method of least squares widely used in processing the experimental results uses the algorithms that minimize the squared deviations from the tabulated values of the sample $M_i(\mathbf{x}_i, \Phi_i)$ from the unknown empirical dependence $\Phi = \Phi(\mathbf{x}, a_1, a_2, \dots, a_M)$, so this method is more preferable than the mean-value method. The mathematical formulation of the problem of selecting the coefficients $a_1, a_2, \dots, a_M$ in the empirical formula $\Phi = \Phi(\mathbf{x}, a_1, a_2, \dots, a_M)$ in this method is stated as follows:

- let us determine the value of the coefficients $a_1, a_2, \dots, a_M$ included in the functional dependence $\Phi = \Phi(\mathbf{x}, a_1, a_2, \dots, a_M)$, so as to provide the minimal value of the squared deviations $S(a_1, a_2, \dots, a_M)$ of the function Φ from the plurality of the values Φ_i corresponding to the values of the independent argument $\mathbf{x}_i$ and defined as the experiment result

$$\min S(a_1, a_2, \dots, a_M) = \sum_{i=1}^{I} (\Phi(\mathbf{x}_i, a_1, a_2, \dots, a_M) - \Phi_i)^2 \qquad (3.70)$$

The solution of the problem can be accomplished using the optimization methods, where the function $S(a_1, a_2, \dots, a_M)$ is the objective function (see Chapter 4). However, with a relatively simple design of the function $\Phi = \Phi(\mathbf{x}, a_1, a_2, \dots, a_M)$ the problem (3.70) can be reduced to solving the system of M equations of the form

$$\frac{\partial S(a_1, a_2, \dots, a_M)}{\partial a_m} = 0, \qquad m = 1, M \qquad (3.71)$$

Let assume that the functional dependence $\Phi = \Phi(\mathbf{x}, a_1, a_2, \dots, a_M)$ of the unknown parameters $a_1, a_2, \dots, a_M$ can be represented as

$$\Phi(\mathbf{x}, a_1, a_2, \dots, a_M) = \varphi_0(\mathbf{x}) + \sum_{m=1}^{M} a_m \varphi_m(\mathbf{x}) \qquad (3.72)$$

In this case, the objective function $S(a_1, a_2, \dots, a_M)$ is defined by the expression

$$S(a_1, a_2, \dots, a_M) = \sum_{i=1}^{I} (\sum_{m=1}^{M} a_m \varphi_m(\mathbf{x}_i) - y_i)^2$$

Here $y_i = \Phi_i - \varphi_0(\mathbf{x}_i)$.

Taking into account the Eqs. (3.72), the problem (3.71) can be re-written as follows:

$$\frac{\partial S}{\partial a_1} = 2\sum_{i=1}^{I}(\varphi_1(\mathbf{x}_i)(\sum_{m=1}^{M} a_m\varphi_m(\mathbf{x}_i) - y_i)) = 0$$

$$\frac{\partial S}{\partial a_2} = 2\sum_{i=1}^{I}(\varphi_2(\mathbf{x}_i)(\sum_{m=1}^{M} a_m\varphi_m(\mathbf{x}_i) - y_i)) = 0$$

$$\ldots$$

$$\frac{\partial S}{\partial a_M} = 2\sum_{i=1}^{I}(\varphi_M(\mathbf{x}_i)(\sum_{m=1}^{M} a_m\varphi_m(\mathbf{x}_i) - y_i)) = 0 \qquad (3.73)$$

Based on the definition of the scalar product, let us denote

$$(\varphi_k(\mathbf{x})\cdot\varphi_j(\mathbf{x})) = \sum_{i=1}^{I}\varphi_k(\mathbf{x}_i)\varphi_j(\mathbf{x}_i)$$

$$(\varphi_k(\mathbf{x})\cdot y) = \sum_{i=1}^{I}\varphi_k(\mathbf{x}_i)y_i$$

With the recent designations the equation system (3.73) takes the form

$$a_1(\varphi_1(\mathbf{x})\cdot\varphi_1(\mathbf{x})) + a_2(\varphi_1(\mathbf{x})\cdot\varphi_2(\mathbf{x})) + \ldots + a_M(\varphi_1(\mathbf{x})\cdot\varphi_M(\mathbf{x})) = (\varphi_1\cdot y)$$

$$a_1(\varphi_2(\mathbf{x})\cdot\varphi_1(\mathbf{x})) + a_2(\varphi_2(\mathbf{x})\cdot\varphi_2(\mathbf{x})) + \ldots + a_M(\varphi_2(\mathbf{x})\cdot\varphi_M(\mathbf{x})) = (\varphi_2\cdot y)$$

$$\ldots\, a_1(\varphi_M(\mathbf{x})\cdot\varphi_1(\mathbf{x})) + a_2(\varphi_M(\mathbf{x})\cdot\varphi_2(\mathbf{x})) + \ldots + a_M(\varphi_M(\mathbf{x})\cdot\varphi_M(\mathbf{x})) = (\varphi_M\cdot y)$$

The latter equation system is greatly simplified if the system of orthogonal functions is used as the basic functions $\varphi_i(\mathbf{x})$ in the Eq. (3.72). In this case, the equations for determining the coefficients $a_1, a_2, \ldots, a_M$ take the form

$$a_1(\varphi_1(\mathbf{x})\cdot\varphi_1(\mathbf{x})) = (\varphi_1(\mathbf{x})\cdot y)$$

$$a_2(\varphi_2(\mathbf{x})\cdot\varphi_2(\mathbf{x})) = (\varphi_2(\mathbf{x})\cdot y) \qquad (3.74)$$

$$\ldots$$

$$a_M(\varphi_M(\mathbf{x})\cdot\varphi_M(\mathbf{x})) = (\varphi_M(\mathbf{x})\cdot y)$$

In the Eqs. (3.74) the property of orthogonal functions (3.65)) is considered, and the coefficients a_m are determined by the equation

$$a_m = \frac{(\varphi_m(\mathbf{x}) \cdot y)}{(\varphi_m(\mathbf{x}) \cdot \varphi_m(\mathbf{x}))} \tag{3.75}$$

3.3.6 REGRESSION ANALYSIS

With the limited experimental results the use of the mean-value method or least-squares method becomes difficult. In this case, the models of regression analysis are proved to be useful. In the simplest case these are the models of the linear regression analysis, mathematical problem formulation for which is as follows:

- let there be the functional relation $\Phi(\mathbf{x})$ presented in the form of tabular sample values $M_i(\mathbf{x}_i, \Phi_i)$ and built to $i = \overline{1, I}$, where I – number of experiments. Let us develop the empirical formula $\Phi = \mathbf{A}\mathbf{x} + \Phi_0$, where the values $\mathbf{A}$ and Φ_0 should be selected on the basis of the table sample values $M_i(\mathbf{x}_i, \Phi_i)$.

Consider the solution of the problem in the case when the independent variable $\mathbf{x}$ is scalar. In this case, the solution of the linear regression analysis reduces to determining the coefficients a, Φ_0 in the equation

$$\Phi = ax + \Phi_0 \tag{3.76}$$

If we assume that the Eq. (3.76) is correct for any i^{th} experiment, we can write down the following system of two algebraic equations

$$\Phi_i = ax_i + \Phi_0$$

$$\Phi_i x_i = ax_i^2 + \Phi_0 x_i$$

Note that the second equation is obtained from the first one by multiplying all the members by the value x_i. Sum both equations by all tabulated values for $i = \overline{1, I}$:

$$\sum_{i=1}^{I} \Phi_i = \sum_{i=1}^{I} ax_i + \sum_{i=1}^{I} \Phi_0$$

$$\sum_{i=1}^{I} \Phi_i x_i = \sum_{i=1}^{I} ax_i^2 + \sum_{i=1}^{I} \Phi_0 x_i$$

Let us use the adopted designations and definitions

$$\bar{\Phi} = \frac{1}{I}\sum_{i=1}^{I}\Phi_i, \quad \bar{x} = \frac{1}{I}\sum_{i=1}^{I}x_i$$

$$\sigma_\Phi^2 = \frac{1}{I}\sum_{i=1}^{I}(\Phi_i - \bar{\Phi})^2, \quad \sigma_x^2 = \frac{1}{I}\sum_{i=1}^{I}(x_i - \bar{x})^2$$

$$v_{11} = \frac{1}{I}\sum_{i=1}^{I}x_i\Phi_i, \quad \mu_{11} = v_{11} - \bar{x}\,\bar{\Phi}$$

Taking into account the written relations, the original equation system to determine the values a and Φ_0 can be re-written as

$$\Phi_0 + a\bar{x} = \bar{\Phi},$$

$$\Phi_0\bar{x} + a(\sigma_x^2 + \bar{x}^2) = \mu_{11} + \bar{\Phi}\,\bar{x}$$

The second equation of the system can be re-written as

$$\left(\Phi_0\bar{x} + a\bar{x}^2 - \bar{\Phi}\,\bar{x}\right) + \sigma_x^2 = \mu_{11}$$

In view of the first equation, the expression $\Phi_0\bar{x} + a\bar{x}^2 - \bar{\Phi}\,\bar{x} = 0$ and, eventually, the sought values of parameters a, Φ_0 can be found from the equations

$$a = \frac{\mu_{11}}{\sigma_x^2}$$

$$\Phi_0 - \bar{\Phi} = \frac{\mu_{11}}{\sigma_x^2} + (x - \bar{x}) \tag{3.77}$$

The Eq. (3.76) taking into account (3.77) takes the final form

$$\Phi - \bar{\Phi} = \frac{\mu_{11}}{\sigma_x^2}(x - \bar{x}) \tag{3.78}$$

In the original formulation of the problem, the values x and Φ are equal. Therefore, under the symmetry the Eq. (3.78) can be written in the form

$$x - \bar{x} = \frac{\mu_{11}}{\sigma_\Phi^2}(\Phi - \bar{\Phi}) \tag{3.79}$$

3.3.7 APPLICATION OF THE EXPERIMENT PLANNING METHOD

The above-discussed possibility to construct the linear regression model assumes the availability of information about at least two tests, in which we measured the independent parameter x and function Φ. The greater the volume of the tests, the more accurate is the approximation of test results of the constructed model. In practical applications, especially if the cost of the experiment is high, the construction of mathematical dependencies on the test results is carried out in accordance with the experiment planning theory. Only the preliminary information relating to the experiment planning theory is presented below. The systematic description of the theory and examples of its application are given in professional literature (e.g., [68, 88]).

Suppose that in the tests functional dependence Φ on several independent parameters $x_1, x_2, ..., x_I$ is defined. In the experiment planning theory the independent parameters are called factors, and the formulated problem relates to the problems of the multi-factor (I-factorial) experiment. Testing each of the independent parameters x_i can be changed within the established boundaries from the minimum $x_{i\,\min}$ to the maximum $x_{i\,\min}$ ($x_{i\,\min} \leq x_i \leq x_{i\,\max}$). If in the tests all the independent factors x_i take only minimum $x_{i\,\min}$ or maximum values $x_{i\,\max}$, then such an experiment is called two-level. Whenever possible, the tests can be carried out for intermediate values x_i, particularly when $x_i = \dfrac{x_{i\,\min} + x_{i\,\max}}{2}$. In this formulation, the experiment is called three-level. The number of levels that can take the independent variable x_i, can be more than three. For convenience, the dimensionless variables are used instead of the dimensional independent variables x_i

$$X_i = \frac{2x_i}{x_{i\,\max} - x_{i\,\min}} - 1 \tag{3.80}$$

the change of which occurs in the interval $-1 \leq X_i \leq +1$. With the two-level experiment the dimensionless variable X_i takes the values -1, $+1$. With the three-level experiment: -1, 0, $+1$.

When performing the tests, we should define the dependence Φ from I factors, and each factor in tests takes values on m levels. In this case, the maximum possible number of tests C can be established by the relation $C = I^m$. The problem in the experiment planning theory is to build the functional relations $\Phi = \Phi(x_1, x_2, ..., x_N)$ approximating the experimental results with the minimum number of tests $C_{\min} < I^m$. The list of planned experiments is recorded in the table called a matrix. As an example, Table 3.1 shows the example of the matrix for the two-factor two-level (2^2) full-scale experiment in which the dependence is defined

$$\Phi = \Phi(X_1, X_2) = a_0 + a_1 X_1 + a_2 X_2 + a_{12} X_1 X_2 \tag{3.81}$$

There are four unknown coefficients a_0, a_1, a_2, a_{12} in the Eq. (3.81). Four experimental results $\Phi_{11}, \Phi_{12}, \Phi_{21}, \Phi_{22}$ allow conclusively establishing their values. From the above we can make the conclusion that if we assume that the experimental dependence (3.81) is correct, then the minimal number of performed experiments should not be less than four. However, the experiment planning theory allows answering the additional question – what is the contribution of each coefficient a_0, a_1, a_2, a_{12} to the functional dependence (3.81) and, in particular, the coefficient a_{12} characterizing the interaction of the factors X_1 and X_2?

The Eq. (3.81) will be written in the matrix form

$$\begin{pmatrix} +1 & -1 & -1 & +1 \\ +1 & +1 & -1 & -1 \\ +1 & -1 & +1 & -1 \\ +1 & +1 & +1 & +1 \end{pmatrix} \cdot \begin{pmatrix} a_0 \\ a_1 \\ a_2 \\ a_{12} \end{pmatrix} = \begin{pmatrix} \Phi_{11} \\ \Phi_{21} \\ \Phi_{12} \\ \Phi_{22} \end{pmatrix}$$

The solution of the matrix equation is the following

$$a_0 = \frac{1}{4}(\Phi_{11} + \Phi_{12} + \Phi_{21} + \Phi_{22})$$

$$a_1 = \frac{1}{4}(-\Phi_{11} - \Phi_{12} + \Phi_{21} + \Phi_{22})$$

$$a_2 = \frac{1}{4}(-\Phi_{11} + \Phi_{12} - \Phi_{21} + \Phi_{22})$$

$$a_{12} = \frac{1}{4}(\Phi_{11} - \Phi_{12} - \Phi_{21} + \Phi_{22})$$

The above algorithm for solving the problem of selecting the functional empirical relation can be extended to the case, in which the experiments with

TABLE 3.1 Matrix of the Full-Scale Experiment $\mathbf{2^2}$

Test number	X_1	X_1	Φ
1	–1	–1	Φ_{11}
2	+1	–1	Φ_{21}
3	–1	+1	Φ_{12}
4	+1	+1	Φ_{22}

the same combination of the varied independent variables are performed repeatedly. The convenience of the experiment planning methods becomes noticeable if the number of varied independent variables I and levels of their variation m are high (I>2, m>2). In this case, it is possible to build the functional dependence of the form (3.62) using the outline of non-full-scale I-factorial with the change of each factor on m levels (plan I^m). The matrices of this experiment are designed and can be found in the professional literature on methods for planning the experiment (e.g., [88]). There you can find estimates that allow determining the significance of individual coefficients in the empirical Eq. (3.62).

3.4 DERIVATION OF ANALYTICAL RELATIONS BY SIMPLIFYING THE ORIGINAL PROBLEM

In engineering applications at early stages, relatively simple analytical expressions are used that evaluate the object designing quality with accuracy up to 5 % … 10% during the project calculations. Such analytical relations can be obtained not only experimentally, but also by simplifying the mathematical models based on the application of the fundamental laws of physics. The simplification methods used in mathematics will be discussed in details in Chapter 7. Below we will discuss as examples only certain particular problems related to the calculation of the temperature of materials that do not pretend to any generalizations.

3.4.1 DETERMINATION OF THE TEMPERATURE STATE OF HOMOGENEOUS MATERIAL

In most cases, the frameworks of SFRE elements are made of metal or plastic materials. First of them have high heat-conducting and accumulating thermal characteristics. Plastic materials can be characterized as relatively low values of heat conductivity coefficients and relatively high values of specific heat. The high values of heat conductivity coefficients, especially at small thicknesses of framework, lead to the uniform heating of the material by its thickness. At the same time, for materials with poor heat conductivity we should consider the distribution of the temperature T on the heated up material layer of the framework. Consider some approximate methods of estimating the temperature state of materials (scheme area is presented in Figure 3.10) based on the solution of the heat conductivity equation written on the following assumptions:

- thermophysical characteristics of the material – heat conductivity coefficient λ, specific heat c, thermal diffusivity coefficient a – do not depend on the temperature level of the material;
- heat flow coming from the combustion products affects the material inner surface; the heat flow equals zero at the material outer edge;
- heat transfer coefficient on the inner surface of the material does not change the entire heating period.

The adopted assumptions allow writing the heat conduction equation in the form

$$\frac{\partial T}{\partial t} = a\frac{\partial^2 T}{\partial x^2} \tag{3.82}$$

The initial and boundary conditions:

$$t = 0 - T(x,0) = T_0$$

$$x = 0 - \alpha\left(T_г - T_w\right) = -\lambda\frac{\partial T}{\partial x}$$

$$x = \Delta - \lambda\frac{\partial T}{\partial x} = 0$$

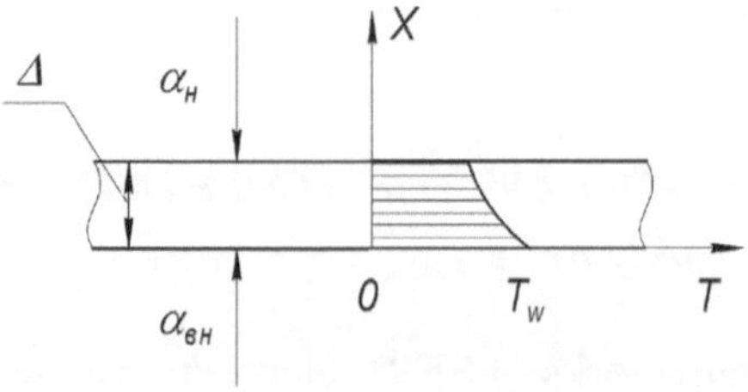

FIGURE 3.10 Design model for the problem on heating the homogeneous material.

Here t – process time, x – spatial coordinate along which the process of heat conduction is examined, Δ – material thickness, $T_г$, T_0, T_w – temperature of combustion products, initial temperature of material, current temperature on inner (heated) surface of material.

Let us consider the special cases, when the analytical solutions of the heat conduction equations can be obtained.

3.4.2 DETERMINATION OF THICKNESS-AVERAGE TEMPERATURE OF MATERIAL

For the design purpose it is necessary to know the framework thickness-average temperature of material (or suppose that the difference between the temperature of the material on its thickness is non-essential – $T(x)\approx$const).

Let us integrate both parts of heat conduction equation by the thickness Δ

$$\int_0^{\Delta} c\rho \frac{\partial T}{\partial t} dx = \lambda \int_0^{\Delta} \frac{\partial}{\partial x} \frac{\partial T}{\partial t} dx$$

After substituting the temperature T for its median value $\overline{T}$ in the expressions for derivatives, the latter equation can be rewritten as follows

$$c\rho\Delta \frac{\partial T}{\partial t} = \lambda \frac{\partial T}{\partial x}\bigg|_0^{\Delta} = 0 - \frac{\partial T}{\partial x}\bigg|_{x=0} = \alpha\left(T_г - T_w\right), \quad T_w \approx \overline{T}$$

This equation can be easily integrated after the separation of variables and takes the final form

$$c\rho\Delta \frac{d\overline{T}}{T_г - \overline{T}} = \alpha\, dt$$

$$\overline{T} = T_г - (T_г - T_0)\exp(-\frac{\alpha}{c\rho\Delta}t) \tag{3.83}$$

It is easy to make sure that the obtained solution corresponds to the physical notion of heating materials. Indeed, when $t=0$, the material average temperature coincides with the initial temperature $-\overline{T} = T_0$, and when $t\rightarrow\infty$, the average temperature approaches the temperature of the combustion products, an increase in the material thickness at the same values α leads to heating at the lower level temperature $\overline{T}$.

3.4.3 DETERMINATION OF TEMPERATURE ON THE INNER SURFACE OF THE HEATED MATERIAL

For the purposes of design, we need to know the temperature at the inner surface of the heated material. If the previous problem is of interest in assessing the thermal condition of the materials the loss of which might be in the work of SFRE, then in this case, the operability of materials whose loss of surface layer is prohibited (e.g., the surface of nodes of SFRE placed in the nozzle block) is evaluated.

In the practice of design, to determine the temperature condition of the materials with the same thermal characteristics, the solutions written in the sum of series, each member of which is represented by the transcendental algebraic function, are used. For the convenience of use of these correlations from the initial heat conductivity Eq. (3.82) we should go to the equation written in dimensionless variables – for criteria of Bio $Bi = \frac{\alpha\Delta}{\lambda}$, Fourier – $Fo = \frac{at^2}{\Delta^2}$, for the temperature simplex $\theta = \frac{T_z - T}{T_z - T_0}$ and dimensionless coordinate $\bar{x} = \frac{x}{\Delta}$.

The transition from the initial equations to the equations in dimensionless variables will be implemented step-by-step:

- transition from the variable T to the dimensionless temperature θ

$$T = T_z - (T_z - T_0)\theta$$

$$\frac{\partial T}{\partial t} = \frac{\partial}{\partial t}\left[T_z - (T_z - T_0)\theta\right] = -(T_z - T_0)\frac{\partial \theta}{\partial t}$$

$$\frac{\partial T}{\partial x} = \frac{\partial}{\partial x}\left[T_z - (T_z - T_0)\theta\right] = -(T_z - T_0)\frac{\partial \theta}{\partial x}$$

$$\frac{\partial^2 T}{\partial x^2} = (T_z - T_0)\frac{\partial^2 \theta}{\partial x^2}$$

- transition to the dimensionless time

$$\frac{\partial \theta}{\partial t} = \frac{\partial \theta}{\partial Fo}\frac{\partial Fo}{\partial t} = \frac{a}{\Delta^2}\frac{\partial \theta}{\partial Fo};$$

- transition to the dimensionless coordinate

$$\frac{\partial \theta}{\partial x} = \frac{\partial \theta}{\partial \overline{x}} \frac{\partial \overline{x}}{\partial x} = \frac{1}{\Delta} \frac{\partial \theta}{\partial \overline{x}}$$

$$\frac{\partial}{\partial x} \frac{\partial \theta}{\partial x} = \frac{1}{\Delta^2} \frac{\partial^2 \theta}{\partial \overline{x}^2}$$

Finally, the heat conductivity equation and boundary conditions in dimensionless variables are written as

$$\frac{\partial \theta}{\partial F_0} = \frac{\partial^2 \theta}{\partial \overline{x}^2}$$

$$\left. \frac{\partial \theta}{\partial \overline{x}} \right|_{\overline{x}=0} = -Bi\theta_w$$

$$\left. \frac{\partial \theta}{\partial \overline{x}} \right|_{\overline{x}=0} = \theta$$

$$\theta_0 = 1 \tag{3.84}$$

The general solution is represented in the form $\theta = \theta(Fo, Bi, \overline{x})$ or as series [124]

$$\theta = \sum_{i=1}^{\infty} \left\{ \frac{2\sin\Phi_i}{\Phi_i + \sin\Phi_i \cos\Phi_i} \exp(-\Phi_i^2 F_0) \cos\left[\Phi_i (1-\overline{x})\right] \right\} \tag{3.85}$$

Here Φ_i is found from the solution of the periodic transcendental equation

$$\Phi_i tg\Phi_i = Bi, \quad i = 1...\infty$$

The index i = 1 corresponds to the solution of Φ_1 placed in the interval $0<\Phi_1<\pi$, $i = 2$ – in the interval $\pi<\Phi_2<2\pi$, etc. It is found that using only one term of series (i = 1) to determine θ and with the value of dimensionless time $Fo \geq 0.3$ the computational error θ will not exceed 1%. The analysis shows that calculating the metal materials under the conditions typical to the nodes of SFRE, it is possible to limit to one member of the series in the solution (3.85).

Calculating the temperature θ_w of the inner wall of the heated material it is advisable to use the ratio

$$\theta_w = \frac{2\sin\Phi_1}{\Phi_1 + \sin\Phi_1 \cos\Phi_1} \cos\Phi_1 \exp(\Phi_1^2 Fo)$$

for the thickness-average wall temperature $\bar{\theta}$

$$\bar{\theta} = \frac{1}{\Delta}\int_0^{\Delta} \theta(x)dx = \int_0^{t} \bar{\theta}(x)d\bar{x} = \frac{2\sin\Phi_1}{\Phi_1 + \sin\Phi_1 \cos\Phi_1} \frac{\sin\Phi_1}{\Phi_1} \exp(\Phi_1^2 Fo)$$
$$= M \exp(\Phi_1^2 Fo)$$

You can use the approximate relations for the temperatures $\theta_w, \bar{\theta}$. First of all, the value M is in the interval 0.81…1.0. In this case, changing Bi in the interval Bi = 0.4…4.0, the following approximate relations are correct:

$$\Phi_1^2 = 0,7B_i^{0,7}, \quad M \approx M_{cp}$$

$$\ln\left(\frac{2\sin\Phi_1 \cos\Phi_1}{\Phi_1 + \sin\Phi_1 \cos\Phi_1}\right) = 0,3Bi$$

Then the equation for $\theta_w, \bar{\theta}$ will be

$$\theta_w = \exp(-0,3Bi - \Phi_1^2 Fo) \tag{3.86}$$

$$\bar{\theta} = M_{cp} \exp(-0,7Bi^{0,7} Fo)$$

3.4.4 DETERMINATION OF THE TEMPERATURE STATE OF THE TWO-LAYER MATERIAL

Estimates of the heat stresses realized in the structural elements of SFRE show that in the constructions of propulsion systems it is necessary to use heat-resistant materials. Passive materials can be used as heat-resistant materials; these are materials that keep their original geometry for the whole operation period. These materials combine high thermal capacity, high temperature of destruction and relatively low thermal diffusivity and thermal conductivity. The use of these materials is due to the structural elements placed in the nozzle block, especially in the vicinity of the critical section, the height of which is not permitted. The heat-resistant effect of the use of this group of materials is only in the implementation of the accumulating effect and high values of temperatures when the destruction of these materials starts. The group of passive materials includes refractory materials, graphite, some metals oxides, borides, nitrides, carbides and zirconates of some metals [136].

In the most general case, to determine the thickness of the heat-resistant material we should solve the heat problem of heating two- or multi-layer wall by the flow of the given intensity, variable or constant in time, for the specified period of time. The problem of determining the minimal allowable thickness of coating is the problem related to the class of inverse problem, whose solution, generally speaking, is not trivial. One way to solve such problems is to determine any considerations of preliminary approximate thicknesses of the heat-resistant material which is further refined using more accurate heat models.

Let us consider the design model given in Figure 3.11. Solving the problem of heating the two-layer material, let us assume the following:

- two-layer wall has a changeless geometric shape;
- heat flow affects only the wall of inner surface from the side of heat-resistant material, the value of heat emission coefficient α at the inner surface is unchangeable during the considered period of heating;
- thermal characteristics of coating materials and framework materials do not depend on the level of temperature and other thermal factors;
- heat conductivity coefficient of the framework material is much larger than the heat conductivity coefficient of coating that leads to approximately uniform heating of the framework material through its thickness.

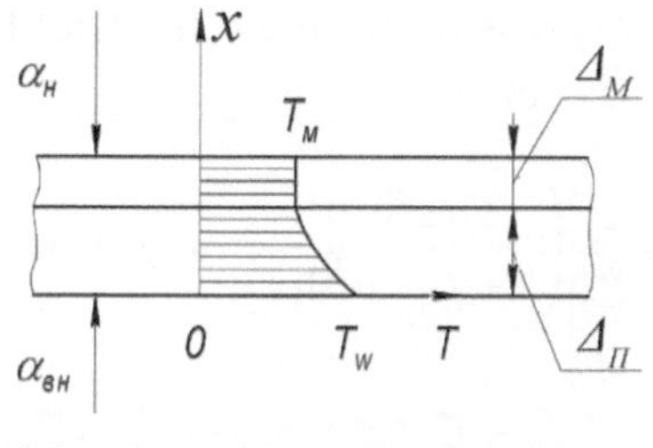

FIGURE 3.11 Design model to the problem of heating the two-layer material.

With the written assumptions the thermal problem for the two-layer wall can be stated as follows:

- equation of the heating process (heat conductivity equation)

$$\frac{\partial T}{\partial t} = a_n \frac{\partial^2 T}{\partial x^2} \tag{3.87}$$

- boundary conditions

$$x = 0 \quad -\lambda_n \frac{\partial T}{\partial t} = \alpha\left(T_г - T_w\right) \tag{3.88}$$

$$x = \Delta \quad -\lambda_n \frac{\partial T}{\partial x} = c_m \rho_m \Delta_m \frac{\partial T}{\partial t}$$

- initial conditions

$$t = 0 \qquad T(x,0) = T_0 = const \tag{3.89}$$

In the Eqs. (3.87)–(3.89) the index n corresponds to the parameters of covering, and the index m – to parameters of the framework (metal).

Let us introduce the following designations:

$$Fo = \frac{a_n t}{\Delta_n^2}, \qquad Bi = \frac{\alpha \Delta_n}{\lambda_n}$$

$$M = \frac{\rho_n c_n \Delta_n}{\rho_M c_M \Delta_M}, \qquad \mu = \frac{1}{Bi} + \frac{1}{M} + \frac{1}{Bi\,M}$$

$$\theta = \frac{T_г - T}{T_г - T_0}, \quad \overline{\mathrm{M}} = \frac{M}{\Delta_n} \tag{3.90}$$

Then, taking into account the known solution of the formulated problem [124], we can write the expression for the dimensionless temperature of the framework material in Fourier series

$$\theta = \sum_{i=1}^{\infty}\left(\frac{2\mu \sec \Phi_i^2}{1 + \mu + \mu^2 \Phi_i} \exp\left(-\Phi_i Fo\right)\right)$$

$$\Phi_i tg\Phi_i = \frac{1}{\mu} \tag{3.91}$$

In [124], it is shown that the Eqs. (3.91) for the problem can be accurately approximated by the following correlation

$$\lg\theta = \lg\theta_0 - \frac{A}{\mu + c} Fo \tag{3.92}$$

Here $\lg\theta_0 = 0,0212; A = 0,45; c = 0.40$.

This dependence in the case of μ, taking the values $\mu = 0.2\ldots20$, provides the defining of the temperature of the framework material with the error less than 2%, and when $\mu\to0$ – no more than 12%.

If in the latter equation we substitute the values Fo, μ calculated by the ratios (3.90), then after transformations we can obtain the expression for the thickness of the heat-protective passive coating providing the temperature of the framework material equal to $\theta_{\partial on}$ by the end of operation:

$$\Delta_n = -F + \sqrt{F^2 - G - H\frac{t}{\lg\theta_{\partial on} - \lg\theta_0}} \tag{3.93}$$

Here $F = \frac{1}{2c}\left(\frac{\lambda_n}{\alpha} + \frac{1}{\overline{M}}\right)$, $G = \frac{1}{c}\frac{\lambda_n}{\alpha\,\overline{M}}$, $H = \frac{A}{c}a_n$.

3.5 APPLICATION OF COMPUTER TECHNOLOGIES TO PROCESS THE EXPERIMENTAL RESULTS

Some functions included in the mathematical system *MathCad* are given in the Table 3.2 [86,141]. Some subprograms and functions included in the *IMSL* library of *Intel Visual Fortran* compiler are given in Table 3.3 [22]. The following are examples and comments on the application of functions and subprograms given in Tables 3.2 and 3.3.

The first group of functions is presented in Table 3.2 that allows calculating the vector of random numbers in the interval $a \le x \le b$. The distribution law of random numbers is uniform (function ***runif*** (*m, a, b*)) or normal (function ***rnorm*** (*m, a, b*)). The system MathCad contains functions that provide the calculation of randomly distributed numbers for various laws of distribution – binomial, χ^2, Fisher, Cauchy, Poisson, Student, Weibull, etc. As a part of compilers program *FORTRAN* there is the subprogram random_ number providing the calculation of ***random number*** or vector uniformly distributed in the interval $0 \le x \le 1$.

TABLE 3.2 Examples of the Function *MathCad* Providing the Construction of Regression Dependencies

No	Function name	Purpose of function	Description of function arguments
1	***dunif*** (*x, a, b*); ***runif*** (*m, a, b*); ***dnorm*** (*x, a, b*); ***rnorm*** (*m, a, b*);	probability density and vector of generated *m* random numbers, with a uniform distribution; the same but for normally distributed random quantity	*x* – calculated value of the probability density; *a, b* – boundary interval where the random quantity is calculated; *m* – number of independent numbers calculated according to the law of random quantity distribution
2	***line*** (*x, y*) ***medfit*** (*x, y*) ***lnfit*** (*x, y*)	regression of the table function *y(x)*, linear, median-linear $y = a + bx$ and logarithmic $y = a\ln x + b$ functions, respectively	*x, y* – vectors with the values of argument and function
3	***regress*** (*x, y, n*)	vector (function) containing the values of the coefficients that ensure the construction of approximating polynomial	*x, y* – vectors with the values of argument and function; *n* – degree of the approximating polynomial
4	***expfit*** (*x, y, g*) ***sinfit*** (*x, y, g*) ***pwrfit*** (*x, y, g*) ***lgsfit*** (*x, y, g*) ***logfit*** (*x, y, g*)	regression of the table function *y(x)*, of the form $y = ae^{bx} + c$, $y = a\sin(x + b)\,e^{bx} + c$, $y = ax^b + c$, $y = \frac{a}{1+be^{-cx}}$, $y = a\ln(x + b) + c$, respectively	*x, y* – vectors with the values of argument and function; *g* – initial approximations of the coefficients a, b, c
5	***medsmooth*** (*y, n*) ***ksmooth*** (*x, y, b*) ***supsmooth*** (*x, y*)	transformed vector y smoothed with the algorithms of shifting medians with the window of width *n*, smoothed using Gaussian kernel to calculate the weighted average elements from *y*, smoothed by the adaptive least-square method is the result of calling to the function	*x, y* – vectors with the values of argument and function; *n* – width of the window over which the smoothing occurs (*n* – odd number); *b* – parameter controlling the window of smoothing (should be greater than the interval value between the points of the argument *x*)

TABLE 3.3 Examples of the Subprogram of *IMSL* Library Providing the Construction of Regression Dependencies

No	Call to the subprogram or function	Purpose of the subprogram or function	Description of arguments of the subprogram or function
1	*call **rline** (nobs, xdata, ydata, b0, b1, stat);*	to the set of points defined on the plane, the smoothing straight line ($y = b_0 + b_1 x$) is built	*nobs* – number of observations; *xdata, ydata* – vector of the size *nobs* containing the values of observations; *b0, b1* – coefficients in the equations for the smoothing straight line; *stat(12)* – average values of *xdata* and *ydata*, dispersion *xdata* and *ydata*, correlation, errors in estimating the coefficients *b0* and *b1*, degree of regression freedom, sum of squared deviations for regression, degree of error freedom, sum of squared deviations for an error, number of unspecified values *xdata, ydata* in the initial data
2	*call **rcurv** (nobs, xdata, ydata, ndeg, b, sspoly, stat)*	to the set of points, defined on the plane, the smoothing polynomial of the form $y = \sum_{k=0}^{K} b_k x^k$ is built	*nobs* – number of observations; *xdata, ydata* – vector of the size *nobs* containing the values of observations; *ndeg* – maximum value of the exponent in the polynomial *y (ndeg=K); b, sspoly* – arrays of dimension K+1; b – contains the coefficients b_k, $k = 0,K$; *sspoly* – contains the sum of squared deviations for *i=1, K* and for the mean values of *x; stat(10)* – average values of *xdata* and *ydata*, dispersion *xdata* and *ydata*, statistics R^2 [22], degree of regression freedom, sum of squared deviations for regression, degree of error freedom, sum of squared deviations for an error, number of unspecified values *xdata, ydata* in the initial data
3	*call **fnlsq**(f, intcep, nbasis, ndata, xdata, fdata, iwt, weight, a, sse)*	to the set of points, defined on the plane, the smoothing polynomial using the set of user basic functions is built	*f* – user set of basic functions; *intcep* – criterion indicating the variant of approximating the vector *fdata* by the set of basic functions; *nbasis* – number of basic functions; *ndata* – number of points for approximation; xdata, fdata – abscissa and ordinate points; *iwt* – criterion defining the values of the weight function; *weight* – weight coefficients for all points; *a* – vector containing the coefficients of the approximating function; *sse* – error equal to the sum of deviations

The next group of functions available as a part of *MathCad* allows you to create the regression dependencies that are based on tabular dependences *y(x)* obtained, for example, as the experiment result. Because of the errors of the function y the use of interpolation dependencies discussed in Chapter 2 is unacceptable. You can build the linear regression of linear type, logarithmic, exponential, polynomial ones, etc.

To choose the right regression dependence of the functions available in *MathCad*, you should have the experience of experimental data processing. In particular, Figure 3.12 shows the nature of power dependencies of the form $y = ax^b$. In this dependence, when $x = 1$ the function value is $y = a$. In the special case ($b = 1$), the power dependence becomes linear and passes through the origin of coordinates (0,0). For values $b > 1$ at the point $x = 0$ the curve touches the abscissa axis. The dependencies where $b > 0$, are sometimes called parabolic. The dependencies where $b < 0$, are sometimes called hyperbolic. If the experimental dependence *y(x)* has the form similar to the one shown in Figure 3.12, then we can use the functions of *MathCad* ***lnfit*** *(x, y)* or ***pwrfit*** *(x, y, g)* in constructing the regression dependence.

Figure 3.13 demonstrates the functional dependencies typical to the exponential functions of the form $y = ae^{bx}$. The variants where $b = 0$ are trivial and out of practical interest. The variants where $a \neq 0$ and $b \neq 0$ can be frequently used in engineering. In *MathCad* the function ***expfit****(x, y, g)* allows obtaining the regression with the use of exponential function.

As a part of the library *IMSL* in *FORTRAN* compilers the regression dependencies can be obtained using, for example, programs ***rline***, ***rcurv***, ***fnlsq***. These programs provide the construction of linear (***rline***) or polynomial

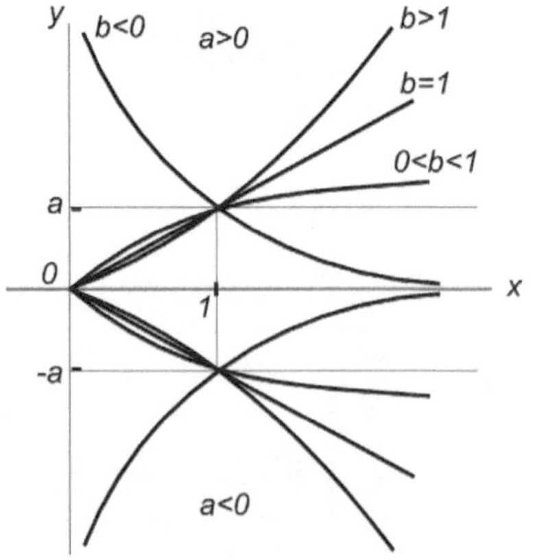

FIGURE 3.12 Character of functional dependencies of the form $y = ax^b$.

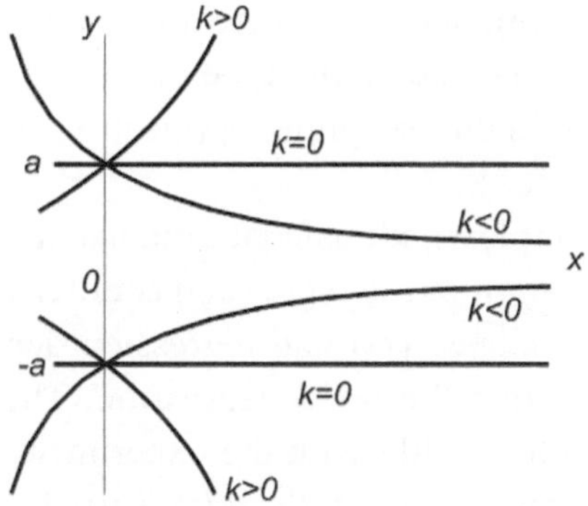

FIGURE 3.13 Character of functional dependencies of the form $y = ar^{bx}$.

(***rcurv***) regression. The program ***fnlsq*** construction of regression dependence is carried out using the user set of basic functions.

The functions and programs given above provide the writing of the approximating function constructed on the results of experiments in the explicit form. However, it is not always necessary. Many functions and programs are available in the system *MathCad*, and in the library *IMSL* of compiler *FORTRAN*, where on the basis of originally given tabular function $y(x)$ the new function $f(x)$ if built, where the systematic and random errors available in the original dependence $y(x)$ are absent or their influence is minimized. The procedure of constructing the dependence $f(x)$ on the basis of "noisy" dependence $y(x)$ is called smoothing. Some smoothing algorithms are available, for example, in [112,230].

Table 3.2 lists the three functions (***medsmooth****(y, n)*, ***ksmooth****(x, y, b)*, ***supsmooth****(x, y)*) ensuring the implementation of the smoothing procedure. The difference between these functions is in the used numerical algorithms of smoothing. The first function ***medsmooth****(y, n)* implements the algorithm of "traveling" medians, the prerequisite of which is the uniformity of the distribution of empirical points on the abscissa. The second function – ***ksmooth****(x, y, b)* implements the smoothing algorithm on Gaussian method. The prerequisite for using this feature is the requirement that the width of the smoothing window *b* should equal the total value of several gaps separating adjacent points in the sample. The third function ***supsmooth****(x, y)* performs smoothing using the adaptive algorithm based on the least squares method, in which the calculation of the smoothed value is carried out taking into account the mutual position of the point under consideration and the nearest to it. The number of the used nearest values of the original dependence is defined by the specific behavior of the graph of this dependence.

In addition to the mentioned ones, there are also the other functions that provide smoothing with the use of any type of splines [86,141]. Many programs of smoothing based on the splines application are in the library *IMSL* of compiler *FORTRAN* [22].

Below are some examples made using the mathematical system *MathCad*.

Figures 3.14 (calculation of the regression coefficients) and 3.15 (graphics of the original dependence *y(x)* and regression dependence *F(x)*) show the example of constructing the linear regression. The initial data (dependence *y(x)*) is taken from the file with the experimental data. To construct the regression, two functions of *MathCad* are used – ***intercept****(x, y)* and ***slope****(x, y)*. The first function allows you to set the coordinate of the intersection point of the regression line with the ordinate axis. The second function sets a slope of the regression line to the abscess axis. The results of work with these functions are shown in Figure 3.14.

Figure 3.15 shows the results of calculating the linear regression which are derived in the form of graphic dependencies. The numeral *1* indicates the initial dependence *y(x)*. Numeral *2* indicates the linear regression dependence built using the functions ***intercept****(x, y)* and ***slope****(x, y)*.

Figures 3.16 and 3.17 show the calculation example of the special regression in the form of the logarithmic curve. The function ***logfi***t(*x, y, g*) is

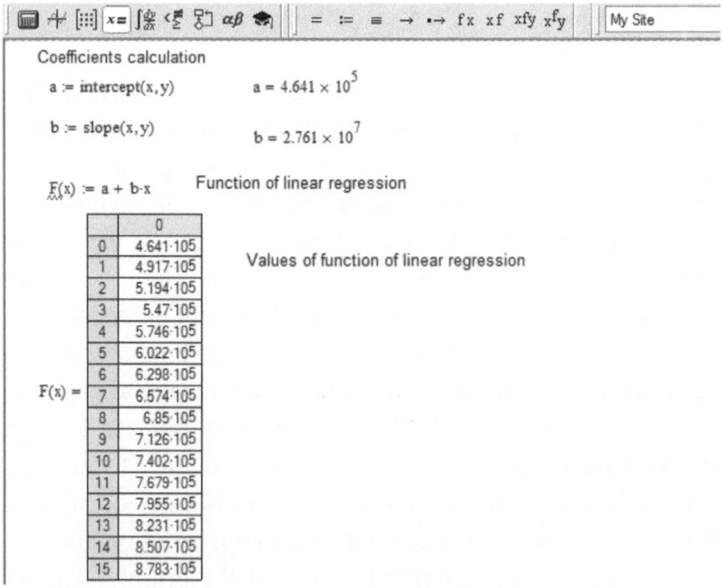

FIGURE 3.14 Construction of the linear regression (program part).

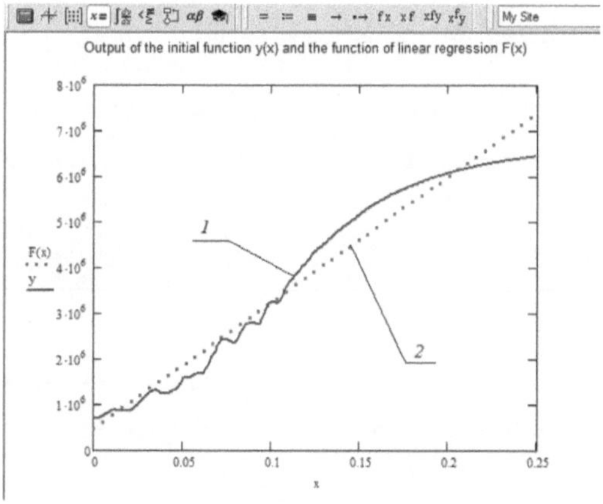

FIGURE 3.15 Construction of the linear regression (graphic dependencies) (1 – initial dependence *y(x)*; 2 – linear regression *F(x)*).

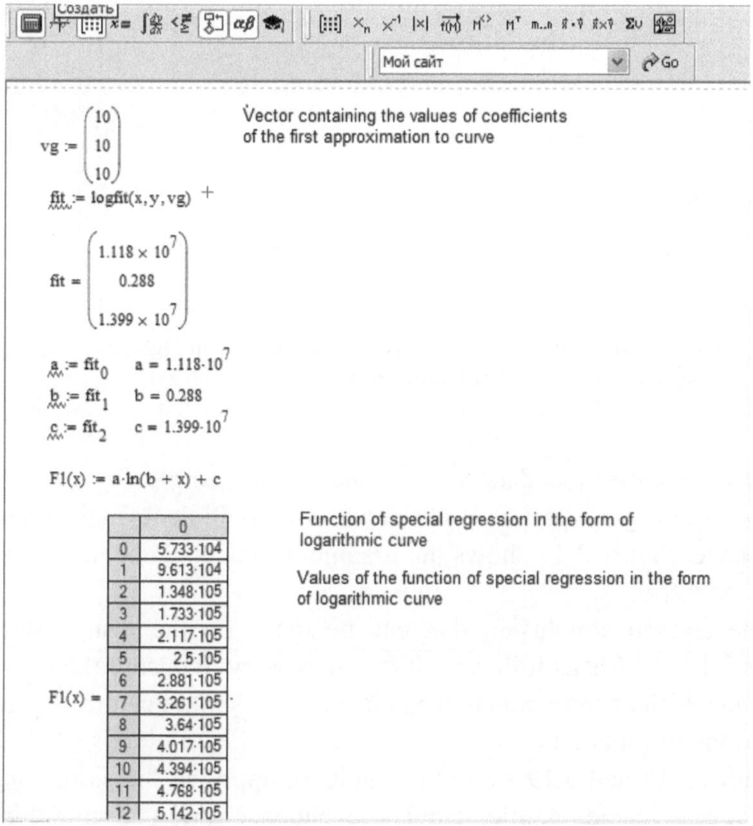

$$vg := \begin{pmatrix} 10 \\ 10 \\ 10 \end{pmatrix}$$

Vector containing the values of coefficients of the first approximation to curve

fit := logfit(x, y, vg)

$$fit = \begin{pmatrix} 1.118 \times 10^7 \\ 0.288 \\ 1.399 \times 10^7 \end{pmatrix}$$

$a := fit_0 \quad a = 1.118 \cdot 10^7$

$b := fit_1 \quad b = 0.288$

$c := fit_2 \quad c = 1.399 \cdot 10^7$

$F1(x) := a \cdot \ln(b + x) + c$

Function of special regression in the form of logarithmic curve

F1(x) =

	0
0	5.733·104
1	9.613·104
2	1.348·105
3	1.733·105
4	2.117·105
5	2.5·105
6	2.881·105
7	3.261·105
8	3.64·105
9	4.017·105
10	4.394·105
11	4.768·105
12	5.142·105

Values of the function of special regression in the form of logarithmic curve

FIGURE 3.16 Construction of the logarithmic regression (program part).

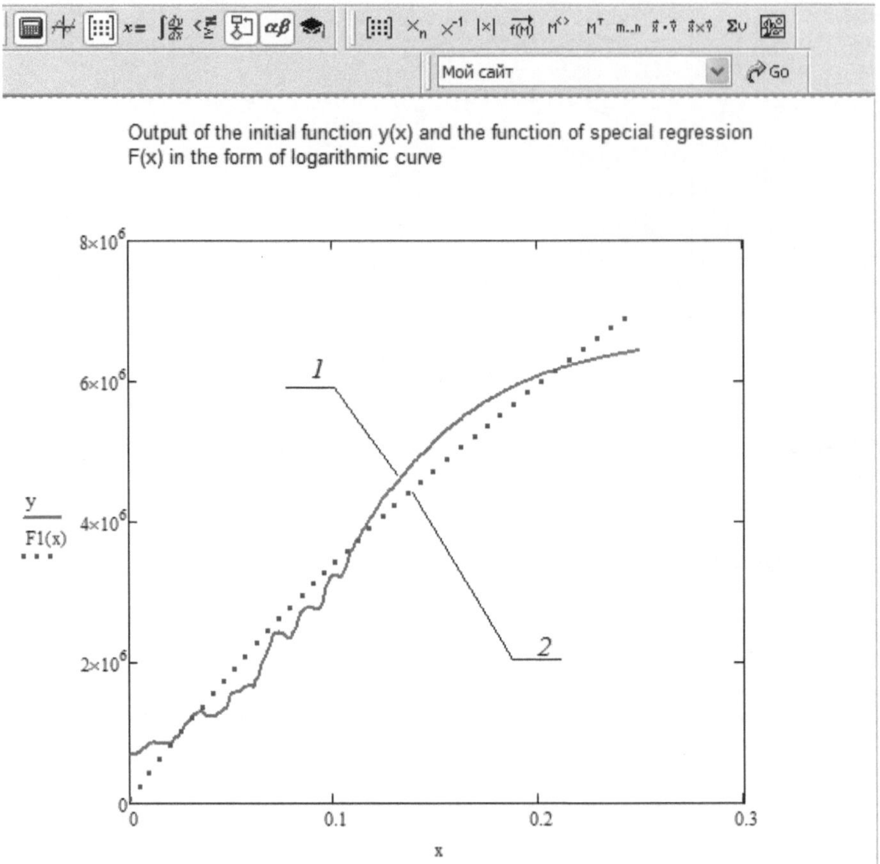

FIGURE 3.17 Construction of the logarithmic regression (graphic dependencies) (1 – initial dependence *y(x);* 2 – logarithmic regression *F1(x)*).

applied to the same input data *y(x)* as in the above example. Figure 3.16 demonstrates the program part which calculates the coefficients in the regression dependence. Figure 3.17 shows the graphic dependencies – initial *y(x)* and regression *F1(x)*.

The general conclusion that can be made on the results shown in Figures 3.14–3.17 is as follows – if it is necessary to ensure the qualitative processing of the experimental dependencies, we should carefully select the approximating functions.

Figures 3.18 and 3.19 show the results of applying the smoothing procedure for the already applied tabular dependence *y(x)*. The smoothing was performed using the function ***supsmooth***(*x*, *y*). According to the results, it can

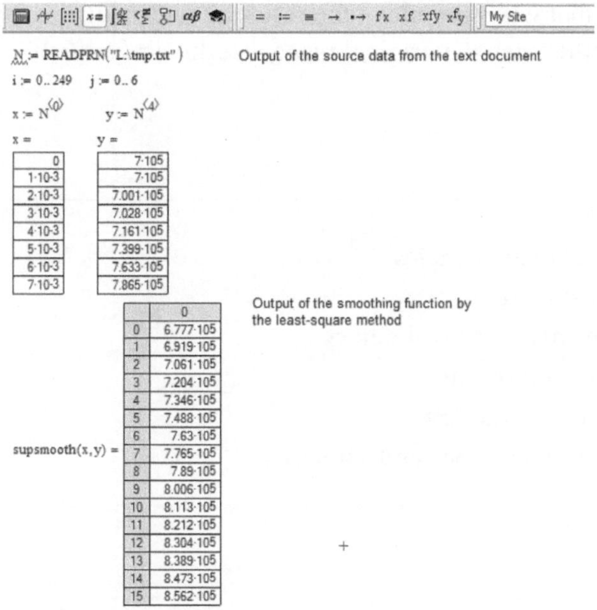

FIGURE 3.18 Construction of the smoothing function (program part).

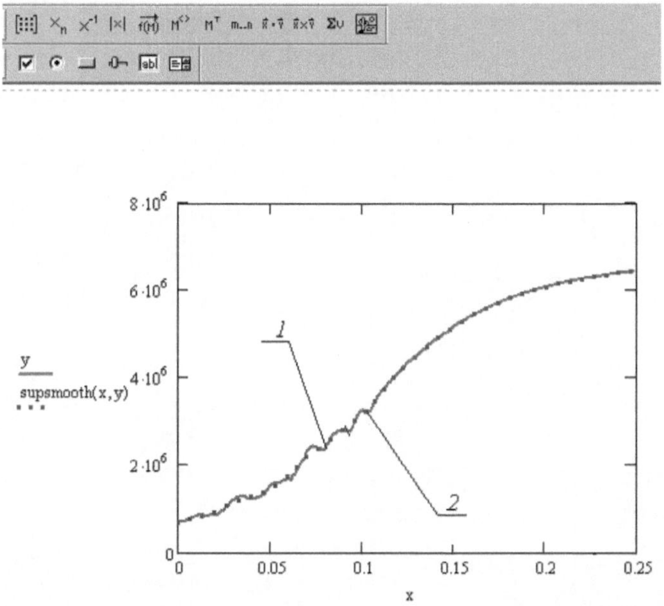

FIGURE 3.19 Construction of the smoothing function (graphic dependencies) (1 – initial dependence *y(x);* 2 – scheme of smoothed function).

be concluded that with smoothing the resulting dependence ***supsmooth***(x, y) is much smoother than the original piecewise-linear function.

KEYWORDS

- **analytical dependencies**
- **analytical relations**
- **behavior of a technical object**
- **experimental results**
- **mathematical models**
- **theory of dimensions and similarity**

CHAPTER 4

MATHEMATICAL MODELS AS SELECTION PROBLEMS

CONTENTS

4.1 FORMULATION OF THE SELECTION PROBLEMS AS OPTIMIZATION PROBLEMS

One of the main problems encountered in technical applications is the selection problems. In mathematical models the feature of the selection problem is the fact that in addition to the basic model of technical object or process, the relations written in the form of one or more linear and non-linear equations are added. The solution of these equations is the solution of the selection problem.

The selection problems can be, for example, direct or inverse problems of technical object projection, problems which involve the application of various game models, etc. Generally speaking, selection is an action that attaches purposefulness to the activity, subjecting of this activity to one goal or set of goals. Solution of the selection problem includes a number of actions with the set of alternatives. As a result, in the problem solving the reduction of the original set of alternatives is carried out and some subset is formed. The reduction of set is possible only if criteria of comparative analysis of alternatives or criteria of preferences are defined.

Definitions of the selection problems vary. Methods for solving such problems also vary. Selection problems may differ both in the original formulation and process of solution.

Here are some options assumed as the basis for the classification of selection problems [34, 154, 155, 166, 215, 225, 253].

1. Many alternatives, among which the most preferred are sought, can be finite, countable or potential. In the first case, all possible

solutions of the selection problem (alternatives) can be numbered and discussed separately. In the second case, the alternative solutions can be numbered, but their quantity is infinite, and the analysis of each of them is impossible. In the third case, the condition of potentiality of alternatives allows us to use the differential apparatus for solving the selection problem.

2. Alternatives can be evaluated by one or several criteria of preference – with qualitative or (and) quantitative nature. Preference criteria will be discussed below. We note only that there are procedures that allow us to reduce the multi-criteria selection problem to the one-criteria.
3. Selection can be single or repeated. An example of such problem can be the problem of selecting the trajectory of antiaircraft missile flight. Indeed, the missile flight trajectory can be repeatedly adjusted on the basis of information analysis about the problem solution of missile and target meeting. As approaching the target, the value of predictable failure should be reduced.
4. Implications of selection can be accurately known, have probabilistic nature or to be unknown. With a limited set of alternatives, the implications of solving the selection problem can be predicted with sufficient accuracy using deterministic models. With a large number of alternatives the analysis of the selection outcomes can be time consuming or unprofitable. An example can be the problem of estimating the quality of ammunition produced. The reliable estimate of quality can be obtained by testing each product that is absurd. The way out is the random selection of samples for tests with the quality assessment of production by the methods of probability theory and mathematical statistics. The most difficult cases are the selection problems when the information about the results of selection is absent (selection is made under conditions of uncertainty). Such problems can be solved using heuristic methods.
5. Responsibility for the selection can be unilateral (individual) or multilateral (group). The simplest examples of such problems are sports – personal (chess, tennis, etc.) and team (football, volleyball, etc.), operation of computer-aided manufacturing, military operations, etc.
6. Level of the goals consistency in multi-selection can be opposite (conflict selection), coinciding (cooperative selection) or intermediate (e.g., coalition selection).

Conflict selection accompanies most of sports. Game example of the coincident selection can be so-called cooperative chess. The goal for both participating parties is the checkmate to the white king. Examples of intermediate selection are political and military and technical solutions in monitoring the conflict between two or more conflicting parties, etc.

It was noted above that the solution of selection problems is possible if the criteria of comparative analysis which is sometimes called criteria of quality, preference, utility function, or simply criterion. Defining the problem, a single criterion can be determined or the set of valuable criteria can be listed.

The most developed are the selection problems solved with one criterion. If we denote $\boldsymbol{x}$ – an arbitrary alternative in the set $\boldsymbol{X}$ ($\boldsymbol{x} \in \boldsymbol{X}$), and $q(x)$ – a value of criterion q for the selected alternative x, then the selection problem can be reduced to the search of alternative x^* that provides the maximum (or minimum) value of the criterion $x^* = \arg\max\ q(x),\ x \in \mathbf{X}$.

There are several examples from engineering, when it is possible to define the single criterion for the solution of the selection problem (e.g., the selection of the optimal design of a technical object). However, the survey of various experts may show that the single selected criterion is a subjective concept. Indeed, with the same technical object at different stages of its existence various subjects (users), who do not need from this object the coincidence of one or more specific properties out of the whole set, are dealt with. Here are some examples [9, 116, 128, 174, 231].

4.1.1 TACTICAL CRITERIA

Tactical criteria can be those which to the greatest extent characterize the properties of the studied (projected) object. In technical applications tactical criteria used in the design or operation of object can include, for example:

- capacity and flight range of the aircraft (for solving the problems of transport aviation design);
- flight speed and flight altitude of the aircraft (for solving the problems of military fighters design);
- payload weight and height of the rocket (for solving the problems of meteorological rockets design);
- depth and speed of the submarine (for solving the naval tactical or strategic problems, and also the problem of submarine design), etc.

4.1.2 MASS CRITERIA

The mass of a technical object can be regarded as an independent criterion if it characterizes the studied object in full. Note that in this case it is correct to examine the mass as a special case of tactical criterion. Many indicators and features of a technical object are related to the mass. For example, an increase of aircraft or car mass certainly leads to the increase of fuel consumption, increase of the complexity of maintenance and operation of vehicle. Production of technical object with a greater mass, usually, requires more material or human resources costs.

4.1.3 LEAD TIME AS A CRITERION

Lead time as a criterion can be taken as an independent criterion if the costs for creating the object are not taken into account. Such situations are real in solving any strategic problems related to the military equipment (in Russian history it is the creation of atomic and thermonuclear bombs, intercontinental missiles, etc.). Other examples are real for the objects of mass production, when in the condition of severe competition an object with high consumer quality is created and product market is won.

At the same time, it should be noted that the lead time is economically related to the costs of technical object construction.

4.1.4 ECONOMIC CRITERIA

The use of economic criteria is the most attractive for solving the problems of selecting the options for the designed technical object [84, 122, 156]. In most cases, economic criteria are reduced to the solution of the selection problem, in which a subset of objects that have minimum value at certain stages of creation or (and) object operation is defined.

4.1.5 MULTI-CRITERION PROBLEMS OF SELECTION

The solution of selection problems becomes more complicated if you cannot define the only criterion of preference. Of many examples, illustrating this case, we will give the following:

- purchase of consumer goods (price, product design, packaging design, consumer properties, etc. influencing the choice);
- creation of a new technical object (the main criterion changes as the new elements of the created object appear), etc.

  One way to solve multi-criterion selection problems is to reduce the multi-criterion problem to one-criterion. For this case, it is important to develop the algorithm of super-criteria selection. The simplest of them are:
- peer review as a sum of N factors using the weighing coefficients

$$q_0 = \sum_{n=1}^{N} \alpha_n \frac{q_n}{S_n} \tag{4.1}$$

  In the formula (4.1) S_n – coefficient ensuring the zero-dimension of the criterion q_n; α_n – relative influence of the criterion q_n on the super-criterion value q_0;
- peer review as a product of N factors using the weighing coefficients

$$q_0 = 1 - \prod_{n=1}^{N} (1 - \beta_n \frac{q_n}{S_n}) \tag{4.2}$$

  In the formula (4.1), S_n – coefficient ensuring the zero-dimension of the criterion q_n; α_n – relative influence of the criterion q_n on the super-criterion value q_0;

  In the formula (4.2), β_n – relative influence of the criterion q_n on the super-criterion value q_0;
- method of "pulling of the most lagging"

$$x^* = \arg\max \left\{ \min (\alpha_n \frac{q_n(x)}{S_n}) \right\}, \quad n = 1, N \tag{4.3}$$

- change of the criterion happens at each stage of solving the selection problem, at the same time, at each stage the value n changes in the additional algorithm

$$x^* = \arg \left\{ \max q_{n1}(x) \middle| q_{n2}(x) \leq C_{n2}, \forall n2 \neq n1 \right\} \tag{4.4}$$

In analyzing complex systems (objects) that are difficult to formalize and select the single criterion of preference, other methods are used that can be referred to heuristic. The difficulty and inconsistence of solving the problems related to the complex systems can be confirmed with a well-known impossibility theorem [172] (Arrow paradox), the simple meaning of which is the following

- if there are two random system statuses x_1 and x_2, then there is always the solution of the selection problem, where the system status can be transferred from x_1 to x_2.

In other words, the essence of the paradox is that it is always possible to offer an algorithm where the arbitrary (or given) value $x \in X$ will be the solution of the selection problem.

Another difficulty of solving the problem related to complex systems can be the fact that seeking the result is the function of random numbers. This type of problems is stochastic selection problems. Stochastic selection problems can also be the problems where the selection result can be estimated with the given degree of certainty (probability of the selected result is given), or random distribution law is unknown.

Of all the variety of selection problems, we sort out the problems of mathematical programming which are generally formulated as follows:

let there be the set X, any point of which satisfies the limit series $f_m(x) \geq 0,\ m = \overline{1, M}$. Find in the set X the point $x^* \in X$ that ensures the ratio $\varphi(x^*) = \min \varphi(x)$. If there is such point, then let us find the exact lower boundary of the function $\varphi_* = \inf \varphi(x)$.

In the problems of mathematical programming, the function $\varphi(x)$ is called the objective function, and the set $\mathbf{X}$, where the point x^* is found, is the search parameters set. The value x^* providing the optimal value of the objective function $\varphi(x)$ is called the stationary point.

In the theory of mathematical programming [70, 110, 155], the following problems are singled out:

- *problems of linear programming,* – an objective function here is linear, and the set, where the extremum of objective function is calculated, is limited to the system of linear equalities and inequalities;
- *problems of non-linear programming,* – an objective function or limits here are non-linear functions. In non-linear programming we can single out the problems of *convex programming* (objective function is convex and set where the extreme problem is solved is also convex), *quadratic programming* (objective function is quadratic, and limits are linear equalities and inequalities), *integer programming* (integrity condition is superimposed on the searched variables);

- *problems of geometric programming,* – objective function and limits here are presented in the form of posynomial (in posynomial, unlike the polynomial, the index can be any real number [216])

$$\varphi(x) = \sum_{t=1}^{T} C_t \prod_{m=1}^{M} x^{\alpha_{tm}} \tag{4.5}$$

where C_t– posynomial coefficients (C_t>0), α_{tm}– indexes (α_{tm}– any real number). Methods of geometric programming problems solution are based on the correctness of the geometrical ratio which can be written as follows

$$\alpha_1 x_1 + \alpha_2 x_2 + \ldots + \alpha_N x_N \geq x_1^{\alpha_1} x_2^{\alpha_2} \cdot \ldots \cdot x_N^{\alpha_N},$$

$$\alpha_1, \alpha_2, \ldots, \alpha_N > 0, \tag{4.6}$$

$$\alpha_1 + \alpha_2 + \ldots + \alpha_N = 1$$

An advantage of the use of geometrical ratios is that in some cases they allow receiving the answer for two important questions without the solution: Under what conditions is the optimal value of the objective function ensured? and What is the value of this optimum? In accordance with the theory of geometric programming the optimal values of unknown variables x_i are ensured only if values of all components in the first Eq. (4.6) coincide

$$\alpha_1 x_1 = \alpha_2 x_2 = \ldots = \alpha_N x_N \tag{4.7}$$

The latter system of equations allows us to simplify the right side of the first Eq. (4.6). Indeed, taking into account (4.7), the expression for the minimum of the unknown objective function can be re-written as

$$\min \varphi(x_1, x_2, \ldots, x_N) = x_1^{\alpha_1} x_2^{\alpha_2} \cdot \ldots \cdot x_N^{\alpha_N} = x_1^{\alpha_1} \left(\frac{\alpha_1 x_1}{\alpha_2}\right)^{\alpha_2} \cdot \ldots \cdot \left(\frac{\alpha_1 x_1}{\alpha_N}\right)^{\alpha_N}$$

$$= \frac{(\alpha_1 x_1)^{\sum_{n=1}^{N} \alpha_n}}{\alpha_2^{\alpha_2} \alpha_3^{\alpha_3} \cdot \ldots \cdot \alpha_N^{\alpha_N}}$$

Finally, using the third Eq. (4.6) we can write down

$$\min \varphi(x_1, x_2, ..., x_N) = \frac{\alpha_1}{\alpha_2^{\alpha_2} \alpha_3^{\alpha_3} \cdot ... \cdot \alpha_N^{\alpha_N}} x_1$$

Note that in [216] there are numerous elegant examples of geometric programming methods for technical applications.

Let us consider the following example. The use of mathematical models of simplified type to analyze the operation of the propulsion system assumes that the definition of the set of concordant coefficients, correct values of which can be obtained by comparing the calculations with experimental results. Particularly, calculating the initial phase of solid fuel engine running the matching (identification) of the results with the experiment can be performed by the coefficient value that determines the relative weight of tablets departing from the initiator frame by the pressure of jet membrane destruction, and by parameters determining the fuel firing, etc. The whole set of these parameters by means of which the matching between the mathematical model and experiment will be performed, can be referred to the searching ones. The limits of search parameters changing are easily determined from the problem formulation. Particularly, the relative weight of tablets departing from the initiator frame is in the range (0, 1). The pressure of the jet membrane destruction exceeds the initial pressure in the engine chamber, but it is less than its quasi-steady level, etc.

The formulation of the problem of mathematical programming in this case will be carried out in full, if an objective function will be formalized. As an objective function degree of distinction between the values of calculated pressure in the engine chamber $p(t)$ (pressure $p(t)$) can be determined by solving the equations of internal ballistics [244]) and values of pressure $p^*(t)$ calculated in the experiment should be taken. The objective function can be written, for example, as $\Phi(\mathbf{x}) = \int_{t_0}^{t_k} (p(t) - p^*(t))^2 dt$. Note that without limiting the problem generality, it can be solved under the experimental information about the change of $p(t)$ in different engine sections or when processing several (N) experimental results. In this case, the objective function can be as follows

$$\Phi(\mathbf{x}) = \sum_{n=1}^{N} \int_{t_0}^{t_k} (p_n(t) - p_n^*(t))^2 dt$$

4.2 METHODS OF REDUCING THE PROBLEMS OF MATHEMATICAL PROGRAMMING TO THE PROBLEMS OF UNCONSTRAINED OPTIMIZATION

The characteristic feature of mathematical programming problems is limitations which complicate the search of a stationary point providing the extreme value of the objective function. Several methods for solving this class of problems is to change the objective function in the way as to exclude all the limits (conditions) existing in the problem. Let us consider some known methods of reducing the problems of mathematical programming to the problems of unconstrained optimization [5, 45, 58, 70, 110, 233].

4.2.1 LAGRANGIAN MULTIPLIER METHOD

The method is suitable for the problems of mathematical programming of the following form:

- to find the stationary point with coordinates $x_1, x_2, \ldots x_N$ providing the extreme of the objective function $F(x_1, x_2, \ldots x_N)$ with the M limits of the form

$$Q_m(x_1, x_2, \ldots x_N) = 0, \; m=1, M \tag{4.8}$$

The objective function F in this method is changed to the objective function Φ as follows:

$$\Phi(x_1, x_2, \ldots x_N, \lambda_1, \lambda_2, \ldots, \lambda_M) = F(x_1, x_2, \ldots x_N) + \sum_{m=1}^{M} \lambda_m Q_m(x_1, x_2, \ldots x_N) \tag{4.9}$$

In the objective function (4.9), the coefficients λ_m, $m=1, M$ – are Lagrangian multipliers which, in general, should be regarded as unknown. It is easy to verify that the change from the problem of mathematical programming to the problem of unconstrained optimization is correct. Indeed, in the optimum point (stationary point), in accordance with the original problem statement, the following conditions should be satisfied

$$\frac{\partial F}{\partial x_n} = 0, \quad n = 1, N \tag{4.10}$$

$$Q_m(x_1, x_2, ..., x_N) = 0, m = \overline{1, M}$$

For a new objective function the following conditions should be hold:

$$\frac{\partial \Phi}{\partial x_n} = \frac{\partial F}{\partial x_n} = 0, \quad n = \overline{1, N} \tag{4.11}$$

$$\frac{\partial \Phi}{\partial \lambda_m} = Q_m(x_1, x_2, ..., x_N) = 0, \quad m = \overline{1, M}$$

which are equivalent to the relations (4.10).

4.2.2 PENALTY FUNCTION METHOD

In the penalty function method the solution of problems with any kind of limitations is possible. The objective function F is changed to the new function Φ of the type

$$\Phi = F + \delta(x/\Omega) \tag{4.12}$$

$$\delta(x/\Omega) = 0 \text{ when } x \in \Omega, \ \delta(x/\Omega) \longrightarrow \infty \text{ when } x \notin \Omega$$

It is supposed that the original problem assumes the extending of the search area of the unknown point on the set Ω. The new objective function extends the set to the whole area, but the objective function value in the limited areas may indefinitely increase (when determining the maximum of the objective function, $\delta(x/\Omega)$ is chosen as negative). The constructive form of the function $\delta(x/\Omega)$ called indicatrix allows the change of the original objective function for the limited value, but a large one to ensure the convergence of solving the problem of unconstrained optimization to the same stationary point as in the original problem solution. There are methods of internal penalty functions and external penalty functions which differ by the method of indicatrix implementation.

In the method of internal penalty functions the constructive form of indicatrix is selected from the conditions:

- from the first iterations the penalty value set for overrunning the accessible computational range is large, and as you get closer to the

stationary point (with the increase of the numbers of iterations) the penalty is reduced;
- initial approximation is selected to satisfy all the limits. In subsequent iterations the next approximation of the unknown stationary point is also within the allowed area.

In the case of calculating the minimum of the objective function $F(x)$ with the limitations of the following form

$$\varphi_m(x) \geq 0, \; m = 1, M \tag{4.13}$$

the constructive form of the new objective function when using the interior penalty function method is as follows:

$$\Phi(x) = F(x) - r_k \sum_{m=1}^{M} \ln[\varphi_m(x)] \tag{4.14}$$

or

$$\Phi(x) = F(x) - r_k \sum_{m=1}^{M} \frac{1}{\varphi_m(x)} \tag{4.15}$$

In the Eqs. (4.14) and (4.15) r_k – penalty parameter whose value depends on the iteration number. In the method of internal penalty functions r_k – sequence of positive numbers monotonically decreasing to zero.

In the external penalty function, the indicatrix constructive form is selected from the conditions:

- penalty value set for overrunning the accessible computational range is initially minimal, but as you get closer to the stationary point (corresponding to the numbers of iterations) it significantly increases;
- initial and the following approximations do not necessarily belong to the area allowed by limitations set.

In case of calculating the minimum of the objective function $F(x)$ with the limits of the form (4.12), the constructive form of new objective function can be as

$$\Phi(x) = F(x) + r_k \sum_{m=1}^{M} \varphi_m(x) \tag{4.16}$$

$$\varphi_m(x) = \min(0, \varphi_m(x))$$

For the limits in the equalities form

$$\varphi_m(x) = 0, \ m = \overline{1, M} \tag{4.17}$$

the objective function can be constructed as follows:

$$\Phi(x) = F(x) + r_k \sum_{m=1}^{M} |\varphi_m(x)| \tag{4.18}$$

or

$$\Phi(x) = F(x) + r_k \sum_{m=1}^{M} (\varphi_m(x))^2 \tag{4.19}$$

In the Eqs. (4.16), (4.18), and (4.19) the penalty parameter r_k– is steadily increasing sequence of positive numbers.

4.3 MATHEMATICAL METHODS FOR SOLVING OPTIMIZATION PROBLEMS

The above information about the formulation of selection problems indicates that its solution is reduced to the solution of one or more algebraic equations (determined by the number of searching parameters) of linear or non-linear form (determined by the type of objective function). Chapter 2 discusses the solution of the systems of linear algebraic equations. Below we discuss the methods of solving the non-linear algebraic equations and systems of equations.

Non-linear systems of equations for the unknowns $x_1, x_2, x_3, \dots, x_N$ will be the systems of the following form

$$\varphi_m(x_1, x_2, x_3, \dots, x_N) = 0, \quad m = 1, \dots, M \tag{4.20}$$

In the special case, the values of N and M in the system of Eqs. (4.20) can coincide in the equation system φ_m – arbitrary algebraic or transcendental function of the variables $x_1, x_2, x_3, \dots, x_N$. Further, for simplification the unknown variables x_n will be presented in the vector form – $\mathbf{x} = (x_1, x_2, x_3, \dots, x_N)$.

In engineering, this class of problems takes a significant place not only in solving the selection problems. Such equations or equation systems can occur alone or as a part of more complex problems containing linear algebraic equations and (or) differential equations. The feature of this class of problems is the possibility of several independent solutions existence. Moreover, complex numbers can be the solution of the equations. The application of computational methods for solving the non-linear equations requires the preliminary analysis (definition of the nature of functions derivatives change in the left side of equations), intervals definition where the roots can be placed. The more carefully the preliminary analysis of the equation system will be carried out, the better solution obtained with the use of computers will be.

The objective function determined in solving the selection problem (e.g., the formulas (4.8), (4.9), (4.15), etc.) differs from the non-linear Eqs. (4.20). However, the differentiation of the objective function for each of search parameters lets us to obtain the equivalent system of linear or (and) non-linear equations.

Indeed, let us consider the optimization problem (seeking the functional $\Phi(\mathbf{x})$) minimum of the form

$$\Phi(\mathbf{x}) = \min \sum_{m=1}^{M} (F_m(x_1, x_2, x_3, \dots, x_N))^2 \qquad (4.21)$$

In the Eq. (4.21) $\mathbf{x}$ – the vector of search parameters – $\mathbf{x} = (x_1, x_2, x_3, \dots, x_N)$.

Conditions of extremum existence (minimum or maximum) of the function $\Phi(\mathbf{x})$ (Eq. (4.21)) can be written as the system of N equations (in equations $n = \overline{1, N}$):

$$\frac{\partial \Phi(x_1, x_2, \dots, x_n, \dots, x_N)}{\partial x_n} = 2 \cdot \sum_{m=1}^{M} \left(F_m(x_1, x_2, x_3, \dots, x_N) \frac{\partial F_m(x_1, x_2, x_3, \dots, x_N)}{\partial x_n} \right) = 0$$

or

$$\varphi_m(x_1, x_2, x_3, \dots, x_N) = \sum_{m=1}^{M} \left(F_m(x_1, x_2, x_3, \dots, x_N) \frac{\partial F_m(x_1, x_2, x_3, \dots, x_N)}{\partial x_n} \right) = 0 \qquad (4.22)$$

The system of N Eqs. (4.22) is the system of algebraic equations to the solution of which the standard computational methods for solving the systems of non-linear equations are acceptable. On the other hand, note that for solving optimization problems there are other solution methods, which can be used for solving the systems of non-linear equations of the form (4.20). To this effect, we should build the functional $\Phi(\mathbf{x})$

$$\Phi(\mathbf{x}) = \min \sum_{m=1}^{M} (\varphi_m (x_1, x_2, x_3, \ldots, x_N))^2 \tag{4.23}$$

reducing the problem of solving the system of non-linear equations to the problem of unconstrained optimization (minimization).

In addition, the important conclusion should be noted, reducing the problem of solving the system of non-linear Eq. (4.20) to the problem of unconstrained optimization (4.23) is an effective method to solve the problem where the number of equations does not necessarily coincide with the number of unknowns ($M \neq N$).

There are many methods for solving the equations and systems of non-linear equations. All these methods are of iteration type (the solution is calculated in several approximations). Particularly, these are successive iteration method, half-interval method and chord methods, Newton method and secant method, descent method and pseudoviscosity method, etc. Below we discuss some of these methods. For simplification in some cases, the presented methods are illustrated by the solution of relatively simple equations and their geometric interpretation.

4.3.1 METHOD OF SUCCESSIVE ITERATION

Method of successive iteration can be applied in solving the systems of non-linear equations in case when the number of equations and number of unknown variables coincide. In this method, the original equation system (4.20) is reduced to the form

$$x_n = G_i (x_1, x_2, x_3, \ldots, x_N), \quad n = 1, \ldots, N \tag{4.24}$$

Let us assume that the initial approximation for all the unknown variables in the form: $x_n = x_n^{(0)}$, $(n = 1, N)$. Then the first approximation can be determined by the formulas

$$x_n^{(1)} = G_n(x_1^{(0)}, x_2^{(0)}, x_3^{(0)}, \ldots, x_N^{(0)}), \quad n = 1, \ldots, N$$

All subsequent iterations are defined by similar formulas. Thus, for the iteration with the number t the values of the variables $x_n^{(t)}$ are determined by the values $x_n^{(t-1)}$ calculated at the previous iteration:

$$x_n^{(t)} = G_n(x_1^{(t-1)}, x_2^{(t-1)}, x_3^{(t-1)}, \ldots, x_N^{(t-1)}), \quad n = 1, \ldots, N$$

Iteration process, described above, is not always reduced to solve the problem (4.20). We can prove a theorem that the iteration process coincides if finding a solution is in vicinity, where $\left|\frac{\partial G_n(\mathbf{x})}{\partial x_n}\right| < 1 \ \forall n$. Let us introduce the function $\rho(\mathbf{f}_1, \mathbf{f}_2) = \sqrt{\sum_n (f_{1n} - f_{2n})^2}$ which is the distance between the vectors $\mathbf{f}_1$ and $\mathbf{f}_2$ (in particular case, the arguments $\mathbf{f}_1$ and $\mathbf{f}_2$ can be scalars). This function is useful in the analyzing the convergence of the successive iteration method. Let us consider two functions of distance. In the first one, as the arguments $\mathbf{f}_1$ and $\mathbf{f}_2$ we consider the values of the functions $G_n(\mathbf{x}^{(t)}), G_n(\mathbf{x}^{(t-1)})$ calculated, respectively, at iterations with the numbers t and *t-1*. In the second function, the distance is determined between the values of the variables $\mathbf{x}^{(t)}, \mathbf{x}^{(t-1)}$ calculated at iterations with the numbers t and *(t-1)*. It can be argued that if in the iteration process the condition $\rho(G_n(\mathbf{x}^{(t)}), G_n(\mathbf{x}^{(t-1)})) \le q \cdot \rho(\mathbf{x}^{(t)}, \mathbf{x}^{(t-1)})$ is satisfied for any n and t and with the values $q < 1$ (this condition is called a contracting mapping), then the iteration process converges to the solution of the original problem (4.20). In practice, the simplified condition of iteration process of the Eq. (4.20) solution is often used. The solution is considered as found if after the iteration with the number t the condition $|\mathbf{x}^{(t)} - \mathbf{x}^{(t-1)}| \le \varepsilon$ is satisfied. ε is the given accuracy of the unknowns' $\mathbf{x} = (x_1, x_2, x_3, \ldots, x_N)$determination.

The foregoing demonstrates that the rate of iteration process convergence significantly depends on the type of the selected functions $G_n(x_1, x_2, x_3, \ldots, x_N)$, $n = 1, \ldots, N$. As an example, let us consider possible solutions of the quadratic equation $x^2 + x - 2 = 0$ by the iteration. In the Eq. (4.20) designations the written quadratic equation can be re-written as

$$F(x) = x^2 + x - 2 = 0 \tag{4.25}$$

The Eq. (4.25) has two exact solutions – $x_1 = -2$; $x_2 = 1$. The iteration process in solving this equation can be constructed by selecting the function $G(x)$, for example, in the following ways:

$G_{(1)}(x) = x^2 + 2x - 2$ (corresponds to solving the equation $x = x^2 + 2x - 2$)

$G_{(2)}(x) = \pm\sqrt{2-x}$ (corresponds to solving the equation $x = \pm\sqrt{2-x}$)

$G_{(3)}(x) = \frac{2-x}{x}$ (corresponds to solving the equation $x = \frac{2-x}{x}$)

The functions $G(x)$ correspond to the derivatives $G'(x)$:

$$G'_{(1)}(x) = 2x + 2$$

$$G'_{(2)}(x) = \mp\frac{1}{2\cdot\sqrt{2-x}}$$

$$G'_{(3)}(x) = -\frac{2}{x^2}$$

For certainty the first solution will be sought in the interval $x \in (-100, -1)$. Derivatives values on the interval boundaries of search $G'(x)$ give the following results – $G'_{(1)}(-100) = -198$, $G'_{(1)}(-1) = 0$; $G'_{(2)}(-100) \approx -0.0495$, $G'_{(2)}(-1) \approx -0.289$; $G'_{(3)}(-100) \approx -0.0002$, $G'_{(3)}(-1) = -2$. The calculations show that the condition $\left|\frac{\partial G(x)}{\partial x}\right| < 1$ satisfies the original equation re-written as

$$x = -\sqrt{2-x} \tag{4.26}$$

Now, if during the iterations the values of the desired unknown do not go beyond the interval $x \in (-100, -1)$, then the iteration process will converge to the desired solution. Let us perform several iteration of the Eq. (4.26). As the initial approximation we will take $x^{(0)} = -100$. Sequentially we have:

$$x^{(1)} = -\sqrt{2 - x^{(0)}} = -\sqrt{2-(-100)} = -\sqrt{102} = -10{,}0995;$$

$$x^{(2)} = -\sqrt{2 - x^{(1)}} = -\sqrt{2-(-10{,}0995)} = -\sqrt{12{,}0995} = -3{,}4784;$$

$$x^{(3)} = -\sqrt{2 - x^{(2)}} = -\sqrt{2 - (-3{,}4784)} = -\sqrt{5{,}4784} = -2{,}3406;$$

$$x^{(4)} = -\sqrt{2 - x^{(3)}} = -\sqrt{2 - (-2{,}3406)} = -\sqrt{4{,}3406} = -2{,}0834;$$

$$x^{(5)} = -\sqrt{2 - x^{(4)}} = -\sqrt{2 - (-2{,}0834)} = -\sqrt{4{,}0834} = -2{,}0207;$$

$$x^{(6)} = -\sqrt{2 - x^{(5)}} = -\sqrt{2 - (-2{,}0207)} = -\sqrt{4{,}0207} = -2{,}0052;$$

$$x^{(7)} = -\sqrt{2 - x^{(6)}} = -\sqrt{2 - (-2{,}0052)} = -\sqrt{4{,}0052} = -2{,}0013;$$

$$x^{(8)} = -\sqrt{2 - x^{(7)}} = -\sqrt{2 - (-2{,}0013)} = -\sqrt{4{,}0013} = -2{,}0003$$

The calculations show that after the eighth iteration the solution of the Eq. (4.26), and, therefore, the Eq. (4.25), is achieved with the accuracy of 10^{-4}. Furthermore, it should be noted that rounding errors when performing the iterations almost have no influence on the final result. Similarly, you can make sure that to define the positive root ($x=1$) the Eqs. (4.25) should be solved in the form $x = \sqrt{2 - x}$, and as the initial approximations we should take, for example, $x^{(0)} = 100$.

Figure 4.1 shows the possible diagrams of the iteration process convergence (diagrams *a* and *b*), and the possible instability of the computational process (diagrams *c* and *d*). The example, discussed above, corresponds to the convergence shown in Figure 4.1*a*.

The above method for solving the Eq. (4.25) in the form (4.26) is not unique. There are other methods [23]. Particularly, in some cases, the selection of such functions $H_n(y_1, y_2, y_3, \ldots, y_N)$, that $H_n(y_1, y_2, y_3, \ldots, y_N) = 0$ only in the case, when $y_1, y_2, y_3, \ldots, y_N = 0$, is recommended. If the iteration process of solving the non-linear equations converges, then in the limit of $t \to \infty$ the value is $y_n = x_n^{(t)} - x_n^{(t-1)} \to 0$. In this case, the iteration process can be constructed by the formula

$$H_n(x_1^{(t)} - x_1^{(t-1)}, x_2^{(t)} - x_2^{(t-1)}, \ldots, x_N^{(t)} - x_N^{(t-1)}) = F(x_1^{(t-1)}, x_2^{(t-1)}, \ldots, x_N^{(t-1)}) \quad (4.27)$$

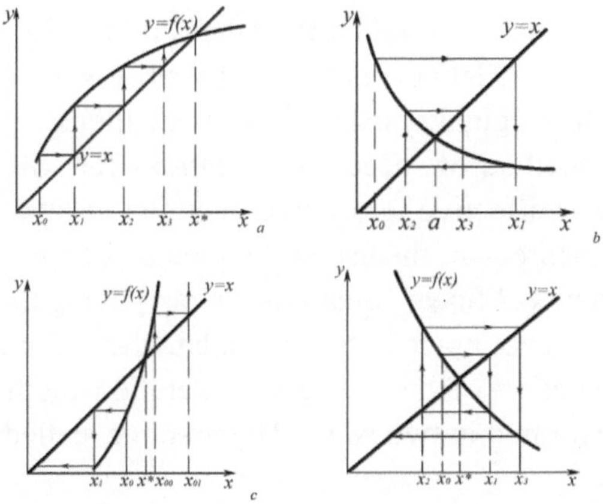

FIGURE 4.1 Development of the iteration process (*a* – iteration process converges to the solution from the right edge of the interval; *b* – iteration process converges to the solution even on both boundaries of the interval; *c* – iteration process diverges toward the right edge of the interval of search interval; *d* – iteration process diverges beyond the search interval).

Another variant of the iteration process construction is to write the function $G_n(x_1, x_2, x_3, \ldots, x_N)$ as

$$G_n(x_1, x_2, x_3, \ldots, x_N) = x_n + \tau(x_1, x_2, x_3, \ldots, x_N) F(x_1, x_2, x_3, \ldots, x_N) \quad (4.28)$$

The coefficient $\tau(x_1, x_2, x_3, \ldots, x_N)$ is given at each iteration step from the condition $\left|\frac{\partial G_n(x)}{\partial x_n}\right| < 1$. Particularly, $\tau(x_1, x_2, x_3, \ldots, x_N) = const$ can be accepted.

It should be noted that in Eqs. (4.26) and (4.27) the functions $G_n(x_1, x_2, x_3, \ldots, x_N)$ can be linear. This allows us to use the method of successive iteration for the solutions of the systems of linear equations. In this case, the system of linear equations should be given in the form (4.21), and the condition of the iteration process stability will be $\left|\frac{\partial G_n(x)}{\partial x_n}\right| < 1$. For the systems of linear equations given in the form (4.21), it means that in the right side of the equation for the unknown x_n

$$x_n = g_{i1}x_1 + g_{i2}x_2 + \ldots + g_{ii}x_i + \ldots + g_{iN}x_N, \qquad n = 1, N$$

values of the coefficients g_{in} should satisfy the condition $|g_{in}| < 1$. Obviously, the implementation of the latter condition is possible by completing the simple conversion of the original system of linear equations.

Note that the method of successive iterations for solving the systems of linear equations is effective for a large numbers of unknowns (from 10^4 to 10^7 unknowns). Moreover, the use of the method of successive iterations for solving the systems of linear equations (as for solving the system of nonlinear equations) does not practically accumulate the computational errors.

The implementation of the successive iteration method for equation systems can be performed in two ways. The first one, called Jacobi method, is written down as

$$x_n^{(t)} = G_n(x_1^{(t-1)}, x_2^{(t-1)}, ..., x_{n-1}^{(t-1)}, x_n^{(t-1)}, x_{n+1}^{(t-1)}, ..., x_N^{(t-1)}), \quad n = 1, ..., N \quad (4.29)$$

The second one, called Seidel method, is written down as

$$x_n^{(t)} = G_n(x_1^{(t)}, x_2^{(t)}, ..., x_{n-1}^{(t)}, x_n^{(t-1)}, x_{n+1}^{(t-1)}, ..., x_N^{(t-1)}), \quad n = 1, ..., N \quad (4.30)$$

The difference between Seidel and Jacobi methods is that the calculation of the next approximation for $x_n^{(t)}$ takes into account the values $x_j^{(t)}$ obtained on the tth iteration calculated for *j=1, n–1*. The convergence of Seidel method is higher than Jacobi method. The conditions of the computational process stability in both cases are the same – $\left|\frac{\partial G_n(x)}{\partial x_n}\right| < 1$.

Let us illustrate the foregoing by the example of solving the system of three linear equations

$$10x_1 + x_2 + x_3 = 12$$

$$2x_1 + 10x_2 + x_3 = 13$$

$$2x_1 + 2x_2 + 10x_3 = 14$$

To use the successive iteration method, let us convert the original equation system of s as follows:

$$10x_1 = 12 - x_2 - x_3;$$

$$10x_2 = 13 - 2x_1 - x_3;$$

$$10x_3 = 14 - 2x_1 - 2x_2;$$

And further

$$x_1 = 1,2 - 0,1x_2 - 0,1x_3;$$

$$x_2 = 1,3 - 0,2x_1 - 0,1x_3;$$

$$x_3 = 1,4 - 0,2x_2 - 0,2x_3$$

As the initial approximation we use free members of the latter equation system $- x_1^{(0)} = 1,2;\ x_2^{(0)} = 1,3;\ x_3^0 = 1,4$.

The calculation results at the first and subsequent iterations when implementing the Jacobi method are written as follows:

$$\begin{aligned} x_1^{(1)} &= 1,2 - 0,1x_2^{(0)} - 0,1x_3^{(0)} = 1,2 - 0,1\cdot 1,3 - 0,1\cdot 1,4 \approx 0,93 \\ x_2^{(1)} &= 1,3 - 0,2x_1^{(0)} - 0,1x_3^{(0)} = 1,3 - 0,2\cdot 1,2 - 0,1\cdot 1,4 \approx 0,92 \\ x_3^{(1)} &= 1,4 - 0,2x_1^{(0)} - 0,2x_2^{(0)} = 1,4 - 0,2\cdot 1,2 - 0,2\cdot 1,3 \approx 0,90 \end{aligned};$$

$$\begin{aligned} x_1^{(2)} &= 1,2 - 0,1x_2^{(1)} - 0,1x_3^{(1)} = 1,2 - 0,1\cdot 0,92 - 0,1\cdot 0,90 \approx 1,018 \\ x_2^{(2)} &= 1,3 - 0,2x_1^{(1)} - 0,1x_3^{(1)} = 1,3 - 0,2\cdot 0,93 - 0,1\cdot 0,90 \approx 1,024 \\ x_3^{(2)} &= 1,4 - 0,2x_1^{(1)} - 0,2x_2^{(1)} = 1,4 - 0,2\cdot 0,93 - 0,2\cdot 0,92 \approx 1,030 \end{aligned};$$

$$\begin{aligned} x_1^{(3)} &= 1,2 - 0,1x_2^{(2)} - 0,1x_3^{(2)} = 1,2 - 0,1\cdot 1,024 - 0,1\cdot 1,030 \approx 0,9946 \\ x_2^{(3)} &= 1,3 - 0,2x_1^{(2)} - 0,1x_3^{(2)} = 1,3 - 0,2\cdot 1,018 - 0,1\cdot 1,030 \approx 0,9934 \\ x_3^{(3)} &= 1,4 - 0,2x_1^{(2)} - 0,2x_2^{(2)} = 1,4 - 0,2\cdot 1,018 - 0,2\cdot 1,024 \approx 0,9916 \end{aligned};$$

$$\begin{aligned} x_1^{(4)} &= 1,2 - 0,1x_2^{(3)} - 0,1x_3^{(3)} = 1,2 - 0,1\cdot 0,9934 - 0,1\cdot 0,9916 \approx 1,00150 \\ x_2^{(4)} &= 1,3 - 0,2x_1^{(3)} - 0,1x_3^{(3)} = 1,3 - 0,2\cdot 0,9946 - 0,1\cdot 0,9916 \approx 1,00192 \\ x_3^{(4)} &= 1,4 - 0,2x_1^{(3)} - 0,2x_2^{(3)} = 1,4 - 0,2\cdot 0,9946 - 0,2\cdot 0,9934 \approx 1,00240 \end{aligned}$$

After the fourth iteration the error of solving the original system of equations does not exceed 0.24%.

The calculation results at the first and subsequent iterations when implementing Seidel method with the same initial approximations $x_1^{(0)}, x_2^{(0)}, x_3^{(0)}$ will take the form:

$$\begin{aligned} x_1^{(1)} &= 1{,}2 - 0{,}1x_2^{(0)} - 0{,}1x_3^{(0)} = 1{,}2 - 0{,}1 \cdot 1{,}3 - 0{,}1 \cdot 1{,}4 \approx 0{,}93 \\ x_2^{(1)} &= 1{,}3 - 0{,}2x_1^{(1)} - 0{,}1x_3^{(0)} = 1{,}3 - 0{,}2 \cdot 0{,}93 - 0{,}1 \cdot 1{,}4 \approx 0{,}974 \qquad ; \\ x_3^{(1)} &= 1{,}4 - 0{,}2x_1^{(1)} - 0{,}2x_2^{(1)} = 1{,}4 - 0{,}2 \cdot 0{,}93 - 0{,}2 \cdot 0{,}974 \approx 1{,}0192 \end{aligned}$$

$$\begin{aligned} x_1^{(2)} &= 1{,}2 - 0{,}1x_2^{(1)} - 0{,}1x_3^{(1)} = 1{,}2 - 0{,}1 \cdot 0{,}974 - 0{,}1 \cdot 1{,}0192 \approx 1{,}00068 \\ x_2^{(2)} &= 1{,}3 - 0{,}2x_1^{(2)} - 0{,}1x_3^{(1)} = 1{,}3 - 0{,}2 \cdot 1{,}00068 - 0{,}1 \cdot 1{,}0192 \approx 0{,}99794 \\ x_3^{(2)} &= 1{,}4 - 0{,}2x_1^{(2)} - 0{,}2x_2^{(2)} = 1{,}4 - 0{,}2 \cdot 1{,}00068 - 0{,}2 \cdot 0{,}99794 \approx 1{,}00028 \end{aligned}$$

The calculations show that Seidel method provides the same accuracy for two iterations as Jacobi method provides for four iterations.

4.3.2 HALF-INTERVAL METHOD AND CHORD METHOD

To solve the non-linear equations the reliable method of root solving, called the half-interval method, can be used. Sometimes this method is called the dichotomy method. Let us consider the scalar equation $F(x) = 0$ whose unique solution is in the interval $x \in (x_-^{(0)}, x_+^{(0)})$ $(x_-^{(0)} < x < x_+^{(0)})$. In this case, the inequality $F(x_-^{(0)}) \cdot F(x_+^{(0)}) < 0$ (while crossing the solution function $F(x)$ changes the sign) is correct. We select the point x in the interval $(x_-^{(0)}, x_+^{(0)})$. This point can be selected by dividing the interval $(x_-^{(0)}, x_+^{(0)})$ into two – $x^{(1)} = \dfrac{x_-^{(0)} + x_+^{(0)}}{2}$. If $F(x^{(1)}) \cdot F(x_+^{(0)}) < 0$, then $x_-^{(1)} = x^{(1)}$; $x_+^{(1)} = x_+^{(0)}$. Otherwise – $x_-^{(1)} = x_-^{(0)}$; $x_+^{(1)} = x^{(1)}$.

For any iteration with the number t the following algorithm is correct:

- if after the iteration with the number *(t–1)* the equation solution $F(x) = 0$ is in the interval $x_-^{(t-1)} < x < x_+^{(t-1)}$, then the next approximation is found by the formula $x^{(t)} = \dfrac{x_-^{(t-1)} + x_+^{(t-1)}}{2}$;
- if the condition $F(x^{(t)}) \cdot F(x_+^{(t-1)}) < 0$ is performed, then the interval boundaries where the solution of the original equations is, are taken

as follows – $x_-^{(t)} = x^{(t)}$; $x_+^{(t)} = x_+^{(t-1)}$. Otherwise – $x_-^{(t)} = x_-^{(t-1)}$; $x_+^{(t)} = x^{(t)}$;

- computational procedure continues until the interval $(x_-^{(t-1)},\ x_+^{(t-1)})$ is reduced to the value comparable with the accuracy ε defined in the problem – $\left|x_-^{(t-1)} - x_+^{(t-1)}\right| < \varepsilon$.

The graphical interpretation of the half-interval method is shown in Figure 4.2. The algorithm of the method is the same as when $F(x\_) < 0$, and when $F(x\_) > 0$. At each step of the computational process in the half-interval method, the initial interval (x_-, x_+), where the solution of the equation $F(x) = 0$ is sought, is halved. With ten approximations the initial interval of search (x_-, x_+) is reduced about a thousand times ($2^{10} = 1024 \approx 10^3$).

Let us consider the application of the half-interval method for solving the equation $F(x) = x^2 + x - 2 = 0$. As in the above-discussed examples, let us limit with the root search in the interval $x \in (-100, -1)$. Let us denote $a = -100$, $b = -1$. For the interval (a, b) we have – $F(a) = -100^2 + 100 - 2 = 10098 > 0$, $F(b) = -1^2 + 1 - 2 = -2 < 0$. The calculation results of the equation root $x = -2$ are presented in Table 4.1.

After ten iterations it is determined that the values of equation root are placed in the interval $x \in (-2.0635, -1.9668)$. As the solution, we can take the middle of the denoted interval – $x \approx -2.0151$, at the same time, the error does not exceed 1.5%.

To accelerate the convergence process of the presented method, the next approximation $x^{(t)}$ should be calculated by the ratio

$$x^{(t)} = x_-^{(t-1)} - (x_+^{(t-1)} - x_-^{(t-1)}) \cdot \frac{F(x_-^{(t-1)})}{F(x_+^{(t-1)}) - F(x_-^{(t-1)})} \tag{4.31}$$

This modification is called the chord method. The geometric interpretation of the chord method is shown in Figure 4.3. The formula (4.35) comes out of the condition that the curve $F(x) = 0$ in the interval (x_-, x_+) is replaced

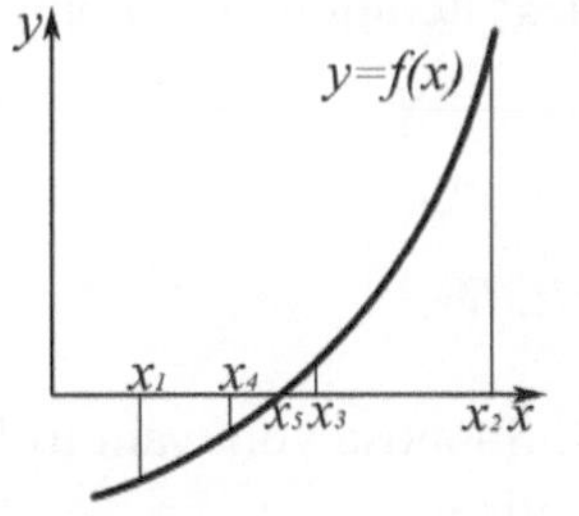

FIGURE 4.2 Geometric interpretation of the half-interval method.

TABLE 4.1 Calculation of the Root of the Equation $x^2 + x - 2 = 0$ by the Half-Interval Method

No	x	F(x)	x_-	x_+	$F(x_-)$	$F(x_+)$
0	–100	10098	–100	–1	10098	–2
1	–50.5000	2498.0	–50.5000	–1	2498	–2
2	–25.7500	635.3	–25.7500	–1	635.3	–2
3	–13.3750	163.5	–13.375	–1	163.5	–2
4	–7.1875	42.5	–7.1875	–1	42.5	–2
5	–4.0938	10.70	–4.0938	–1	10.70	–2
6	–2.5469	1.94	–2.5469	–1	1.94	–2
7	–1.7734	–0.63	–2.5469	–1.7734	1.94	–0.63
8	–2.1602	0.51	–2.1602	–1.7734	0.51	–0.63
9	–1.9668	–0.098	–2.1602	–1.9668	0.51	–0.098
10	–2.0635	0.195	–2.0635	–1.9668	0.195	–0.098

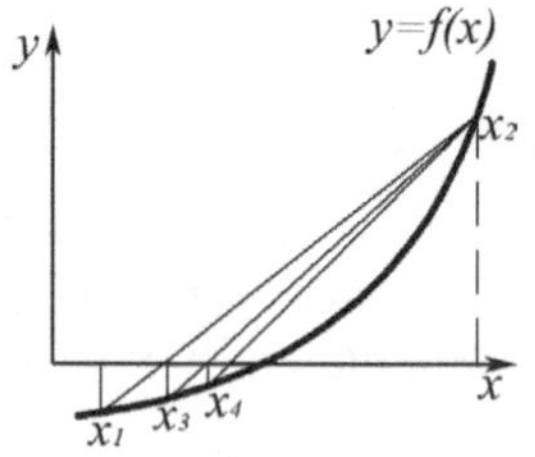

FIGURE 4.3 Geometric interpretation of the chord method.

by the line segment connecting the points with the coordinates $(x_-, F(x_-))$ and $(x_+, F(x_+))$.

The half-interval and chord methods can be used for solving the systems of non-linear equations. In this case, for reliability of the computational process convergence for the next iteration one should specify the value of only one unknown $x_n, n = \overline{1, N}$.

4.3.3 *NEWTON METHOD*

Let us consider Figure 4.4. Suppose you want to find the intersection point of the curve *F(x)* with the axis *x*.

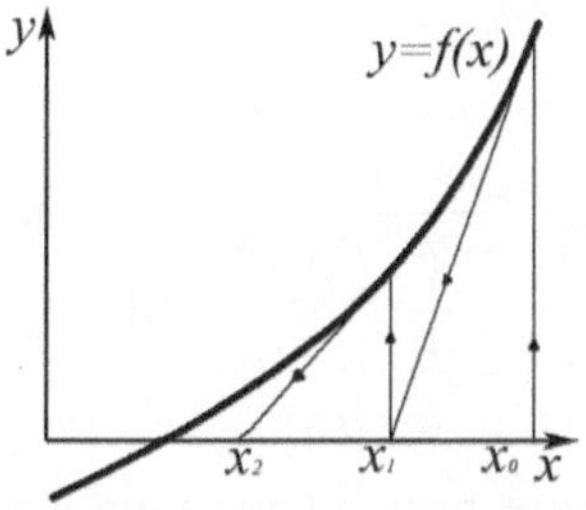

FIGURE 4.4 Geometric interpretation of Newton method.

Let us take the value $x = x_0$ as the initial approximation. Draw the tangent to the curve *F(x)* at the point $x = x_0$. Find the intersection point of the tangent and axis x ($x = x_1$). After that again draw the tangent to the curve *F(x)* at the point $x = x_1$ and find the point of intersection with the axis x.

Continuing this process, the intersection point of the curve *F(x)* with the axis x (otherwise – to provide the solution of $F(x) = 0$) can be found with any desired accuracy. This is the graphical interpretation of the powerful method for solving non-linear equations (and equation systems) called Newton method.

Newton method is based on the possibility to use the following differential ratio with small values of Δx_n, $(n = 1, N)$

$$F_n(x_1, x_2, ..., x_n + \Delta x_n, ..., x_N) \approx F_n(x_1, x_2, ..., x_N) + \frac{\partial F_n(x_1, x_2, ..., x_N)}{\partial x_n} \Delta x_n \approx 0$$

If we denote the values $x_n^{(t-1)} \equiv x_n$ as the approximation obtained at the iteration (approximation) with the number (t-1), then we can assume that $x_n^{(t)} \equiv x_n^{(t-1)} + \Delta x_n$. This allows us to use the above equation for the arrangement of successive approximations as follows:

$$F_n(x_1^{(t-1)}, x_2^{(t-1)}, ..., x_n^{(t)}, ..., x_N^{(t-1)}) + \frac{\partial F_n(x_1^{(t-1)}, x_2^{(t-1)}, ..., x_n^{(t-1)}, ..., x_N^{(t-1)})}{\partial x_n}(x_n^{(t)} - x_n^{(t-1)}) \approx 0$$

or

$$x_n^{(t)} = x_n^{(t-1)} - \frac{F_n(x_1^{(t-1)}, x_2^{(t-1)}, ..., x_n^{(t-1)}, ... x_N^{(t-1)})}{\left(\dfrac{\partial F_n(x_1^{(t-1)}, x_2^{(t-1)}, ..., x_n^{(t-1)}, ... x_N^{(t-1)})}{\partial x_n}\right)}$$

or as the vector

$$x_n^{(t)} = x_n^{(t-1)} - \frac{F_n(\mathbf{x}^{(t-1)})}{\left(\dfrac{\partial F_n(\mathbf{x}^{(t-1)})}{\partial x_n}\right)}, \quad n = 1, N \tag{4.32}$$

If there is a solution of one non-linear equation with one unknown, the formula (4.32) can be re-written as

$$x^{(t)} = x^{(t-1)} - \frac{F(x^{(t-1)})}{F'(x^{(t-1)})}$$

The latter formulas show that in comparison with the method of successive iteration in Newton method at each iteration step not only the values of function, but also the values of its derivatives should be calculated. If it is difficult to present the derivatives $\frac{\partial F_n(\mathbf{x}^{(t-1)})}{\partial x_n}$ in explicit (analytical) form, the application of Newton method requires using any computational procedures to estimate the value of these derivatives. In the compact form the values of the derivatives $\frac{\partial F_n(\mathbf{x}^{(t-1)})}{\partial x_n}$ can be presented as matrix **J** called Jacobian matrix:

$$\mathbf{J} = \left\{ \begin{array}{l} \dfrac{\partial F_1(\mathbf{x})}{\partial x_1} \; \dfrac{\partial F_1(\mathbf{x})}{\partial x_2} \cdots \dfrac{\partial F_1(\mathbf{x})}{\partial x_N} \\ \dfrac{\partial F_2(\mathbf{x})}{\partial x_1} \; \dfrac{\partial F_2(\mathbf{x})}{\partial x_2} \cdots \dfrac{\partial F_2(\mathbf{x})}{\partial x_N} \\ \cdots\cdots\cdots\cdots\cdots\cdots\cdots\cdots \\ \dfrac{\partial F_N(\mathbf{x})}{\partial x_1} \; \dfrac{\partial F_N(\mathbf{x})}{\partial x_2} \cdots \dfrac{\partial F_N(\mathbf{x})}{\partial x_N} \end{array} \right\} \tag{4.33}$$

The advantage of Newton method is its quick convergence to the solution, but only in case of selecting the successful initial approximation for the calculated unknowns. In addition, the method convergence depends essentially on the values of Jacobian norms $\|\mathbf{J}\|$ and their inversion $\|\mathbf{J}^{-1}\|$. The values of the second derivatives $\frac{\partial^2 F_n(\mathbf{x})}{\partial x_n^2}$ calculated at the stationary point, impact on the method convergence. The strict evaluation of Newton method convergence can be found in [21].

Using Newton method in the optimization problem for the next approximation calculation, the Eq. (4.29) can be written as

$$x^{(t+1)} = x^{(t)} - \alpha_n \cdot \frac{F(x^{(t-1)})}{F'(x^{(t-1)})} = x^{(t)} - \alpha_n \cdot \frac{\Phi'(x^{(t-1)})}{\Phi''(x^{(t-1)})}$$

From the latter equation it means that using Newton method for solving the optimization problem at each iteration step one should calculate the values of the first and second derivatives of the objective function, and to analyze the convergence of the iteration in this case one should know the third derivative of the objective function.

The issue of Newton method convergence is not the only problem. Another problem of the method is the need to calculate Jacobian **J** and its inverse form $\mathbf{J}^{-1}$ at each step. There are modifications of Newton method consisting in the fact that the value $\mathbf{J}^{-1}$ is calculated only once before the first approximation. The modifications where the value $\mathbf{J}^{-1}$ is periodically re-calculated after several iterations are possible. The interesting modification called the secant method is used for solving scalar non-linear equations (one equation with one unknown [21]). In the secant method the partial derivative $\frac{\partial F_n(\mathbf{x}^{(t-1)})}{\partial x_n}$ is defined by the following approximate relation

$$\frac{\partial F_n(\mathbf{x}^{(t-1)})}{\partial x_n} \approx \frac{F_n(\mathbf{x}^{(t-1)}) - F_n(\mathbf{x}^{(t-2)})}{\mathbf{x}^{(t-1)} - \mathbf{x}^{(t-2)}}$$

In this case, the calculation of the next approximation in the secant method is performed according to the formula

$$\mathbf{x}^{(t)} = \mathbf{x}^{(t-1)} - \frac{F(\mathbf{x}^{(t-1)}) \cdot (\mathbf{x}^{(t-1)} - \mathbf{x}^{(t-2)})}{F(\mathbf{x}^{(t-1)}) - F(\mathbf{x}^{(t-2)})} \tag{4.34}$$

From (4.34) it follows that to calculate the next approximation for the unknown value $x^{(t)}$ it is necessary to know the values of x and $F(x)$ at iterations with the numbers *(t-1)* and *(t-2)*.

As an example, let us consider Newton method implementation for solving the equation $F(x) = x^2 + x - 2 = 0$. For this equation $F'(x) = 2x + 1$ and formula for iteration can be written as

$$x^{(t)} = x^{(t-1)} - \frac{(x^{(t-1)})^2 + x^{(t-1)} - 2}{2(x^{(t-1)} + 1)}$$

Let us take $x^{(0)} = -100$ as the initial approximation. Then the iteration process will continue as follows

$$x^{(1)} = x^{(0)} - \frac{(x^{(0)})^2 + x^{(0)} - 2}{2x^{(0)} + 1} = -100 - \frac{100^2 - 100 - 2}{-2 \cdot 100 + 1} = -49.7588$$

$$x^{(2)} = x^{(1)} - \frac{(x^{(1)})^2 + x^{(1)} - 2}{2x^{(1)} + 1} = -49.7588 - \frac{49.7588^2 - 49.7588 - 2}{-2 \cdot 49.7588 + 1} = -24.7266$$

$$x^{(3)} = x^{(2)} - \frac{(x^{(2)})^2 + x^{(2)} - 2}{2x^{(2)} + 1} = -24.7266 - \frac{24.7266^2 - 24.7266 - 2}{-2 \cdot 24.7266 + 1} = -12.6822$$

$$x^{(4)} = x^{(3)} - \frac{(x^{(3)})^2 + x^{(3)} - 2}{2x^{(3)} + 1} = -12.6822 - \frac{12.6822^2 - 12.6822 - 2}{-2 \cdot 12.6822 + 1} = -6.6898$$

$$x^{(5)} = x^{(4)} - \frac{(x^{(4)})^2 + x^{(4)} - 2}{2x^{(4)} + 1} = -6.6898 - \frac{6.6898^2 - 6.6898 - 2}{-2 \cdot 6.6898 + 1} = -3.7759$$

$$x^{(6)} = x^{(5)} - \frac{(x^{(5)})^2 + x^{(5)} - 2}{2x^{(5)} + 1} = -3.7759 - \frac{3.7759^2 - 3.7759 - 2}{-2 \cdot 3.7759 + 1} = -2.4816$$

$$x^{(7)} = x^{(6)} - \frac{(x^{(6)})^2 + x^{(6)} - 2}{2x^{(6)} + 1} = -2.4816 - \frac{2.4816^2 - 2.4816 - 2}{-2 \cdot 2.4816 + 1} = -2.0586$$

$$x^{(8)} = x^{(7)} - \frac{(x^{(7)})^2 + x^{(7)} - 2}{2x^{(7)} + 1} = -2.0586 - \frac{2.0586^2 - 2.0586 - 2}{-2 \cdot 2.0586 + 1} = -2.0011$$

$$x^{(9)} = x^{(8)} - \frac{(x^{(8)})^2 + x^{(8)} - 2}{2x^{(8)} + 1} = -2.0011 - \frac{2.0011^2 - 2.0011 - 2}{-2 \cdot 2.0011 + 1} = -2.0000$$

Nine iterations are enough to solve the equation with four digits after the decimal point. Similarly, one can make sure that it is possible to use Newton method for finding the positive root of the original equation $F(x) = x^2 + x - 2 = 0$.

4.3.4 GRADIENT METHODS

Gradient methods are used mainly to solve the optimization problems. The next approximation when searching the stationary point is performed by the ratio

$$x_n^{(t)} = x_n^{(t-1)} - \alpha_t \cdot \frac{\partial \Phi(x^{(t-1)})}{\partial x_n}, \quad n = 1, N, \quad \alpha_t > 0 \tag{4.35}$$

Here n – number of current point coordinates, N – number of coordinates, t – iteration number, α_t – coefficient whose values can be variable from iteration to iteration.

The gradient methods are the methods of the first-order accuracy. The method modifications are in selecting the values of the coefficient α_t:

a) method with the step fragmentation – the value α_t is selected to ensure the inequality

$$\Phi(x_n^{(t)}) = \Phi(x_n^{(t-1)} - \alpha_t \cdot \frac{\partial \Phi(x_n^{(t-1)})}{\partial x_n}) < \Phi(x_n^{(t-1)}) \leq \varepsilon \cdot \alpha_t \cdot \left\| \Phi'(x^{(t-1)}) \right\| \tag{4.36}$$

Here ε characterizes the accuracy of the problem solution ($0 < \varepsilon < 1$); The geometric interpretation of the gradient descent with the step fragmentation is shown in Figure 4.5. It shows the lines of the level of the function $\Phi(x)$ having the minimum in the point x^*, and $C_1 > C_2 > C_3$..., and some zigzag trajectory $x_0 x_1 ... x_k$, orthogonal at each point $x_0, x_1, ..., x_k$ corresponding to the lines of the levels and resulting from the initial point x_0 to the minimum point x^*. The broken line

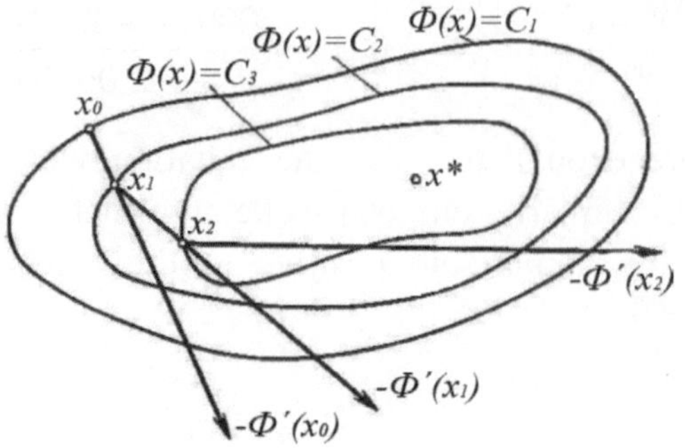

FIGURE 4.5 Geometric interpretation of the gradient descent with the step fragmentation.

$x_0x_1...x_k$ approximates the so-called gradient curve which satisfies the differential equation

$$\frac{dx}{d\alpha} = -\Phi'(x_k)$$

b) steepest descent method – the values α_t are selected from the condition $\frac{\partial \varphi_n(\alpha_t)}{\partial \alpha_t} = 0$, where the function $\varphi_n(\alpha_t)$ is defined by the relation

$$\varphi_n(\alpha_t) = \Phi(x_n^{(t-1)} - \alpha_t \cdot \Phi'(x_n^{(t-1)}); \tag{4.37}$$

In this method, unlike an ordinary gradient descent, the direction from the point x_k touches the line level in the point x_{k+1}. The sequence of points $x_0, x_1, x_2, ..., x_k, ...$ in zigzag fashion approaches the minimum point x^*, and the zigzag sections are orthogonal to each other (Figure 4.6);

c) coordinate descent method differs from the methods with the step fragmentation and steepest descent method in the implementation of the next iteration along one of the coordinates (Figure 4.7). In subsequent iterations the descent is performed by other coordinates. In this method, different modifications can be applied.

4.3.5 METHODS OF CONJUGATE GRADIENTS (DIRECTIONS)

Two vectors $\mathbf{x}, \mathbf{y}$ are called H – conjugate if their scalar product is zero – $(\mathbf{x} \cdot H\mathbf{y}) = 0$.

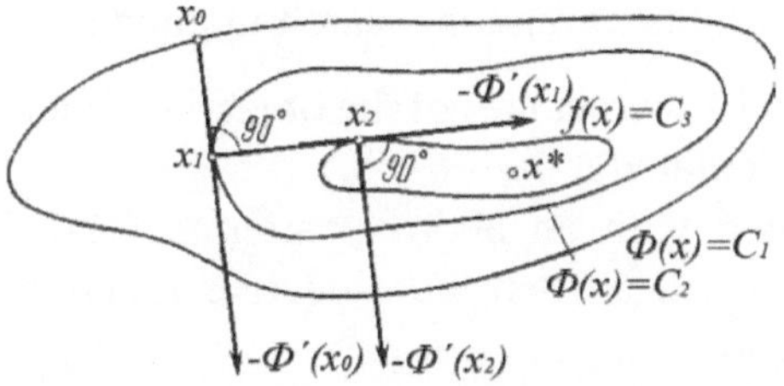

FIGURE 4.6 Geometric interpretation of the steepest descent method.

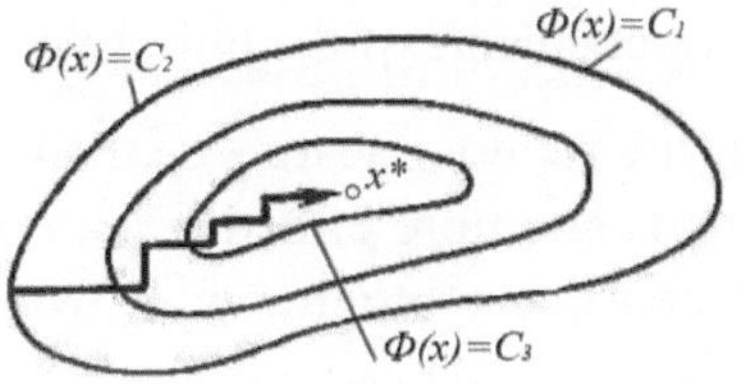

FIGURE 4.7 Geometric interpretation of the coordinate descent method.

Let us select the system $\mathbf{S}_1, \mathbf{S}_2, ..., \mathbf{S}_t -$ H-conjugate vectors ($(\mathbf{S}_i \cdot H\mathbf{S}_j) = 0,\ i \neq j$). Then the next approximation for the stationary point can be defined by the ratio

$$\mathbf{x}^{(t)} = \mathbf{x}^{(t-1)} + \alpha_t \cdot \mathbf{S}_t \tag{4.38}$$

Here $\alpha_t = -\dfrac{(\Phi'(x^{(t-1)}) \cdot \mathbf{S}_t)}{(\mathbf{S}_t \cdot H\mathbf{S}_t)}$

The conjugate gradient method is the method of the first-order accuracy. The advantages of the conjugate gradient method should include the convergence rate close to the quadratic. In addition, in gradient methods the number of arithmetic operations is less than in Newton method. The disadvantage of gradient methods is the need of good initial approximation when finding the optimal solution. The abovementioned is correct for the frequently used algorithms – Fletcher – Reeves and Davidon – Fletcher – Powell.

4.3.6 RANDOM SEARCH METHODS

The calculations by the random search method are carried out as in the gradient methods by the relations (4.35), the difference is in the parameter α_t selection. In random search methods the value α_t is random subject to any statistic distribution law, for example, normal one and lying in the interval *(0,1)*.

The following variants of random search method are used:

- pair test method, – two values of the objective function with $\alpha_t = \alpha$ and $\alpha_t = -\alpha$ are calculated;
- returned method with an abortive attempt (abortive attempt corresponds to the variant when the objective function is changed in the opposite direction);
- linear extrapolation method formed by the attempts previously performed;
- training method where the ban is to use α_t resulting in the previous attempts to the incorrect change of the objective function.

As gradient methods, the random search methods have the first-order convergence. These methods are more effective than the previous ones for solving the problems of non-linear programming with a large number of unknown parameters.

4.3.7 GENETIC ALGORITHMS

Effectiveness of random search method can be improved if to order the selection of coefficients α_t at the next iteration. Partially, this problem is solved by the random search method with learning. But the group of methods, importing genetic knowledge, has been rapidly developing recently [211]. Genetic algorithms are based on the random search method where "learning" is performed by the constructing analogy with laws disclosed in biology while studying the evolution and origin of species. The development of any population of living organisms occurs random, but in the evolution the fittest organisms survive. With the development, the adjustment of population increases allowing it to survive better in changing environment.

First, such algorithm was offered by John Holland at the University of Michigan in 1975. It received the name of "Holland's reproductive plan" and was the basis for almost all variants of genetic algorithms. The main ideology of genetic algorithms is based on the following premises:

- evolution of living organisms' population is in its expansion and domination in the constantly changing living conditions;
- specific population of living organisms can be described by properties that distinguish it from other populations. The set of these properties is called a phenotype;

- information about the main phenotype properties is fixed in chromosomal genes in a living organism belonging to the examined phenotype. The chromosome set of genes is called genotype;
- any gene belonging to the genotype is reflected in the phenotype and vice versa;
- population evolution happens due to the random pair (binary) crossing of animal units, meanwhile, the units with a weak set of genes disappear (become extinct), individuals with a newly acquired set of genes obtained in a pair crossing take their place;
- animal units newly formed in pair crossing inherits the genetic features of their parents, the set of genes in the newly formed individuals is the result of concatenating the genes of both parents (new individual chromosome contains the part of genetic code of the first parent and the part of genetic code of the second parent);
- chromosome in the new animal unit can mutate, and it is shown as a distortion (defect) of a newly acquired genetic code.

The algorithm of unconstrained optimization problem solution repeats the listed above laws of evolution of living organisms' population. Suppose we want to solve the problem (4.21) of the objective function $\Phi(\mathbf{x})$ minimum search depending on the N variables x_n $(n = \overline{1, N})$. Let us list the main steps of genetic algorithm applied to this problem. The final goal of evolutionary process in solving the mathematical problem is to find the optimum of the objective function (maximum or minimum is inessential). The genotype in the problem is the set of independent variables of the objective function x_n $(n = \overline{1, N})$. Each of the variables is a particular gene. The chromosome γ is a vector containing the specific values of the variables x_n. The beginning of the problem solving corresponds to the population consisting of M individuals, each of them has the unique chromosome $\gamma_m = \gamma(x_1^{(m)}, x_2^{(m)}, ..., x_N^{(m)})$, $m = \overline{1, M}$. Populations of M individuals are regenerated by means of crossings. When crossing, the chromosomes γ_m are re-calculated by an algorithm called a crossover. The crossover is constructed so that to take into account the chromosomes values involved in crossing, and values of the objective functions corresponding to these chromosomes. At the next stage of the algorithm, the content of population is re-viewed. From the content, the weak chromosomes are removed and newly acquired strong (in terms of the objective function value) chromosomes are added.

Nowadays a large number of numerical algorithms for genetic methods is developed. In these methods a new gene included in the newly calculated unique chromosome $\gamma = \gamma(x_1, x_2, ..., x_N)$ at the next iteration is calculated as a linear combination of genes included in the chromosomes with the numbers *m* and *t*

$$x_n = a \cdot x_n^{(m)} + b \cdot x_n^{(t)} \tag{4.39}$$

Values of the coefficients *a* and *b* can be defined, for example, by the formulas [57]:

$$a = 1 + \alpha - u(1 + 2\alpha), \quad b = u(1 + 2\alpha) - \alpha; \quad \alpha \in [0,1], \quad u \in (0,1) \tag{4.40}$$

$$a = \frac{1+z}{2}, \quad b = \frac{1-z}{2}; \quad z = \begin{cases} (2u)^{\frac{1}{1+\beta}}, & u \le 0,5; \\ (2(1-u))^{\frac{-1}{1+\beta}}, & u > 0,5; \end{cases} \tag{4.41}$$

$$a = 1 - 2^{\xi - n}, \quad b = 2^{\xi - n}; \quad \xi \in [0, n] \tag{4.42}$$

In the Eqs. (4.40), (4.41), and (4.42) α, u, ξ are random numbers, and the value $\beta > 1$. Moreover, in all these equations the condition $a + b = 1$ is satisfied.

To learn more about the theory of genetic methods and its applications we should refer to [211].

4.3.8 DEFORMABLE POLYHEDRON METHOD

The deformable polyhedron method [233] is characterized by high efficiency in solving the large-scale problems. The method is easily implemented when performing the calculations on a computer; first, it was used by Nelder and Mead (University of Texas, USA) in 1966.

The method under study is operated by the set of *N+1* points $\mathbf{x}_1$, $\mathbf{x}_2$,..., $\mathbf{x}_{N+1}$ sorted to perform the inequalities $\Phi_{N+1} \ge \Phi_N \ge ... \ge \Phi_2 \ge \Phi_1$ for the corresponding values of the objective function (the problem of minimizing the values of the objective function is solved). The set of points $\mathbf{x}_1$, $\mathbf{x}_2$,..., $\mathbf{x}_{N+1}$ can be interpreted as the polyhedron vertexes in *N*-dimensional space.

At each iteration the current polyhedron is replaced by the new one following the rule – "the worst" vertex $\mathbf{X}_{N+1}$ (corresponds to the largest value of the objective function) is rejected, and the new point calculated according to the algorithm is introduced to the set instead of the previous one. In iteration algorithms the information about partial derivatives of the objective function is not used.

Let us discuss the main steps of computational algorithm of deformable polyhedron at the following iteration:

- coordinates of the gravity center N of the best polyhedron vertexes $\mathbf{x}_1, \mathbf{x}_2, \ldots, \mathbf{x}_N$
- are calculated

$$\mathbf{c} = \frac{1}{N}\sum_{n=1}^{N}\mathbf{x}_n \tag{4.43}$$

- coordinate of the new vertex of polyhedron $\mathbf{x}_r$ is calculated and the value of the objective function Φ_r is calculated

$$\mathbf{x}_r = \mathbf{c} + \alpha \cdot (\mathbf{c} - \mathbf{x}_{N+1}), \quad \alpha > 0 \tag{4.44}$$

- renewal of the coordinate $\mathbf{x}_{N+1}$ is performed on the following conditions

$$\mathbf{x}_{N+1} = \mathbf{x}_r, \text{ if } \Phi_1 \le \Phi_r \le \Phi_N,$$

$$\mathbf{x}_{N+1} = \begin{cases} \mathbf{c} + \beta(\mathbf{x}_r - \mathbf{c}), & at\ \Phi_c < \Phi_r; \\ \mathbf{x}_r, & at\ \Phi_c \ge \Phi_r; \end{cases}, \quad \text{if } \Phi_r \le \Phi_1 \tag{4.45}$$

$$\mathbf{x}_{N+1} = \begin{cases} \mathbf{c} + \gamma(\mathbf{x}_{N+1} - \mathbf{c}), & at\ \Phi_r \ge \Phi_{N+1}; \\ \mathbf{c} + \gamma(\mathbf{x}_r - \mathbf{c}), & at\ \Phi_r < \Phi_{N+1}; \end{cases}, \quad \text{if } \Phi_r > \Phi_N$$

In the above equations $\beta > 1,\ 0 < \gamma < 1$. At each iteration, there can be several compressions or stretchings with different coefficients β, γ.

Apart from the usual iterations with reflections, stretchings and compressions, the discussed method involves the iterations of the forced substitution of the current polyhedron for the appropriate polyhedron performed with a certain frequency. These substitutions are called recoveries. Periodic recoveries are needed to eliminate the unnecessary irregularity of vertexes placement arising as a result of several successive stretchings. In such recoveries, only two best vertexes of the latter polyhedron are saved. The distance between them becomes the length of each side of the newly generated correct figure.

The deformable polyhedron method refers to minimization methods based on the comparison of the function values, which are often called direct search methods. These methods are heuristic, they are characterized by a large number of settings, and the method success is often completely determined by how well the values of its parameters are selected. Thus, in the polyhedron method, the following settings are used: α – reflection coefficient, β – coefficient of stretch, γ – compression ratio. In most cases, it is recommended to use the following coefficient values – $\alpha=1$; $\beta=0.5$; $\gamma=2$.

4.3.9 METHODS OF SEQUENTIAL ANALYSIS OF VARIANTS

Methods of sequential analysis of variants are appropriate to apply in integer programming problems, when it is possible to enumerate all possible combinations of the variables. Sometimes, these methods are used for problems with continuous parameters. The application of the methods is useful when there is a relatively small number of required parameters, otherwise, the complexity of solving the problem may be excessive. Indeed, if each parameter can take N values, and the number of unknown parameters is m, the number of possible variants that should be examined in the problem solving is N^m. Already for $N = 10$ and $m = 4$ the complexity of the full search of all possible variants can make to reject the use of these methods.

Various heuristic approaches that reduce the number of viewed variants solving the problem are developed. In particular, analogies to biological laws (e.g., eagle hunting algorithm) allow us to reduce the viewed variants in optimization in more than two times. However, the probability of finding the global optimum using heuristic approaches is definitely decreasing.

4.3.10 REDUCING OPTIMIZATION PROBLEMS TO THE PROBLEMS ON THE SOLUTION OF ORDINARY DIFFERENTIAL EQUATIONS

Analysis of the use of optimization methods shows that to the present time there is no universal algorithm that would solve any practical problem in a reasonable number of iterations. That is why the number of methods for solving optimization problems is large and growing. Relatively new methods for solving this class of problems are the methods where the optimization problem is reduced to the problem of solving systems of ordinary differential equations. The use of such methods, in general, increases the complexity of solving the original problem, but it does increase the reliability of its solutions. The approach, where the solution of systems of nonlinear equations is reduced to Cauchy problem solution, is called the relaxation method in computational mathematics.

The problem can be formulated as follows:

To find the solution $\mathbf{x} = (x_1, x_2, x_3, \ldots, x_N)$ for values $t \to \infty$ that satisfies the equation

$$\frac{d\mathbf{x}}{dt} + \text{grad}\,\Phi(\mathbf{x}) = 0 \tag{4.46}$$

if it is known that when $t = 0$, the vector $\mathbf{x}(t)$ values take the values $\mathbf{x}(0) = (x_1^{(0)}, x_2^{(0)}, \ldots, x_n^{(0)}, \ldots, x_N^{(0)},)$.

It can be shown that if the stationary solutions of the Eq. (4.46) exist, these solutions correspond to $t \to \infty$ and coincide with the solutions of the problem (4.21) of finding the functional minimum.

In the case of slow convergence the following equation can be solved instead of the Eq. (4.46)

$$\frac{d^2\mathbf{x}}{dt^2} + \gamma \cdot \frac{d\mathbf{x}}{dt} + \text{grad}\,\Phi(\mathbf{x}) = 0 \tag{4.47}$$

In this case the convergence rate is largely determined by the successful choice of the parameter γ (the slow convergence corresponds to $\gamma \to 0$ and $\gamma \to \infty$).

Issues related to the numerical methods of solving the systems of ordinary differential equations will be discussed in Chapter 5.

4.4 SOLUTION OF NONLINEAR EQUATIONS AND SYSTEMS OF NONLINEAR EQUATIONS USING COMPUTERS

4.4.1 *SOFTWARE USED TO SOLVE THE SYSTEMS OF NONLINEAR EQUATIONS BY MEANS OF MATHCAD AND FORTRAN*

Tables 4.2–4.4 show some programs available as the part of MathCad and IMSL libraries of compiler FORTRAN and also used for solving nonlinear equations, systems of nonlinear equations and optimization problems [22, 141].

In the package MathCad the solution of non-linear equations with one unknown can be performed, for example, using the function ***root***. A solvable equation can be given before the call to the function in the analytical form by a separate operator or directly in the function call (e.g., as ***root*** (x^2+2x+1, x)). The function allows you to calculate all the roots, real and complex. Before the call to the function you should determine an initial approximation to the root. If an additional root is wanted, you should change the initial approximation. When finding the complex root, the initial approximation should be written down in the complex form. The function ***root*** calculates the roots using the method of secants. If the solution cannot be found by this method, the special Muller method is used.

The function ***polyroots*** (*y*) allows us to calculate all the polynomial roots $a_0+a_1x+a_2x^2+...+a_nx^n$ of n degree. The vector y should have the length (n+1) and contain the information about the polynomial coefficients $a_0, a_1, a_2, ..., a_n$.

The functions ***Find***, ***Minerr*** allow us to solve the system of non-linear equations. The ***given*** equation system should be enclosed in a computing unit opened by the directive Given. The equation system, restrictive conditions for unknown variables and calls to function are sequentially written inside the unit. Before the computing unit the initial values for unknown variables should be given. The more versatile function is ***Minerr***. Unlike the function ***Find***, the function ***Minerr*** allows us to calculate the roots $x_1, x_2, ..., x_n$ in a way to satisfy the limits for the unknown roots to the best advantage. Particularly, if there is no accurate solution of the equation system, the use of the function ***Minerr*** still offer the root values where the difference between the right sides of equations and initially written ones will be minimal.

To solve the systems of non-linear equation in the MathCad package, the functions ***Maximize*** and ***Minimize*** can be used. However, these functions are most effective when solving the optimization problems.

TABLE 4.2 Examples of *MathCad* Functions Providing the Solution of Non-Linear Equations and Systems of Non-Linear Equations

No	Function name	Purpose of the function	Function arguments description
1	***root*** ($f(x)$, x);	roots of the non-linear equation of the form $f(x)=0$ are calculated	$f(x)$ – function whose root is calculated; x – name of the argument function
2	***polyroots*** (y)	polynomial roots of power n are calculated	y – vector of dimension ($n+1$) containing the coefficient values $a_0, a_1, a_2, \ldots, a_n$
3	***find*** $(x_1, x_2, \ldots, x_n)$	calculation of the roots of the system of non-linear equations	$x_1, x_2, \ldots, x_n$ – variables (unknown) whose values are defined by solving the system of non-linear equations
4	***minerr*** $(x_1, x_2, \ldots, x_n)$	calculation of roots of the system of non-linear	$x_1, x_2, \ldots, x_n$ – variables (unknown) whose values are defined by solving the system of non-linear equations
5	***minimize*** (y, x)	determining the values x, ensuring the minimum value of the function $y(x)$	y – function whose minimum is searched; x – function argument
6	***maximize*** (y, x)	determining the values x, ensuring the maximum value of the function $y(x)$	y – function whose maximum is searched; x – function argument

In IMSL – FORTRAN library to solve non-linear equation the programs ***zbren***, ***zreal***, ***zanly*** are used (Table 4.3). The latter program allows us to obtain the solution in problems where the roots are complex numbers. The program ***zbren*** uses a combined method for roots calculation, programs ***zreal***, ***zanly*** use Muller method [22]. To solve equations presented by polynomials the programs ***zplrc***, ***zporlrc***, ***zplrc*** can be used.

Out of the large number of programs allowing to calculate roots of the system of non-linear equations, the program ***neqnf*** seems the simplest and most effective. This program uses the computer algorithm based on the modified method of Powell [22]. Jacobi matrix, necessary for calculating the next approximation for the unknown variables, is calculated within the program. As a part of the library, the program ***neqnj*** exists in which Jacobian matrix is computed in the user program. The programs ***neqbf***, ***neqbj***, IMSL

TABLE 4.3 Examples of the Subprograms of *IMSL* Library Providing the Solution of Non-Linear Equations and Systems of Non-Linear Equations

No	Call to subprogramor function	Purpose of the subprogram or function	Description of the function arguments or subprogram arguments
1	*call* ***zbren*** *(f, errabs, errel, a, b, maxfn), call* ***zreal*** *(f, errabs, errel, eps, eta, nroot, itmax, xguess, x, info); call* ***zanly*** *(f, errabs, errel, nknown, nnew, nguess, zinit, itmax, z, info)*	Calculation of the roots of the non-liner equation $f(x)=0$ by the combined method (***zbren***), Muller method with the calculation of real roots (***zreal***), Muller method with the calculation of complex roots (***zanly***)	*f* – user function given by the operator *external*; *errabs, errel* – criterion of finishing the calculations and relative error; *a, b* – interval within the roots; *maxfn* – maximum allowable number of function calculations; *eps, eta* – criteria of roots location; *nroot* – number of roots which the program *zreal* should find; *itmax* – maximum allowable number of iterations in the calculation of the next root; *xguess, x* – vectors of *nroot* size containing the initial approximations and final values of the equation roots; *info* – vector containing the information about the number of iterations used in the calculation of each root of equation; *nknown, nnew* – number of known roots, number of roots to be calculated; *nguess, itmax* – number of initial approximations and the maximum number of iterations; *zinit, z* – initial and final approximations of all complex roots

TABLE 4.3 Continued

No	Call to subprogramor function	Purpose of the subprogram or function	Description of the function arguments or subprogram arguments
2	*call* ***zplrc*** *(ndeg, coeff, root); call* ***zporlrc*** *(ndeg, coeff, root); call* ***zplrc*** *(ndeg, coeff, root);*	Defining the roots of polynomial with real coefficients by Laguerre method (***zplrc***), Jenkins-Traub method and polynomial with complex coefficients (***zporlrc***)	*ndeg* – degree of polynomial; *coeff* – vector of *ndeg*+1 size containing the coefficients of polynomial $a_0, a_1, \ldots a_{\text{ndeg}}$; *root* – vector of ndeg size containing the calculated polynomial roots
3	*call* ***neqnf*** *(fcn, errel, n, itmax, xguess, x, fnorm)*	Calculation of the roots of the system *n* of non-linear equations	*fcn* – user program where the functions of the system of non-linear equations are calculated; *errel* – criterion of the calculation stopping; *itmax* – maximum allowable number of iterations; *xguess, x* – initial approximations and final values of the equation roots; *fnorm* – scalar containing the assessment of the accuracy

TABLE 4.4 Examples of the Subprograms of *IMSL* Library Providing the Solution of Optimization Problems

No	Call to the subprogram or function	Purpose of the subprogram or function	Description of the function arguments or subprogram arguments
1	*call **uminf** (fcn, n, xguess, xscale, fscale, iparam, rparam, x, fvalue);* *call **umcgf** (fcn, n, xguess, xscale, gradtl, maxfn, dfpred, x, g, fvalue);* *call **umpol** (fcn, n, xguess, s, ftol, maxfn, x, fvalue);*	Determination of the minimum of n variables by quasi-Newton method (***uminf***), conjugate gradient method (***umcgf***) and deformable polyhedron method (***umpol***)	*fcn* – user program where the minimization problem is given; *xguess* – initial approximation; *xscale* – scaling vector for the variables; *fscale* – scaling vector for the function; *iparam* – vector containing the service information from six numbers; *rparam* – vector containing the service information from seven numbers; *x* – vector containing the problem solution; *fvalue* – minimized function value calculated at the final stage; *gradtl* – convergence criterion; *maxfn* – maximum allowable number of the function evaluations; *dfpred* – estimate of the expected function decrease; *g* – vector containing the gradient components; *s* – initial simplex in the method of deformable polyhedron; *ftol* – operating parameter used to generate the convergence criterion;

TABLE 4.4 Continued

No	Call to the subprogram or function	Purpose of the subprogram or function	Description of the function arguments or subprogram arguments
2	*call **bconf** (fcn, n, xguess, ibtype, xlb, xub, xscale, fscale, iparam, rparam, x, fvalue);* *call **bcpol** (fcn, n, xguess, ibtype, xlb, xub, ftol, maxfcn, x, fvalue)*	Determination of the minimum of *n* variables with simple constraints by quasi-Newton method (***bconf***) and deformable polyhedron method (***bcpol***)	*fcn* – user program where the minimization problem is given; *xguess* – initial approximation; *ibtype* – scalar specifying the kind of restrictions on the variables; *xlb, xub* – vectors containing the lower and upper boundaries of the variables; *xscale* – scaling vector for the variables; *fscale* – scaling vector for the function; *iparam* – vector containing the service information from six numbers; *rparam* – vector containing the service information from seven numbers; *x* – vector containing the problem solution; *fvalue* – minimized function value calculated at the final stage; *ftol* – operating parameter used to generate the convergence criterion; *maxfn* – maximum allowable number of the function evaluations

libraries provide the solution to the problem of the roots of non-linear equations by the intersecting method, etc.

To solve the problems of unconstrained optimization the functions ***uminf***, ***umcgf***, ***umpol*** (Table 4.4) seem convenient. The determination of the function minimum in these programs is provided by quasi-Newton method, conjugate gradient method and method of deformable polyhedron, respectively.

In simple constraints (constraints of the form $a_i \leq x_i \leq b_i$, where $i=1, n$) the programs ***bconf*** or ***bcpol*** can be applied. In these programs quasi-Newton method and method of deformable polyhedron, respectively, are used.

Additionally, it should be noted that the IMSL library has programs that provide the solution of problems of linear programming, quadratic programming and non-linear programming [22].

4.4.2 SELECTION PROBLEMS IN HEAT ENGINES DESIGN

4.4.2.1 Selecting the Fuel Charge Geometry

The problem of solid charge design for gas producer plant or rocket engine, in some cases, can be non-trivial. As an example, there are two designs of fuel charges in Figures 4.8 and 4.9.

The first design (Figure 4.8) is the charge of a set-in type with a circular grooving at the ends [136]. The outer surface part is protected from burning by the coating. The change of the area *S* of the fuel charge burning surface is presented by the dependence on quantity of burnt arch $e = \int_t u_m dt$ (u_m – solid fuel burning rate, *t* – time). The actual dependence *S(e)* is determined by geometric relations (we use the condition that the fuel burns by the normal to the surface, and the burning rate u_m is the same in any section of the charge) and is the non-linear function of the selected geometric charge sizes (numerical values of coordinates $x_i, r_i,\ i = 1,20$). According to the technical specification, the law of the surface area variation as a function of the burnt arch is given – $S^*(e)$. The projection problem is the selection of such values $x_i, r_i,\ i = 1,20$ which would provide the actual law of the burning surface variation *S(e)* maximally close to the law $S^*(e)$ given in the technical specification.

The problem of selecting the sizes of fuel charge under study can be formulated as a mathematical programming problem [110, 155]

- to determine the geometric sizes of the fuel charge $x_i, r_i,\ i = 1,20$; $R_1,...,R_4, h_1,...,h_4$ providing the minimum of the objective function Φ which is defined by the relation

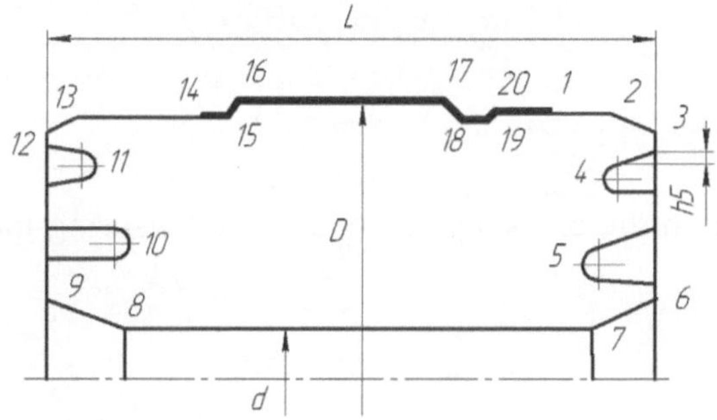

FIGURE 4.8 Charge with a circular groove.

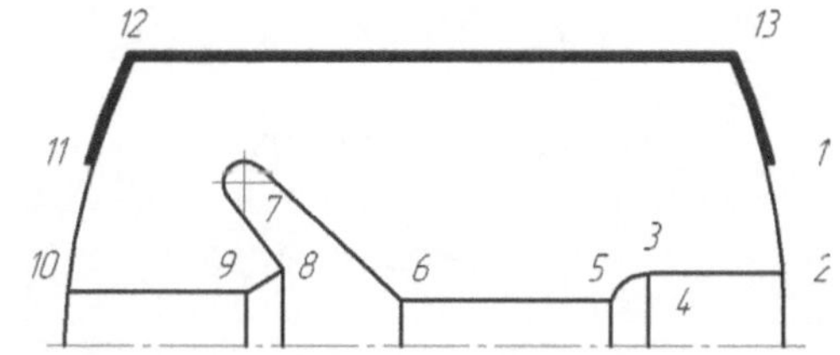

FIGURE 4.9 Scheme of the charge with an annular slot.

$$\Phi(x_1,...,x_{20},r_1,...,r_{20},R_1,...,R_4,h_1,...,h_4) = \min \int_0^{e_{\max}} \left(\frac{S(e)-S^*(e)}{S^*(e)} \right)^2 de \quad (4.48)$$

under the following constraints on geometric sizes of the fuel charge:

- constraints on the longitudinal coordinate ensuring the charge shape constancy

$$x_1<x_2;\ x_3>x_2;\ x_4<x_3;$$

$$x_5<x_4;\ x_6=x_3;\ x_7<x_6;$$

$$x_8<x_7;\ x_9<x_8;\ x_{12}=x_9;$$

$$x_{10}>x_9;\ x_{11}>x_{12};\ x_{13}>x_{12};$$

$$x_{14}<x_{15};\ x_{14}>x_{15};\ x_{16}>x_{15};$$

$$x_{17}>x_{16};\ x_{18}>x_{17};\ x_{19}>x_{18};$$

$$x_{20}>x_{19};\ x_{1}>x_{20};\ x_{6}-x_{9}=L;$$

- constraints on the cross-section coordinate ensuring the charge shape constancy

$$r_{1}=r_{2};\ r_{8}=r_{7};\ r_{17}=r_{16};$$

$$r_{13}=r_{14};\ r_{19}=r_{18};\ r_{20}=r_{1};$$

$$r_{2}>r_{3};\ r_{3}>r_{4};\ r_{15}=r_{14};$$

$$r_{5}>r_{6};\ r_{6}>r_{7};\ r_{4}>r_{5};$$

$$r_{10}>r_{9};\ r_{11}>r_{10};\ r_{9}>r_{8};$$

$$r_{13}>r_{12};\ r_{16}>r_{15};\ r_{12}>r_{11};$$

$$r_{17}>r_{18};\ r_{7}=d_{2};\ r_{19}-r_{7}=e_{max};$$

$$r_{20}>r_{19};\ r_{16}=D/2;$$

- other constructive constraints – geometric conditions, condition on the work of attachment point of charge, condition on non-erosive burning, on ensuring the given volume of charge, etc. (listed conditions are discussed in [136])

$$R_{4}+h_{4}<r_{3};\quad R_{5}+h_{5}<R_{4-}h_{4};\quad R_{5}-h_{5}>r_{6};$$

$$R_{11}+h_{11}<r_{12};\quad R_{11}-h_{11}<R_{10};\quad (x_{4}-x_{19})^{2}+(R_{4}-R_{19})^{2}>e_{max};$$

$$\frac{S(0)}{\pi r_{7}^{2}}\leq\kappa^{*};\quad \int_{0}^{e_{\max}} S(e)de=W$$

Similarly, we can formulate the problem of determining the charge size of case-bonded type with an annular slot (Figure 4.9). Unlike the previous problem, the charge design problem cannot be separated from the problem of internal ballistics. This is due to the fact that the fuel charge surface is deformed from the intrachamber pressure impact. The deformation degree defines the real area of solid fuel burning surface. The objective function in this problem is reasonable to build not on the difference of the planned and actual areas of the fuel charge burning surface, on the difference of actual pressure *p(t)* in the combustion chamber from the given technical specification of the law of variation $p^*(t)$. Thus, the calculation of the objective function in solving the problem of fuel charge design requires the joint solution of the geometric problem and problem of internal ballistics.

The problem of designing the charge of case-bonded type can be formulated as follows

- to determine the geometric size of the fuel charge $x_i, r_i,\ i = 1,13$; R_2, R_5, R_7, R_{10} providing the minimum of the objective function Φ which is defined by the relation

$$\Phi(x_1,...,x_{13},r_1,...,r_{13},R_1,R_5,R_7,R_{10}) = \min \int_0^{t_{\max}} \left(\frac{p(t)-p^*(t)}{p^*(t)} \right)^2 dt \quad (4.49)$$

  under the following constraints on the geometric sizes of the fuel charge:
- constraints on the longitudinal coordinate ensuring the charge shape constancy

$$x_1<x_2;\ x_2>x_3;\ x_3=x_4;$$

$$x_5<x_4;\ x_6<x_5;\ x_8<x_6;$$

$$x_7<x_8;\ x_9<x_8;\ x_{10}<x_9;$$

$$x_{10}>x_{11};\ x_{12}>x_{11};\ x_{13}>x_{12};$$

$$x_1>x_{13};$$

- constraints on the cross-section coordinate ensuring the charges shape constancy

$$r_1>r_2;\ r_2=r_3;\ r_4=0;$$

$$r_5=r_6;\ r_7>r_6;\ r_7>r_8;$$

$$r_8>r_9;\ r_9=r_{10};\ r_{11}>r_{10};$$

$$r_{12}>r_{11};\ r_{13}=r_{12};\ r_{13}=D/2;$$

- other constructive constraints – geometric conditions, condition on the work of attachment point of charge, condition on non-erosive burning, on ensuring the given volume of charge, etc. (listed conditions are discussed in [136])

$$R_4+h_4<r_3; \quad R_5+h_5<R_{4-} h_4; \quad R_5-h_5>r_6;$$

$$R_{11}+h_{11}<r_{12}; \quad R_{11}-h_{11}<R_{10}; \quad (x_4-x_{19})^2 + (R_4-R_{19})^2>e_{max};$$

$$\frac{S(0)}{\pi r_5^2} \leq \kappa^*; \qquad \int_0^{e_{\max}} S(e)de = W \text{ и др}$$

The formulated problems of designing the fuel charge (4.48) and (4.49) are non-linear quadratic programming problems (quadratic objective function and constraints are linear equalities and inequalities). Both problems can be reduced to the problem of unconstrained programming. For this purpose, the objective function should be changed to a new one, different from the above-written form (Eqs. (4.48) and (4.49)) in that it would include "penalties." "Penalties" should be written for each constraint written in the form of equalities and inequalities. The total number of penalties will be numbered in the tens.

It should be noted that the considered problems of selecting the geometric sizes of the fuel charge formulated as a mathematical programming problem are essentially non-linear problems. The objective function calculation is a consuming task, especially in the problem on the charge of case-bonded type (Figure 4.9). In addition, the calculation of partial derivatives of the objective function for each search parameter in the analytical form is almost an unrealizable problem. Therefore, there is a computational method for solving problems with the use of partial derivatives, the values of the derivatives should be determined numerically. However, on a modern PC the solution of the considered problems is performed in reasonable time (a few seconds).

As an example, below we discuss the solution of the problem of geometric sizes selection of the solid fuel charge of a relatively simple geometric shape (Figure 4.10). To solve the problem, the length of the charge and its outer diameter were taken as invariable. As search parameters, the values

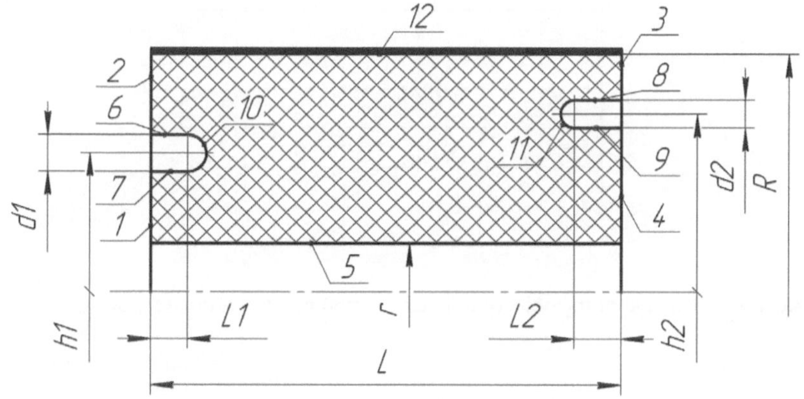

FIGURE 4.10 Scheme of the charge with a circular groove.

r_0, h_1, h_2, L_1, L_2, r_1, r_2 were accepted. Defined by the technical specification the law of surface area variation of the solid fuel combustion was adopted in a simplified form $S^*(e) = \frac{W}{e_{\max}} = const$. The value of the maximum burning arch $e_{\max}$ was not limited. The calculation of the dependence *S(e)* under the known geometric sizes was performed in accordance with [136].

The problem of unconstrained optimization was solved by the methods of coordinate descent, random search method and the use of genetic algorithm. As the initial approximation was taken: $L = 0{,}4$ м; $L_1 = 0{,}007$ м; $L_2 = 0{,}005$ м; $r_0 = 0{,}016$ м; $r_1 = 0{,}004$ м; $r_2 = 0{,}006$ м; $h_1 = 0{,}042$ м; $h_2 = 0{,}055$ м. The calculations showed that the initial values of the search parameters insignificantly affect the solution convergence. The main results of the calculations are presented in Figures 4.11–4.16. The figures demonstrate the dependence of the objective function change from the number of iterations (Figures 4.11, 4.13 and 4.15) and relations *S (e)* (Figures 4.12, 4.14 and 4.16), corresponding to the original data and the data obtained in the solution of mathematical programming problems are compared.

The calculations indicate that the method of coordinate descent (Figures 4.11 and 4.12) using a wide range of initial configurations of geometric parameters of charge allows us to reduce the objective function value, on the average of 3–5 times from its initial value. The drawback of the method for solving the considered problem (and other multiparameter optimization problems) is a slow convergence to the solution.

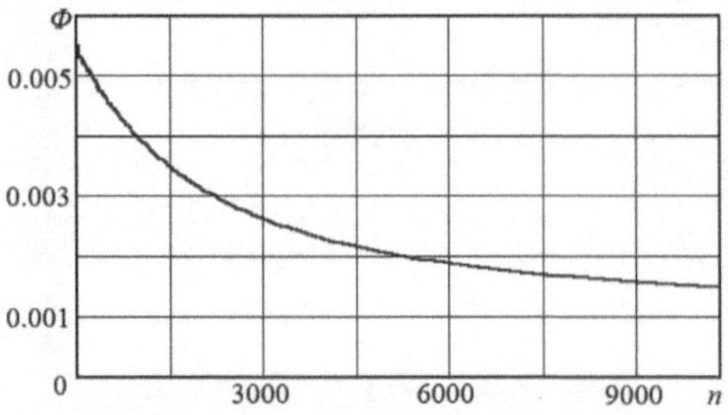

FIGURE 4.11 Change of the objective function (method of coordinate descent).

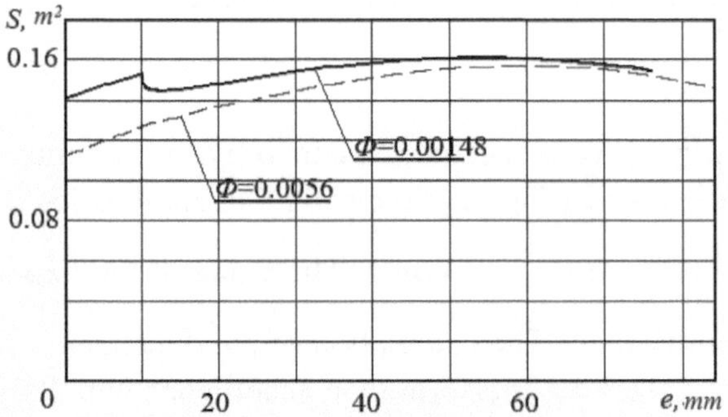

FIGURE 4.12 Dependence of the burning area surface of the solid fuel charge on the burnt arch value (method of coordinate descent) (Φ=0.0056 – initial approximation, Φ=0.00148 – final approximation).

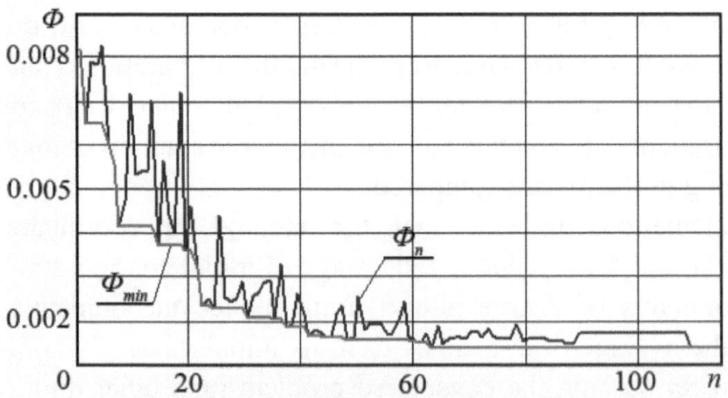

FIGURE 4.13 Change of the objective function (random search method) (Φ=0.0056 – initial approximation, Φ=0.000362 – final approximation).

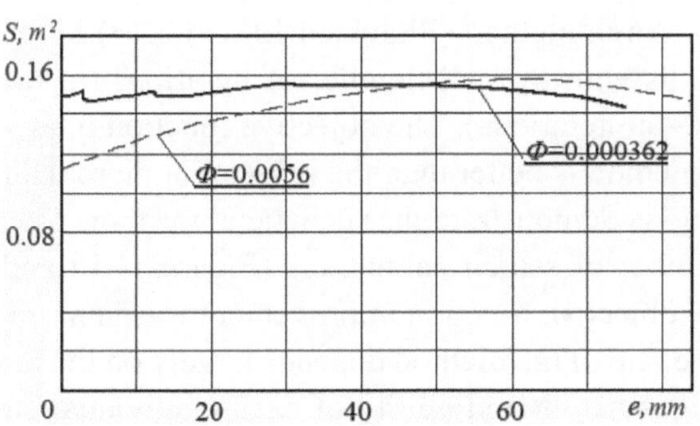

FIGURE 4.14 Dependence of the burning area surface of the solid fuel charge on the value of the burnt arch (random search method).

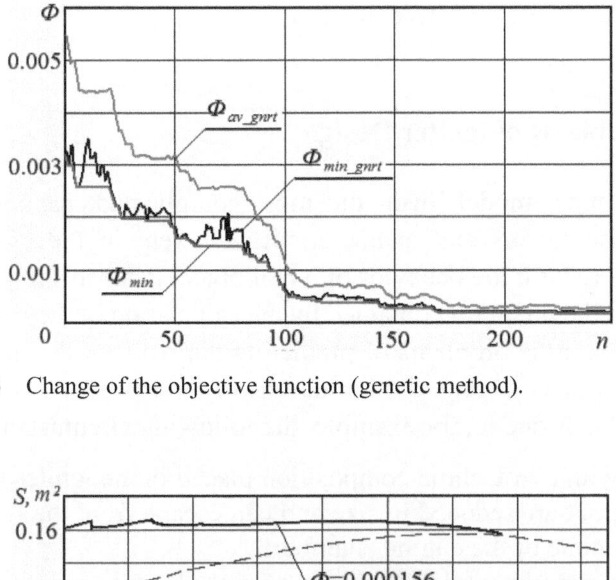

FIGURE 4.15 Change of the objective function (genetic method).

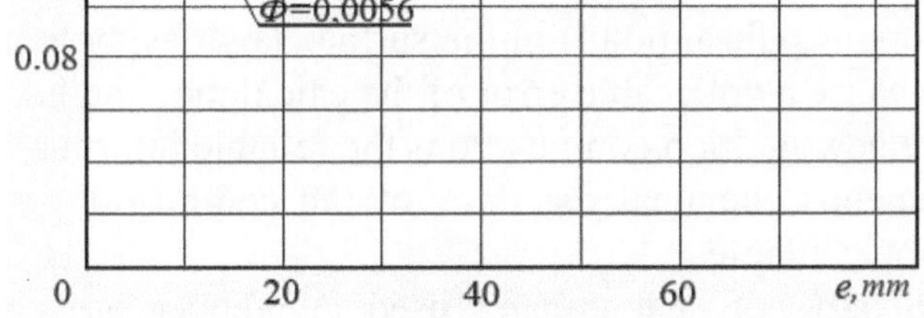

FIGURE 4.16 Dependence of the burning area surface of the solid fuel charge on the value of the burnt arch (genetic method) (Φ=0.0056 – initial approximation, Φ=0.000156 – final approximation).

The random search method (Figures 4.13 and 4.14) in most cases gives satisfactory results with a small number of iterations (in the solved problem these are 50–60 iterations). The objective function final value using the random search method is better than the original in more than 15 times.

The calculations demonstrate that genetic algorithms are most effective with a large number of search parameters (Figures 4.15 and 4.16). In our calculations, the objective function improvement was in more than 25 times. The convergence rate of the method depends largely on the number of "individuals" in the generation and ability of each "individual" to mutate from generation to generation. It may be noted that the method converges better with a large number of "individuals" in each generation. Figure 4.15 shows: Φ_{av} – average value of the objective function in the "generation," $\Phi_{\min}$ – minimum value of the objective function in the "generation," $\Phi_{\min}$ – reached the minimum value of the objective function.

4.4.2.2 Problems of Igniter Design

Any mathematical model, even the most complicated, cannot take into account all the factors and characteristics inherent in the real designed object. In this regard, the behavior of a real object is always to some extent different from the prediction obtained by the calculation of the mathematical problem. Solving the problems of predicting the work of the initial part of the solid fuel rocket engine, the limitations of the mathematical model are natural, and this is due to, for example, the following circumstances:

- tablets with pyrotechnic composition placed in the igniter body, when burning out are reduced in size and can escape from the igniter to the inner volume of the engine chamber;
- existing experimental techniques do not allow to accurately set the ignition time of the solid fuel, and the use of obtaining the given level of temperature (flash point) in the surface layer as the ignition condition can cause a noticeable error of the calculation method;
- using the new ignition compositions the reliable information about the heterogeneous chemical reactions of the combustion products with fuel is not available;
- empirical laws of heat transfer used in practice are not universal, and therefore, cannot be applied to engines of individual structural schemes.

The written above is illustrated by the calculations performed using the software package [10] on the calculation of internal ballistics of solid propellant in the initial part of its work. The calculations were performed using the models discussed [48, 244]. Figures 4.17 and 4.18 present the results of calculations of a large engine (Figure 4.5), the mass of which is ~ 50 tons. The engine length is about 5 meters, the weigh mass used in the igniter body is 5 kg.

Figure 4.17 demonstrates the dependence of the calculated pressure in the volume of the igniter body as the time function calculated under the assumption that weigh part of the ignition composition flows from the igniter body into the combustion chamber volume. The metering characteristic ***G(t)*** describing the flow of the combustion products from the igniter body is given by the formula

$$G(t) = \rho_в u_в S_в(t) \cdot \varpi \cdot (1 - \exp(- t/3t_*))$$

In the above formula it is indicated:

- t, t_* – current process time and time determining the duration of the escape of ignition composition tablets from the igniter body;
- $\rho_в, u_в, S_в$ – density of the ignition composition, its burning rate and the current value of the burning surface of the ignition composition;
- ϖ – empirical coefficient in the law for the metering characteristics.

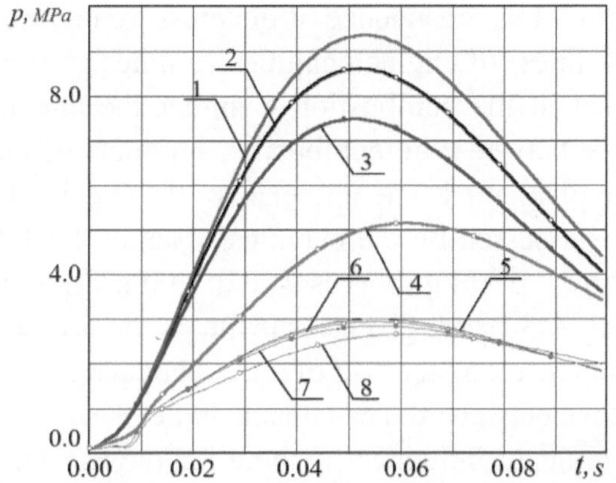

FIGURE 4.17 Dependence of the combustion product pressure in the engine chamber on time with different values of time of tablets escape from the igniter body.

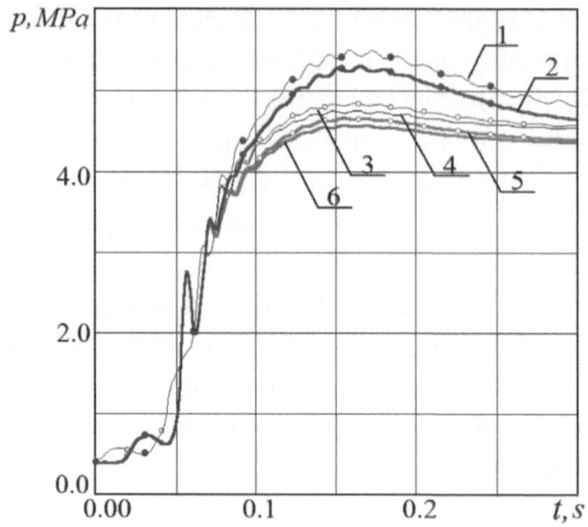

FIGURE 4.18 Dependence of the combustion product pressure in the engine chamber on time with different values of the chemical reaction thermal effect.

Figure 4.18 shows the dependence on the pressure time *p (t)* in the igniter body (thick lines) and in the engine chamber (thin lines) obtained with the values t_* =0.100 sec (curves *1, 2*); 0.050 sec (curves *2_1, 2_2*); 0.025 sec (curves *3_1, 3_2*); 0.001 sec (curves *4_1, 4_2*). In all calculations it is assumed that $\varpi = 0,4$. The analysis of the results presented in Figure 4.18 shows that the tablet escape is an important factor that has a significant influence on the dependence *p (t)* in the igniter body and, as a consequence, its metering characteristics.

Figure 4.18 shows the dependence of the pressure in the front (thin lines) and plenum (thick lines) of engine chamber volumes *p (t)* calculated under the assumption that in the combustion chamber the chemical reaction of heterogeneous type between the combustion products of the igniter weigh and solid fuel is implemented. The influence of the chemical reaction of heterogeneous type is assessed by the chemical reaction heat ΔQ. The results for three values ΔQ – 0 kJ/kg (curves *1, 22*); 200 kJ/kg (curves *1_1, 1_2*); 400 kJ kg (curves *3_1, 3_2*) are given. As in the previous case, the uncertainty in the initial data on the thermal effect of the chemical reaction can lead to the significant difference between the calculated and experimental results.

Due to the significant influence of some unknown parameters included in the mathematical model of the processes, the selection problem of these parameters values arises, for which, to the maximum extent, the coincidence between the calculated dependencies of internal ballistics characteristics of solid propellant and experimental data are provided. This problem can be solved using mathematical programming. The formulation of mathematical programming problem involves the assignment of the objective function, definition of the list of search parameters and limits of its search (minimal and maximal values), selection of the algorithm for calculating the objective function and selection of the method for solving the constrained optimization problem.

In practice, when conducting the experiments, pressure is the main measured parameter. Therefore, defining the problem of mathematical programming, we will use the value determined by the following expression as a target function

$$\Phi(\mathbf{x}) = \min \int_{t_0}^{t_k} (p(t) - p^*(t))^2 \cdot dt$$

In the latter formula the vector **x** corresponds to the set of search parameters, ***p*(*t*)** corresponds to the calculated pressure in the engine chamber, and ***p**(*t*)** – pressure determined as the experiment result. Note that without the problem generality constraints, it is possible to solve the problem in the presence of experimental information about the change ***p*(*t*)** in different sections of the engine or in the processing of multiple (e.g., *N*) experimental results. In this case, the objective function can be adopted as

$$\Phi(\mathbf{x}) = \min \sum_{n=1}^{N} \int_{t_0}^{t_k} (p_i(t) - p_i^*(t))^2 dt$$

As the search parameters, various values can be taken at different stages of solid propellant work. Calculating the initial part of the engine operation, three characteristic stages of work can be distinguished – period prior to fuel ignition, fuel ignition period, period after the fuel ignition to the engine output to the completion of the quasi-stationary mode.

In the period prior to the fuel ignition the coincidence between the calculated results with the experiment can be carried out on the coefficient value that determines the relative weight of the tablets escaping the igniter body, on the duration of the connection to the burning of all fractions of the initiation composition, on the pressure of nozzle cover destruction.

In the fuel ignition period, the coincidence is possible by the parameters determining the fuel ignition (e.g., the impact of the chemical reagent which is a part of the initiation composition on the fuel ignition temperature and its combustion rate) and pressure of the nozzle cover destruction.

At the final stage, the coincidence should be based on the parameters that determine the chemical reaction between the products of combustion initiating weigh and solid fuel (additional heat generation, the chemical reaction rate). In addition, with the compliance there can be the set the corrections to the coefficients values defining the fuel burning rate (coefficients u_1 and v), coefficients in law defining the heat exchange between the combustion products of the ignition composition and solid fuel surface.

The calculation of the objective function $\Phi(\mathbf{x})$ at each iteration step is performed using the software package [10]. To find the next approximation for the wanted parameters Newton method and method of deformable polyhedron are used. Solving the selection problem, the matching coefficients in the calculation of the initial section of solid propellant are provided after 10–20 iterations. In Figure 4.19, the example of calculating the formulated mathematical programming problem using a personal computer (attached picture from the screen) is given as the illustration. The bold line in the diagram $p(t)$ corresponds to the experimental curve.

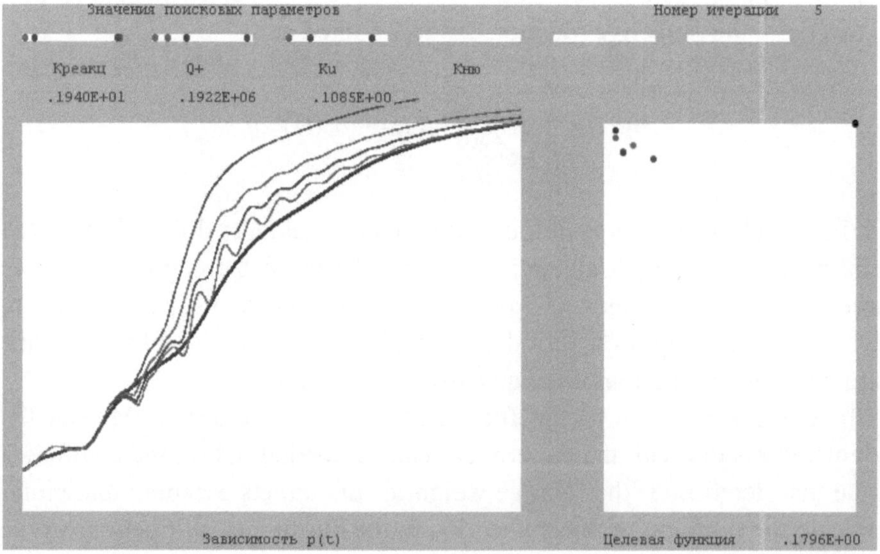

FIGURE 4.19 Example of calculating the mathematical programming problem for the propulsion system.

The developed tools of coincidence of computational models with the experimental results are related to the assumptions adopted in the models construction and contain four ways to identify the mathematical models with experimental data:

1. Identification is performed on the coefficient values allowing us to calculate the burning rate of the ignition composition fractions, and on the time value during which the ignition composition fraction is connected to the burning. The mathematical models allow you to determine the coefficient values for the three laws of burning rate – power, linear and two-tier power. In this identification variant, the largest influence of the listed values on the change of internal ballistics parameters is developed prior to the fuel ignition;
2. Identification is performed by the fuel ignition temperature values for the three cases – when there is no catalyst in the combustion products, when there is a catalyst in the amount up to 50% of the maximum possible concentration and when there is a catalyst in the maximum possible amount. Furthermore, the pressure of nozzle cover distraction and relative mass of particles escaping the igniter body can be included in the search parameters. In this identification variant, the largest influence of the listed values on the change of internal ballistics parameters is developed during the fuel ignition;
3. Identification is performed for the case when the ignition weigh formulation contains a catalyst and it is required to set the parameters characterizing additional chemical reactions caused by the catalyst presence. The identification is performed on the values of chemical reaction rate, the quantity of the additional heat generation, and the coefficients in the law for the combustion rate – corrections to the values u_1 and v are calculated. In this identification variant, the objective function calculation can be performed during the entire period of the engine output to the stationary operation mode;
4. Standard identification variant. The identification is performed on the value of the relative mass of the particles escaping the igniter body, the duration of connection to the combustion of the main weigh fraction of the ignition composition, and the fuel ignition temperature.

The listed variants do not fully solve the problem of selecting the identification parameters which provide an ideal coincidence of calculation and experimental dependencies of internal ballistics parameters. In particular, when calculating the large solid propellant engines, the values of flame spread rate on the fuel surface located in the central channel are overrated. The analysis of the calculation results shows that the flame spread rate through the central channel determined, to the greatest extent, by the functional dependence of the heat flow from the combustion products into the surface layer of the fuel charge. In the software package to calculate the heat flow in the flow channels the formula of Dyunze-Zhimolokhin which is correct for pyrotechnic compositions [244] is used

$$Nu = 0.485 \cdot (\mathrm{Re} \cdot \mathrm{Pr})^{0.63} \cdot (1 + \frac{x}{d})^{-0.59}$$

It should be noted that the written formula (it is obtained in the experiments for the flow channels) can hardly claim to universality. First, this is due to the fact that the channels of modern engines are not necessarily round. Second, the formulation of modern fuels and modern ignition compositions differ significantly from the formulations applied in the experiments by M. F. Dyunze and V. G. Zhimolokhin on defining the heat flows in the combustion chamber. The formula for the heat flow can be refined using the identification variant with the use of mathematical programming. We assume that the heat flow values can be found from the dependence

$$Nu = a \cdot \mathrm{Re}^{n_1} \cdot \mathrm{Pr}^{n_2} \cdot \left(1 + \frac{x}{d}\right)^{n_3} \tag{4.50}$$

In this dependence, the coefficients a, n_1, n_2, n_3 are unknown and can be determined as a result of search by the algorithms of mathematical programming methods.

Similarly, one can determine the values of the heat flows in the volumes in the case of thermodynamic models (combustion chamber rate is close to zero). In this case, to determine the dependence of the heat flow the following dependence is applied

$$Nu = b \cdot (Gr \cdot \mathrm{Pr})^{m_1}$$

In this relation, the unknowns are values of the coefficients b, m_1.

Figure 4.20 shows the dependencies between pressure and time in the front volume (shown with thick lines) and nozzle volume (shown with thin lines). In the figure, the pressure values obtained when calculating the heat flow in the central channel using the formula of Dyunze-Zhimolokhin (curves *1, 2*)

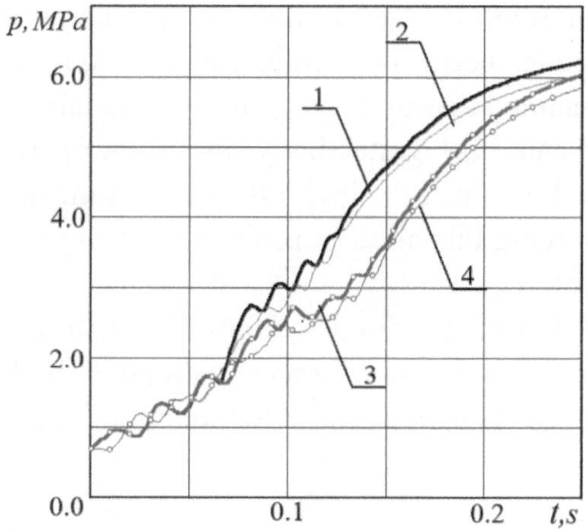

FIGURE 4.20 Pressure dependence in the combustion chamber for different laws of heat change in the charge central channel.

and pressure values obtained when calculating the heat flow in the central channel by formula (4.50), where $a = 0,485;\ n_1 = 0,63;\ n_2 = 0,63;\ n_3 = -1,0$ (curves *3,4*) are compared. The analysis demonstrates that the coefficient n_3 significantly effects the dependence *p(t)* (particularly, the coefficient n_3 affects the rise rate of the pressure in the chamber upon the fuel ignition). The calculations show that for $n_3 = -1,0$ (curves *3,4*) the pressure growth rate in the chamber during the period from 0.065 sec to 0.130 sec (corresponds to the flame spread period through the central channel) is noticeably shorter than for $n_3 = -0.59$. The calculations of the problem of coefficients a, n_1, n_2, n_3 selection using the methods of mathematical programming for the large solid propellant showed that the best coincidence between the calculated curves *p(t)* and experimental results corresponds to the variants in which it is assumed that the formula of Dyunze-Zhimolokhin is inapplicable for the heat exchange calculation in the central channel of the charge. The formula of heat exchange in the form of $Nu = 0.25 \cdot \mathrm{Re}^{0.67}\,\mathrm{Pr}^{0.8}(1 + x / D)^0$ allows you to get good results.

In conclusion, we should note a high efficiency of applying the mathematical programming methods for identifying the calculation and experimental results. The developed method allows you to determine the coefficient values, the exact values of which cannot be known by the designer. Thus, the calculations showed that using the ignition composition weigh in the tablet form, the mathematical models of intrachamber processes should be introduced in the corrections taking into account the escape of the part of tablets from the igniter body. The optimal coincidence in this case is provided when it is assumed that from 20 to 40% of the tablets of the pyrotechnic composition escape the igniter body during its operation. In addition, the calculations demonstrated the high efficiency of the deformable polyhedron method for solving this class of problems. The application of Newton method is less effective.

It should be noted that in the discussed formulation the problem of selecting the parameters of the gas generator device with pellet fuel was successfully solved in the design of the gas generator for automobile airbags [12].

KEYWORDS

- **Chord method**
- **conjugate gradients method**
- **FORTRAN**
- **gradient method**
- **half-interval method**
- **iteration method**
- **Lagrangian multiplier method**
- **linear system**
- **MathCad**
- **Newton method**
- **non-linear system**
- **penalty function method**

CHAPTER 5

DYNAMIC SYSTEMS

CONTENTS

5.1 TECHNICAL OBJECT AS A DYNAMIC SYSTEM

5.1.1 GENERAL INFORMATION ON DYNAMIC SYSTEMS

The forecast of dynamic features of a technical object functioning is an important task to be solved by an engineer on the object designing stage. As a rule, the tasks to be solved on the object designing stage are formulated in a technical task. Thus, for instance, a rocket missile design assumes the setting up of maximum permissible g-loads acting upon it during its movement along the planned trajectory [1, 83, 130, 168, 177, 203]. When designing a missile, it is necessary to provide its movement stability along the flight trajectory [175, 188]. The forecast of missile stability quality can be assessed when solving a flight dynamics problem [242]. A particular place in designing problems belongs to those on providing a rocket missile guided flight [179, 234].

The foregoing could be applied to any other vehicle – automobile, locomotive, aircraft, etc. The performance of any heat engine – rocket, gas-turbine, piston – also requires the solution of a number of problems on dynamics and control [36, 97, 218, 222]. It can be asserted that the necessity to solve similar problems arises only when analyzing the functioning of technical objects. The dynamic development of objects and processes is observed when analyzing physical and chemical systems [237], in biologic objects on macro- and micro-levels [105]. The regional economy development in one or another country, commonwealth of countries is also objects, the change in dynamics of which needs to be modeled and analyzed [236].

Under dynamic systems in engineering or natural science we take any object or process for which the initial state defined in the form of the initial data population is unambiguously fixed, and the law determining the object or process development in time is set up.

Mathematical models, the mathematical record of which is based on differential or (and) integral transformations, table or geographic description, etc., can be applied to describe dynamic systems. Dynamic systems can develop with time discretely (cascades) or continuously (fluences). At discrete development, we say that the dynamic system is described by the sequence of states. In the systems with continuous changes with time we say about fluences during the system evolutionary development.

The model represented in the form of the system of ordinary differential equations is a spread out way of mathematical description

$$\frac{d\mathbf{y}}{dt} = \mathbf{F}(t,\mathbf{y}) \tag{5.1}$$

For the dynamic system modeled by the Eqs. (5.1), the following definitions and characteristics are applied:

- value vector $\mathbf{y} = (y_1, y_2, \ldots, y_N)$ is called a phase (image) point in N– dimensional space in which y_i – coordinates of this point;
- set of phase points defining the space of states composes the phase space;
- movement of phase point in time is called phase trajectory;
- vector $\mathbf{F}(t,\mathbf{y})$ composed of the right sides of the Eq. (5.1).

$\mathbf{F}(t,\mathbf{y}) = (f_1(t, y_1, y_2, \ldots, y_N), f_2(t, y_1, y_2, \ldots, y_N), \ldots, f_N(t, y_1, y_2, \ldots, y_N))$ is called velocity vector;

- dynamic system is called conservative if it possesses the energy stock unchangeable with time;
- dynamic system is called dissipative if its energy decreases with time. If the system energy increases with time, this system refers to systems with negative dissipation;
- dynamic system is called autonomous, if the vector of right sides in the Eq. (5.1) does not contain forces dependent on time.

The simplest variants of recording mathematical models of technical systems are, as a rule, based on multiple (depending on degrees of freedom of the mechanical system investigated) application of Newton's second law written down in differential form. Thus, for example, in [217] the issues connected with the vibrations of transport moving along an uneven road are discussed. In [120] the issues connected with the dynamics of internal combustion engines are considered. In [38, 81, 175] the vibrations of dynamic systems found in problems on aircraft flight analysis are investigated.

In practice even more complicated mathematical models containing derivatives of high orders can be used $\frac{d^n y_i}{dt^n}$, $n > 2$. Such models can be arranged with experimental material that allows assessing the contribution of derivatives of high order on the evolution of vector $\mathbf{y}$. In this case, the class of the problem describing the dynamic system behavior does not mathematically change, since the equation for the derivative $\frac{d^n y_i}{dt^n}$ can be deduced to the system n of ordinary differential equations of the first order.

In technical applications connected with the analysis of dynamics systems other types of problems can be found, in particular:

- to arrange the automatic change of one or another parameter characterizing the technical object work process in accordance with the law preliminary chosen;
- to arrange the automatic selection of actions upon the technical object providing its functioning in the mode corresponding to the control aim set up.

In the theory of automatic control [19, 31] the first problem is called the problem of automatic regulations, and the second – the problem of automatic control. The mathematical modeling of automatic regulation systems, as a rule, comes to writing down the systems of ordinary differential equations of linear or (and) nonlinear types.

As an example, let us consider the dynamic system with the regulation demonstrated in Figure 5.1 [19].

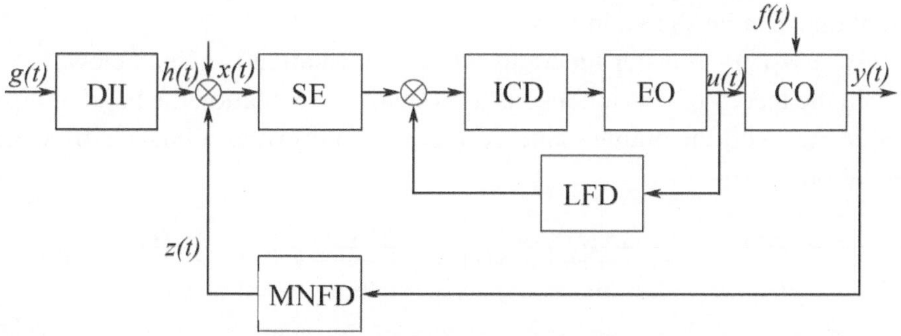

FIGURE 5.1 Scheme of closed system of automatic regulation (DII – device of instruction input; SE – sensing element; ICD – intensive-converter device; EO – executive organ; CO – control (regulation) object; LFD – local feedback device; MNFD – main negative feedback device).

The system works as follows. The control object (CO) produces the parameter $y(t)$ whose value can differ from the program value $y_{pr}(t)$ due to the disturbing effect $f(t)$. To provide the correlation $y(t) = y_{pr}(t)$, the regulation (control) system is used containing the device of instruction input (DII), sensing element (error sensor) (SE), intensive-converter device (ICD), executive organ (EO), local feedback device (LFD), main negative feedback device (MNFD).

The regulation (control) is provided as follows.

The device of instruction input (DII) receives the input signal $g(t)$ whose value corresponds to the value of controlling instruction action $g(t) = y_{np}(t)$. At DII output the signal value $g(t)$ can change to the value $g(t)$. In a particular case $h(t) = g(t)$. The signal $h(t)$ is received by the error sensor (summation unit). The same sensor receives the feedback signal $z(t)$ which is the converted output signal $y(t)$ after the main negative feedback device (MNFD). In a particular case $z(t) = y(t)$. The output signal $y(t)$ is a regulated variable whose value needs to coincide with the program set value $y(t) = y_{np}(t) = g(t)$. The error sensor produces the signal $x(t)$ defined as the difference $x(t) = h(t) - z(t)$. The regulating action $u(t)$ produced by the executive organ (EO) and acting upon the control (regulation) object (CO) is the function depending on the error ratio $x(t)$. In particular, if $x(t) \to 0$, then $u(t) \to 0$. It should be taken into consideration that the signal $x(t)$ formed by the error sensor will change on all devices available in the regulation (control) system – on sensing element, intensive-converter device, executive organ, etc.

Bonds and transformations, to which the signals on each device comprised by the regulation (control) system are subordinated, are demonstrated in the structural regulation schemes. Thus, the foregoing scheme, as a structural one, can be shown in Figure 5.2.

Here $W_i(s)$ – transfer functions whose information is given below.

In the theory of continuous linear systems of regulation the bond of input value $x(t)$ with the output value $y(t)$ is written down as a linear differential equation

$$a_n \frac{d^n y(t)}{dt^n} + \ldots + a_1 \frac{dy(t)}{dt} + a_0 y(t) = b_m \frac{d^m x(t)}{dt^m} + \ldots + b_1 \frac{dx(t)}{dt} + b_0 x(t). \tag{5.2}$$

The Eqs. (5.2) can be written down in the operating format in which the differentiation operator $\frac{d}{dt}$ is designated as $s = \frac{d}{dt}$

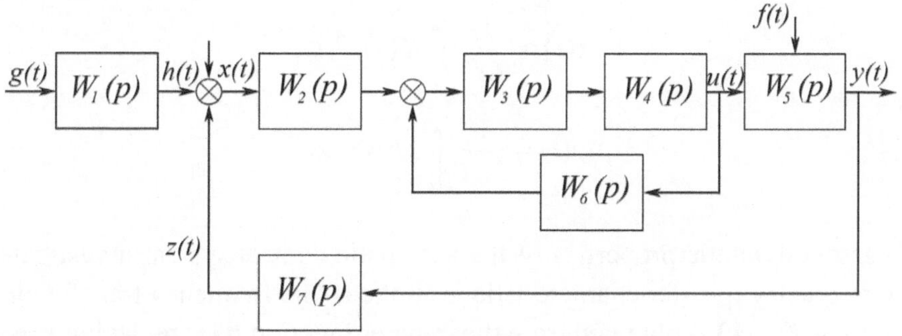

FIGURE 5.2 Structural scheme of the closed system of automatic regulation.

$$(a_n s^n + \ldots + a_1 s + a_0)y(t) = (b_m s^m + \ldots + b_1 s + b_0)x(t) \quad (5.3)$$

Let us use Laplace transformation [202]

$$X(s) = \int_0^{\infty} x(t)e^{-st}dt$$

$$Y(s) = \int_0^{\infty} y(t)e^{-st}dt$$

In this case, the Eqs. (5.3) can be written down as follows:

$$(a_n s^n + \ldots + a_1 s + a_0)Y(s) = (b_m s^m + \ldots + b_1 s + b_0)X(s)$$

or

$$Y(s) = W(s)X(s) \quad (5.4)$$

where $W(s)$ – transfer function defined by the following expression

$$W(s) = \frac{b_m s^m + \ldots + b_1 s + b_0}{a_n s^n + \ldots + a_1 s + a_0} = \frac{B(s)}{A(s)}$$

A differential equation constraining the output $x(t)$ and input $x(t)$ values can be restored from the operator Eq. (5.4) using the reverse Laplace transformation.

$$x(t) = \frac{1}{2\pi i}\int_{-i\infty}^{i\infty} X(s)e^{st}ds$$

$$y(t) = \frac{1}{2\pi i}\int_{-i\infty}^{i\infty} Y(s)e^{st}ds$$

To assess dynamic properties of the automatic control system investigated, the frequency transfer characteristic is applied. The frequency transfer characteristic $W(j\lambda)$ is obtained from the transfer function $W(s)$ replacing $s = j\lambda$ (j – imaginary unit, λ – angular oscillation frequency). Taking the foregoing into account, the transfer function can be demonstrated as follows:

$$W(j\lambda) = \frac{\alpha(\lambda) + j\beta(\lambda)}{\gamma(\lambda) + j\delta(\lambda)} = P(\lambda) + jQ(\lambda) = A(\lambda)\cdot\exp(j\Theta(\lambda)) \qquad (5.5)$$

In the Eq. (5.5) $P(\lambda), Q(\lambda)$ – real and imaginary components of frequency characteristic, $A(\lambda), \Theta(\lambda)$ – amplitude and phase characteristics. Their values are set up by the following correlations:

$$P(\lambda) = \frac{\alpha\gamma + \beta\delta}{\gamma^2 + \delta^2}; \quad Q(\lambda) = \frac{\beta\gamma - \alpha\delta}{\gamma^2 + \delta^2};$$

$$A(\lambda) = \sqrt{\frac{\alpha^2 + \beta^2}{\gamma^2 + \delta^2}}; \quad \Theta(\lambda) = \mathrm{arctg}\left(\frac{\beta\gamma - \alpha\delta}{\alpha\gamma + \beta\delta}\right)$$

The amplitude characteristic $A(\lambda) = \frac{y(\lambda)}{x(\lambda)}$ – the ratio of dimensionless amplitudes of oscillations at the output and input of the dynamic system as the function of frequency of harmonic oscillations at the system input. $\Theta(\lambda)$ – dependence of phase shift of oscillations at the output on the frequency of harmonic oscillations at the system input.

Frequency transfer functions are convenient to apply to the analysis of the functioning of dynamic systems. In particular, they allow setting up resonance frequencies λ at which the amplitude frequency characteristic $A(\lambda)$ (or logarithmic amplitude frequency characteristic $\ln A(\lambda)$) becomes unacceptably large. The knowledge of such frequencies helps preventing the instability of dynamic system performance inserting special filters into their composition. It has been discussed earlier that amplitude and frequency characteristics are used when developing the algorithms of regulation system performance, in particular, for setting up the values of resonance frequencies.

The models of dynamic systems are extremely important for practice. Therefore the increased attention has been paid to their investigation. The dynamic system model can contain several differential equations (e.g., the problem on the dynamics of a solid body movement), but can also comprise millions of differential equations (e.g., problems on the movement and interaction of molecules and atoms being of interest in chemistry, physics, nano-engineering). Practices, methods and regularities reasonable for relatively small dynamic systems can be also reasonable for large dynamic systems.

The following problems on the behavior of dynamic systems are interesting for practice:

- character of the development of evolutionary processes;
- asymptomatic behavior of a dynamic system (at $t \to \infty$);
- stability of a dynamic system;
- resonance phenomena in dynamic systems, etc.

The development of evolutionary processes in a dynamic system is set up by solving the system of differential Eqs. (5.1), at that both analytical and numerical methods of their solution can be applied. It should be pointed out that many properties of the evolutionary development of a dynamic system can be correctly forecasted by the type of the system of differential equations whose properties were investigated in numerous literature (e.g., [108]). Large dynamic systems with nonlinear right parts require the application of computational methods.

The solution of the system of differential equations allows setting up the dependence of dynamic parameters y_i on time $y_i = y_i(t)$. However, when investigating dynamic systems, the dependencies interconnecting all parameters y_i with each other – $\phi(y_1, y_2, \ldots, y_i, \ldots, y_n)$ built up with different values of initial data are also applied. Such dependencies are called the phase portrait of a dynamic system and can be arranged for different time moments $t \in (0, \infty)$. The limiting set of phase portraits of a dynamic system corresponding to $t \to \infty$ is called an attractor. The areas of one or another behavior of the dynamic system can be set up by the graphic representation of phase portraits, and also its stability can be evaluated, etc. Unfortunately with a large number of differential equations describing the dynamic system behavior, the application of phase portraits loses its visualization.

The analysis of asymptomatic behavior of a dynamic system allows answering the following questions important for practice:

- Does the stationary state exist for the system parameters?
- Is this stationary state stable?

The simplest way of investigating the asymptomatic behavior of a dynamic system is the solution of the system of ordinary differential Eqs. (5.1). Actually if there is a stationary state and it is stable, it will be obtained by solving the system (5.1) at large values of the process time ($t \to \infty$). However, other ways of investigating stationary states are also applied. In particular, in nonlinear systems instead of the system of ordinary differential Eqs. (5.1) the system of nonlinear algebraic equations obtained from the system (5.1) by the assumption that $\frac{dy}{dt} = 0$ is solved

$$
\begin{gathered}
f_1(y_1, y_2, \ldots, y_3, \ldots, y_n) = 0 \\
f_2(y_1, y_2, \ldots, y_3, \ldots, y_n) = 0 \\
\ldots \\
f_i(y_1, y_2, \ldots, y_3, \ldots, y_n) = 0 \\
\ldots \\
f_n(y_1, y_2, \ldots, y_3, \ldots, y_n) = 0
\end{gathered}
\tag{5.6}
$$

The system of nonlinear Eqs. (5.6) can have any improvised character. The methods discussed in Chapter 4 can be applied for solving the system. Here, the method of equation reduction (5.6) to the problem of mathematical programming with further solution of this problem, for example, with genetic methods is the most effective.

The issues of the stability of dynamic systems are most easily solved by the analysis of the solution stability of the systems of ordinary differential equations [202].

The issues of resonance are topical when creating any new technical objects. In some cases, the phenomena of resonance are useful in the object designed, but in other cases, they are dangerous and should be avoided. Nevertheless, for the majority of technical objects the information on the resonance possibility with the indication of certain frequencies corresponding to the resonance is important and useful. The issues connected with resonance phenomena are partly covered by the mathematical apparatus based on the application of transfer functions (amplitude, frequency and phase characteristics). Another approach consists in analyzing own values of the system of ordinary differential Eqs. (5.1).

Some of the issues listed will be considered in more details in the next chapters.

5.1.2 THERMODYNAMIC MODEL OF SOLID FUEL GAS GENERATOR PERFORMANCE

Let us consider the model of gas generator performance with the charge of solid fuel at the initial stage of its functioning as an example of a dynamic system (Figure 5.3). The gas generator performance starts with an initiator operation. The initiator contains the weigh of tablets of igniting mixture the combustion of which provides the flow of hot products into the internal chamber volume. The expansion of hot products inside the gas generator internal volume is heating the body and fuel charge. The body is inert and the fuel, after being heated up to a certain temperature, gets involved into combustion, the burning speed is defined by the combustion products pressure in the gas generator volume.

Physical processes accompanying the gas generator operation are described in detail in monographs [2, 8, 36, 48, 49, 91–94, 104, 107, 182, 204, 205, 244, 246–248]. The modeling of installation operation at the initial non-stationary period of time assumes the consideration of the following processes [244]:

- initiator performance;
- expansion of the combustion products of the initiator igniting composition and solid fuel (after its ignition) through the gas generator free volume (gas-dynamic processes);
- computational domain of deformation;

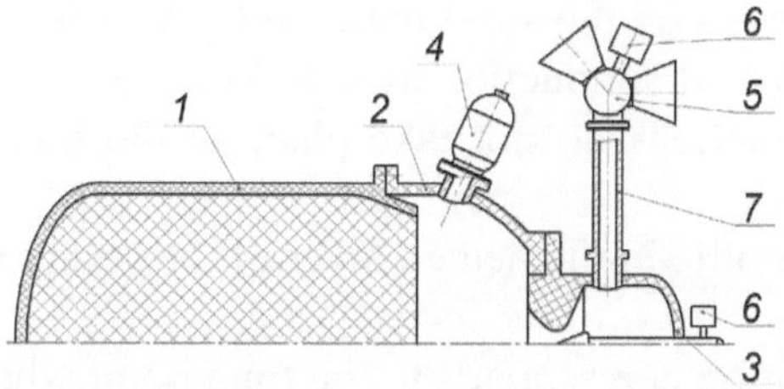

FIGURE 5.3 Gas generator construction diagram (1 – igniter, 2 – solid fuel).

- heat release from the gaseous phase to the body and fuel charge surfaces;
- heating of the body material and fuel, fuel ignition and its combustion after the ignition.

Let us list the main assumptions taken by the processes indicated:

- flow of combustion products in the gas generator combustion chamber is considered in the setting averaged by internal volume, at that the pressure values in the gas generator combustion chamber and igniter body volume coincide;
- it is assumed that the weigh of igniting composition placed inside the igniter gets involved into the combustion during the specified time period by the exponential law;
- combustion surface of the igniting composition weigh changes during combustion by the geometric law corresponding to the spherical shape of the igniting composition tablets;
- combustion products placed inside the combustion chamber are the chemically nonreactive mechanical mixture containing "cold" gas (air, nitrogen, etc.) that initially filled the space inside the chamber, combustion products of the igniting composition and solid fuel. The combustion products are considered an ideal compressible gas (viscosity influence and mixture heat conductivity are neglected). The mass concentration of the components indicated changes with time;
- change in the combustion chamber internal volume is conditioned only by the solid fuel burning;
- heat exchange between the combustion products and solid boundaries of the computational area (fuel surface, gas generator body, etc.) is performed in the combined mode (convective, radiant and conductive components of the heat flow are taken into account) and is set up by the experimental criterial correlations;
- radiate properties of the combustion products correspond to the properties of "grey" gas and radiate properties of solid materials – to the properties of "grey" material;
- heat exchange inside the solid material (fuel or body material) is performed due to heat conduction mechanism;
- solid-phase reactions do not take place in the fuel charge during its heating;
- temperature profile in the heated layer of the solid material is of exponential type;
- solid fuel ignition corresponds to the time point when its surface temperature becomes critical which is called the ignition temperature;

- after the ignition, the fuel is burning with the rate defined by the pressure level of combustion products in the gas generator chamber and set up by the experimental dependence.

Taking the assumptions into consideration, the equations of thermodynamic processes in the gas generator combustion chamber will be as follows:

$$\frac{dW}{dt} = u_f S_{f0}$$

$$\frac{d\rho W}{dt} = G_{\rho i} + G_{\rho f} - G_c$$

$$\frac{d\rho \alpha_i W}{dt} = G_{pi} - G_c \alpha_i$$

$$\frac{d\rho \alpha_f W}{dt} = G_{\rho f} - G_c \alpha_f$$

$$\frac{d\rho WE}{dt} = G_{Ei} + G_{Ef} - kG_c E$$

$$\frac{dz}{dt} = \frac{u_i}{e_{\max}}, \quad z \le 1 \tag{5.7}$$

$$\frac{d(T_m - T_0)^2}{dt} = \frac{2}{c_m \rho_m \lambda_m} |q_m| q_m$$

$$\frac{d(T_f - T_0)^2}{dt} = \frac{2}{c_f \rho_f \lambda_f} |q_f| q_f$$

$$p = \rho(k-1)E$$

$$k = {}^{c_p}\!/\!{}_{c_v}$$

$$c_p = c_{pi}\alpha_i + c_{pf}\alpha_f + c_{p0}(1 - \alpha_i - \alpha_f)$$

$$c_v = c_{vi}\alpha_i + c_{vf}\alpha_f + c_{v0}(1 - \alpha_i - \alpha_f)$$

$$T = {}^{E}\!/\!{}_{c_v}$$

$$G_c = A_c p \cdot (\varphi_p F)$$

$$\begin{cases} A_c = \sqrt{\dfrac{k}{\phi_{\hat{e}} RT}} \left(\dfrac{2}{k+1} \right)^{\frac{k+1}{2(k-1)}}, & at \quad \dfrac{p_a}{p} \le \left(\dfrac{2}{k+1} \right)^{\frac{k}{k-1}}, \\ A_c = \sqrt{\dfrac{2k}{\phi_{\hat{e}} (k-1) RT} \left(\left(\dfrac{p_a}{p} \right)^{\frac{2}{k}} - \left(\dfrac{p_a}{p} \right)^{\frac{k+1}{k}} \right)}, & at \quad \left(\dfrac{2}{k+1} \right)^{\frac{k}{k+1}} < \dfrac{p_a}{p} < 1. \end{cases}$$

$$G_{\rho i} = \rho_i S_{i0} (1-z)^2 u_i$$

$$G_{Ei} = G_{\rho i} H_i$$

$$u_i = u_{1i} \left(\frac{p}{0.98 \cdot 10^5} \right)^{\nu_i}$$

$$G_{\rho f} = \rho_f S_f u_f$$

$$G_{Ef} = G_{\rho f} H_f$$

$$\begin{cases} u_f = u_{1f} \left(\dfrac{p}{0.98 \cdot 10^5} \right)^{\nu_f}, & at \quad T_f \ge Ti, \\ u_f = 0 \quad at \quad T_f < T_i \end{cases}$$

$$q_f = q_{fc} + q_{fr}$$

$$q_{fc} = Nu \frac{\lambda}{D} (T - T_f)$$

$$q_{fr} = \sigma_0 \frac{1}{1/\varepsilon + 1/\varepsilon_f - 1} (T^4 - \grave{O}_f^4)$$

$$q_m = q_{mc} + q_{mr}$$

$$q_{mc} = Nu \frac{\lambda}{D} (T - T_m)$$

$$q_{mr} = \sigma_0 \frac{1}{1/\varepsilon + 1/\varepsilon_m - 1} (T^4 - T_m^4)$$

$$Q_f = q_f S_f$$

$$Q_m = q_m S_m$$

$$\frac{1}{\lambda} = \frac{\alpha_i}{\lambda_i} + \frac{\alpha_f}{\lambda_f} + \frac{1-\alpha_i-\alpha_f}{\lambda_0}$$

$$Nu = AGr^{\gamma}$$

$$Gr = \frac{g\rho\beta_t D^3 (T - T_m)}{\mu^2}$$

In the system of Eqs. (5.7) the ordinary differential equations are successively written down for the combustion chamber volume value W, the conservation equation for the total mass of the products placed in the combustion chamber volume, the equations for mass conservation of combustion products of igniting composition and fuel, the equation of energy conservation, the equation for calculating the relative combustion crown z of the igniting composition grains, the equations for temperature determination on the solid fuel surface T_f and the body material surface T_m. Additionally the algebraic correlations that allow calculating the right sides of the system of ordinary differential equations – the equations for thermodynamic and thermophysical characteristics of the mixture (pressure p, adiabatic exponent k, specific heat capacity of the mixture under the constant pressure c_p and constant volume c_v, temperature T), the equation for calculating the consumption G_c of the combustion products from the chamber into the environment (environment pressure p_a), the equations for determining mass (G_{pi}, $G_{\rho f}$) and energy input (G_{Ei}, G_{Ef}) from the weigh of igniting composition and solid fuel are written down.

The following designations are given in the Eqs. (5.7):

- α_i, α_f – mass concentration of combustion products of the weigh of igniting composition and solid fuel;
- $c_{pi}, c_{vi}, c_{pf}, c_{vf},$ – values of specific heat capacities for the combustion products of the weigh of igniting composition and solid fuel;
- c_{p0}, c_{v0} – values of specific heat capacities, respectively under constant pressure and constant volume, for air (or gas that initially filled the internal volume of the gas generator or engine chamber);
- $\varphi_p F$ – consumption complex determined through the product of consumption coefficient φ by the hole area F, through which the combustion products exhaust into the environment;

- A_c – coefficient of the exhaust of combustion products from the igniter body;
- φ_κ – coefficient of heat loss in the engine chamber (coefficient values are in the range from 0.90 up to 0.95);
- R – gas constant;
- S_{i0} – area of the combustion surface of the weigh of igniting composition at the initial time point;
- H_i, H_B – heat content of the combustion products of the igniter weigh;
- q_f, q_{fc}, q_{fr}, q_m, q_{mc}, q_{mr} – total, convective and radiate heat flows coming into the fuel charge and body material of the gas generator, respectively;
- Q_f, Q_m, T_0 – amount of heat coming into the fuel surface and body material surface, initial temperature of the materials;
- λ – heat conduction coefficient of combustion products;
- Nu, Gr – criteria of Nusselt and Grashof conformities;
- σ_0 – Stefan-Boltzmann constant ($\sigma_0 \approx 5.75 \cdot 10^{-8}$ W/m$^2 \cdot$ K^4);
- $\varepsilon, \varepsilon_f, \varepsilon_m$ – coefficients of the blackness of combustion products, solid fuel surface and body material surface, respectively;
- D – characteristic dimension of the combustion chamber free volume;
- A, γ – experimental coefficients;
- g – acceleration of free falling ($g = 9.81$ m/s);
- β_t – temperature coefficient ($\beta_t = 1/273$ 1/K);
- μ – dynamic viscosity of combustion products.

The Eqs. (5.7) correspond to the Eqs. (5.1), at the same time, the following should be understood under vector **y**:

$$\mathbf{y} = \begin{pmatrix} W \\ \rho W \\ \rho \alpha_в W \\ \rho \alpha_m W \\ \rho W E \\ z \\ (T_м - T_0)^2 \\ (T_m - T_0)^2 \end{pmatrix}$$

The right sides of ordinary differential equations are calculated by the foregoing algebraic equations.

The system of Eqs. (5.7) is solved under the initial conditions

$$\mathbf{y}_0 = \begin{pmatrix} W_0 \\ \rho_0 W_0 \\ 0 \\ 0 \\ \rho_0 W_0 E_0 \\ 0 \\ 0 \\ 0 \end{pmatrix}$$

Let us consider the gas generator system that differs from the previous one (Figure 5.3) by the availability of the control block. The gas generator shown in Figure 5.4 additionally comprises the control block containing the control actuator 9, nozzle block 8, flow control of the combustion products 6 and gas duct 5.

The combustion products coming into the chamber volume 2 are distributed into the nozzle blocks 8. Starting from some given time point the consumption of the combustion products through the nozzle blocks 8 is controlled by the actuator 9. The control is provided by the change in the minimal section area in the nozzle blocks and it is constantly performed until the finish of the gas generator operation in accordance with the given law of pressure control in the combustion chamber $p_{pr}(t)$ (Figure 5.5). The program pressure in the chamber contains the zones of increased and decreased pressure. The transition from one mode to another is performed by the given law, not necessarily linear.

In accordance with the theory of controllable schemes [19, 31, 67, 99, 121, 168], the gas generator block diagram can have the appearance demonstrated, in particular, in Figure 5.6.

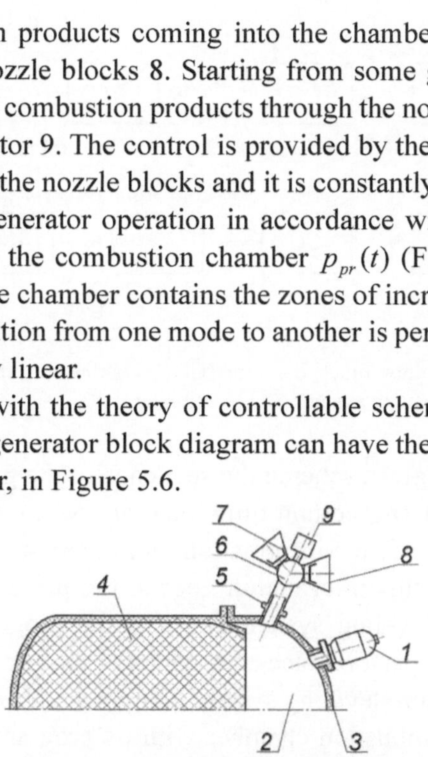

FIGURE 5.4 Controllable gas generator construction diagram (1 – igniter; 2 – combustion chamber free volume; 3 – chamber body; 4 – fuel charge; 5 – gas duct; 6 – flow control of the combustion products; 7 – nozzle cover; 8 – nozzle block; 9 – control actuator).

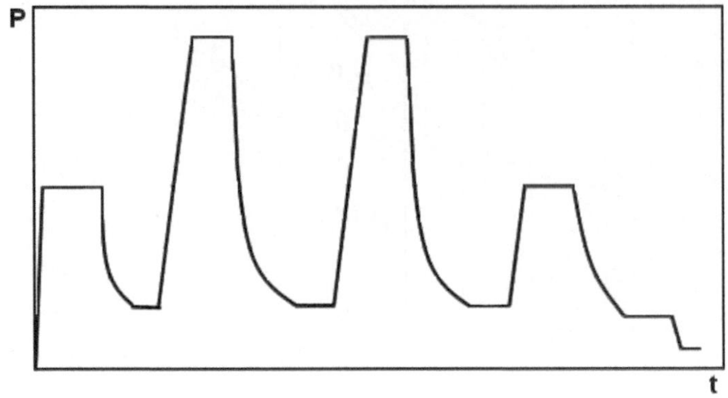

FIGURE 5.5 Typical scheme of the pressure change in the gas generator combustion chamber as a time function.

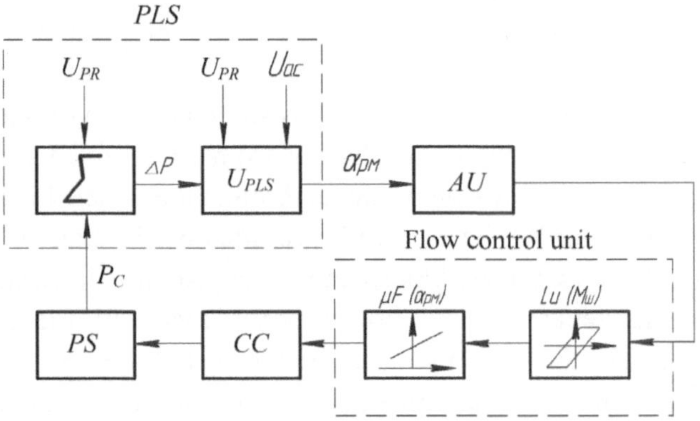

FIGURE 5.6 Gas generator block diagram (CC – combustion chamber; PS – pressure sensor; PLS – pressure leveling system; AU – actuating unit).

According to the given scheme the sequence of the gas generator operation is as follows. After a certain time interval (cycle), the pressure sensor PS is required to provide the information on the pressure value in the combustion chamber p_c, this information goes to the pressure leveling system PLS. The pressure leveling system contains the programmed cyclogram of the gas generator performance that contains the program pressure values $p_{pr}(t)$. The pressure leveling system compares the actual pressure p_c in the gas generator combustion chamber with its program value p_{pr}, and by the error ratio $\Delta p = p_c - p_{pr}$ the control signal U_{kont} is produced expressed in the rotation angles of the actuating unit AU shaft. The value U_{kont} is set up taking into account the values of program signal U_{pr} and stabilization controller signal U_{ac}. In accordance with the incoming signal, the actuating unit changes through kinematics the location of shafts of the flow control units of all nozzle control blocks by the angle $\zeta_{\text{рм}}$. As a result, the consumption complex $(\varphi_p F)$ changes to the value providing the condition of the pressure error at the level $\Delta p \approx 0$.

In Figure 5.7 you can see the scheme of practicable flow control units [65, 165, 180–182, 210, 218, 252]. In this scheme the combustion products enter through the inlet from the engine chamber, and flow into the nozzle blocks through the outlets *1* and *2*. When the shaft of the flow control unit rotates (its rotation is provided by the actuating unit shaft rigidly connected with the shaft of the flow control unit), the part of outlets either open or close. In particular, the complete opening of the outlet *1* can be provided with the completely closed outlet *2*, and vice versa.

The dependence of the consumption complex $(\varphi_p F)$ on the turning angle ζ_{pm} of the actuating unit can be found experimentally and demonstrated by the table or functional relationship. However, when modeling the processes to control the consumption characteristics of the gas generator, it is absolutely important to know not only the listed dependencies but also the dependencies for values of the derivative $\dfrac{d(\varphi_p F)}{d\zeta_{pm}}$. In some cases – also the second derivative $\dfrac{d^2(\varphi_p F)}{d\zeta_{pm}^2}$.

The shaft with the flow control unit is activated by the actuating unit. The actuating unit operates in the pulse mode with the set up time interval τ_0 between the impulses. The angle, at which the control unit turns at the

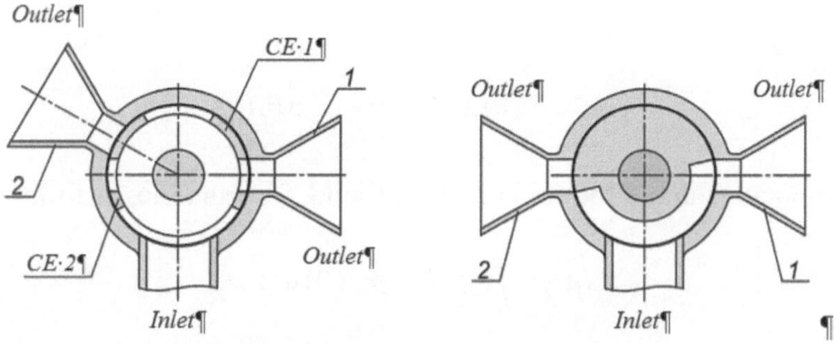

FIGURE 5.7 Scheme of the flow control unit of the combustion products (1, 2 – outlets of nozzle blocks; CE 1, CE 2 – control elements).

beginning of each cycle, is determined by the signal value δ_s produced by the control system. The signal level is limited, therefore, the angle at which the control unit shaft can be rotated through the regular operation cycle of the actuating unit is limited as well. The connection between the angle ζ_{pm} of the turn of the flow control unit shaft and the signal level δ_s transferred by the control system to the actuating unit is determined by the relation:

$$k_1 k_2 \frac{d^2\zeta_{pm}}{dt^2} + (k_1 + k_2)\frac{d\zeta_{pm}}{dt} + \zeta_{pm} = K\,\delta_c(t) \qquad (5.8)$$

In the Eq. (5.8) – $K, k_{1,} k_2$ – coefficients conditioned by the design of the actuating unit and control unit shaft whose values are identified by the experimental results.

The gas generator in question is the unit of automatic control, therefore, the approaches applied in the theory of dynamic systems and theory of automatic control are entirely applicable to it [19, 31, 67, 99, 121, 237]. In accordance with the theory basic guidelines, a gas generator can be referred to closed systems. For such systems the evaluated algorithms for the control of output parameters are available, including those evaluated under the action of perturbing factors on the system. These algorithms can also be applied when creating the control systems for solid fuel gas generators. Thus the Eq. (5.8) can be rewritten as follows:

$$W_{\zeta\delta}(s) = \frac{\zeta_{pm}(s)}{\delta_c(s)} = \frac{K}{(k_1 s + 1)(k_2 s + 1)} = \frac{K}{k_1 k_2 (s + 1/k_1)(s + 1/k_2)}$$

The dependence by which the value of the signal transferred to the actuating unit is calculated can contain components of proportional and integral control [31]

$$\delta_c(t) = \delta_c(0) + a_1\, I_1(t) + a_0\, \Delta P_f(t) \qquad (5.9)$$

The components in the formula (5.9) are found, for example, as follows:

$$I_1(t) = \int_{\tau_*}^{t} ((p(t) - p_{pr}(t))\, dt\,;$$

$$a_0 = a_2 + a_3\, \tau_0 \int_{\tau_*}^{t} p_{pr}(t)dt\,;$$

$$\Delta P_f(t) = p_f(t) - p_{pr}(t);$$

$$\frac{p_f(t)}{p(t)} = W_f(s)$$

In the foregoing equations the following designations are used:

- t, τ_*, τ_0 – process current time, time corresponding to the moment of control system switching on, time interval during which the control impulse is processes;
- $p(t)$, $P_{pr}(t)$ – current pressure value in the engine chamber and its program value;
- $\delta_c(0), a_1, a_2, a_3$ – experimentally found coefficients whose values are found by the pressure $P_{pr}(t)$ (are set in the table form as initial data);
- $\Delta P_f(t) = P_f(t) - P_{pr}(t)$ – difference of the phase pressure (pressure set by the pressure sensor PS, Figure 5.6) from the program level;
- $W_f(s)$ – transfer function by pressure. Till the time moment $t \le \tau_*$ – $W_f(s) = 1$. After this time point its values are determined by the formula:

$$W_f(s) = \frac{T_1^2 s^2 + 2\xi_1 T_1 s + 1}{T_2^2 s^2 + 2\xi_2 T_2 s + 1}$$

In the formula for $W_f(s)$ the values of empirical coefficients T_1, T_2, ξ_1, ξ_2 are set in the table form as pressure functions or as constants for the whole operation period of the solid fuel controllable engine unit. Taking into account the transfer function $W_f(s)$, the formula for pressure $p_f(t)$ can be written down as follows:

$$T_2^2 \frac{d^2 p_f}{dt^2} + 2\xi_2 T_2 \frac{dp_f}{dt} + p_f = T_1^2 \frac{d^2 p}{dt^2} + 2\xi_1 T_1 \frac{dp}{dt} + p \qquad (5.10)$$

In the last equation the value of the right side is calculated when integrating the equations of thermogasdynamics for the internal volume of the combustion chamber. Therefore it can be considered that the values of the right side of the equations are known at any time point. In particular case, it can be considered that the transfer function $W_f(s) \equiv 1$ during the entire gas generator operation. In this case, the following formula is true during the entire gas generator operation $p_f(t) \equiv p(t)$.

The control model analysis demonstrates that its application in the composition of the gas generator functioning model is possible only with the known values of coefficients $\delta_c(0)$, a_1, a_2, a_3 and T_1, T_2, ξ_1, ξ_2. The values of these coefficients can be found using the identification on the results of model or natural experiments.

The Eqs. (5.8)–(5.10) can be rewritten in the form (5.1) and solved together with the Eqs. (5.7) as the system of ordinary differential equations for the dynamic system. The system solution will be the only one at setting up the initial conditions for the derivatives in the Eqs. (5.8)–(5.10).

5.1.3 KINETIC PROCESSES IN CHEMICAL REACTORS

Problems of chemical kinetics take a special place among the problems referred to dynamic systems. Such problems can be found when analyzing processes in chemical reactors, heat engines operated on chemical fuel, in problems on the operation of furnace systems, etc. The topicality of solving problems of chemical kinetics grows and this is connected with the increased attention of the society to ecological issues.

Let us consider some problems on kinetic processes in chemical reactors found in practice [43, 61, 139, 147, 148, 206, 212, 219].

1. Non-reversible chemical reaction A→B (substance A decomposes with the formation of substance B)

Let us consider the reactor in which there are n chemical substances with mass concentrations α_i, $i=1,n$. Chemical substances β_j, $j=1,m$ are formed as a result of the chemical reaction flowing in the reactor. From the problem condition it follows that

$$\alpha_1+\alpha_2+\ldots+\alpha_i+\ldots+\alpha_{n-1}+\alpha_n=1,\quad 0\le\alpha_i\le1,$$

$$\beta_1+\beta_2+\ldots+\beta_i+\ldots+\beta_{n-1}+\beta_n=1,\quad 0\le\beta_i\le1$$

The equations for changing mass concentrations in the process of the chemical reaction can be written down as follows:

$$\frac{d\alpha_i}{dt}=-k\alpha_1^{a_1}\alpha_2^{a_2}\ldots\alpha_i^{a_i}\ldots\alpha_n^{a_n},\quad i=1,n,$$

$$\frac{d\beta_i}{dt}=k\alpha_1^{a_1}\alpha_2^{a_2}\ldots\alpha_i^{a_i}\ldots\alpha_n^{a_n},\quad j=1,m$$

Here k – constant of the reaction rate, $a_1, a_2, \ldots, a_n$ – reaction orders in relation to the components $\alpha_1, \alpha_2, \ldots, \alpha_n$. As a rule, the values of the coefficients specified are the functions of pressure and temperature in the chemical reactor.

The recording of the foregoing equations for the chemical reactor can be simplified. Actually there is no need to solve similar equations $n+m$. It is easier to proceed from the assumption that there are two substances in the chemical reactor with mass concentrations α_1 and α_2 ($\alpha_1 + \alpha_2 = 1$), respectively. The first substance is the collection of all reagents contained in the reactor before the chemical reaction. The second substance is the collection of all reagents contained in the reactor after the chemical reaction. In this case, the chemical composition of the combustion products in the reactor is described by the system of equations

$$\frac{d\alpha_1}{dt} = -k\alpha_1^{a_1}$$
$$\frac{d\alpha_2}{dt} = k\alpha_1^{a_1} \tag{5.11}$$

or

$$\frac{d\alpha_1}{dt} = -k\alpha_1^{a_1}$$
$$\alpha_1 + \alpha_2 = 1 \tag{5.12}$$

2. Non-reversible parallel reaction $A \frac{\to B}{\to C}$ *(material A decomposes into two materials B and C)*

In contrast to the previous problem, we will assume that there are two chemical reactions in the chemical reactor. The initial substance with the mass concentration α_1 is transformed into two new ones with the mass concentrations α_2, α_3 ($\alpha_1 + \alpha_2 + \alpha_3 = 1$). The reaction rates are set up by the coefficients k_{12}, a_{12} and k_{13}, a_{13}, respectively.

The equations of the kinetic process are written down as follows:

$$\frac{d\alpha_1}{dt} = -k_{12}\alpha_1^{a_{12}} - k_{13}\alpha_1^{a_{13}}$$
$$\frac{d\alpha_2}{dt} = k_{12}\alpha_1^{a_{12}} \tag{5.13}$$

$$\frac{d\alpha_3}{dt} = k_{13}\alpha_1^{a_{13}}$$

3. Non-reversible consecutive multistep reaction (material A decomposes following the scheme: $A \rightarrow B \rightarrow C \rightarrow D$)

Let there be four substances in the chemical reactor with mass concentrations $\alpha_1, \alpha_2, \alpha_3, \alpha_4$, respectively. The availability of the first substance produces the formation of the second one (reaction parameters k_{12}, a_{12}). The third one is formed from the second substance (reaction parameters k_{23}, a_{23}). The fourth substance is formed from the third one (reaction parameters k_{34}, a_{34}).

The equations determining the kinetic process for this situation are written down as follows:

$$\begin{aligned}
\frac{d\alpha_1}{dt} &= -k_{12}\alpha_1^{a_{12}} \\
\frac{d\alpha_2}{dt} &= k_{12}\alpha_1^{a_{12}} - k_{23}\alpha_2^{a_{23}} \\
\frac{d\alpha_3}{dt} &= k_{23}\alpha_2^{a_{23}} - k_{34}\alpha_3^{a_{34}} \\
\frac{d\alpha_4}{dt} &= k_{34}\alpha_3^{a_{34}}
\end{aligned} \tag{5.14}$$

4. Chemical reaction proceeding in forward and backward directions (reaction scheme: $A \leftrightarrow B$)

In the reactor chemical reactions can proceed in two directions. Let the forward reaction proceed with the rate k_f, and reaction order – a_f. The backward reaction rate – k_b, and reaction order – a_b. In this case, the Eq. (5.11) can be rewritten as follows:

$$\begin{aligned}
\frac{d\alpha_1}{dt} &= -k_f\alpha_1^{a_f} + k_b\alpha_2^{b} \\
\frac{d\alpha_2}{dt} &= k_f\alpha_1^{a_f} - k_b\alpha_2^{b}
\end{aligned} \tag{5.15}$$

Let us point out that the limiting values of the concentrations α_1, α_2 in the reactor correspond to the condition $\frac{d\alpha_1}{dt} = 0, \frac{d\alpha_2}{dt} = 0$. In this case, the following correlation is true

$$K_c = \frac{k_{np}}{k_{обр}} = \frac{\alpha_2^{a_{обр}}}{\alpha_1^{a_{np}}} \tag{5.16}$$

Coefficient K_c is called the equilibrium constant of the chemical reaction. Its value is the function of pressure and temperature in the chemical reactor and can be found based on theoretical assumptions or experimentally.

5. Non-reversible reaction of two substances with the formation of the third substance (reaction scheme: $A + B \rightarrow C$)

The chemical reactor can contain three substances with the mass concentrations $\alpha_1, \alpha_2, \alpha_3$, respectively. The first two substances interact (reaction parameters k, a) with the formation of the third one.

The equations determining the kinetic process are written down as follows:

$$\frac{d\alpha_1}{dt} = -k\alpha_1^a \alpha_2^a$$

$$\frac{d\alpha_2}{dt} = -k\alpha_1^a \alpha_2^a$$

$$\frac{d\alpha_3}{dt} = k\alpha_1^a \alpha_2^a \tag{5.17}$$

6. Non-reversible reaction of two substances with the successive formation of the third and fourth substances (reaction scheme: $A + B \rightarrow C \rightarrow D$)

The chemical reactor contains four substances with the mass concentrations $\alpha_1, \alpha_2, \alpha_3, \alpha_4$, respectively. The interaction of the first and second substances results in the formation of the third one (reaction parameters k_{12}, a_{12}). The fourth substance is formed from the third one (reaction parameters k_3, a_3).

The kinetic equations are written down as follows:

$$\frac{d\alpha_1}{dt} = -k_{12}\alpha_1^{a_{12}} \alpha_2^{a_{12}}$$

$$\frac{d\alpha_2}{dt} = -k_{12}\alpha_1^{a_{12}} \alpha_2^{a_{12}}$$

$$\frac{d\alpha_3}{dt} = k_{12}\alpha_1^{a_{12}}\alpha_2^{a_{12}} - k_3\alpha_3^{a_3}$$
$$\frac{d\alpha_4}{dt} = k_3\alpha_3^{a_3} \tag{5.18}$$

7. Combined chemical reaction of four materials implemented following the scheme $\begin{pmatrix} A+B\rightarrow C \\ C+B\rightarrow D \end{pmatrix}$

The chemical reactor contains four substances with the mass concentrations $\alpha_1, \alpha_2, \alpha_3, \alpha_4$, respectively. The interaction of the first and second substances results in the formation of the third one (reaction parameters k_{12}, a_{12}). The second substance interacts with the newly formed third one (reaction parameters k_{23}, a_{23}), at the same time, the final product – the fourth substance – is formed.

The kinetic equations are written down as follows:

$$\frac{d\alpha_1}{dt} = -k_{12}\alpha_1^{a_{12}}\alpha_2^{a_{12}}$$
$$\frac{d\alpha_2}{dt} = -k_{12}\alpha_1^{a_{12}}\alpha_2^{a_{12}} - k_{23}\alpha_2^{a_{23}}\alpha_3^{a_{23}}$$
$$\frac{d\alpha_3}{dt} = k_{12}\alpha_1^{a_{12}}\alpha_2^{a_{12}} - k_{23}\alpha_2^{a_{23}}\alpha_3^{a_{23}}$$
$$\frac{d\alpha_4}{dt} = k_{23}\alpha_2^{a_{23}}\alpha_3^{a_{23}} \tag{5.19}$$

We could continue with building up kinetic equations, however, the examples given are sufficient to write down the equation system for more complicated practicable cases of chemical interaction of reacting materials. Let us point out that in practice the chain of fuel chemical reacting to final combustion products can comprise over a hundred simplest chemical reactions of the types considered.

The following peculiarities when modeling chemical reactors can be pointed out:

- in the most general case the systems of kinetic equations are the systems of nonlinear ordinary differential equations. If the reactions between chemical reagents are of the first order ($a_i = 1$), separate equation systems (e.g., the Eqs. (5.11)–(5.14)) can be referred to the systems of nonlinear ordinary differential equations;

- values of different constants of reaction rates within the system of equations of chemical kinetics can differ by orders. Thus, for instance, in hydrogen and oxygen reactions the constant rate of the reaction $(H_2 + O = OH + H)$ has the order ~10^4 s^{-1}, and reaction $(O_2 + H = OH + O)$ – the order ~10^{14} s^{-1} [43].

Both peculiarities mentioned significantly complicate the solution of the considered class of ordinary differential equations. Due to speedily evolving solutions, such equation systems are referred to stiff differential equations [112, 221, 227, 250].

It should be pointed out that the foregoing models of kinetic processes are true for reactors in which the reaction product rates can be neglected. However, in practice, there are many technical objects that can be referred to chemical reactors but the rates of gas products cannot be neglected. In particular, these are heat engines (piston and combined internal combustion, gas-turbine, rocket, etc.), furnace systems and flue systems they comprise, purification systems for chemical reaction products, etc. The models of the above systems functioning are much more complicated than those discussed before. The basic notions connected with building up such models, their analysis and solution methods are considered, for example, in Refs. [7, 147, 148, 152, 164, 206].

5.2 PROPERTIES OF DYNAMIC SYSTEMS: PHASE PORTRAITS

One of the effective means of investigating the properties of dynamic systems is the building up of phase portraits of these systems. However, the application of phase portraits in practice is usually limited by the problems on dynamic systems whose mathematical model contains two or three ordinary differential equations. Besides, the visualization of graphic expression of phase portraits is provided in problems on the development of autonomous dynamic systems (in autonomous systems the right sides of differential equations do not depend on time).

Let us consider how the basic regularities in the development of dynamic systems are represented on phase portraits. For simplicity, the analysis will be carried out on the example of the dynamic system consisting of two differential equations

$$\frac{dy_1}{dt} = f_1(y_1, y_2)$$

$$\frac{dy_2}{dt} = f_2(y_1, y_2) \tag{5.20}$$

For the indicated system the phase portraits will be represented in the plane y_1y_2 called the phase plane. They will be built up by solving the system of differential Eqs. (5.20) for the derivatives y_1 and y_2 with further formation of the dependence $y_2(y_1)$. The dependencies $y_2(y_1)$ need to be built up many times for all the possible collections of initial conditions $y_1(t_0) = y_0$ and $y_2(t_0) = y_{20}$. The phase trajectories of all the possible solutions of the system of differential Eqs. (5.20) in the phase plane y_1y_2 satisfy the following differential equation

$$\frac{dy_2}{dy_1} = \frac{f_2(y_1, y_2)}{f_1(y_1, y_2)} \tag{5.21}$$

The features of the Eq. (5.21) will be represented when the following conditions are fulfilled

$$f_1(y_1, y_2) = 0 \tag{5.22}$$

$$f_2(y_1, y_2) = 0 \tag{5.23}$$

The points in which the conditions (5.22) or (5.23) are carried out in the phase plane are called "special points." As a rule, in the special points the dynamic system is in equilibrium, stable or unstable. It should be pointed out that the limiting states of the dynamic system investigated corresponding to $t \to \infty$ and called "trajectories" are of interest in practice.

The examples of phase portraits implemented for typical dynamic systems are demonstrated in Figure 5.8 [105, 214]. The phase portraits of dynamic systems with a special point called "node" are shown in Figures 5.8*a* and 5.8*b*. The stable node corresponds to the problems in which the dynamic system tends to stable equilibrium. The unstable node – to absolutely unstable systems. The dynamic systems with a special point "focus" (stable or unstable, Figures 5.8*c*, 5.8*d*) are characterized by the oscillating character of the changes in parameters. The oscillating systems without dissipation have a special point "center" (Figure 5.8*e*). A special point "saddle" (Figure 5.8*f*) determines the area of bifurcational (catastrophic) change in the dynamic system state (from one stable state into another stable one).

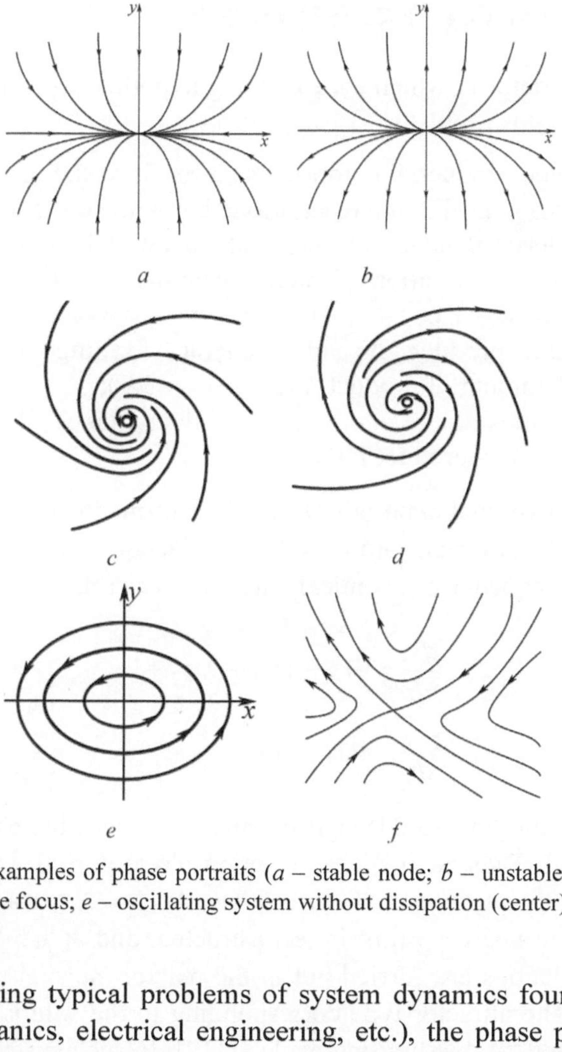

FIGURE 5.8 Examples of phase portraits (*a* – stable node; *b* – unstable node; *c* – stable focus; *d* – unstable focus; *e* – oscillating system without dissipation (center); *f* – saddle).

When solving typical problems of system dynamics found in practice (e.g., in mechanics, electrical engineering, etc.), the phase portraits come down to those shown in Figure 5.8. It is discussed in [214] that practically all the above special points can belong to the following equation

$$a\frac{d^2x}{dt^2}+b\frac{dx}{dt}+cx+d\left(x^2+(\frac{dx}{dt})^2\right)\frac{dx}{dt}=0$$

at different values of the coefficients a,b,c,d (this equation can come down to the system of two ordinary differential equations of the first order).

Let us specify other known problems of dynamic systems whose investigation results are discussed, for instance, in [78, 112, 214].

5.2.1 PROBLEM ON "BRUSSELATOR"

The chemical reactor in which the following hypothetical reactions proceed is considered in this problem:

- intermediate product X is formed as a result of the decomposition of substance A placed in the reactor together with substance B ($A \to X$);
- initial material B enters the chemical reaction with intermediate product X with the formation of intermediate material Y and final product D ($B + X \to D + Y$);
- intermediate products X and Y interact forming, as a result, the increased amount of product X ($2X + Y \to 3X$);
- final decomposition of intermediate product X takes place with the formation of a new product E ($X \to E$).

As a result, two final materials D and E are formed from the initial materials A and B. The system of ordinary differential equations corresponding to the considered sequence of chemical reactions is as follows:

$$\frac{dy_1}{dt} = -(\mu + 1)y_1 + y_1^2 y_2 + 1$$
$$\frac{dy_2}{dt} = \mu y_1 - y_1^2 y_2 \qquad (5.24)$$

Parameter μ in the Eqs. (5.24) significantly influences the character of the results obtained. Thus, at $\mu < 2$ the solution phase portrait is implemented with a special point "node" (Figure 5.9a). At $\mu = 2$ (parameter bifurcational value) the phase portrait is reconstructed, and at $\mu > 2$ the constant chemical oscillations are carried out in the reactor (special point "center," Figure 5.9b). The attractor type corresponding to the solution of the latter problem is called "stable limiting cycle" [117]. In such systems auto-oscillating modes take place regardless of the initial conditions.

5.2.2 EQUATIONS OF LOTKA-VOLTERRA

Equations of Lotka-Volterra are considered in many problems connected with the analysis of dynamic systems. These equations can be applied in the problems of kinetics [214], problems on the dynamics of biological populations [112], problems on modeling military-historical situations (conflicts, mobilization processes [150]), etc.

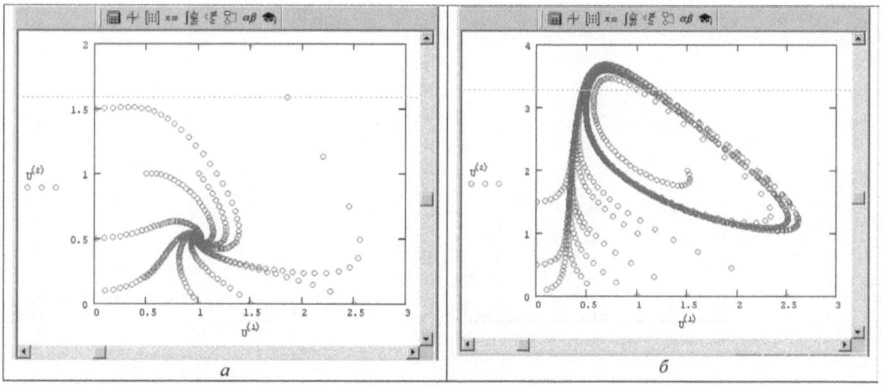

FIGURE 5.9 Phase portraits of "brusselator" at μ = var.

Equations of Lotka-Volterra are written down as follows:

$$\frac{dy_1}{dt} = k_1 a_1 y_1 - k_2 y_1 y_2$$

$$\frac{dy_2}{dt} = -k_3 a_2 y_2 + k_2 y_1 y_2$$

$$a - \mu = 0{,}5\,;\, b - \mu = 2{,}5 \tag{5.25}$$

In the Eqs. (5.25) k_1, k_2, k_3, a_1, a_2 – constants whose values are determined by the conditions of the problem being solved (or the parameters of chemical reactions, or the rate of population propagation and food stock, or the parameters characterizing a conflict development, etc.).

The solutions of the Eq. (5.25) can have the appearance shown in Figure 5.10 [112]. The figure demonstrates the typical decision for the derivatives y_1 and y_2, as well as the solution phase portrait. The given figures show that the system does not have a stationary solution, but the solution has an auto-oscillating character of non-harmonic type. The oscillation period, oscillation amplitudes, minimum and maximum values of the derivatives y_1 and y_2 are determined by the coefficients k_1, k_2, k_3, a_1, a_2.

5.2.3 DYNAMIC LORENTZ MODEL

The Lorentz model was proposed in 1963 as a simplified model of convective turbulent flow of liquid in a heated toroidal-shaped vessel. This model consists of three ordinary differential equations of the following type

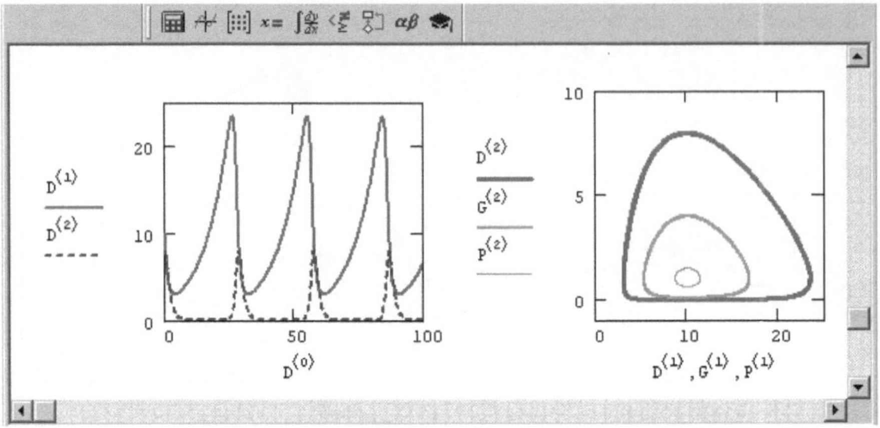

FIGURE 5.10 – Solution graph (on the left) and phase portrait (on the right) of the system "predator-prey".

$$\frac{dy_1}{dt} = k_1(y_2 - y_1)$$
$$\frac{dy_2}{dt} = -y_1 y_3 + k_2 y_1 - y_2 \quad (5.26)$$
$$\frac{dy_3}{dt} = y_1 y_2 - k_3 y_3$$

The coefficients k_1, k_2, k_3 are the parameters determining the equation system solution (5.26). The coefficient k_2 is the most significant, its change is in the range $k_2 < k_*$ (k_* – parameter bifurcational value) has an attractor with a stable special point (see Figure 5.8*c*). At $k_2 > k_*$ the solution phase portrait will have the appearance demonstrated in Figure 5.11. It should be noted that there are three unknown functions in the Lorentz model, therefore the system phase portrait should be determined not on the plane but in 3D space. For the simplicity and visualization the phase portrait is represented in the projection onto one of the planes of the phase space.

In accordance with the figure, there are two stable solutions. The system transition from one stable solution to another stable solution takes place in the mode which could be considered stochastic. However, it follows from the Eq. (5.26) that this transition is established by the system of determinate differential equations. The attractor corresponding to the Eq. (5.26) at $k_2 > k_*$ is called "strange attractor."

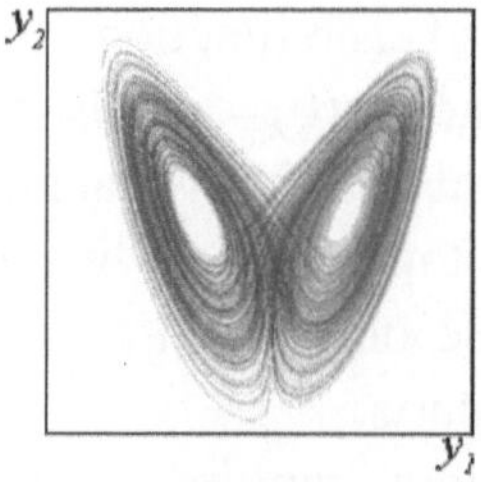

FIGURE 5.11 Projection of the phase portrait of Lorentz equation system.

More complicated examples of dynamic systems and their regularities can be found in special literature, for example [54, 77, 78, 105, 114, 115, 158, 214]. It can be additionally pointed out that recently the modeling in problems of nanotechnology has been intensively developing. The solution of such problems is also the solution of problems of dynamic systems comprising over a million of equations describing the movement of molecules and atoms.

5.3 STABILITY OF DYNAMIC SYSTEMS

5.3.1 STATIC AND DYNAMIC STABILITY

Stability is an important property of any dynamic system. At the same time, both static and dynamic stability are important. Under the static stability we understand the stability of the system being in equilibrium under unchangeable external conditions. Under the dynamic stability we understand the stability of the system at an unspecified point of time under the action of external conditions changing with time.

To assess the static and dynamic stability of the dynamic systems of the type (5.1), the mathematical apparatus of the stability of the systems of ordinary differential equations can be applied [108, 117, 173, 202].

The basic theorem of stability is as follows:

- solution $y_i = y_i(t)$, $(i = 1, n)$ of the system of ordinary differential equations of the following type

$$\frac{dy_i}{dt} = f_i(t, y_1, y_2, \ldots, y_n), \quad (i = 1, n) \tag{5.27}$$

is called stable by Lyapunov, if each function $y_i(t)$ is a uniformly continuous function from all initial values $y_i(0)=y_{i0}$ $(i=1,n)$ in the interval $0 \le t < \infty$ and little change in the initial conditions cannot cause significant changes in the solution. In particular, the system of linear differential equations with constant coefficients is stable only when all the roots of its characteristic equation have negative real parts.

The theorem of equilibrium stability (static stability) is formulated as follows:

The equilibrium solution $y_i = y_{i0}$, $(i=1,n)$ corresponding to such stationary point $(y_{10}, y_{20}, \ldots, y_{n0})$ that

$$f_i(t, y_{10}, y_{20}, \ldots, y_{n0}) = 0 \quad (i=1,n) \tag{5.28}$$

is stable if it is such for the linearized system

$$\frac{dy_i}{dt} = \sum_{k=1}^{n} \frac{\partial f_i}{\partial y_k}(y_k - y_{k0}) \quad (i=1,n) \tag{5.29}$$

where partial derivatives are taken at $y_1 = y_{10}$, $y_2 = y_{20}$, $\ldots$, $y_n = y_{n0}$. This takes place when all the roots λ_i $(i=1,n)$ of characteristic equation

$$\det\begin{pmatrix} \frac{\partial f_1(t,\mathbf{y})}{\partial y_1} - \lambda & \cdots & \frac{\partial f_1(t,\mathbf{y})}{\partial y_i} & \cdots & \frac{\partial f_1(t,\mathbf{y})}{\partial y_n} \\ \cdots & \cdots & \cdots & \cdots & \cdots \\ \frac{\partial f_i(t,\mathbf{y})}{\partial y_1} & \cdots & \frac{\partial f_i(t,\mathbf{y})}{\partial y_i} - \lambda & \cdots & \frac{\partial f_i(t,\mathbf{y})}{\partial y_n} \\ \cdots & \cdots & \cdots & \cdots & \cdots \\ \frac{\partial f_n(t,\mathbf{y})}{\partial y_1} & \cdots & \frac{\partial f_n(t,\mathbf{y})}{\partial y_i} & \cdots & \frac{\partial f_n(t,\mathbf{y})}{\partial y_n} - \lambda \end{pmatrix} = 0 \tag{5.30}$$

have negative real parts. The equilibrium will be unstable if the Eq. (5.30) has at least one root with a positive real part. If there are no roots with a positive real part but there are purely imaginary roots, the additional investigation is required.

The stability conditions by Lyapunov defined above are a powerful apparatus which can be applied to analyze the stability of technical systems both at the initial and unspecified time points:

- to analyze the stability at an unspecified time point, the initial equation system (5.27) should be linearized and reduced to the form, similar to (5.29);
- stability conditions resulting from the theorem of the stability of solutions by Lyapunov should be provided at an unspecified time point for the dynamic system under investigation, at the same time, it is necessary to the solve the characteristic equation of the type (5.30) at each time point.

It should be pointed out that the stability analysis of a dynamic system should be also carried out in the situation when the system is affected by the disturbances of random character.

5.3.2 APPLICATION OF LINEARIZATION FOR THE ANALYSIS OF SYSTEM STABILITY

The stable development of processes in dynamic systems is characterized by the fact that if a light action of disturbing factor is observed at an unspecified time point, after some time the system will return to its stable state. The disturbance development with time, in stable or unstable process, can be determined by solving the equations for the deviation y_i' $(i=1,n)$ found as the difference between the actual value of this magnitude $\tilde{y}_i$ and its equilibrium level y_i $(y_i'=\tilde{y}_i-y_i)$. The linearization method can be applied only when the disturbance value y_i' $(i=1,n)$ is significantly less than the actual value y_i $(i=1,n)$.

This corresponds to the following conditions

$$\left|\frac{y_i'}{y_i}\right| << 1 \quad (i=1,n); \tag{5.31}$$

$$\frac{y_i'}{y_i}\frac{y_j'}{y_j} \approx 0 \quad (i \neq j) \tag{5.32}$$

When solving technical problems in practice, it is believed that the linearization method can be applied, if $\left|\frac{y_i'}{y_i}\right| < 0{,}1$ $(i=1,n)$.

Let us see how the linearization procedure can be applied for the system of ordinary differential equations of the type (5.27). In accordance with [201], the following correlations for the disturbed values of the function $\tilde{y}_i$ can be written down

$$\tilde{y}_i = y_i + y_i'$$

$$\frac{d\tilde{y}_i}{dt} = \tilde{f}_i, \quad i = 1, N$$

$$\tilde{f}_i = f_i + f_i' = f_i + \sum_{j=1}^{M} \varphi_j \cdot \frac{\partial F_i}{\partial \varphi_j} \cdot \frac{\varphi_j'}{\varphi_j} = f_i + \sum_{j=1}^{M} f_{ij} \cdot \frac{\varphi_j'}{\varphi_j}$$

$$f_{ij} = \varphi_j \cdot \frac{\partial f_i}{\partial \varphi_j}$$

In the foregoing equations $\varphi_j, j = 1, M$ – external disturbance factors. In special cases, the values of the function f_{ij} take on zero values.

The differential equation for the disturbed value $\tilde{y}_i$ can be rewritten in one of the following forms convenient for application

$$\frac{dy_i\left(1 + \frac{y_i'}{y_i}\right)}{dt} = f_i + \sum_{j=1}^{M} f_{ij} \cdot \frac{\varphi_j'}{\varphi_j}, \quad i=1,\ N; \tag{5.33}$$

$$\frac{dy_i\left(\frac{y_i'}{y_i}\right)}{dt} = \sum_{j=1}^{M} f_{ij} \cdot \frac{\varphi_j'}{\varphi_j}, \quad i=1,\ N; \tag{5.34}$$

$$\frac{dy_i\left(\frac{y_i'}{y_i}\right)}{dt} = f_i', \quad i=1,\ N \tag{5.35}$$

5.3.3 LINEARIZATION OF THE EQUATIONS OF INTERNAL BALLISTICS FOR SOLID FUEL GAS GENERATOR

The system of algebraic and ordinary differential equations discussed above (5.7) is quite applicable to analyze processes in a solid fuel gas generator, and for this system the methods and approaches applied in the theory of dynamic systems can be used to the full extent. Thus, in particular, it is known (e.g., [136, 204, 205, 210]) that if the index of power in the law for the solid fuel combustion rate $0 < \nu_m < 1$, the solid fuel gas generator is stable as a dynamic sýstem. The phase portrait of the processes in such gas generator (e.g., in the plane "pressure-temperature") will correspond to the case with the special point "stable node." The stability of more complicated models of the functioning of solid fuel gas generators can be investigated when using the linearized analog together with the main model. Hereinafter under the parameters y_i we will mean, for instance, the density values of combustion products ρ, pressure p, energy E, rate u and others.

The problem on changing parameters in the internal volume of the gas generator by the Eqs. (5.7) is solved in the determinate setting assuming that all the unknown variables and initial data are precise numbers. However, it is not so in reality. For example, the actual values of fuel thermophysical characteristics can differ from the value set up in initial data and can be characterized as stochastic values. In this case, the unknown functions should also be considered as stochastic when solving the problem. Another difficulty when solving the considered class of problems consists in the lack of knowledge of precise values of some group of parameters which cannot be referred to stochastic (e.g., some parameter cannot be measured). Such uncertainty forces to solve the problem and process its solution results when varying the unknown parameters in all the possible range of their change.

The approach attractive for practice is the one when the determinate solution of the problem on the process development in a heat engine is considered as the determination of the mathematical expectation of unknown parameters and to determine the deviations from the mathematical expectation, the basic system of the equations is supplemented. It should be pointed out that in such approach it is also possible to solve the problem on the stability of internal processes in a heat engine.

The problem will be solved for the initial part of operation of solid fuel gas generator (Figure 5.3), the processes in which are described by the Eqs. (5.7). The physical picture of the process developed in the gas generator is as follows. First, the initiator combustion products (located on the gas generator left boundary) flow into the gas generator internal volume. Filling the internal volume, the products heat up the fuel surface layer, which ignites after heating above the critical value. In some time after the fuel ignition and beginning of the combustion product outflow from the internal volume, the parameters characterizing the processes in the gas generator are established.

Let us rewrite the equation system (Eq. (5.7)) in a simplified form

$$\frac{d}{dt}W = f_w$$

$$\frac{d}{dt}\rho W = f_\rho$$

$$\frac{d}{dt}\rho WE = f_E$$

$$p = \rho(k-1)E \tag{5.36}$$

For simplification the right sides of the Eq. (5.36) are designated as f_w, f_ρ and f_E. The constructive form of the right sides can be determined by [244].

Taking into account the Eqs. (5.31) and (5.32), the bond of equilibrium y_i and disturbed parameters $\tilde{y}_i$ with the deviations y_i' for the parameters of internal ballistics are written down as

$$\tilde{W} = W\left(1 + \frac{W'}{W}\right)$$

$$\tilde{\rho}\tilde{W} = \rho W\left(1 + \frac{\rho'}{\rho} + \frac{W'}{W}\right)$$

$$\tilde{\rho}\tilde{W}\tilde{E} = \rho W E\left(1 + \frac{\rho'}{\rho} + \frac{W'}{W} + \frac{E'}{E}\right)$$

$$\tilde{p} = p(1 + \frac{p'}{p}) = p(1 + \frac{\rho'}{\rho} + \frac{E'}{E} + \frac{k'}{k-1})$$

$$\Delta f_w = \tilde{f}_w - f_w$$

$$\Delta f_\rho = \tilde{f}_\rho - f_\rho$$

$$\Delta f_E = \tilde{f}_E - f_E$$

The execution of transformations for the equation of mass conservation allows consecutively obtaining the following equations

$$\rho W \frac{d}{dt}\frac{\rho'}{\rho} + \frac{\rho'}{\rho}\frac{d}{dt}\rho W = \Delta f_\rho - \frac{d}{dt}\rho W \frac{W'}{W}$$

$$\rho W\frac{d}{dt}\frac{\rho'}{\rho}+\frac{\rho'}{\rho}f_1=\Delta f_\rho-\frac{d}{dt}\rho W\frac{W'}{W},$$

$$\frac{d}{dt}\frac{\rho'}{\rho}+\frac{\rho'}{\rho}\frac{f_\rho}{\rho W}=\frac{\Delta f_\rho}{\rho W}-\frac{1}{\rho W}\frac{d}{dt}\rho W\frac{W'}{W}$$

The execution of transformations for the equation of energy conservation allows consecutively obtaining the following equations

$$\rho WE\frac{d}{dt}(\frac{\rho'}{\rho}+\frac{E'}{E})+(\frac{\rho'}{\rho}+\frac{E'}{E})\frac{d}{dt}\rho WE=\Delta f_E-\frac{d}{dt}\rho WE\frac{W'}{W},$$

$$\rho WE\frac{d}{dt}(\frac{\rho'}{\rho}+\frac{E'}{E})+(\frac{\rho'}{\rho}+\frac{E'}{E})f_E=\Delta f_E-\frac{d}{dt}\rho WE\frac{W'}{W},$$

$$\frac{d}{dt}(\frac{\rho'}{\rho}+\frac{E'}{E})+(\frac{\rho'}{\rho}+\frac{E'}{E})\frac{f_E}{\rho WE}=\frac{\Delta f_E}{\rho WE}-\frac{1}{\rho WE}\frac{d}{dt}\rho WE\frac{W'}{W},$$

$$\frac{d}{dt}\frac{E'}{E}+\frac{E'}{E}\frac{f_E}{\rho WE}+\frac{\rho'}{\rho}\left(\frac{f_E-f_\rho E}{\rho WE}\right)=\frac{(\Delta f_E-\Delta f_\rho E)}{\rho WE}-\frac{1}{\rho WE}\left(\frac{d}{dt}\rho WE\frac{W'}{W}-E\frac{d}{dt}\rho W\frac{W'}{W}\right)$$

The transformations for the state equation are written down as follows:

$$\frac{p'}{p}=\frac{\rho'}{\rho}+\frac{E'}{E}+\frac{k'}{k-1},$$

$$\frac{d}{dt}(\frac{\rho'}{\rho}+\frac{E'}{E})+(\frac{\rho'}{\rho}+\frac{E'}{E})\frac{f_E}{\rho WE}=\frac{\Delta f_E}{\rho WE}-\frac{1}{\rho WE}\frac{d}{dt}\rho WE\frac{W'}{W},$$

$$\frac{d}{dt}\frac{p'}{p}+\frac{p'}{p}\frac{f_E}{\rho WE}=\frac{\Delta f_E}{\rho WE}-\frac{1}{\rho WE}\frac{d}{dt}\rho WE\frac{W'}{W}+\frac{k'}{k-1}\frac{f_E}{\rho WE}+\frac{d}{dt}\frac{k'}{k-1}$$

The right sides of the equations can be also simplified using the initial equations for the undisturbed stream

$$\frac{d}{dt}\rho W\frac{W'}{W}=\rho W\frac{d}{dt}\frac{W'}{W}+f_\rho\frac{W'}{W};$$

$$\frac{d}{dt}\rho\, WE\frac{W'}{W} = \rho\, WE\frac{d}{dt}\frac{W'}{W} + f_E\frac{W'}{W};$$

$$\frac{d}{dt}\rho\, WE\frac{W'}{W} - E\frac{d}{dt}\rho\, W\frac{W'}{W} = (f_E - Ef_\rho\frac{W'}{W})$$

Eventually after the transformations in the compact view the equations will be written down as follows:

$$\frac{d}{dt}\frac{W'}{W} = \mathrm{K}^w$$

$$\frac{d}{dt}\frac{\rho'}{\rho} + \mathrm{K}_\rho^\rho\frac{\rho'}{\rho} = \mathrm{K}^\rho$$

$$\frac{d}{dt}\frac{E'}{E} + \mathrm{K}_E^E\frac{E'}{E} + \mathrm{K}_\rho^E\frac{\rho'}{\rho} = \mathrm{K}^E$$

$$\frac{d}{dt}\frac{p'}{p} + \mathrm{K}_p^p\frac{p'}{p} = \mathrm{K}^p \tag{5.37}$$

Here the following is designated:

$$\mathrm{K}^w = \frac{\Delta f_w}{W}$$

$$\mathrm{K}_\rho^\rho = \frac{f_\rho}{\rho W}$$

$$\mathrm{K}^\rho = \frac{\Delta f_\rho}{\rho\, W} - \frac{W'}{W}\frac{f_\rho}{\rho\, W} - \frac{d}{dt}\frac{W'}{W}$$

$$\mathrm{K}_E^E = \frac{f_E}{\rho WE}$$

$$\mathrm{K}_\rho^E = \frac{f_E - f_\rho E}{\rho WE}$$

$$\mathrm{K}^E = \frac{(\Delta f_E - \Delta f_\rho E)}{\rho WE} - \frac{W'}{W}\frac{f_E - Ef_\rho}{\rho WE}$$

$$K_p^p = \frac{f_E}{\rho WE}$$

$$K^p = \frac{\Delta f_E}{\rho\, WE} - \frac{W'}{W}\frac{f_E}{\rho\, WE} - \frac{d}{dt}\frac{W'}{W} + \frac{k'}{k-1}\frac{f_E}{\rho\, WE} + \frac{d}{dt}\frac{k'}{k-1}$$

When using the non-stationary one-dimensional models of gas-dynamic processes, the system of basic equations (conservation of mass, amount of movement, energy, as well as the equation of state) can be written down as follows:

$$\frac{\partial}{\partial t}\rho F + \frac{\partial}{\partial x}\rho Fu = f_1,$$

$$\frac{\partial}{\partial t}\rho Fu + \frac{\partial}{\partial x}\rho Fu^2 + F\frac{\partial p}{\partial x} = f_2,$$

$$\frac{\partial}{\partial t}\rho FE + \frac{\partial}{\partial x}\rho FEu + \frac{\partial}{\partial x}Fup = f_3, \qquad (5.38)$$

$$p = \rho\cdot(k-1)\cdot\left(E - \frac{1}{2}u^2\right)$$

The bond between the disturbed and non-disturbed gas-dynamic parameters by the analogy with the foregoing correlations for the thermodynamic correlations can be written down as follows:

$$\tilde{F} = F\left(1+\frac{F'}{F}\right),\quad \tilde{k} = k\left(1+\frac{k'}{k}\right),\quad \tilde{\rho} = \rho\left(1+\frac{\rho'}{\rho}\right),\quad \tilde{p} = p\left(1+\frac{p'}{p}\right),$$

$$\tilde{E} = E\left(1+\frac{E'}{E}\right),\quad \tilde{u}^2 = u^2\left(1+2\frac{u'}{u}\right),\quad \tilde{\rho}\tilde{F} = \rho F\left(1+\frac{\rho'}{\rho}+\frac{F'}{F}\right),$$

$$\tilde{\rho}\tilde{F}\tilde{u} = \rho Fu^2\left(1+\frac{\rho'}{\rho}+\frac{F'}{F}+\frac{u'}{u}\right),\quad \tilde{\rho}\tilde{F}\tilde{u}^2 = \rho Fu^2\left(1+\frac{\rho'}{\rho}+\frac{F'}{F}+2\frac{u'}{u}\right),$$

$$\tilde{\rho}\tilde{F}\tilde{E} = \rho FE\begin{pmatrix}1+\dfrac{\rho'}{\rho}+\dfrac{F'}{F}\\ +\dfrac{E'}{E}+\dfrac{u'}{u}\end{pmatrix}\quad \tilde{\rho}\tilde{F}\tilde{E}\tilde{u} = \rho FEu\begin{pmatrix}1+\dfrac{\rho'}{\rho}+\dfrac{F'}{F}\\ +\dfrac{E'}{E}+\dfrac{u'}{u}\end{pmatrix}$$

Besides, let us designate $\Delta f_1 = \tilde{f}_1 - f_1$, $\Delta f_2 = \tilde{f}_2 - f_2$, $\Delta f_3 = \tilde{f}_3 - f_3$.

After the transformations the linearized equations for the deviations of gas-dynamic parameters are as follows:

$$\frac{\partial}{\partial t}\frac{\rho'}{\rho} + u\frac{\partial}{\partial x}\frac{\rho'}{\rho} + K_{ux}^{\rho}\frac{\partial}{\partial x}\frac{u'}{u} + K_{\rho}^{\rho}\frac{\rho'}{\rho} + K_{u}^{\rho}\frac{u'}{u} = K^{\rho},$$

$$\frac{\partial}{\partial t}\frac{u'}{u} + u\frac{\partial}{\partial x}\frac{u'}{u} + K_{px}^{u}\frac{\partial}{\partial x}\frac{p'}{p} + K_{\rho}^{u}\frac{\rho'}{\rho} + K_{u}^{u}\frac{u'}{u} + K_{p}^{u}\frac{p'}{p} = K^{u},$$

$$\frac{\partial}{\partial t}\frac{E'}{E} + u\frac{\partial}{\partial x}\frac{E'}{E} + K_{ux}^{E}\frac{\partial}{\partial x}\frac{u'}{u} + K_{px}^{E}\frac{\partial}{\partial x}\frac{p'}{p}$$

$$+K_{\rho}^{E}\frac{\rho'}{\rho} + K_{u}^{E}\frac{u'}{u} + K_{E}^{E}\frac{E'}{E} + K_{p}^{E}\frac{p'}{p} = K^{E}, \tag{5.39}$$

$$\frac{\partial}{\partial t}\frac{p'}{p} + K_{\rho x}^{p}\frac{\partial}{\partial x}\frac{\rho'}{\rho} + K_{ux}^{p}\frac{\partial}{\partial x}\frac{u'}{u} + K_{px}^{p}\frac{\partial}{\partial x}\frac{p'}{p} + K_{Ex}^{p}\frac{\partial}{\partial x}\frac{E'}{E} + K_{\rho}^{p}\frac{\rho'}{\rho}$$

$$+K_{u}^{p}\frac{u'}{u} + K_{E}^{p}\frac{E'}{E} + K_{p}^{p}\frac{p'}{p} = K^{p}$$

Here we designate:

$$K^{\rho} = \frac{1}{\rho F}\left(\Delta f_1 - \frac{F'}{F}f_1\right)$$

$$-\left(\frac{\partial}{\partial t}\frac{F'}{F} + u\frac{\partial}{\partial x}\frac{F'}{F}\right);\quad K_{\rho}^{\rho} = \frac{f_1}{\rho F};\quad K_{u}^{\rho} = \frac{1}{\rho F}\frac{\partial}{\partial x}\rho F u;\quad K_{ux}^{\rho} = u;$$

$$K^{u} = \frac{1}{\rho F u}\begin{pmatrix}\Delta f_2 - u\Delta f_1 \\ +\dfrac{F'}{F}(uf_1 - f_2)\end{pmatrix};\quad K_{\rho}^{u} = \frac{1}{\rho F u}\begin{pmatrix} f_2 - uf_1 \\ -F\dfrac{\partial p}{\partial x}\end{pmatrix};$$

$$K_{u}^{u} = \frac{1}{\rho F u}\begin{pmatrix} f_2 + \rho F u\dfrac{\partial u}{\partial x} \\ -F\dfrac{\partial p}{\partial x}\end{pmatrix};\quad K_{p}^{u} = \frac{F\dfrac{\partial p}{\partial x}}{\rho F u};\quad K_{px}^{u} = \frac{F p}{\rho F u};\quad K_{ux}^{E} = \frac{Fup}{\rho F E};$$

$$K^{E} = \frac{1}{\rho F E}\begin{pmatrix}\Delta f_3 \\ -E\Delta f_1\end{pmatrix} + \frac{F'}{F}\frac{1}{\rho F E}\begin{pmatrix}Ef_1 \\ -f_3\end{pmatrix} - \frac{Fup}{\rho F E}\frac{\partial}{\partial x}\frac{F'}{F};\quad K_{p}^{E} = \frac{1}{\rho F E}\frac{\partial}{\partial x}Fup;$$

$$K_{\rho}^{E} = \frac{1}{\rho FE}\begin{pmatrix} f_3 - Ef_1 \\ -\frac{\partial}{\partial x}Fup \end{pmatrix}; \quad K_{u}^{E} = \frac{1}{E}\begin{pmatrix} u\frac{\partial E}{\partial x} \\ +\frac{1}{\rho F}\frac{\partial}{\partial x}Fup \end{pmatrix}; \quad K_{px}^{E} = \frac{Fup}{\rho FE};$$

$$K_{E}^{E} = \frac{1}{\rho FE}\begin{pmatrix} f_3 \\ -\frac{\partial}{\partial x}Fup \end{pmatrix}; \quad K^{p} = K^{\rho} + \frac{1}{1-\frac{u^2}{2E}}\cdot\left(\frac{u^2}{2E}\cdot K^{u} + K^{E}\right) - \frac{\partial}{\partial t}\frac{k'}{k-1};$$

$$K_{\rho}^{p} = K_{\rho}^{\rho} + \frac{1}{1-\frac{u^2}{2E}}\cdot\left(\frac{u^2}{2E}\cdot K_{\rho}^{u} + K_{\rho}^{E}\right);$$

$$K_{E}^{p} = \frac{K_{E}^{E}}{1-\frac{u^2}{2E}} + \frac{\partial}{\partial t}\frac{1}{1-\frac{u^2}{2E}};$$

$$K_{u}^{p} = K_{u}^{\rho} + \frac{1}{1-\frac{u^2}{2E}}\cdot\left(\frac{u^2}{2E}\cdot K_{u}^{u} + K_{u}^{E}\right) + \frac{\partial}{\partial t}\frac{\frac{u^2}{2E}}{1-\frac{u^2}{2E}};$$

$$K_{p}^{p} = \frac{1}{1-\frac{u^2}{2E}}\cdot\left(\frac{u^2}{2E}\cdot K_{p}^{u} + K_{p}^{E}\right);$$

$$K_{\rho x}^{p} = u; \; K_{ux}^{p} = K_{ux}^{\rho} + \frac{1}{1-\frac{u^2}{2E}}\cdot\left(\frac{u^2}{2E}\cdot u + K_{ux}^{E}\right);$$

$$K_{Ex}^{p} = \frac{u}{1-\frac{u^2}{2E}}; \quad K_{px}^{p} = \frac{1}{1-\frac{u^2}{2E}}\cdot\left(\frac{u^2}{2E}\cdot K_{px}^{u} + K_{px}^{E}\right)$$

The foregoing equation system (5.39) gives the possibility to determine the values of deviations of the process main characteristics from their mathematical expectations at each stage of integrating the basic equation system (5.38). Both equation systems should be supplemented by initial and boundary conditions.

Additionally some applied aspects which are true for the Eqs. (5.38) and (5.39) should be pointed out. During the operation of the heat engine in the normal mode the solution for the parameter values ρ, p, E, u is set up at any time point by solving the Eq. (5.38). If a random factor acts at some time point (e.g., the appearance of a crack in the propellant charge of the solid fuel gas generator), the normal development of the process is disturbed, and together with the Eqs. (5.38) and (5.39) need to be solved. In this case, the initial conditions for the latter equation system will equal zero. The unstable heat engine operation can occur, to some extent, at any moment even with zero initial values for the disturbances ρ', p', E', u' and zero values of the right sides of the Eqs. (5.38). The analysis of the occurrence of unstable operation modes of the heat engine and development of disturbances with time can be reduced to the matrix analysis composed of the coefficients $K_{\varphi}^{\phi}, K_{\varphi x}^{\phi}$ in the left sides of the Eqs. (5.39). It is important that these coefficients are the functions of variables whose values are found when solving the basic system of Eqs. (5.38) and only by them.

Let us additionally mention the following. In a working technical device (including any heat engine) there are always disturbing factors (variations in the fuel combustion rate, inaccuracies in producing or assembling some parts and units of the structure, etc.). The action of these disturbing factors will result in the following: the actual parameters of the operation process of the technical object investigated will differ from their determinate values. This is also true for controllable heat engines (Figure 5.4). Some papers, for instance [153, 220], are dedicated to the analysis of heat engines operation under the action of disturbing factors. They present the ways of analyzing the stability of engine functioning and the development of dispersions of inter-ballistic parameters under the action of disturbing factors on operation processes.

All the disturbing factors acting upon the operation of heat engine or gas generator can be divided into the following groups:

- initial conditions for pressure and temperature;
- parameters determining the ignition system operation;
- parameters determining the variations in solid fuel properties;
- parameters determining the variations in internal volume geometry.

The first group of parameters should comprise the initial pressure value in the engine chamber p_0, temperature of gas T_0 filling the engine internal volume and initial temperature of solid fuel T_{T0}. Variations in the environment parameters (pressure, temperature, etc.) are insignificant and, as a rule, are not taken into account.

The second group of parameters comprises:

- thermophysical and thermodynamic characteristics of the igniting composition combustion products – heat capacities c_{pi}, c_{vi}, heat conductivity coefficient λ_i, dynamic viscosity μ_i, molecular mass M_i, temperature of combustion products T_i, mass concentration of condensed particles in the combustion products of igniting composition α_{ki};
- information on the igniting composition weigh in solid phase – weigh mass m_i, igniting composition density ρ_i;
- characteristics of the igniting composition combustion – combustion rate u_i and enthalpy of combustion products H_i;
- value of the hole area in the igniter body F_i.

The third group of parameters comprises:

- thermophysical and thermodynamic characteristics of the fuel combustion products – heat capacities c_{pF}, c_{vF}, heat conductivity coefficient λ_F, dynamic viscosity μ_F, molecular mass M_F, temperature of combustion products T_F, mass concentration of condensed particles in the fuel combustion products α_{kF};
- information on the fuel in solid phase – fuel density ρ_F, heat capacity c_F;
- characteristics of the fuel combustion – combustion rate u_F;
- value of the parameters characterizing the fuel ignition – ignition temperature T_* and (or) depth of the chemical reaction in the fuel surface layer during which the fuel β_* ignites.

The fourth group of parameters comprises the geometry characteristics of the engine calculated – volume value W and area of the fuel combustion surface S, and for the nozzle block – area of the nozzle minimal section $F_{\min}$. Besides, the last group of parameters comprises the values of plug decomposition pressure p_{noz} and duration of the plug decomposition t_{noz}. When performing the calculations for all the foregoing parameters, it is necessary to set up the information on their average values, as well as the value of variations of these parameters from the average value.

The method for evaluating the stability of intrachamber process given in Ref. [153] for the conditions of the controllable solid fuel engine unit is quite easily applied, however, it has a significant shortcoming. The unstable operation modes can only be set at the joint numerical solution of the system of equation of internal ballistics and equations for disturbances

(Eqs. (5.38) and (5.39)). The possibilities of assessing the stability of operating processes using simpler mathematical models are interesting in practice. One of the methods of stability assessment is the representation of disturbing factors influencing the internal ballistics of the engine unit in the form of disturbances of operating pressure level in the combustion chamber.

The random pressure deviations are conditioned by the foregoing disturbing factors. The pressure variations are mostly influenced by the variations of solid fuel combustion rate value, small cracks in the solid fuel charge body, etc. The action of disturbing factors can be taken into account by the modification of the control Eqs. (5.9), in which the disturbed pressure value $p^*(t) = p(t)\cdot(1+\delta'(t))$ is used in the combustion chamber of the engine unit instead of the pressure values *p(t)*. The values $\delta'(t)$ determining the level of random pressure deviations $p(t)$ are found by the correlations, which are true for the normal distribution law. At the same time, the mathematical expectation is taken as $v_1(\delta'(t)) = 0$, and the value of standard deviation $\sigma_\delta = \sigma(\delta')$ is set up in initial data. The algorithm of the calculations of the values $\delta'(t)$ corresponding to the normal distribution law is described in the papers [198, 230]. Besides, the options of modeling the pressure disturbances obeying the periodic law can be considered

$$p^*(t) = p(t)\cdot(1+\delta'(t));$$
$$\delta'(t) = A\cdot\sin(2\pi\, N\, t) \tag{5.40}$$

When analyzing the action of disturbances on the dynamic controllable systems, other approaches are also applied which allow assessing the system asymptomatic behavior by analytical methods. In particular, the calculation of amplitude and phase frequency characteristics of non-stationary processes is widely spread [19, 31, 36, 168, 176]. This approach allows additionally finding the possibility of occurrence of resonance oscillations in the combustion chamber of the controllable solid fuel engine unit (periodic pressure fluctuations occur in the combustion chamber which can coincide with the own fluctuations of the intrachamber volume).

5.3.4 MODELING OF THE ACTION OF STOCHASTIC DISTURBANCES ON DYNAMIC SYSTEMS

Generally speaking, the solution of problems on processes in dynamic systems in determinate setting does not meet the requirements of accuracy in forecasting the development of these processes. And this is conditioned by the fact that there are always disturbing factors in the working technical device, the action of which results in the situation that the actual parameters of the operation process of the technical object investigated will differ from their determinate values. The influence of disturbing factors on the processes in the system can be assessed, for example, by the following methods.

The first method consists in solving the linearized Eq. (5.39) together with the Eq. (5.38). The linearized equations are solved repeatedly, at the same time, the action randomness of one or another factor set up in initial data, initial and boundary conditions is taken into account in each calculation. The combined action of disturbing factors is assessed by the apparatus of probability theory and mathematical statistics. The considered method of calculating the evolution of disturbances has shortcomings. First of all, it is conditioned by the fact that when deriving the linearized equations we make the assumption on the limitation of deviations of gas-dynamic parameters (the deviation value φ of some gas-dynamic parameter φ is significantly less than the absolute value of the non-disturbed parameter – $\left|\frac{\varphi'}{\varphi}\right| << 1$). However, for instance, in the calculations of processes in gas generators and heat engines this condition cannot be fulfilled when analyzing the variations in the rates of combustion products. In particular, in the areas where the rate values are close to zero ($u \approx 0$). Besides, in the method considered it is necessary to create separate calculation algorithms to solve the Eq. (5.38) and to solve the Eq. (5.39).

Another method of calculating disturbances consists in multiple solution of the Eq. (5.38) with different initial data whose values are set up taking their random character into account. At the first stage, the Eq. (5.38) is solved for non-disturbed right sides and non-disturbed boundary and initial conditions. At the second stage, the same equation system is solved but for "disturbed" parameters. When solving the problem, the right sides and (or) boundary and initial conditions are changed taking the acting disturbance factors into account. The values of parameter deviations are found by the difference of the first (corresponds to the non-disturbed case) and second (corresponds to the case with disturbed right sides of equations and disturbed initial and boundary conditions) solutions. Finally, at such approach there is a possibility to calculate the deviations of the parameters corresponding to any disturbance of initial and boundary conditions or any disturbance of the right sides of differential equations. At such approach, the deviation values corresponding to the combined action of several or all disturbing factors can be calculated (it should be pointed out that such approach is more appropriate to the conditions under which the natural experiment is carried out).

The shortcoming of the approach considered is the necessity to repeatedly solve the Eq. (5.38). The advantage is that at such calculation the computational program does not practically change; only the executive program will change insignificantly. The deviation values calculated repeatedly at different disturbing factors can be considered as general regularities of stochastic values and allow calculating the values of dispersions and mean-square deviations important for the statistic analysis of the investigated object operation [66, 68, 178].

Further we will consider the latter method for calculating dispersions and mean-square deviations.

In probability theory it is mentioned that the deviations of random numbers from their mathematical expectations obey certain regularities. In particular, these regularities show up through the integral functions of the distribution of random numbers. In actual processes connected with random factors a limited number of variants of integral distribution factors is implemented. Normally distributed and binomially distributed integral functions (distributions) are most widely spread in engineering. Random values have deviations from the actual value distributed by the normal law if they satisfy the conditions of Chebyshev theorem [178, 251].

Hereinafter the following additional designations are accepted:

- P – event probability written down in curved brackets;
- N – number of experiments (or tests);
- $\overline{x}_i$ – mathematical expectation of the variables x_i;
- σ_x – mean-square deviation of the random value from its mathematical expectation;
- μ_i – central moment of the random value of the order i;
- $\rho_{x_i x_j}$ – paired coefficient of the correlation of i and j arguments.

In Chapter 3, there are regularities which are true for the normally distributed random variable x:

- density of the probability distribution $f(x)$ (Eq. (3.38));
- probability of the event that the measured variable x is in the range from a to b $(a < x \le b)$ (Eq. (3.39));
- probability of the events at different values of mean-square deviations (Eqs. (3.40));

- conditions under which the random variable x satisfies the normal distribution law (conditions for initial and central moments of distribution of the Eq. (3.44)), etc.

Further, when solving specific technical problems, we believe that the normal law of the distribution of random values is true. Other laws of the distribution of random numbers are also implemented in engineering. However, the analysis demonstrates that if during the object analysis we assume that the distribution of random value deviations from its actual value obeys the normal distribution law, the final error will not be significant. Besides, the following circumstance needs to be pointed out. If during the analysis of random values their probabilistic distributions change with time or random values depends on each other, then, in accordance with the definition [66] we can speak of the analysis of random processes. Random processes can be stationary and this definition of random processes is true for the case when no distribution of probabilities for all random variables changes when the argument (time) changes. Gauss random processes are those in which all their distributions of probabilities are normal for all random variables determining the random process.

The determination of parameters of the distribution of random values, at the normal distribution law as well, can be carried out using [9, 68, 178]. In accordance with these works when calculating the distribution parameters using the method of linearization of random functions, it is assumed that the distribution function can be characterized by the magnitudes of average value (mathematical expectation), dispersion, asymmetry and excess. Let the functional dependence be as follows: $\Phi = \Phi(x_1, x_2, ..., x_N)$. In this case, the average value of the determining characteristic is found with the help of the following formula

$$\bar{\Phi} = \Phi\left(\bar{x}_1, \bar{x}_2 \ldots \bar{x}_i, \ldots \bar{x}_k\right) + \frac{1}{2}\sum_{i=1}^{k}\left(\frac{\partial^2 \Phi}{\partial x_i^2}\right)_{x_i} \sigma_{x_i}^2 + \sum_{i>j}^{k}\left(\frac{\partial^2 \Phi}{\partial x_i \partial x_j}\right)_{x_i x_j} \rho_{x_i x_j} \sigma_{x_i} \sigma_{x_j}$$

In general, dispersion σ_Φ of value $\bar{\Phi}$ is found by the formula

$$\sigma_\Phi^2 = \sum_{i=1}^{k}\left(\frac{\partial \Phi}{\partial x_i}\right)_{x_i}^2 \sigma_{x_i}^2 + \frac{1}{2}\sum_{i=1}^{k}\left(\frac{\partial^2 \Phi}{\partial x_i^2}\right)_{x_i} \sigma_{x_i}^4$$

$$+ \sum_{i>j}^{k}\left(\frac{\partial^2 \Phi}{\partial x_i \partial x_j}\right)_{x_i x_j}^2 \sigma_{x_i}^2 \sigma_{x_j}^2 + 2\sum_{i>j}^{k}\left(\frac{\partial \Phi}{\partial x_i}\right)_{x_i}\left(\frac{\partial \Phi}{\partial x_j}\right)_{x_j} \rho_{x_i x_j} \sigma_{x_i} \sigma_{x_j}$$

The values of asymmetry and excess of the determining characteristic are found by the formulas

$$A = \sum_{i=1}^{k} \left(\frac{\partial \Phi}{\partial x_i} \right)_{x_i}^{3} \mu_3(x_i)$$

$$E = \sum_{i=1}^{k} \left(\frac{\partial \Phi}{\partial x_i} \right)_{x_i}^{4} \mu_4(x_i) + 6 \sum_{i>j}^{k} \left(\frac{\partial \Phi}{\partial x_i} \right)_{x_i}^{2} \left(\frac{\partial \Phi}{\partial x_j} \right)_{x_j}^{2} \sigma_{x_i}^2 \sigma_{x_j}^2$$

The average value of the variable and its dispersion can be determined by the following formulas with the accuracy sufficient for practical purposes

$$\bar{\Phi} = \Phi(\bar{x}_1, \bar{x}_2 \ldots . \bar{x}_i, \ldots \bar{x}_k)$$

$$\sigma_{\Phi}^2 = \sum_{i=1}^{k} \left(\frac{\partial \Phi}{\partial x_i} \right)_{x_i}^{2} \sigma_{x_i}^2$$

The criteria of permissibility of using the assumption on the truthfulness of the normal law of the distribution of random values (or on the minimal number of tests k fulfilled when solving the problem (5.1) at random values of initial data and initial conditions) are the following conditions

$$A \approx 0, \quad E \approx 3$$

The dispersions and mean-square deviations are calculated using the method of statistic tests [68, 251]. According to this method the statistics is accumulated for all the calculated inter-ballistic parameters. The statistics is formed carrying out a series of calculations, and when setting up the initial data their random character for each of the calculations performed is taken into account. Certain values of the initial data (let us designate them as $x_i, i = 1, N$) having a random character are calculated using the algorithm of generating random numbers by the following formula

$$x_i = \bar{x}_i + K \cdot \alpha_i \cdot \sigma_{x_i}$$

Here α_i – random number located in the interval $-1 \leq \alpha_i \leq 1$, K – coefficient taking into account the uncertainty interval (see the Eq. (3.60)) of the random value. It is accepted on default that $K = 3$.

Random numbers α_i are calculated in the following sequence:

- random numbers γ_{ij}, $j = 1, j$ uniformly distributed in the interval from 0 to 1 are calculated;
- normally distributed random numbers α_i are calculated by the calculated series of computations γ_{ij}, $j = 1, j$ whose mathematical expectations equal zero and the maximal deviation takes on the value set up by the user (e.g., 1).

There are many algorithms for calculating random numbers uniformly distributed in the interval from 0 to 1 [22, 23, 230]. The correctness of the algorithms applied is conditioned by the computer precision. For instance, when using the compiler Intel Visual Fortran for the personal computer, the computation of random numbers uniformly distributed in the interval from 0 to 1 is performed by the embedded program *Random_number* [22]. To operate this program it is necessary to calculate the first random number γ_{i0}. The first random number (it is called "seed") is calculated by the program *Random_seed.* It should be pointed out that the random variable uniformly distributed in the interval from 0 to 1 can be reduced to the range from −1 to 1 by the linear transformation

$$\gamma'_{ij} = 2\gamma_{ij} - 1$$

The random numbers α_i obeying the normal distribution law can be obtained by the values calculated γ_{ij}, $j = 1, j$ in accordance with the formula [198]

$$\alpha_i = \frac{\sum_{j=1}^{J} \gamma_{i,j} - \frac{J}{2}}{\sqrt{\frac{J}{12}}} \cdot \sigma_\alpha + \bar{\alpha}_i$$

The more is the number of tests J ($J \geq 12$ is recommended), the higher is the accuracy of this formula. In the foregoing formula $\bar{\alpha}_i, \sigma_\alpha$ – required average value of the random variable and its required standard deviation. For the rated random variable in the interval $-1 \leq \alpha_i \leq 1$, we should take $\bar{\alpha}_i = 0;\ \sigma_\alpha = 0,707$.

Let us point out the following features of implementing the method of statistic tests:

- in contrast to the classical method of statistic tests all the initial data having the stochastic character change at the recurrent calculation cycle. 200 random numbers are calculated at the recurrent cycle. Such variant of implementing the method of statistic tests models, to the most extent, the natural (physical) modeling, in which the number of experiments is limited and does not exceed several dozen;
- with a large number of experiments (several dozen and more) it is quite possible to apply the random variable uniformly distributed in the interval from 0 to 1 instead of the distributed random variable when performing the calculations.

5.3.5 SPECTRAL ANALYSIS OF DYNAMIC SYSTEMS

Earlier we considered the mathematical approaches that allow assessing the stability of process development in dynamic systems. In particular, the determination of the own values of the matrix composed of the coefficients contained in the right sides of the differential Eq. (5.1) gives the possibility to find the potential oscillation frequencies characteristic for the dynamic system investigated. Some of these frequencies can be dangerous for the technical object that is demonstrated through the resonance phenomena.

There is an approach which allows determining the list of frequencies of the dynamic system oscillations actually implemented. This approach consists in the fact that the solution of the evolution problem (5.1) is represented in the form of series by functions possessing the periodicity property. In particular, series of Fourier, Walsh, Chebyshev, wavelets, etc. [77, 112, 202, 230].

Fourier-spectra are most widely spread in the analysis of dynamic systems, for which the following assertions are true [230]:

- if some periodic function $f(t)$ with the period 2π is determined in the interval $(-\pi, \pi)$, is continuous and integratable in this interval, then the following resolution is true for all values of the argument x

$$f(t) = \frac{a_0}{2} + \sum_{k=1}^{\infty}(a_k \cos kt + b_k \sin kt) \tag{5.41}$$

- if the series (5.41) uniformly reduces to $f(t)$, Fourier formulas are true for the series coefficients

$$a_0 = \frac{1}{\pi}\int_{-\pi}^{\pi} f(t)\,dt$$

$$a_k = \frac{1}{\pi}\int_{-\pi}^{\pi} f(t)\cos kt\,dt$$

$$b_k = \frac{1}{\pi}\int_{-\pi}^{\pi} f(t)\sin kt\,dt \tag{5.42}$$

A stricter formulization of the above problem requires the execution of Dirichlet conditions which are described, for example, in [202].

The representation of the function $f(t)$ in the form (5.41) is called by the function expanding to Fourier series, numbers a_k, b_k determined by the formulas (5.42) are called the coefficients of Fourier function $f(t)$. The value k is called "frequency," and expression $A_k = \sqrt{a_k^2 + b_k^2}$ – frequency power k.

The function $f(t)$ expanding to Fourier series is simplified if it is known that the function possesses the properties of evenness or oddity:

- for the even function $f(t)$ the Eq. (5.41) is written down as follows:

$$f(t) = a_0 + \sum_{k=1}^{\infty} a_k \cos kt \tag{5.43}$$

and coefficients are determined by the following formulas

$$a_0 = \frac{1}{\pi}\int_{0}^{\pi} f(t)\,dx,\ a_k = \frac{2}{\pi}\int_{0}^{\pi} f(t)\cos kt\,dt,\ b_k = 0$$

- for the odd function $f(t)$ the Eq. (5.41) is written down as follows:

$$f(t) = \sum_{k=1}^{\infty} b_k \sin kt \tag{5.44}$$

and Fourier coefficients are determined by the following formulas

$$a_0 = 0,\ a_k = 0,\ b_k = \frac{2}{\pi}\int_{0}^{\pi} f(t)\sin kt\,dt$$

For practice, the expanding to the series of Fourier functions set up in the range $(-\tau, \tau)$ instead of $(-\pi, \pi)$ is interesting. In this case, in accordance with [202], the following formula is true

$$f(t) = \frac{a_0}{2} + \sum_{k=1}^{\infty}(a_k \cos\frac{k\pi\, t}{\tau} + b_k \sin\frac{k\pi\, t}{\tau}) \tag{5.45}$$

in which Fourier coefficients are found by the following correlations

$$a_0 = \frac{1}{\tau}\int_{-\tau}^{\tau} f(t)\, dt$$

$$a_k = \frac{1}{\tau}\int_{-\tau}^{\tau} f(t)\cos\frac{k\pi\, t}{\tau}\, dt$$

$$b_k = \frac{1}{\tau}\int_{-\tau}^{\tau} f(t)\sin\frac{k\pi\, t}{\tau}\, dt \tag{5.46}$$

When analyzing the dynamic systems with the help of computers, the calculation results are presented by the tables, therefore, in the Eqs. (5.41)–(5.46) the integration procedures should be replaced by the summation with a limited number of summands.

Let in the specified terminal $t \in (-\tau, \tau)$ the function $f(t)$ be represented by discrete values *2K*. In this case, the following dependencies are true [230]

$$f(t) = \frac{a_0}{2} + \sum_{k=1}^{K-1}(a_k \cos\frac{k\pi}{K}t + b_k \sin\frac{k\pi}{K}t);$$

$$a_k = \frac{1}{K}\sum_{t_1}^{t_{2K}} f(t)\cos\frac{k\pi}{K}t, \quad (k=0,\ 1,\ ...,\ K); \tag{5.47}$$

$$bk = \frac{1}{K}\sum_{t_1}^{t_2 k} f(t)\sin\frac{k\pi}{K}t, \quad (k=0,\ 1,\ ...,\ K-1)$$

Below is the example which allows assessing the advantages of spectral analysis applied to evaluate the properties of the arbitrary function $f(t)$. For illustration, the function $f(t)$ is taken as the sum of five periodic functions (Figure 5.12).

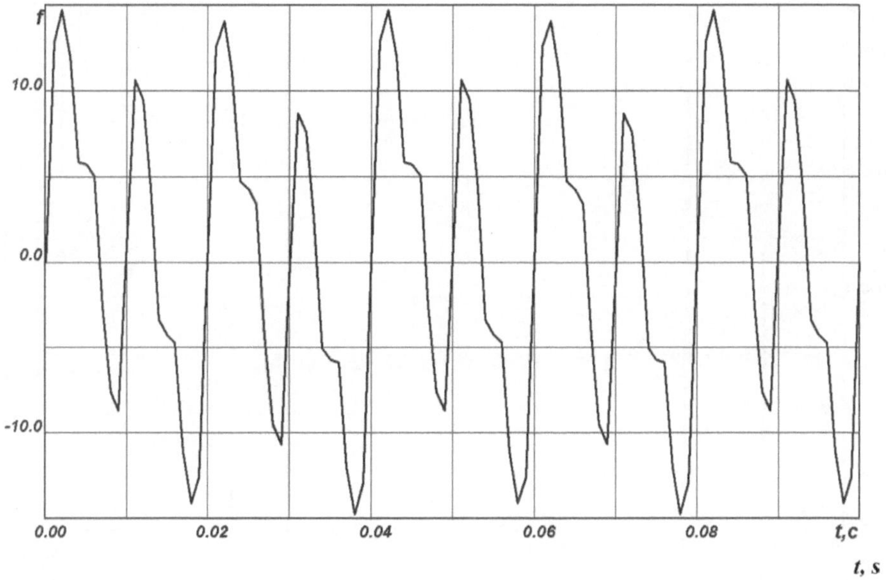

FIGURE 5.12 Dependence of the test function f on time.

$$f(t) = \sin(2\pi N_1 t) + 5\sin(2\pi N_2 t) + 10t^2 \sin(2\pi N_3 t) + 5\sin(2\pi N_4 t) + \sin(2\pi N_5 t)$$

In the latter formula it is accepted that $N_1 = 25.0$, $N_2 = 50.0$, $N_3 = 100.0$, $N_4 = 200.0$, $N_5 = 400.0$. The frequencies N_i in the formula are measured in Hertz.

In Figure 5.13 there is the dependence of the power $\boldsymbol{A}$ on the value of harmonic frequency N. The analysis result presented in Figures 5.12, 5.13 demonstrates that Fourier-analysis correctly determined the frequency spectrum of the function $f(t)$ and correctly ranked these frequencies by their power.

During the numerical finding of Fourier transformations it should be taken into account that the number of harmonics analyzed is limited, although by the initial assumptions their number must be infinite. Naturally, this fact can be reflected in the calculation results and requires their thorough analysis. Errors on the boundaries of minimal and maximal frequencies are most probable. These frequencies are more probable as, in accordance with the determination of the discrete Fourier transformation, the initial function is assumed to equal zero outside the calculated interval. The minimal and maximal frequencies which can be found by the computational algorithms

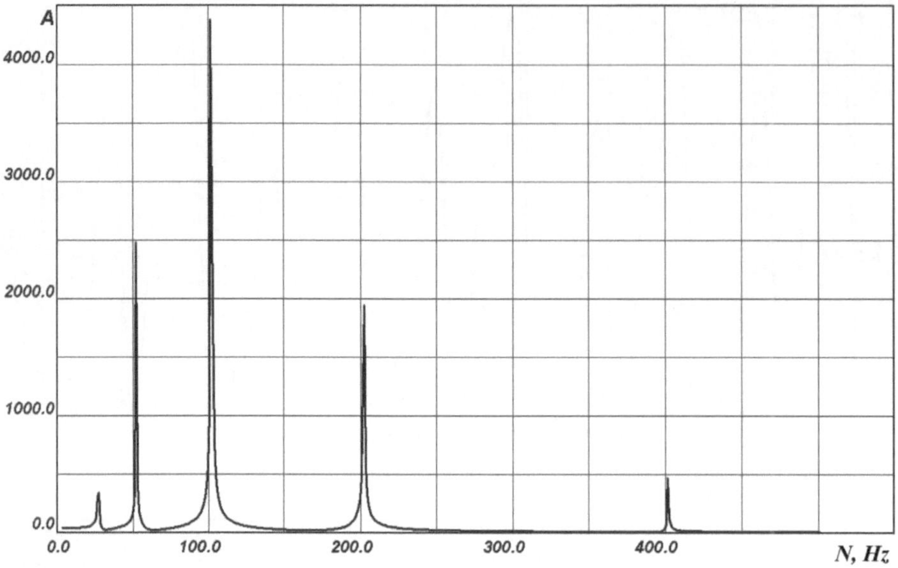

FIGURE 5.13 Dependence of power on the value of harmonic frequency.

of Fourier transformations are called "the boundary frequency and Nyquist frequency," respectively. If the function $f(t)$ is determined in the interval $t \in (\tau_0, \tau_k)$ and calculation results contain K of the data, the boundary frequency k_0 is found by the formula $k_0 = \dfrac{1}{\tau_k - \tau_0}$, and Nyquist frequency k_N – by the formula $k_N = \dfrac{K}{\tau_k - \tau_0}$.

Fourier-analysis can be also applied in problems connected with the solution of differential equations in partial derivatives. The issues connected with the application of Fourier expanding for these cases are considered, for example, in [77, 202, 230].

5.3.6 WAVELET-ANALYSIS OF THE PROCESSES IN DYNAMIC SYSTEMS

Wavelet-transformation (or discrete wave transformation) has been quite recently applied in computational practice. The number of applications grows continuously which is connected with the method high efficiency, in comparison with Fourier-analysis methods as well. Wavelet-forming functions are applied, for example, to compress the digital information (audio, video, etc.) in the problems of financial analysis, analysis of seismic information, study of resonance effects in engineering, etc. The attractiveness of wavelet-analysis in comparison with Fourier-analysis consists in the fact that this method allows not only finding the availability of harmonics in the function $f(t)$, but also determining at what time moment and in what space points these harmonics are displayed.

The basics of wavelet theory (originated from the English word "wave") are discussed, for instance, in [82]. In the same way as in Fourier transformations when applying wavelet-transformations the function $f(t)$ is represented as the series total. Basis functions in this expanding are not the trigonometric functions but the so-called wavelet-forming (soliton-forming, wave) functions. Each of the functions of this basis characterizes a certain frequency and its localization in time. The wavelet-forming functions can be localized in some limited area of its argument and equal zero or insignificant far from it. They can be understood as the method of enforcing the effect studied in some surroundings of the function argument.

In practice, several dozen of different wavelet-forming functions are used. Haar wavelet is the simplest wavelet applied when analyzing the rectangular oscillations

$$\psi(t) = \phi(2t) - \phi(2t-1) \tag{5.48}$$

Here the function φ is called the scaling function and is determined as follows:

$$\phi(t) = \begin{cases} 1, & if\ 0 \le t < 1 \\ 0, & otherwise \end{cases}$$

In Figure 5.14 you can find the examples of wavelet-forming functions obtained when differentiating Gaussian function.

$$\phi(t) = (-1)^m \frac{d^m}{dt^m} \exp(-\frac{t^2}{2}) \tag{5.49}$$

In case $m = 1$ and $m = 2$ (curves 1 and 2 in Figure 14), wavelet-forming functions are as follows, respectively

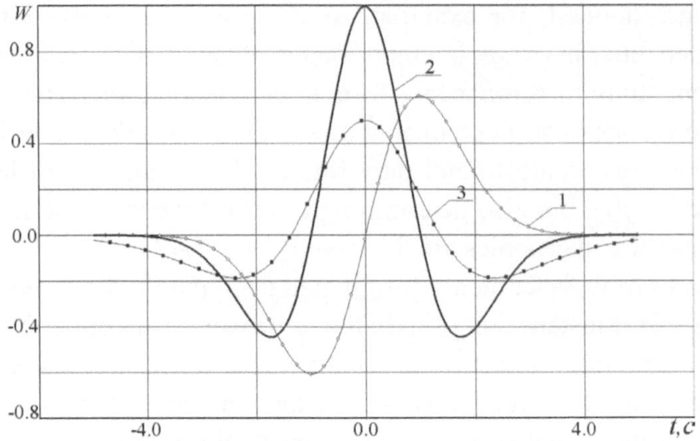

FIGURE 5.14 Comparison of wavelet-forming functions (1 – Gauss wavelet (m=1); 2 – Gauss wavelet (m=2); 3 – DOG-wavelet).

$$\varphi(t) = t \exp\left(-\frac{t^2}{2}\right)$$

$$\varphi(t) = (1 - t^2)\exp(-\frac{t^2}{2})$$

The function corresponding to the value $m = 2$ is called "Mexican hat" or MHAT-wavelet [112]. For comparison, in Figure 5.14 there is a DOG-wavelet (curve 3) determined by the following function

$$\varphi(t) = \exp(-\frac{t^2}{2}) - \frac{1}{2}\exp(-\frac{t^2}{8})$$

In practice, complex wavelet-forming functions are also used. In particular, this is Morlet function

$$\varphi(t) = \exp(2\pi i t)\exp(-\frac{t^2}{2\sigma^2})$$

or Paul wavelet

$$\varphi(t) = (1 - \frac{2\pi i t}{p-1})^{-p}$$

In the latter equations σ and p determine the wavelet length in the frequency space.

When setting up the wavelet-forming function ϕ, the wavelet spectrum is established as the integral

$$\psi(a,b)=\int_{-\infty}^{\infty} f(t)\phi(\frac{t-b}{a})\,dt \tag{5.50}$$

or

$$\psi(a,b)=\frac{1}{t_{\max}-t_{\min}}\int_{t_{\min}}^{t_{\max}} f(t)\phi(\frac{t-b}{a})\,dt$$

As follows from the Eq. (5.50), the wavelet spectrum $\psi(a,b)$ is the function of two arguments. There is some freedom in selecting these arguments. In particular, for the non-stationary processes in the quality of the argument a, the value t_* characterizing the oscillation period can be applied. In this case, the oscillation frequency N will be determined by the formula $N=\frac{1}{t_*}$. The physical sense of the parameter b can be connected with the value of the phase shift $\varphi=\frac{t-b}{t}$ or $\varphi=\operatorname{arctg}\left(\frac{t-b}{t}\right)$. The results of the calculations of wavelet-functions ψ are represented in the form of the level lines on the planes (a, b), (a, t), (b, t), etc.

5.4 COMPUTATIONAL METHODS IN THE PROBLEMS ON THE ANALYSIS OF DYNAMIC SYSTEMS

5.4.1 METHODS OF SOLVING THE SYSTEMS OF ORDINARY DIFFERENTIAL EQUATIONS

The systems of ordinary differential equations considered in Section 5.1 are referred to the class of problems called Cauchy problems. The total solution of the systems of n ordinary differential equations (or ordinary differential equation of order n) is as follows [108]:

$$\mathbf{y}=\mathbf{y}(t;C_1,C_2,\ldots,C_n)$$

where $C_1,C_2,\ldots,C_n$ – integration constants.

In Cauchy problem the partial solution of one or several ordinary equations is found, for which n initial conditions of the type $\mathbf{y} = \mathbf{y}(0)$ are set up and by which n of the constants $C_1, C_2, ..., C_n$ are found.

It should be pointed out that together with Cauchy problem the boundary value problem is of interest for practice, in which n conditions are set up on the boundaries of function change – $\mathbf{y} = \mathbf{y}(t_a)$ and $\mathbf{y} = \mathbf{y}(t_b)$ $(t_a \le t \le t_b)$.

In the most general case, the solution of Cauchy problem for the equation systems describing dynamic systems does not have the analytical solution. Such problems can be solved with numerical methods. It was pointed out before (Section 2.3) that the numerical solution of differential equations can be based on applying the procedures of numerical differentiation or numerical integration. The accuracy of the solution of differential equations depends on the precision of the selected approximation of derivatives or integrals. In practice, one-step and multi-step methods are applied to solve the systems of ordinary differential equations with numerical methods [75, 112, 230].

5.4.1.1 One-Step Methods for Solving Differential Equations

One-step numerical methods are the simplest, and therefore, they are most frequently used in practice. One-step methods for determining the regular functional approximation use only one previous point $\mathbf{y}(t_i)$. The computational algorithm called Euler method is the simplest in this group of methods. For the system of differential equations $\frac{d\mathbf{y}}{dt} = \mathbf{f}(t, \mathbf{y})$ with the initial conditions $\mathbf{y}(t_0) = \mathbf{y}_0$ at i integration step Euler method is implemented by the following formula

$$\frac{\mathbf{y}(t_i + \Delta t_i) - \mathbf{y}(t_i)}{\Delta t_i} \approx \mathbf{f}(t_i, \mathbf{y}(t_i)) \tag{5.51}$$

or

$$\mathbf{y}(t_i + \Delta t_i) \approx \mathbf{y}(t_i) + \mathbf{f}(t_i, \mathbf{y}(t_i)) \cdot \Delta t_i \tag{5.52}$$

Here $\mathbf{y}(t_i)$, $\mathbf{y}(t_i + \Delta t_i)$ – values of the integrated function vector at the time points t_i and $t_{i+1} = t_i + \Delta t_i$. The values of vector $\mathbf{y}(t_i)$ are known at the time point t_i, and the values of $\mathbf{y}(t_{i+1})$ corresponding to the time point t_{i+1} need to be found. Δt_i – step of the integration of differential equations. The value of step Δt_i can be variable and change from step to step. The graphic interpretation of Euler method which is true for the solution of one ordinary differential equation is demonstrated in Figure 5.15*a*. In the figure: f_1 – derivative of the function $y = y(t)$ calculated in the point with the coordinates (t_i, y_i), e_1 – actual error of the integration at the step Δt_i.

Despite of the simplicity, the application of the algorithm (5.51) [or (5.52)] is limited due to the low accuracy of the equation integration (Euler method provides the first accuracy order by the argument t). Higher accuracy orders among one-step methods of solving differential equations are provided by numerical methods of Runge-Kutta type. There is a large number of one-step computational methods of Runge-Kutta type with different accuracy. Let us demonstrate some of them.

5.4.1.2 Modified Euler Method

The analysis of solving the equations with the help of Euler method allows making the conclusion that the algorithm accuracy can be increased if a more precise value of the differential equation right side is used in the formula.

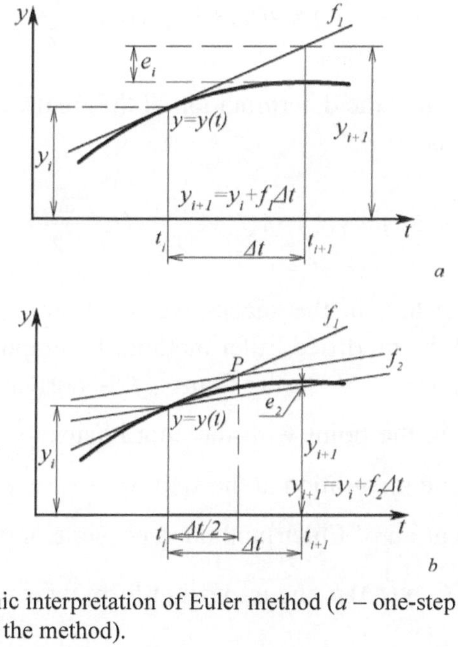

FIGURE 5.15 Graphic interpretation of Euler method (*a* – one-step variant of the method; *b* – two-step variant of the method).

In accordance with the formulas of numerical integration (Section 2.3), the Eq. (5.52) can be rewritten as follows:

$$\mathbf{y}(t_i + \Delta t_i) \approx \mathbf{y}(t_i) + \int_{t_i}^{t_{i+1}} \mathbf{f}(t_i, \mathbf{y}(t_i))dt \tag{5.53}$$

The substitution of integral $\int_{t_i}^{t_{i+1}} \mathbf{f}(t_i, \mathbf{y}(t_i))dt$ for the formula $\int_{t_i}^{t_{i+1}} \mathbf{f}(t_i, \mathbf{y}(t_i))dt = \mathbf{f}(t_i, \mathbf{y}(t_i)) \cdot \Delta t_i$ is the application of the triangle formula (2.33) to calculate the integral in its worst implementation. This implementation corresponds to the algorithm of Euler method. A more accurate implementation – the calculation of the integral by the value of subintegral function in the middle of the segment $t \in (t_i, t_{i+1})$ by the triangle formula.

$$\int_{t_i}^{t_{i+1}} \mathbf{f}(t_i, \mathbf{y}(t_i))dt = \mathbf{f}(t_i + \frac{\Delta t_i}{2}, \mathbf{y}(t_i + \frac{\Delta t_i}{2})) \cdot \Delta t_i \tag{5.54}$$

In this case, at the first stage, the computational algorithm assumes the calculation of the value of $\mathbf{y}(t_i + \frac{\Delta t_i}{2})$, for example, by the formula

$$\mathbf{y}(t_i + \frac{\Delta t_i}{2}) \approx \mathbf{y}(t_i) + \mathbf{f}(t_i, \mathbf{y}(t_i)) \cdot \frac{\Delta t_i}{2} \tag{5.55}$$

and at the second stage – the determination of the final value of the function $\mathbf{y}(t_{i+1})$ by the formula

$$\mathbf{y}(t_i + \Delta t_i) \approx \mathbf{y}(t_i) + \mathbf{f}(t_i + \frac{\Delta t_i}{2}, \mathbf{y}(t_i + \frac{\Delta t_i}{2})) \cdot \Delta t_i \tag{5.56}$$

The algorithm consisting in the successive integration of the Eqs. (5.55) and (5.56) is called the modified Euler method. Its graphic interpretation is demonstrated in Figure 5.15*b*. In the figure: f_2 – derivative of the function $y = y(t)$ calculated in the point with the coordinates $(t_i + \Delta t_i, \frac{1}{2}(y_i + y_{i+1}))$, e_2 – actual error of the integration at the step Δt_i.

Another modification of Euler method is possible, in which for calculating the integral $\int_{t_i}^{t_{i+1}} \mathbf{f}(t_i, \mathbf{y}(t_i))dt$ the trapezium formula is used

$$\int_{t_i}^{t_{i+1}} \mathbf{f}(t_i, \mathbf{y}(t_i))dt = \frac{1}{2}(\mathbf{f}(t_i, \mathbf{y}(t_i)) + \mathbf{f}(t_{i+1}, \mathbf{y}(t_{i+1}))\Delta t_i \quad (5.57)$$

The application of such modification of Euler method is implemented by successively applying the following formulas

$$\tilde{\mathbf{y}}(t_{i+1}) \approx \mathbf{y}(t_i) + \mathbf{f}(t_i, \mathbf{y}(t_i))\,\Delta t_i \quad (5.58)$$

$$\mathbf{y}(t_i + \Delta t_i) \approx \mathbf{y}(t_i) + \frac{1}{2}(\mathbf{f}(t_i, \mathbf{y}(t_i) + \mathbf{f}(t_{i+1}, \tilde{\mathbf{y}}(t_{i+1})) \cdot \Delta t_i \quad (5.59)$$

The computational error while solving the differential equations by the algorithms (5.56), (5.57) and (5.58), (5.59) corresponds to the second order $0(\Delta t^2)$.

5.4.1.3 Runge-Kutta Methods

In Runge-Kutta methods the final value of the function $\mathbf{y}(t_{i+1}) = \mathbf{y}(t_i + \Delta t_i)$ is found after several preliminary approximations calculated by Euler formulas. The final result is obtained by summing up these preliminary approximations. The computational algorithm of Runge-Kutta method is represented by the formation of the following equations:

- schemes of the third accuracy order $0(\Delta t^3)$

$$\mathbf{y}(t_{i+1}) = \mathbf{y}(t_i) + \frac{1}{6}(\mathbf{K}_1 + 4\mathbf{K}_2 + \mathbf{K}_3)\Delta t_i \quad (5.60)$$
$$\mathbf{K}_1 = \mathbf{f}(t_i, \mathbf{y}(t_i)),$$
$$\mathbf{K}_2 = \mathbf{f}\left(t_i + \frac{\Delta t_i}{2}, \mathbf{y}(t_i) + \mathbf{K}_1 \frac{\Delta t_i}{2}\right),$$
$$\mathbf{K}_3 = \mathbf{f}(t_i + \Delta t_i, \mathbf{y}(t_i) + (-\mathbf{K}_1 + 2\mathbf{K}_2)\Delta t_i)$$

or

$$\mathbf{y}(t_{i+1}) = \mathbf{y}(t_i) + \frac{1}{4}(\mathbf{K}_1 + 3\mathbf{K}_3)\Delta t_i \quad (5.61)$$

$$\mathbf{K}_1 = \mathbf{f}(t_i, \mathbf{y}(t_i))$$
$$\mathbf{K}_2 = \mathbf{f}\left(t_i + \frac{\Delta t_i}{3}, \mathbf{y}(t_i) + \mathbf{K}_1 \frac{\Delta t_i}{3}\right)$$
$$\mathbf{K}_3 = \mathbf{f}\left(t_i + \frac{2\Delta t_i}{3}, \mathbf{y}(t_i) + \mathbf{K}_2 \frac{2\Delta t_i}{3}\right)$$

- schemes of the fourth accuracy order $0(\Delta t^4)$

$$\mathbf{y}(t_{i+1}) = \mathbf{y}(t_i) + \frac{1}{6}(\mathbf{K}_1 + 2\mathbf{K}_2 + 2\mathbf{K}_3 + \mathbf{K}_4)\Delta t_i, \tag{5.62}$$

$$\mathbf{K}_1 = \mathbf{f}(t_i, \mathbf{y}(t_i))$$
$$\mathbf{K}_2 = \mathbf{f}\left(t_i + \frac{\Delta t_i}{2}, \mathbf{y}(t_i) + \mathbf{K}_1 \frac{\Delta t_i}{2}\right)$$
$$\mathbf{K}_3 = \mathbf{f}\left(t_i + \frac{\Delta t_i}{2}, \mathbf{y}(t_i) + \mathbf{K}_2 \frac{\Delta t_i}{2}\right)$$
$$\mathbf{K}_4 = \mathbf{f}(t_i + \Delta t_i, \mathbf{y}(t_i) + \mathbf{K}_3 \Delta t_i)$$

or

$$\mathbf{y}(t_{i+1}) = \mathbf{y}(t_i) + \frac{1}{8}(\mathbf{K}_1 + 3\mathbf{K}_2 + 3\mathbf{K}_3 + \mathbf{K}_4)\Delta t_i, \tag{5.63}$$

$$\mathbf{K}_1 = \mathbf{f}(t_i, \mathbf{y}(t_i))$$
$$\mathbf{K}_2 = \mathbf{f}(t_i + \frac{\Delta t_i}{3}, \mathbf{y}(t_i) + \mathbf{K}_1 \frac{\Delta t_i}{3})$$
$$\mathbf{K}_3 = \mathbf{f}(t_i + \frac{2\Delta t_i}{3}, \mathbf{y}(t_i) + (-\frac{1}{3}\mathbf{K}_1 + \mathbf{K}_2)\Delta t_i)$$
$$\mathbf{K}_4 = \mathbf{f}(t_i + \Delta t_i, \mathbf{y}(t_i) + (\mathbf{K}_1 - \mathbf{K}_2 + \mathbf{K}_3)\Delta t_i)$$

5.4.1.4 Kutta-Merson Method

As for the implementation, Kutta-Merson method is very close to the algorithms (5.62), (5.63). The computational algorithm of Kutta-Merson method is represented by the following formation of equations

$$\mathbf{y}(t_{i+1}) = \mathbf{y}(t_i) + \frac{1}{6}(\mathbf{K}_1 + 4\mathbf{K}_3 + \mathbf{K}_4)\Delta t_i \tag{5.64}$$

$$\mathbf{K}_1 = \mathbf{f}(t_i, \mathbf{y}(t_i)),$$
$$\mathbf{K}_2 = \mathbf{f}\left(t_i + \frac{\Delta t_i}{2}, \mathbf{y}(t_i) + \mathbf{K}_1 \frac{\Delta t_i}{2}\right),$$
$$\mathbf{K}_3 = \mathbf{f}\left(t_i + \frac{\Delta t_i}{2}, \mathbf{y}(t_i) + \frac{\mathbf{K}_1 + \mathbf{K}_2}{2} \frac{\Delta t_i}{2}\right),$$
$$\mathbf{K}_4 = \mathbf{f}(t_i + \Delta t_i, \mathbf{y}(t_i) + (2\mathbf{K}_3 - \mathbf{K}_2)\Delta t_i)$$

The computational Kutta-Merson algorithm provides the solution of the system of ordinary differential equations with the fourth accuracy order (the error basic term has the accuracy $O(\Delta t^4)$).

5.4.1.5 Multi-Step Methods of Solving Differential Equations

Apart from one-step methods, multi-step methods are used in computational practice, in which the solution of $\mathbf{y}(t_{i+1})$ at the regular step of approximation $t \in (t_i, t_{i+1})$ is determined by several preceding values of the function $\mathbf{y}(t_i)$, $\mathbf{y}(t_{i-1})$, $\mathbf{y}(t_{i-2})$, etc. The computational algorithm providing the implementation of multi-step method can start only after the preliminary calculation of the missing values of the function $\mathbf{y}(t_{i-1})$, $\mathbf{y}(t_{i-2})$,..., for instance, with one-step methods. The development of computational algorithms in multi-step methods is based on high-precision approximation of the derivatives $\frac{d\mathbf{y}}{dt}$ and integrals $\int_{t_i}^{t_{i+1}} \mathbf{f}(t_i, \mathbf{y}(t_i))dt$ (see Section 2.3.2).

"Predictor-corrector" methods are applied to obtain high-precision results. At the first stage (predictor), the preliminary value of the function $\mathbf{y}(t_{i+1})$ is calculated, at the second stage (corrector), this value is clarified. It should be pointed out that in multi-step methods both predictor and corrector provide the high accuracy of calculation. There is a large number of multi-step algorithms [76, 230, 250]. Some of them are given below. The algorithms are written down in the assumption that the integration step of the equations $\Delta t = h = const$.

1. Computational algorithm in Milne method is performed by the following formulas:
   - predictor

$$\mathbf{y}(t_{i+1}) = \mathbf{y}(t_{i-3}) + \frac{4}{3}h(2\mathbf{f}(t_i, \mathbf{y}(t_i)) - \mathbf{f}(t_{i-1}, \mathbf{y}(t_{i-1})) + 2\mathbf{f}(t_{i-2}, \mathbf{y}(t_{i-2}))) \tag{5.65}$$

- corrector (is based on Simpson formula)

$$\mathbf{y}(t_{i+1}) = \mathbf{y}(t_{i-1}) + \frac{1}{3}h(\mathbf{f}(t_{i+1}, \mathbf{y}(t_{i+1})) + 4\mathbf{f}(t_i, \mathbf{y}(t_i)) + \mathbf{f}(t_{i-1}, \mathbf{y}(t_{i-1}))) \tag{5.66}$$

The computational error of the function $\mathbf{y}(t_{i+1})$ in the formula (5.65) is $\sim \frac{28}{90}h^5 \frac{d^5\mathbf{y}}{dt^5}$, and in the formula (5.66) $\sim -\frac{1}{90}h^5 \frac{d^5\mathbf{y}}{dt^5}$.

2. In Hemming method the predictor is calculated by the formula (5.65), and the corrector is calculated as follows:

$$\begin{aligned}\mathbf{y}(t_{i+1}) = \frac{1}{8}(9\mathbf{y}(t_i) - \mathbf{y}(t_{i-2})) + 3h(\mathbf{f}(t_{i+1}, \mathbf{y}(t_{i+1})) \\ + 2\mathbf{f}(t_i, \mathbf{y}(t_i)) - \mathbf{f}(t_{i-1}, \mathbf{y}(t_{i-1})))\end{aligned} \tag{5.67}$$

The computational error of the function $\mathbf{y}(t_{i+1})$ in the formula (5.67) is $\sim -\frac{1}{40}h^5 \frac{d^5\mathbf{y}}{dt^5}$.

3. In Adams-Bashforth method the computational algorithm is implemented as follows:

- predictor (is based on Newton's backward interpolation formula)

$$\begin{aligned}\mathbf{y}(t_{i+1}) = \mathbf{y}(t_i) + \frac{1}{24}h(55\mathbf{f}(t_i, \mathbf{y}(t_i)) - 59\mathbf{f}(t_{i-1}, \mathbf{y}(t_{i-1})) \\ + 37\mathbf{f}(t_{i-2}, \mathbf{y}(t_{i-2})) - 9\mathbf{f}(t_{i-3}, \mathbf{y}(t_{i-3})))\end{aligned} \tag{5.68}$$

- corrector

$$\begin{aligned}\mathbf{y}(t_{i+1}) = \mathbf{y}(t_i) + \frac{1}{24}h(9\mathbf{f}(t_{i+1}, \mathbf{y}(t_{i+1})) - 19\mathbf{f}(t_i, \mathbf{y}(t_i)) \\ - 5\mathbf{f}(t_{i-1}, \mathbf{y}(t_{i-1})) + \mathbf{f}(t_{i-2}, \mathbf{y}(t_{i-2})))\end{aligned} \tag{5.69}$$

The computational error of the function $\mathbf{y}(t_{i+1})$ in the formula (5.68) is $\sim \frac{251}{720}h^5 \frac{d^5\mathbf{y}}{dt^5}$, d in the formula (5.69) $\sim \frac{19}{720}h^5 \frac{d^5\mathbf{y}}{dt^5}$.

5.4.1.6 Solution of Stiff Systems of Differential Equations

The considered methods of solving ordinary differential equations and equation systems can be ineffective when solving stiff systems of equations. From those discussed in Section 5.1, the equation systems describing chemical kinetics can be referred to such systems (Eqs. (5.11)–(5.19)). The systems of ordinary differential equations, in which the processes with significantly different scales are available, belong to stiff ones. In this case, the integration step by time should be selected in such a way as to provide the calculation accuracy of all the components forming the problem solution. In a number of cases, this requires the selection of an unacceptably small integration step Δt.

In [193] there is the following definition of the stiffness of the system of ordinary differential equations:

The system of n differential equations $\frac{d\mathbf{y}}{dt} = A\mathbf{y}$ with the constant matrix A with the dimensionality $n \times n$ is called stiff in the interval $t \in (t_0, t_{\max})$, if in this interval $\operatorname{Re}\lambda_k(t) < 0$, k=1, 2, …, m, and correlation $s(t) = \frac{\max|\operatorname{Re}\lambda_k(t)|}{\min|\operatorname{Re}\lambda_k(t)|}$ is great.

Here $\lambda_k(t)$ – own values of the matrix A, $\operatorname{Re}\lambda_k(t)$ – their real part, s(t) – stiffness number. The more is the stiffness number, the more is the degree in which the properties of the system stiffness of ordinary differential equations are expressed.

The simplest algorithms that allow solving the problems described by the stiff systems of differential equations are the computational algorithms of inexplicit type being the modification of Euler method

$$\frac{\mathbf{y}(t_i + \Delta t_i) - \mathbf{y}(t_i)}{\Delta t_i} \approx \mathbf{f}(t_{i+1}, \mathbf{y}(t_{i+1})) \qquad (5.70)$$

The Eq. (5.70) differs from the Eq. (5.51) by the right side which should be calculated at the time moment t_{i+1}, for which the function value is not found yet. If the right sides of the system of differential equations are linear $\frac{d\mathbf{y}}{dt} = A\mathbf{y}$, in which the matrix A does not depend on time, the inexplicit Euler scheme (5.70) in the time interval $t \in (t_i, t_{i+1})$ reduces as follows:

$$\mathbf{y}(t_{i+1}) = \mathbf{y}(t_i) + \Delta t_i \cdot A\mathbf{y}(t_{i+1}) \qquad (5.71)$$

and is solved with the known methods applicable to the systems of linear algebraic equations (see Chapters 2 and 8). In the Eq. (5.71) the unknown variables are the values $\mathbf{y}(t_{i+1})$.

In general case, the equation system is of linear type, and its solution requires the application of the corresponding methods (the methods of solving the nonlinear systems are considered in Chapter 4). However, the special methods of solving the stiff systems of ordinary differential equations are also developed. In particular, in [112] Rosenbrock algorithm applicable to the Eq. (5.1) is discussed.

In accordance with Rosenbrock algorithm, the following equation is solved in the time interval $t \in (t_i, t_{i+1})$ instead of the Eq. (5.1)

$$\mathbf{B}\frac{\mathbf{y}(t_{i+1}) - \mathbf{y}(t_i)}{\Delta t_i} = \mathbf{F}(t_i, \mathbf{z}(t_i)) \tag{5.72}$$

Here $\mathbf{B} = \mathbf{E} - \alpha \Delta t_i \mathbf{A} - \beta \Delta t_i^2 \mathbf{A}^2$; $\mathbf{z}(t_i) = \mathbf{y}(t_i) + \gamma \Delta t_i \mathbf{F}(t_i, \mathbf{y}(t_i))$; $\alpha = 1.077$; $\beta = -0.372$; $\gamma = -0.577$.

Matrix **A** is Jacobian composed of partial derivatives

$$\mathbf{A} = \begin{pmatrix} \frac{\partial f_1(t,\mathbf{y})}{\partial y_1} & \cdots & \frac{\partial f_1(t,\mathbf{y})}{\partial y_i} & \cdots & \frac{\partial f_1(t,\mathbf{y})}{\partial y_n} \\ \cdots & \cdots & \cdots & \cdots & \cdots \\ \frac{\partial f_i(t,\mathbf{y})}{\partial y_1} & \cdots & \frac{\partial f_i(t,\mathbf{y})}{\partial y_i} & \cdots & \frac{\partial f_i(t,\mathbf{y})}{\partial y_n} \\ \cdots & \cdots & \cdots & \cdots & \cdots \\ \frac{\partial f_n(t,\mathbf{y})}{\partial y_1} & \cdots & \frac{\partial f_n(t,\mathbf{y})}{\partial y_i} & \cdots & \frac{\partial f_n(t,\mathbf{y})}{\partial y_n} \end{pmatrix}$$

5.4.2 METHODS OF DETERMINING THE EIGEN VALUES

In Chapter 2 the notion of matrix eigen numbers is introduced when considering the elements of matrix calculation. Eigen numbers are sometimes called eigen or characteristic matrix values. The mathematical problem on finding the matrix **A** eigen numbers arises, for example, when seeking nontrivial solutions of the system of differential equations

$$\frac{d\mathbf{Y}}{dt} = \mathbf{A} \cdot \mathbf{Y} \tag{5.73}$$

in the form $\mathbf{Y} = e^{\lambda t}\mathbf{X}$. Here $\mathbf{Y}$ – vector-column of the variables to be found y_i $(i = 1, n)$, $\mathbf{X}$ – corresponding vector-column called the dimensionality eigen vector *n*, **A** – matrix with the dimensionality $n \times n$. After the formula substitution for the solution sought $\mathbf{Y} = e^{\lambda t}\mathbf{X}$ into the Eq. (5.56), the following matrix equation can be obtained

$$\mathbf{AX} = \lambda\mathbf{X}$$

or

$$(\mathbf{A} - \lambda\mathbf{E})\mathbf{X} = 0 \tag{5.74}$$

In the form (5.74) the complete problem of the matrix **A** eigen values is formulated as follows:

- to determine all the formation of eigen pairs $\{\lambda, \mathbf{X}\}$ (eigen numbers λ and eigen vectors X) satisfying the Eq. (5.74) and providing the availability of non-trivial solutions of the system of differential Eq. (5.73).

Instead of the complete problem of the matrix **A** eigen values, the partial problem of eigen values, in which only several eigen pairs $\{\lambda, \mathbf{X}\}$ are found (e.g., several maximal and several minimal values) can be solved.

It was pointed out before that the mathematical problem considered is widely applied in engineering applications. Thus, eigen numbers determine the values of main normal stresses in the problems on material mechanics, and eigen vectors determine the action directions of these stresses. In the problems of acoustics, problems for mechanical systems and problems on electrical engineering the eigen numbers determine the eigen frequencies of oscillations, and eigen vectors characterize the modes of these oscillations.

Three groups of methods are applied to solve the complete or partial problem of eigen values [23, 46, 75]. The first group of methods is based on the solution of characteristic (eigen value) Eq. (2.5). The second group of methods is based on iteration algorithms called the power algorithms. The third group of effective methods consists in reducing the initial matrix to the triangular (diagonal) type with the formation of similar orthogonal transformations.

5.4.2.1 Methods Based on the Solution of Characteristic Equation

The solution of the problem of eigen values of the arbitrary matrix A using the characteristic equation consists of two stages. At the first stage, the type of characteristic equation is determined. For the matrix with the dimensionality $n \times n$ the characteristic equation is the polynomial equation of the power n

$$D(\lambda) = \det(\mathbf{A} - \lambda\mathbf{E}) = \begin{vmatrix} a_{11} - \lambda & a_{12} & a_{13} \dots & a_{1n} \\ a_{21} & a_{22} - \lambda & a_{23} \dots & a_{2n} \\ \dots\dots\dots & & & \\ a_{n1} & a_{n2} & a_{n3} \dots & a_{nn} - \lambda \end{vmatrix} = 0 \quad (5.75)$$

At the second stage, the polynomial algebraic equation is solved

$$\lambda^n - p_1\lambda^{n-1} - p_2\lambda^{n-2} - \dots - p_{n-1}\lambda - p_n = 0 \quad (5.76)$$

corresponding to the Eq. (5.75).

5.4.2.2 A.M. Danilevsky Method

The method is based on the transformation of the initial matrix **A** to Frobenius matrix **P**

$$\mathbf{P} = \begin{pmatrix} p_1 & p_2 & p_3 & \cdots & p_i & \cdots & p_{n-1} & p_n \\ 1 & 0 & 0 & \dots & 0 & \dots & 0 & 0 \\ 0 & 1 & 0 & \dots & 0 & \dots & 0 & 0 \\ \dots & \dots & \dots & \dots & \dots & \dots & \dots & \dots \\ 0 & 0 & 0 & \dots & 0 & \dots & 0 & 0 \\ \dots & \dots & \dots & \dots & \dots & \dots & \dots & \dots \\ 0 & 0 & 0 & \dots & 0 & \dots & 0 & 0 \\ 0 & 0 & 0 & \dots & 0 & \dots & 1 & 0 \end{pmatrix}$$

In this matrix, the elements $a_{1i} = p_i$, $(i = 1, n)$, $a_{i+1,i} = 1$, $(i = 1, n-1)$ and all the other elements equal zero. The matrix **A** reduction to matrix **P** is performed by the similar transformations of the type $\mathbf{P} = \mathbf{S}^{-1}\mathbf{A}\mathbf{S}$, where **S** – arbitrary non-degenerate matrix.

In the matrix form, the algorithm of A.M. Danilevsky method can be written down as follows:

$$\mathbf{P} = \mathbf{M}_1^{-1}\mathbf{M}_2^{-1}...\mathbf{M}_i^{-1}...\mathbf{M}_{n-2}^{-1}\mathbf{M}_{n-1}^{-1}\mathbf{A}\mathbf{M}_{n-1}\mathbf{M}_{n-2}...\mathbf{M}_i...\mathbf{M}_2\mathbf{M}_1 \qquad (5.77)$$

In the Eq. (5.77) the lines successively starting from the n line and finishing with the second line are transformed into the type corresponding to Frobenius matrix. Thus, for instance, the matrixes $\mathbf{M}_{n-1}$, $\mathbf{M}_{n-1}^{-1}$ can be written down as follows:

$$\mathbf{M}_{n-1} = \begin{pmatrix} 1 & 0 & \dots & 0 & \dots & 0 & 0 \\ 0 & 1 & \dots & 0 & \dots & 0 & 0 \\ \dots & \dots & \dots & \dots & \dots & \dots & \dots \\ 0 & 0 & \dots & 1 & \dots & 0 & 0 \\ \dots & \dots & \dots & \dots & \dots & \dots & \dots \\ m_{n-1,1} & m_{n-1,2} & \dots & m_{n-1,i} & \dots & m_{n-1,n-1} & m_{n-1,n} \\ 0 & 0 & \dots & 0 & \dots & 0 & 1 \end{pmatrix};$$

$$\mathbf{M}_{n-1}^{-1} = \begin{pmatrix} 1 & 0 & \dots & 0 & \dots & 0 & 0 \\ 0 & 1 & \dots & 0 & \dots & 0 & 0 \\ \dots & \dots & \dots & \dots & \dots & \dots & \dots \\ 0 & 0 & \dots & 1 & \dots & 0 & 0 \\ \dots & \dots & \dots & \dots & \dots & \dots & \dots \\ a_{n,1} & a_{n,2} & \dots & a_{n,i} & \dots & a_{n,n-1} & a_{n,n} \\ 0 & 0 & \dots & 0 & \dots & 0 & 1 \end{pmatrix}$$

Here the matrix elements $m_{n-1,i}$ are calculated by the following formulas

$$m_{n-1,i} = -\frac{a_{ni}}{a_{n,n-1}} \quad \text{at } i \neq n-1;$$

$$m_{n-1,n-1} = \frac{1}{a_{n,n-1}}$$

The matrix multiplication $\mathbf{M}_{n-1}^{-1}\mathbf{A}\mathbf{M}_{n-1}$ brings the initial matrix $\mathbf{A}$ to the new form – matrix B

$$\mathbf{B} = \mathbf{M}_{n-1}^{-1}\mathbf{A}\mathbf{M}_{n-1} = \begin{pmatrix} b_{11} & b_{12} & \dots & b_{1i} & \dots & b_{1,n-1} & b_{1,n} \\ b_{21} & b_{22} & \dots & b_{2i} & \dots & b_{2,n-1} & b_{2n} \\ \dots & \dots & \dots & \dots & \dots & \dots & \dots \\ b_{i1} & b_{i2} & \dots & b_{ii} & \dots & b_{i,n-1} & b_{in} \\ \dots & \dots & \dots & \dots & \dots & \dots & \dots \\ b_{n-1,1} & b_{n-1,2} & \dots & b_{n-1,i} & \dots & b_{n-1,n-1} & b_{n-1,n} \\ 0 & 0 & \dots & 0 & \dots & 1 & 0 \end{pmatrix}$$

The matrix **B** elements are calculated by the following formulas

$$b_{ij} = c_{ij} \quad \text{at } 1 \le i \le n-2, \quad 1 \le j \le n;$$

$$b_{n-1,j} = \sum_{k=1}^{n} a_{nk} c_{kj} \quad \text{at } 1 \le j \le n;$$

$$c_{ij} = a_{ij} + a_{i,n-1} m_{n-1,j} \quad \text{at } 1 \le i \le n-1, \quad 1 \le j \le n,\ j \ne n-1;$$

$$c_{i,n-1} = a_{i,n-1} m_{n-1,n-1} \quad \text{at } 1 \le i \le n-1$$

The matrix **B** changes in the similar way, at the same time, the line with the number (*n-2*) acquires the form corresponding to Frobenius matrix. For this, the matrixes $\mathbf{M}_{n-2}, \mathbf{M}_{n-2}^{-1}$ are constructed and the matrix product $\mathbf{M}_{n-2}^{-1}\mathbf{M}_{n-1}^{-1}\mathbf{A}\mathbf{M}_{n-1}\mathbf{M}_{n-2}$ is determined. The execution of the foregoing procedure (*n–1*) times brings the initial matrix **A** to Frobenius matrix **P**.

By the number of elementary arithmetic operations performed, A.M. Danilevsky method is one of the most effective among the known ones. However, the method has a number of shortcomings. In particular, when calculating the intermediary matrix coefficients on the iteration with the number *k*, during the calculation of $m_{n-k,i} = -\dfrac{b_{n-k+1,i}}{b_{n-k+1,n-k}}$ there can be the accuracy loss caused by the insignificance of the denominator $b_{n-k+1,n-k} \approx 0$. The overcoming of the above shortcoming complicates the computational algorithm.

Finishing the description of A.M. Danilevsky method, it should be pointed out that the matrixes $\mathbf{M}_1, \mathbf{M}_2, \dots, \mathbf{M}_{n-1}$ calculated when solving the problem allow calculating the matrix **A** eigen vectors. Actually, let all eigen numbers $\lambda_1, \lambda_2, \dots, \lambda_n$ and coefficients of the characteristic equation $p_1, p_2, \dots, p_n$ be already calculated. Besides, the matrixes $\mathbf{M}_1, \mathbf{M}_2, \dots, \mathbf{M}_{n-1}$

are calculated. Let the vector **Y** be the eigen vector of Frobenius matrix **P**. In this case, for the values of λ being the eigen values of Frobenius matrix, the matrix equation $(\mathbf{P}-\lambda\mathbf{E})\mathbf{Y}=0$ is true. In the open form this equation is written down as the equation system

$$(p_1-\lambda)y_1+p_2y_2+p_3y_3+...+p_ny_n=0$$

$$y_1-\lambda y_2=0$$

$$y_2-\lambda y_3=0$$

$$\dots$$

$$y_{n-1}-\lambda y_n=0$$

The vector **Y** is the specific solution of the homogeneous system of equations whose elements are calculated by the formulas $y_i=\lambda^{n-i}$ (*i=1, n*). In this case, the eigen vector **X** corresponding to the eigen value λ is found as a result of the matrix multiplication

$$\mathbf{X}=\mathbf{M}_{n-1}\mathbf{M}_{n-2}...\mathbf{M}_i...\mathbf{M}_2\mathbf{M}_1\mathbf{Y} \tag{5.78}$$

5.4.2.3 Method of Indefinite Coefficients

The method is based on the identity of the Eqs. (5.75) and (5.76). These equations are really true for any values of λ. In particular, they are true for the values $\lambda=0$, $\lambda=1$, ..., $\lambda=n-1$. Inserting these values of λ into the Eq. (5.75), we can calculate the values of the determinants *D*(λ) corresponding to the foregoing values of λ. Further, applying the calculated values of the determinants, let us write down the system of linear equations with the values of the polynomial p_i coefficients based on the Eq. (5.76)

$$0^n+0^{n-1}p_1+0^{n-2}p_2+...+0^{n-i}p_i+...+p_n=D(0),$$

$$1^n+1^{n-1}p_1+1^{n-2}p_2+...+1^{n-i}p_i+...+p_n=D(1),$$

$$2^n+2^{n-1}p_1+2^{n-2}p_2+...+2^{n-i}p_i+...+p_n=D(2),$$

$$\dots$$

$$(n-1)^n+(n-1)^{n-1}p_1+(n-1)^{n-2}p_2+...+(n-1)^{n-i}p_i+...+p_n=D(n-1)$$

The values of the coefficient p_n are found in the first equation of the foregoing system. The other coefficients can be found in the matrix equation

$$\mathbf{CP} = \mathbf{D} \tag{5.79}$$

In (5.79) the matrixes **C**, **D** and **P** are as follows:

$$\mathbf{C} = \begin{pmatrix} 1^{n-1} & 1^{n-2} & \dots & 1^1 \\ 2^{n-1} & 2^{n-2} & \dots & 2^1 \\ \dots & \dots & \dots & \dots \\ (n-1)^{n-1} & (n-1)^{n-2} & \dots & (n-1)^1 \end{pmatrix};$$

$$\mathbf{D} = \begin{Bmatrix} D(1) - D(0) - 1^n \\ D(2) - D(0) - 2^n \\ \dots\dots\dots\dots\dots\dots \\ D(n-1) - D(0) - (n-1)^n \end{Bmatrix}; \quad \mathbf{P} = \begin{Bmatrix} p_1 \\ p_2 \\ \dots \\ p_{n-1} \end{Bmatrix}$$

It is not difficult to solve the matrix Eq. (5.79) even when applying Cramer method ($\mathbf{P} = \mathbf{C}^{-1}\mathbf{D}$) since the reciprocal matrix $\mathbf{C}^{-1}$ can be preliminarily calculated. The main working time of the method of undetermined coefficients comes down to calculating the determinants $D(0), D(1), \dots, D(n-1)$.

5.4.2.4 Le Verrier Method

In comparison with the above, Le Verrier method is considered cumbersome, however, the idea of the computational algorithm applied in it is relatively simple. Besides, the algorithm efficiency considerably increases if the initial matrix eigen values and eigen vectors need to be determined in the process of solving the problem in addition to finding the coefficients of the characteristic equation.

In Le Verrier method the coefficients of characteristic polynomial are found when solving the equation system

$$\begin{gathered} p_1 = -s_1 \\ p_2 = -\frac{1}{2}(s_2 + p_1 s_1) \\ \dots \end{gathered} \tag{5.80}$$

$$p_n = -\frac{1}{n}(s_n + \sum_{k=1}^{n-1} p_k s_{n-k})$$

Here, the coefficients s_i are the traces of Sp (diagonal sums) of the matrixes $\mathbf{A}^i$. The matrixes $\mathbf{A}^i$ are obtained as a result of multiplying the matrix **A** by itself *i* times (*i* – power index of the matrix **A**):

$$s_1 = \lambda_1 + \lambda_2 + ... + \lambda_n = \sum_{i=1}^{n} a_{ii} = \text{Sp}\mathbf{A},$$

$$s_2 = \lambda_1^2 + \lambda_2^2 + ... + \lambda_n^2 = \sum_{i=1}^{n} a_{ii}^{(2)} = \text{Sp}\mathbf{A}^2, \quad (5.81)$$

$$...$$

$$s_n = \lambda_1^n + \lambda_2^n + ... + \lambda_n^n = \sum_{i=1}^{n} a_{ii}^{(n)} = \text{Sp}\mathbf{A}^n$$

The calculation of the matrix **A** power index is a cumbersome procedure, however, $\mathbf{A}^i$ values can be demanded in some methods of calculating eigen numbers and eigen vectors of the matrix **A**.

5.4.2.5 Power Methods for Determining Matrix Eigen Values

For the convenience of further description, let us put in order the indexing of eigen values of the matrix **A** as follows – $|\lambda_1| > |\lambda_2| > ... > |\lambda_n|$. It is frequent in practical applications to be limited by the determination of several eigen values of the matrix and eigen vectors corresponding to these values – one or two maximal by the module λ_1, λ_2 and (or) one or two minimal by the module λ_{n-1}, λ_n. These values can be calculated when solving the Eq. (5.76) by any solution method for non-linear equations (see Chapter 4). Further, other computational algorithms will be additionally considered.

5.4.2.6 Determination of Eigen Values by Matrix Traces

Let us apply the Eq. (5.81) for traces of the matrix $\text{Sp}\mathbf{A}^m$. Let the trace values of the initial matrix $\text{Sp}\mathbf{A}^m$ and $\text{Sp}\mathbf{A}^{m+1}$ be known. In this case, the following equations can be written down

$$\mathbf{SpA}^{m} = \lambda_1^{m} + \lambda_2^{m} + \ldots + \lambda_n^{m},$$

$$\mathbf{SpA}^{m+1} = \lambda_1^{m+1} + \lambda_2^{m+1} + \ldots + \lambda_n^{m+1}$$

After dividing the second equation by the first one and performing additional transformations, we can write down the following:

$$\lambda_1 \cdot \frac{1 + (\lambda_2/\lambda_1)^{m+1} + \ldots + (\lambda_n/\lambda_1)^{m+1}}{1 + (\lambda_2/\lambda_1)^{m} + \ldots + (\lambda_n/\lambda_1)^{m}} = \frac{\mathbf{SpA}^{m+1}}{\mathbf{SpA}^{m}}$$

Since $|\lambda_1| > |\lambda_2| > \ldots > |\lambda_n|$, with the m increase, the value of the fraction $\frac{1 + (\lambda_2/\lambda_1)^{m+1} + \ldots + (\lambda_n/\lambda_1)^{m+1}}{1 + (\lambda_2/\lambda_1)^{m} + \ldots + (\lambda_n/\lambda_1)^{m}}$ will be tending to one, and the value of the unknown eigen number λ_1 will be approximating the real value

$$\lambda_1 \approx \frac{\mathbf{SpA}^{m+1}}{\mathbf{SpA}^{m}} \tag{5.82}$$

5.4.2.7 Power Methods with the Application of Eigen Vectors

Let $\mathbf{X}_1, \mathbf{X}_2, \ldots, \mathbf{X}_n$ be the matrix $\mathbf{A}$ eigen vectors. Let us select the arbitrary vector $\mathbf{Y}^0$ whose expansion in the basis $\mathbf{X}_1, \mathbf{X}_2, \ldots, \mathbf{X}_n$ can be written down as follows:

$$\mathbf{Y}^0 = c_1\mathbf{X}_1 + c_2\mathbf{X}_2 + \ldots + c_n\mathbf{X}_n$$

Let us find the vector $\mathbf{Y}^1$ connected with the vector $\mathbf{Y}^0$ by the formula

$$\mathbf{Y}^1 = \mathbf{AY}^0 = c_1\mathbf{AX}_1 + c_2\mathbf{AX}_2 + \ldots + c_n\mathbf{AX}_n = c_1\lambda_1\mathbf{X}_1 + c_2\lambda_2\mathbf{X}_2 + \ldots + c_n\lambda_n\mathbf{X}_n$$

Similarly, the following powers of the vector $\mathbf{Y}^{m+1} = \mathbf{AY}^{m}$ will be found

$$\mathbf{Y}^{m+1} = c_1\lambda_1^{m+1}\mathbf{X}_1 + c_2\lambda_2^{m+1}\mathbf{X}_2 + \ldots + c_n\lambda_n^{m+1}\mathbf{X}_n$$

Based on the condition that $|\lambda_1| > |\lambda_2| > ... > |\lambda_n|$ in the same way as in the method of determining the eigen number λ_1 by the formula (5.82), starting from some rather large m ($m \to \infty$) it can be taken into account that

$$\frac{y_i^{(m+1)}}{y_i^{(m)}} = \lambda_1 \cdot \frac{1 + \frac{c_2 x_{2i}}{c_1 x_{1i}} \cdot (\frac{\lambda_2}{\lambda_1})^{m+1} + ... + \frac{c_n x_{ni}}{c_1 x_{1i}} \cdot (\frac{\lambda_n}{\lambda_1})^{m+1}}{1 + \frac{c_2 x_{2i}}{c_1 x_{1i}} \cdot (\frac{\lambda_2}{\lambda_1})^{m} + ... + \frac{c_n x_{ni}}{c_1 x_{1i}} \cdot (\frac{\lambda_n}{\lambda_1})^{m}} \approx \lambda_1 \tag{5.83}$$

The eigen vector $\mathbf{X}_1$ corresponding to the eigen number λ_1 can be accepted in accordance with the formula

$$\mathbf{X}_1 \approx \frac{\mathbf{Y}^{m+1}}{\left\| \mathbf{Y}^{m+1} \right\|_E} \tag{5.84}$$

5.4.2.8 Methods of Scalar Products

In contrast to the previous method, the value of λ_1 is found by the relation of two scalar products of the vectors $\mathbf{Y}^m$, $\mathbf{Y}^{m-1}$:

$$\frac{(\mathbf{Y}^m \cdot \mathbf{Y}^m)}{(\mathbf{Y}^m \cdot \mathbf{Y}^{m-1})} = \lambda_1 \cdot \frac{1 + (\frac{c_2}{c_1})^2 \cdot (\frac{\lambda_2}{\lambda_1})^{2m} + ... + (\frac{c_n}{c_1})^2 \cdot (\frac{\lambda_n}{\lambda_1})^{2m}}{1 + (\frac{c_2}{c_1})^2 \cdot (\frac{\lambda_2}{\lambda_1})^{2m-1} + ... + (\frac{c_n}{c_1})^2 \cdot (\frac{\lambda_n}{\lambda_1})^{2m-1}} \approx \lambda_1 \tag{5.85}$$

It should be pointed out that the iteration cycle convergence in the method of scalar products is sufficiently greater than in the power method.

5.4.3 DILATATION METHODS

Dilatation methods are the most effective methods for solving the problem of eigen values. The application of these methods is based on the following properties of matrix calculation:

1. Let the non-degenerate matrixes **A** and **S** with the dimensionality $n \times n$ be set up, and $\{\lambda, \mathbf{X}\}$ – eigen pair of the matrix **A** (matrix **A**

eigen values and eigen vectors). Let us mark the matrix **B** determined by the product $\mathbf{B}=\mathbf{S}^{-1}\mathbf{AS}$ (or $\mathbf{B}=\mathbf{SAS}^{-1}$). The matrix **B** constructed in such way is called similar to the matrix **A**, and the eigen pair of this matrix – $\{\lambda, \mathbf{SX}\}$ (eigen values of the matrixes **A** and **B** coincide, and the eigen vectors are connected with the matrix multiplication);
2. Diagonal elements of diagonal or triangular matrix are its eigen numbers;
3. Let **D** be the diagonal matrix with the matrix **A** eigen values as its elements, and **X** – matrix formed by the matrix **A** eigen vectors. Then the following matrix correlation is true $\mathbf{D}=\mathbf{X}^{-1}\mathbf{AX}$.

The computational algorithm in the methods of similar transformations reduces to the multiple constructions of similar matrixes with the final aim to obtain the diagonal or triangular matrix. Then the complete problem of eigen numbers is solved based on the above properties 2 and 3. When constructing the similar matrixes $\mathbf{B}_i$, the matrixes $\mathbf{S}_i$ are selected in such a way that when multiplying $\mathbf{B}_i\mathbf{S}_i$ one or several elements of the matrix $\mathbf{B}_i$ are nulled. For this, the orthogonal rotation matrixes can be applied as the matrixes $\mathbf{S}_i$. The matrix $\mathbf{B}_{i+1}$, which is similar to the matrixes $\mathbf{B}_i$, $\mathbf{B}_{i-1}$, …, $\mathbf{B}_2$, $\mathbf{B}_1$, as well as the initial matrix **A** is obtained by the additional multiplication by the matrix $\mathbf{S}_i^{-1}$. Thus, the matrix $\mathbf{B}_{i+1}$ is found by the multiplication $\mathbf{B}_{i+1}=\mathbf{S}_i^{-1}\mathbf{B}_i\mathbf{S}_i$. As a result of such multiplication, the part of previously nulled elements of the initial matrix **A** will again take the values different from zero. However, with a rather large number of iterations (with the values $i \to \infty$) the computational algorithm $\mathbf{B}_{i+1}=\mathbf{S}_i^{-1}...\mathbf{S}_2^{-1}\mathbf{S}_1^{-1}\mathbf{AS}_1\mathbf{S}_2....\mathbf{S}_i$ eventually allows obtaining the diagonal or triangular matrix $\mathbf{B}_{i+1}$.

The matrix $\mathbf{B}_{i+1}$ diagonal elements are the eigen values of the matrix **A**, and the matrix $\mathbf{X}=\mathbf{S}_1\mathbf{S}_2....\mathbf{S}_i$ elements are the eigen vectors (elements of the eigen vectors are located along the columns of the final matrix **X**).

The described computational algorithm for eigen values and vectors of the arbitrary matrix is called Jacobi method. Its implementation in computational algorithms is based on the application of **LU**- and **QR** decompositions of the matrix **A**.

Let the matrix $\mathbf{A}=\mathbf{LU}$ decomposition be found. Let us consider the new matrix $\mathbf{A}_1=\mathbf{UL}$ which is obtained when changing the multiplication order of the matrixes **L** and **U**. In this case, the bond between the matrixes **A** and $\mathbf{A}_1$ is written down with the matrix equation $\mathbf{A}=\mathbf{LA}_1\mathbf{L}^{-1}$, from which it follows that the matrixes **A** and $\mathbf{A}_1$ are similar. Similarly, when decomposing the matrix

$\mathbf{A}_1 = \mathbf{L}_1\mathbf{U}_1$, the matrix $\mathbf{A}_2 = \mathbf{U}_1\mathbf{L}_1$ can be constructed and this matrix will be similar to the matrixes **A** and $\mathbf{A}_1$ ($\mathbf{A} = \mathbf{L}\mathbf{A}_1\mathbf{L}^{-1} = \mathbf{L}_1\mathbf{L}\mathbf{A}_2\mathbf{L}^{-1}\mathbf{L}_1^{-1}$). Continuing the computational process in such a way, at the *i* step for the matrix $\mathbf{A}_i$ we find the decomposition $\mathbf{A}_i = \mathbf{L}_i\mathbf{U}_i$ and calculate the matrix $\mathbf{A}_{i+1} = \mathbf{U}_i\mathbf{L}_i$ which is similar to the matrixes **A**, $\mathbf{A}_1$, $\mathbf{A}_2$, …, $\mathbf{A}_i$ ($\mathbf{A} = \mathbf{L}_i...\mathbf{L}_1\mathbf{L}\mathbf{A}_{i+1}\mathbf{L}^{-1}\mathbf{L}_1^{-1}...\mathbf{L}_i^{-1}$). If we assume that with rather large values of *i* ($i \to \infty$) the matrix $\mathbf{A}_{i+1}$ becomes triangular, the diagonal elements of this matrix will be the eigen elements of the matrix $\mathbf{A}_{i+1}$ and, consequently, the initial matrix **A**. In the same way as in Jacobi method, after finding the matrix eigen matrix, its eigen vectors are determined by the product $\mathbf{X} = \mathbf{L}\mathbf{L}_1\mathbf{L}_2....\mathbf{L}_i$.

The shortcomings of the application of the matrix **A LU**-decomposition for the computation of eigen values are the lack of the method absolute stability and accumulation of computational errors during the iteration. To the lesser extent these shortcomings are revealed if similar matrixes are formed by the proposed algorithm using the matrix **A** demonstration in the form of the product of the orthogonal matrix **Q** and right triangular matrix **R** (**QR**-decomposition). The matrix $\mathbf{A}_i$ similar to the matrixes **A**, $\mathbf{A}_1$, $\mathbf{A}_2$, …, $\mathbf{A}_{i-1}$ at the *i* step is displayed in the form of the product $\mathbf{A}_i = \mathbf{Q}_i\mathbf{R}_i$ and the matrix $\mathbf{A}_{i+1} = \mathbf{Q}_i\mathbf{R}_i$ is calculated. In the same way as in the previous method, the matrix $\mathbf{A}_{i+1}$ tends to the triangular type when applying the **QR**-decomposition with rather large value of *i* ($i \to \infty$). The triangular matrix diagonal elements will be the eigen values of the matrix **A**. At the first iterations, Householder method is applied to obtain the **QR**-decomposition to accelerate the computations, and when the matrix **Q** takes the tridiagonal form, the rotation method is applied (see Chapter 2).

5.5 APPLICATION OF COMPUTER TECHNOLOGIES TO ANALYZE DYNAMIC SYSTEMS

5.5.1 SOFTWARE PRODUCTS APPLIED TO INVESTIGATE DYNAMIC SYSTEMS WITH MATHCAD AND FORTRAN

Table 5.1 contains some functions comprised by the mathematical package *MathCad* [86, 141] and applied to analyze the dynamic systems whose models are the systems of ordinary differential equations. In Table 5.2 there are several subprograms comprised by the library *IMSL* of the compiler *Visual Fortran* [22]. Below you can find the examples and comments on the application of the functions and subprograms indicated in Tables 5.1 and 5.2.

TABLE 5.1 Examples of MathCad Functions Providing the Work with Ordinary Differential Equations

No	Function name	Function assignment	Description of function arguments
1	***rkfixed*** (*y, a, b, n, F*); ***Rkadapt*** (*y, a, b, n, F*); ***rkadapt*** (*y, a, b, acc, n, F, k, s*); ***Bulstoer*** (*y, a, b, n, F*); ***bulstoer*** (*y, a, b, acc, n, F, k, s*); ***odesolve*** (*y, a, b, n*);	Functions provide the solution of the system of ordinary differential equations by Runge-Kutta or Bulirsch-Stoer methods. The solution result is the matrix containing the table of values of Cauchy problem solution (table formation can be adjusted with the parameters *k* and *s*. The calculations are performed with the variable step by the argument)	*y* – vector of the functions integrated (contains the initial conditions before looking for the solution); *a, b* – interval boundaries within which the integration is performed; *n* – number of steps within which the integration is performed; *F* – vector of the right sides of differential equations; *acc* – solution accuracy; *k* – maximal number of the solution intermediary points; *s* – minimal permissible interval between the points
2	***Stiffb*** (*y, a, b, n, F, J*); ***stiffb*** (*y, a, b, acc, n, F, J, k, s*); ***Stiffr*** (*y, a, b, n, F, J*); ***stiffr*** (*y, a, b, acc, n, F, J, k, s*); ***Radau*** (*y, a, b, n, F*)	Functions provide the solution of the system of stiff ordinary differential equations by Bulirsch-Stoer method, Rosenbrock method or application of the algorithm RADAU5 which does not require Jacobian implicit set up	*J* – Jacobian composed of partial derivatives of the vector *F* of the right sides of differential equations by the vector *y* elements
3	***qr*** (*A*)	Function provides the matrix *A* decomposition onto the orthonormal *Q* and upper triangular *R* ($A=QR$)	*A, Q, R* – square matrixes
4	***fft*** (*y*), ***ifft*** (*c*); ***cfft*** (*y*), ***cifft*** (*c*)	Functions provide the computation of Furrier row coefficients (***fft***, ***cfft***) for the function *y* and calculation of the function (***ifft***, ***cifft***) by the known values of the coefficients *c*	*y, c* – vectors of the function and Fourier coefficients; Functions ***fft*** (*y*), ***ifft*** (*c*) are applied for real values of *y*, and functions ***cfft*** (*y*), ***cifft*** (*c*) – for real and complex values of *y*

TABLE 5.2 Examples of the Subprogram of *IMSL* Library Providing the Work with Ordinary Differential Equations

No	Call to subprogram or function	Function or subprogram assignment	Description of subprogram or function arguments
1	*call **ivprk** (ido, n, fcn, t, tend, tol, param, y);* *call **ivmrk** (ido, n, fcn, t, tend, y, yprime);* *call **ivpag** (ido, n, fcn, fcnj, a, t, tend, tol, param, y);*	Solution of the system of ordinary differential equations (Cauchy problem) by Runge-Kutta-Werner, Runge-Kutta, Adams-Moulton or Gear methods	*ido* – attribute demonstrating the computation state; *n* – number of differential equations; *fcn* – name of the user's subprogram calculating the right sides of differential equations (values of derivatives); *t, tend* – initial and final values of the argument by which the integration is performed; *tol* – permissible error value during the integration; *param(50)* – work array; *y* – vector of the variables integrated; *yprime* – values of the derivatives calculated for (*t, y*); *fcnj* – name of the user's subprogram calculating Jacobian values; *a* – matrix applied when solving the tacit systems
2	*call **bvpfd** (f1, f2, f3, f4, f5,* *n, nleft, ncupbc,* *tleft, tright,* *pistep, tol,* *ninit, tinit, yinit,* *Ldyini, Linear,* *print, mxgrid,* *nfinal, tfinal,* *yfinal, Ldyfin,* *errest);* *call **bvpms** (f1, f2, f3,*	Solution of the system of ordinary differential equations (boundary problems) by the methods of finite differences and method of multiple shooting	*f1, f2, f3, f4, f5* – user software, in which the derivatives, Jacobian, numerical values of boundary conditions, values of the partial derivatives by the set up parameter, values of partial derivatives on the boundaries are calculated; *n* – number of differential equations; *nleft, ncupbc* – number of initial conditions and number of bound conditions; *tleft, tright* – right and left boundaries of the argument; *pistep* – operation boundaries;

TABLE 5.2 Continued

No	Call to subprogram or function	Function or subprogram assignment	Description of subprogram or function arguments
	n, tleft, tright, *dtol, btol,* *maxit, ninit, tinit, yinit, Ldyini,* *nmax, nfinal,* *tfinal, yfinal,* *Ldyfin)*		*tol, dtol, btol* – tolerances providing the control of calculation errors; *ninit, tinit, yinit* – number of initial points and corresponding argument and function values; *Ldyini, Ldyfin* – operation parameters; *Linear=.true.* – if the linear problem is solved; *print=.true.* – if the intermediary results of the calculations need to be developed; *mxgrid, nfinal* – maximal number of computational points and their final value; *tfinal, yfinal* – final values of the argument and function; *errest* – error of the function calculation; *maxit* – permissible number of iterations; *nmax* – is set up *nmax=ninit*

The first group of functions demonstrated in Table 5.1 allows solving the problem described by the system of ordinary differential equations with the set up initial conditions (Cauchy problem). The simplest of the functions – ***rkfixed*** – solves the problem by Runge-Kutta method applying the integration fixed step. If the variable step needs to be applied, the functions ***Rkadapt*** or ***rkadapt*** should be used. The latter provides a user with more capabilities, in particular, its application allows memorizing the *k* intermediary points of the solution, establishing the permissible interval *s* between points.

When solving slack ordinary differential equations, as a rule, the accuracy provided by the application of the foregoing functions is sufficient, in which the computational algorithms based on Runge-Kutta methods are used. However, the package *MathCad* contains the functions ***Bulstoer*** and ***bulstoer***, in which the computational algorithm is based on Bulirsch-Stoer method.

The latest version of *MathCad* contains the function ***odesolve*** which can be recommended for an inexperienced user. A lot of issues which should be solved when programming the problem with this function are solved automatically (putting equations in the standard forms, selection of the integration step and computational functions from those considered above, etc.).

The following group of functions comprised by *MathCad* allows solving the systems of stiff ordinary differential equations. The functions ***Stiffb*** and ***stiffb*** provide the solution of stiff systems of equations by Bulirsch-Stoer method, and functions ***Stiffr*** and ***stiffr*** – Rosenbrock method. The latest versions of *MathCad* contain the function ***Radau*** which looks simpler in operation (it does not require setting up Jacobian).

Table 5.1 displays the function ***qr(A)*** which allows representing the square matrix A in the form of the product of the orthogonal matrix Q and upper triangular matrix R. Such representation can be used when finding matrix eigen values, including matrixes in the problems with the systems of linear ordinary differential equations.

Table 5.1 demonstrates the functions ***fft** (y), **ifft** (c)*; ***cfft** (y), **cifft** (c)* providing the work with Furrier-decompositions (direct and inverse transformations) for real and complex values of the vector y.

Table 5.2 contains the main software products which help to solve the systems of ordinary differential equations when working with the compiler ***Intel Visual Fortran.***

The first group of subprograms provides the solution of the systems of slack ordinary differential equations arranged as Cauchy problem. This is the software ***ivprk*** in which Runge-Kutta-Werner algorithms of fifth and sixth accuracy orders are applied, software ***ivmrk*** in which Runge-Kutta algorithms with different accuracy are implemented, and software ***ivpag*** providing high accuracy orders due to the application of computational algorithms of Adams-Moulton and Gear. The software ***ivpag*** also allows solving the systems of stiff ordinary differential equations.

The second group of software provides the solution of the systems of ordinary differential equations arranged as boundary problems. The finite-difference methods of solving the boundary problem are applied in the software ***bvpfd***, and the method of multiple shooting is applied in the software ***bvpms***. It should be pointed out that the software ***bvpfd*** and ***bvpms*** contain many formal parameters and their application can appear difficult for inexperienced users.

The library *IMSL* comprises a lot of software which allow working with Furrier rows. In particular, the software providing the procedures of direct and inverse Furrier transformations should be pointed out. These are the software ***FFTRF*** and ***FFTCF*** (computation of Furrier coefficients of real and complex sequence), ***FFTRB*** and ***FFTCB*** (perform the inverse Furrier transformation, respectively, calculating the real and complex sequences by Furrier coefficients). Besides, there are functions providing the fast sine- and cosine-transformations of Furrier, Laplace transformations, etc.

Let us consider the solution of the problem of nitrogen oxide *NO* formation in the chemical reactor containing only oxygen and nitrogen, as an example. The equations describing the kinetic processes refer to stiff systems of ordinary differential equations and look as follows [171]

$$\frac{d\alpha_1}{dt} = -2k_{11}\alpha_1\alpha_1 - k_{12}\alpha_1\alpha_2 - k_{13}\alpha_1\alpha_3 + k_{24}\alpha_2\alpha_4 + k_{35}\alpha_3\alpha_5 + 2k_{45}\alpha_4\alpha_5$$

$$\frac{d\alpha_2}{dt} = -k_{12}\alpha_1\alpha_2 + k_{13}\alpha_1\alpha_3 - k_{24}\alpha_2\alpha_4 + k_{35}\alpha_3\alpha_5$$

$$\frac{d\alpha_3}{dt} = k_{12}\alpha_1\alpha_2 - k_{13}\alpha_1\alpha_3 + k_{24}\alpha_2\alpha_4 - 2k_{34}\alpha_3^2\alpha_4 - k_{35}\alpha_3\alpha_5 + 2k_{44}\alpha_4\alpha_4$$

$$\frac{d\alpha_4}{dt} \approx 0$$

$$\frac{d\alpha_5}{dt} \approx 0$$

The equations are written down in the assumption that the following reactions take place in the reactor

$$O + N_2 \underset{k_{12}}{\overset{k_{35}}{\Leftrightarrow}} NO + N; \quad O_2 + N \underset{k_{13}}{\overset{k_{24}}{\Leftrightarrow}} NO + O;$$

$$O_2 \underset{k_{34}}{\overset{k_{44}}{\Leftrightarrow}} 2O; \quad N_2 + O_2 \underset{k_{11}}{\overset{k_{45}}{\Leftrightarrow}} 2NO$$

The reactions occur in direct and inverse directions, at the same time, the rates of direct reactions equal $k_{35}, k_{24}, k_{44}, k_{45}$, respectively, and inverse reactions – $k_{12}, k_{13}, k_{34}, k_{11}$. It is additionally assumed that there is an ideal mixing in the reactor, the reagent temperature and concentration are similar everywhere in the reaction volume, and the temperature is constant with time, N_2, O_2 concentrations change insignificantly.

These equations can be solved with MathCad facilities. The problem is solved with the initial conditions corresponding to *t*=0

$$\alpha_1 = 0; \quad \alpha_2 = 0; \quad \alpha_3 = 0; \quad \alpha_4 = 0,232; \quad \alpha_5 = 0,756$$

In accordance with [171] the rate values of the chemical reactions have the following values

$$k_{11} = 0,023; k_{12} = 2,8 \cdot 10^{13}; k_{13} = 1,6 \cdot 10^{8}; k_{24} \approx 0;$$

$$k_{34} \approx 0; k_{35} = 16,0; k_{44} \approx 0; k_{45} = 3,7 \cdot 10^{-9}$$

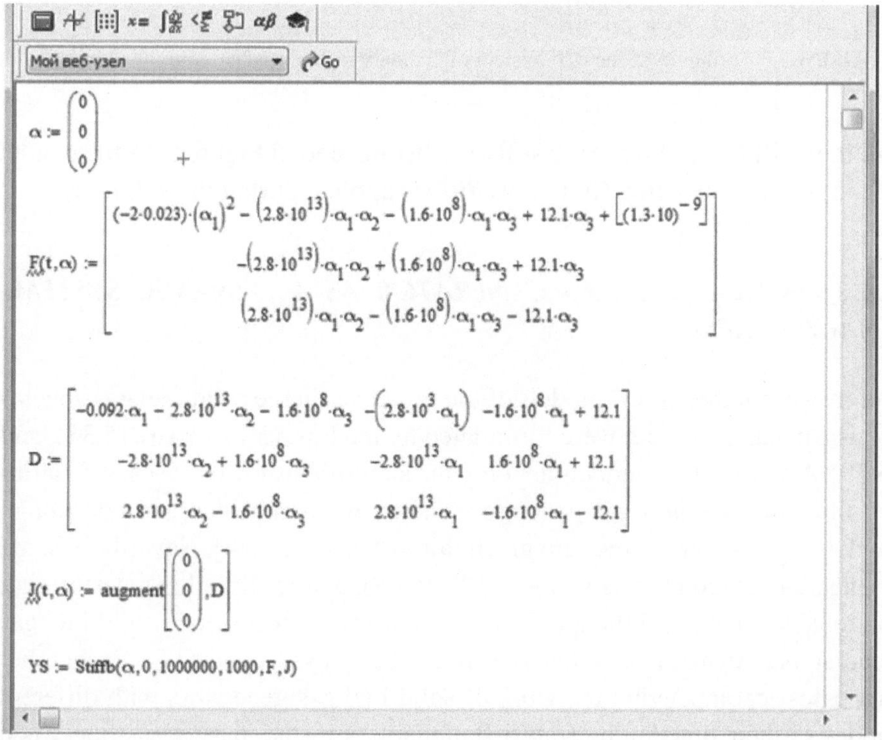

FIGURE 5.16 Listing of the software for the calculation of the stiff system of ordinary differential equations.

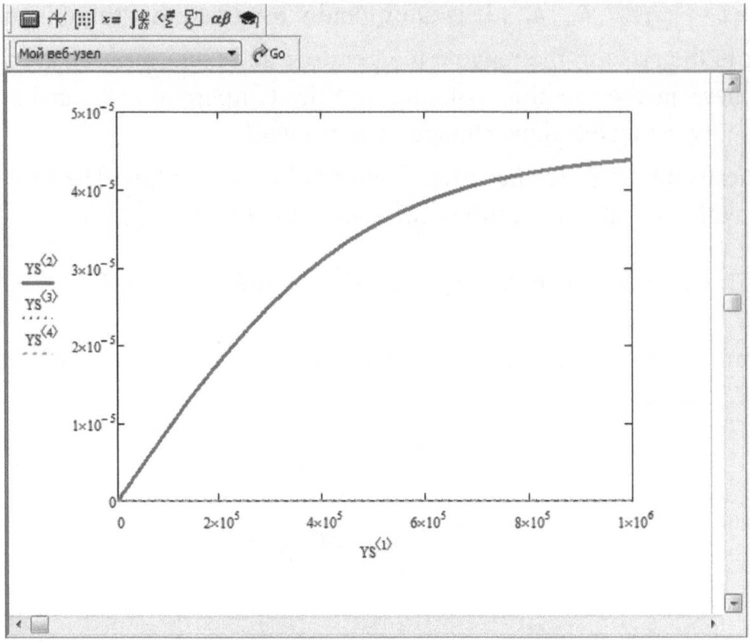

FIGURE 5.17 Solution of the stiff system of ordinary differential equations.

In Figure 5.12 you can see the software listing, and in Figure 5.13 the results of the solution executed (function ***Stiffr*** is applied during the solution).

5.5.2 SOLID FUEL GAS GENERATOR AS A DYNAMIC SYSTEM. ANALYSIS RESULT

Before, the mathematical models of the processes in the solid fuel gas generator as a dynamic system were formulated by the Eqs. (5.7)–(5.10), (5.38), and (5.39). Actually, these models are only the simplified specific cases of describing the processes in a gas generator. The more complicated and conceptual models by their capabilities are given, for instance, in [244]. Nevertheless, the application of the models (5.7)–(5.10), (5.38), and (5.39) allows performing the reliable analysis of the quality of intrachamber processes in solid fuel gas generators as dynamic systems. Below are the analysis results of intrachamber processes accompanying the work of solid fuel gas generators with different modes of their functioning (quasi-stationary, non-stationary, transition when changing the pressure mode in the gas generator combustion chamber, etc.).

5.5.3 ANALYSIS OF THE SCATTERING OF INTER-BALLISTIC PARAMETERS OF INTRACHAMBER PROCESSES IN A SOLID FUEL GAS GENERATOR

Let us consider the gas generator performance whose structural scheme is given in Figure 3.5. The scattering of inter-ballistic parameters in the gas generator combustion chamber can be found by the multiple solutions of the equation systems on internal ballistics with different initial data whose values are set up taking their random character into consideration. In accordance with the algorithm given in Section 5.3.4, at the first stage, the Eq. (5.7) is solved for the undisturbed right sides and undisturbed boundary and initial conditions. At the second stage, the same equation systems are solved for "disturbed" parameters. When solving the problem, the right sides and (or) boundary and initial conditions are changed taking into account the acting disturbing factors. The deviation values of inter-ballistic parameters are determined by the difference of the first (corresponds to the undisturbed case) and second (corresponds to the case with the disturbed right sides of equations and disturbed initial and boundary conditions) solutions. Finally, with such approach it is possible to calculate the deviations of inter-ballistic parameters corresponding to any disturbance of initial and boundary conditions or any disturbance of right sides of differential equations. With such an approach the deviation values corresponding to the combined action of several or all disturbing factors can be calculated. The deviation values multiply calculated at different disturbing factors can be considered as general populations of stochastic values and allow computing the values of dispersion and mean-square deviations important for the statistical analysis of heat engine or gas generator performance.

Below are the computational results of scattering the inter-ballistic parameters in the solid fuel gas generator at the initial stage of its operation. The calculations are performed for the gas generator functioning model written down in the one-dimensional non-stationary setting (Eqs. (5.38), (5.39)).

The main initial data at which the following results are performed taking into account the setting up of the disturbing factors are as follows:

1. *Ignition system*

- mass of the igniting weigh: 4 kg (± 2%);
- weigh material density: 2500 kg/m^3 (± 2%);
- enthalpy of combustion products: 0.300E+7 J/kg (± 2%);

- specific heat capacity of combustion products c_p: 1200J/kg*K (± 2%);
- specific heat capacity of combustion products c_v: 1100 J/kg*K (± 2%);
- molecular mass of combustion products: 50 kg/mol (± 3%).

2. *Solid fuel*

   (a) gas phase parameters:

   - specific heat capacity of the products c_p: 1870 J/kg *K (± 3%);
   - specific heat capacity of the products c_v: 1570 J/kg *K (± 3%);
   - molecular mass: 28.5 kg/mol (± 2%);
   - temperature of combustion products: 3600 K (± 10%);

   (b) solid phase parameters:

   - fuel density: 1800 kg/m^3 (± 2%);
   - specific heat capacity: 1200 J/kg *K (± 2%);
   - coefficients in the combustion law (first model taking erosion into account): 0.004; 0.4 (± 2%);

3. *Information on the gas generator geometry*

   - chamber free volume: 2.2 m^3 (± 1%);
   - square of the fuel combustion surface (first channel): 22.8 m^2 (± 1%);
   - square of the minimal section: 0.135 m^2 (± 2%).

In Figure 5.18 you can see the computational results of pressure change in the gas generator front volume. The values of maximal, nominal and minimal pressures obtained by the method of statistic tests with test number N=200 are given. In the calculations the array of random numbers is formed by the algorithms for the random value uniformly distributed in the interval from −1 to +1.

It should be pointed out that the pressure dispersions in the gas generator combustion chamber are calculated using the linearization method as well. Qualitatively, the computational results coincide with those demonstrated in Figure 5.18, however, the qualitative difference of the results is insignificant. This result confirms the correctness of the assumption that the linearization method application has limitations when calculating the parameter dispersion in internal ballistics and the computational results can contain significant errors. At the same time, the dispersion computation results performed by the method of statistic tests with the test numbers *N=50, N=100, N=200* differ insignificantly.

All the foregoing is confirmed by 5.19, in which the results of the calculations of the pressure mean square deviations obtained by the linearization method are shown (curve *1*) and method of statistic tests (curve *2* corresponds

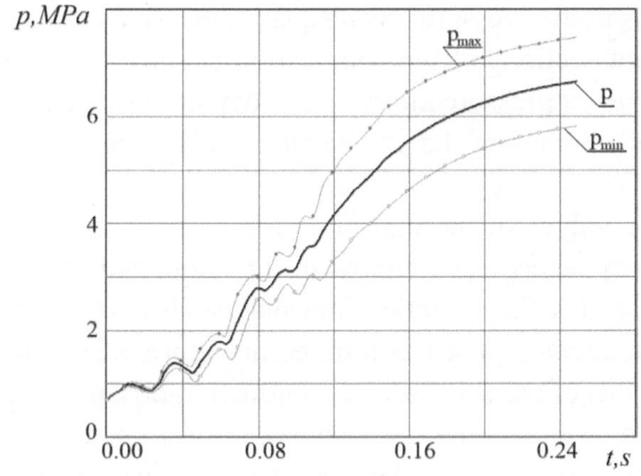

FIGURE 5.18 Dependence of maximal, nominal and minimal pressures in the gas generator combustion chamber on the process time.

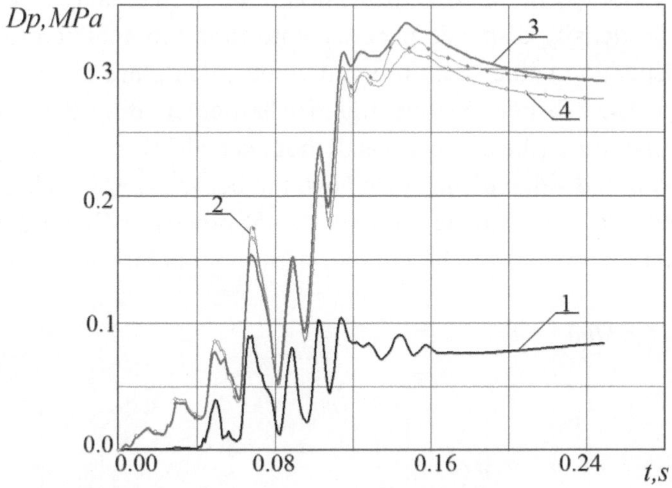

FIGURE 5.19 Dependence of mean square deviations of pressure on time (*1* – linearization method; *2, 3, 4* – method of statistic tests).

to 50 tests, *3*–100 tests, *4*–200 tests). The results obtained by the linearization method differ in more than three times from those obtained by the method of statistic tests. Moreover, the local maximums of mean square deviations (and dispersions) in the calculations by the linearization method and method of statistic tests correspond to different time moments (in the linearization method – ~0.11 s, in the method of statistic tests – ~0.15 s).

The changes in the pressure mean square deviations have the same character in different sections of the combustion chamber. In Figure 5.20, you can see the values of pressure mean square deviations calculated in the front and pre-nozzle volumes of the gas generator chamber with the number of tests $N=100$.

The relative values of the main internal ballistic parameter dispersion (pressure, density, energy and temperature) calculated in the period of the gas generator starting the operation (the analysis is performed by the values of mean square deviations) are compared in Figure 5.21. The values of the relative deviations of the flow rate of combustion products are not given, as at separate time moments the rate values can be close to zero. As a result, the computational accuracy can be lost when calculating the relative deviations [142].

The results given in Figure 5.21 demonstrate that the combustion product density has the greatest deviations in the period of the gas generator starting the operation (with the used dispersion in the initial data). The maximal values of the density dispersions at separate time moments are over 16%. The less dispersion values are observed when calculating the pressures (up to 8%). The least dispersions are found when calculating the combustion product temperature (the dispersions do not exceed 2%).

Changes in the value dispersions have the non-monotonic character that is connected with the sequence of operational processes flowing in the gas generator. At the first stage, the igniter operational parameters influence the

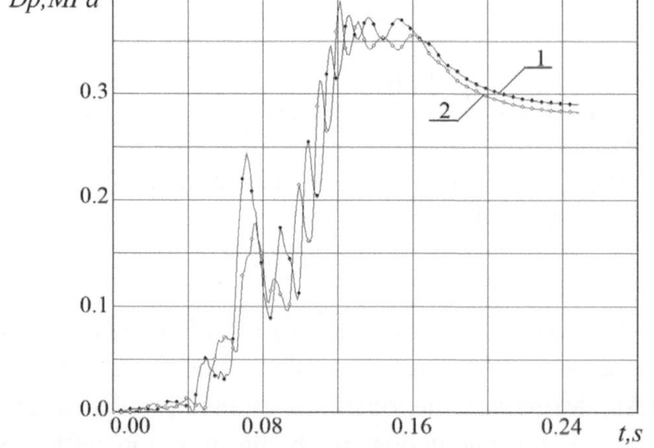

FIGURE 5.20 Dependence of pressure mean square deviations in front (*1*) and pre-nozzle (*2*) volumes on time (N=100).

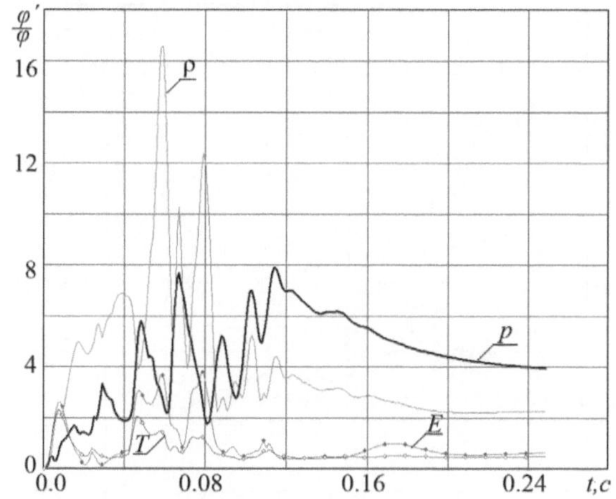

FIGURE 5.21 Dependence of the relative values of mean square deviations of pressure *p*, density ρ, energy *E*, temperature *T* on time.

dispersion values. At the second stage – the parameters determining the nozzle membrane decomposition and beginning of the combustion product outflow from the combustion chamber. At the next time moments – flame expansion, first, along the charge central channel, then along the fuel charge front end-face, along the fuel surface located in the over-nozzle recess and further – in the nozzle end-face. After the gas generator starting the operation, the dispersions decrease conditioned by the lack of disturbing factors connected with the igniter operation.

5.5.4 DETERMINATION OF EIGEN VALUES IN THE GAS GENERATOR CHAMBER

At the initial stage, the eigen values accompanying the gas generator operation are calculated using the models (5.38) and (5.39) following the technique described in Section 5.4. The partial derivatives by the spatial coordinate in the Eq. (5.39) are approximated by finite-difference equations based on the selected numerical method of equation integration (5.38) [153]. The calculations are carried out at the foregoing initial data.

In Figure 5.22 you can see the modules of maximal and minimal eigen values for deviations calculated near the charge channel left boundary (curve ***1***), in its center (curve ***2***), and at the output from the channel to the pre-nozzle volume (right boundary, curve ***3***).

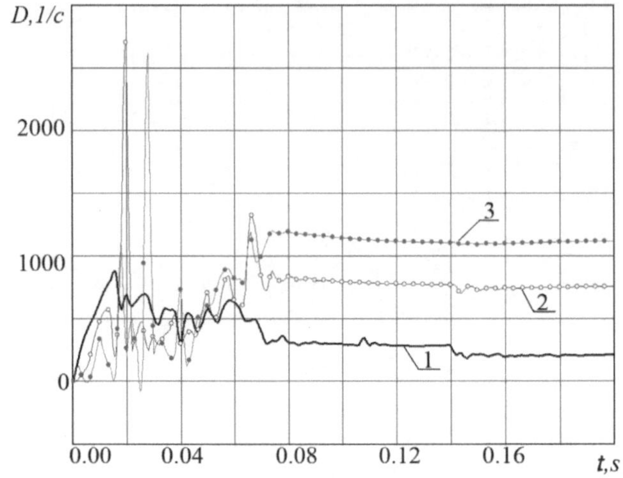

FIGURE 5.22 Dependence of the first eigen value module on time (1 – initial part of the channel, 2 – center of the channel, 3 – output part of the channel).

The analysis demonstrates that the modules of eigen values are maximal at the initial channel part before the fuel ignition, but in the process of flame expansion along the ignited fuel their values decrease and become less than the modules of eigen values at the output from the charge channel (the modules are maximal) and in the center of the charge channel. After the gas generator starting the mode of quasi-stationary operation, the modules of eigen values are fixed at the constant level. The calculations performed demonstrate that the real part of the calculated eigen values is mainly negative. The real part sign can be positive only in short time intervals, which do not exceed 0.005 sec (correspond to the time moments of fuel ignition, end of the igniter system operation). This allows asserting that the initial part of the gas generator performance develops in a stable mode.

The imaginary component of eigen values equals zero in the majority of sections of channel area (the resonance frequencies of oscillations for the deviations of inter-ballistic parameters are not available). The imaginary component of eigen values does not equal zero in the vicinity of the gas generator front volume. The oscillation frequencies corresponding to the maximal module of eigen values propagate to the section with the coordinate $x < 0.1$ m. The oscillation frequencies corresponding to the minimal module of eigen values propagate to the sections with the coordinate $x < 2.0$ m.

The analysis of the calculations performed demonstrates that the eigen frequencies of the oscillations of inter-ballistic parameters are preserved on the channel left boundary during practically all the period of starting the mode of quasi-stationary operation (oscillation frequency is 0–80 Hz). This fact appears to be quite logical as the combustion product rates on the channel left boundary are rather low, especially after the ignition system operation is over. Similar dependencies of the changes in oscillation frequency at $x \approx 0.03$ m correspond also to other eigen values.

In Figure 5.23 you can see the computational results of the frequency of eigen oscillations corresponding to the minimal module of eigen values and calculated in the channel section $x \approx 0.70$ m. The analysis demonstrates that in the section considered the oscillation eigen frequencies appear only during the ignition system operation. Initially the oscillation frequency exceeds 80 Hz. Further, in the process of the gas generator starting the quasi-stationary operation mode, the eigen oscillations decrease in the section considered.

5.5.5 DYNAMICS OF THE PROCESSES IN THE CONTROLLABLE GAS GENERATOR COMBUSTION CHAMBER

The results given below are obtained for the controllable gas generator shown in Figure 5.4 by the mathematical model comprising the Eqs. (5.7)–(5.10). The disturbing factors affect the operational processes with the corrections by the signal value coming to the control system pressure sensor.

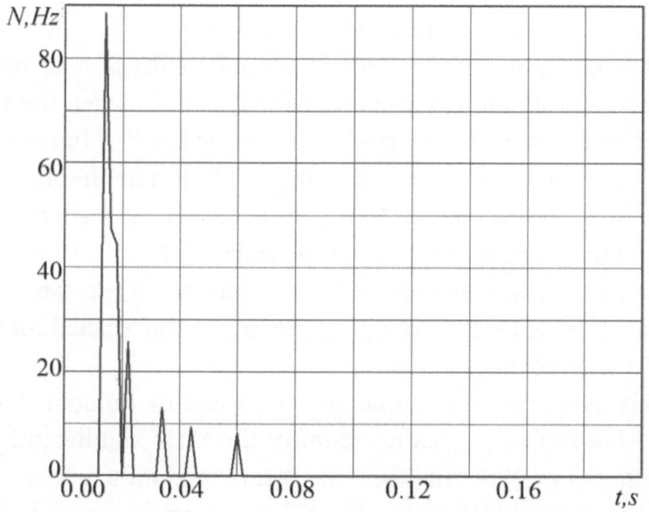

FIGURE 5.23 Dependence of the frequency of eigen oscillations on time.

The pressure "disturbed" value is calculated taking the assumptions (5.31) and (5.32) into account. The pressure deviations from the level of their mathematical expectation found when solving the equation system in thermodynamics (5.7) are conditioned by the aforementioned disturbing factors. The dispersions of the solid fuel combustion rates, availability of small cracks in the solid fuel charge body, etc. mostly influence the pressure dispersions. The values of $\delta'(t)$ determining the level of pressure random deviations $p(t)$ are found by the correlations which are true for the normal distribution law. At the same time, the mathematical expectation $v_1(\delta'(t)) = 0$, and the standard deviation value $\sigma_\delta = \sigma(\delta')$ is set up in the initial data. The value of σ_δ in the calculations is selected to fulfill the condition $|\delta'(t)| \leq \delta_{\max}$. Besides, the pressure disturbances conforming to the periodic law are considered – $\delta'(t) = A \cdot \sin(2\pi N t)$.

The results given below are obtained at the following main initial data:

- initial temperature of the elements of the gas generator structure: 268 K;
- initial air pressure in the combustion chamber: 0.0981 MPa;
- weight of the ignition stick: 0.20 kg;
- ballistite solid fuel with the ignition temperature: 650 K;
- value of the combustion chamber internal volume: 0.04 m^3;
- initial area of the fuel combustion surface: 0.20 m^2;
- start of the control block operation: 0.06 sec.

The optimal variants of the functions in the parameter control law of the solid fuel engine unit are applied in the calculations.

In Figures 5.24 and 5.25, the computational results of inter-ballistic processes in the controllable gas generator for the cases when the pressure in the combustion chamber is set up with the correction δ'. In Figure 5.24 the value $\delta_{\max} = 1\%$, and in Figure 5.25 – $\delta_{\max} = 2\%$. In the figures the curve 1 (solid black line) corresponds to "the non-disturbed" law of *p(t)* (magnitude of the maximal program value of the pressure – $p_{\max} = 5.0$ MPa), curve 2 corresponds to the pressure change in the chamber when the disturbances are available at the same pressure $p_{\max}$, curve 3 is constructed for the variant with pressure disturbances but at $p_{\max} = 9.0$ MPa.

The analysis demonstrates that the influence of random deviations at the selected laws of gas generator control does not significantly influence the dependence *p(t)*. Nevertheless, the disturbance influence on the dependence *p(t)* in various time intervals of the gas generator performance has

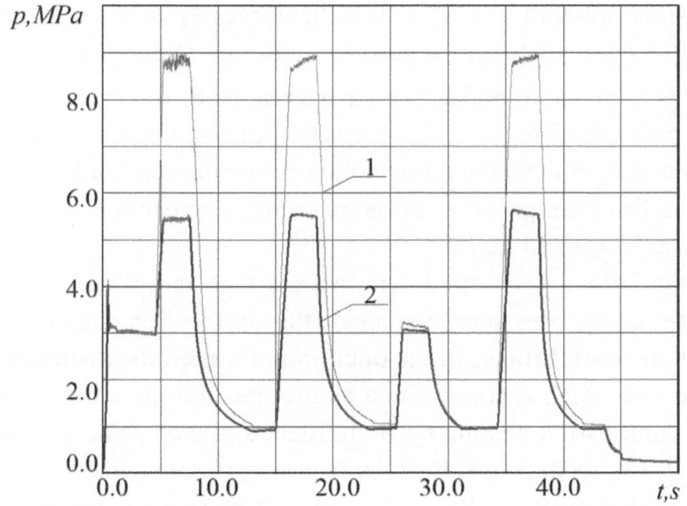

FIGURE 5.24 Pressure dependencies in the combustion chamber on time, $\delta_{max} = 1\%$.

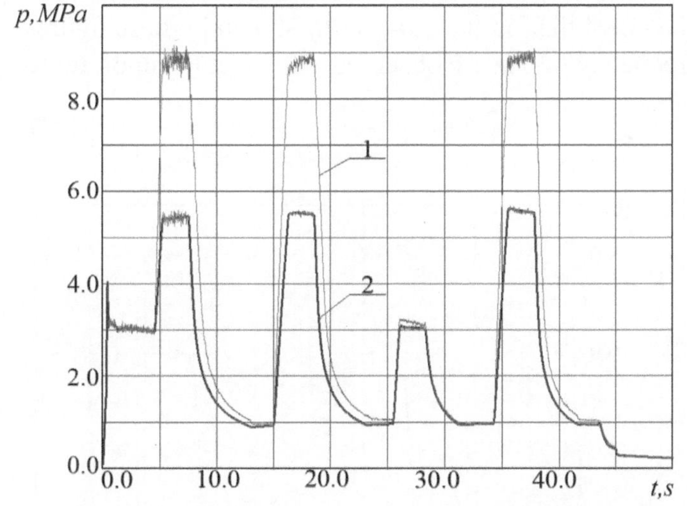

FIGURE 5.25 Pressure dependencies in the combustion chamber on time, $\delta_{max} = 2\%$.

differences. The disturbance influence is mostly significant at the first program stages of the operation process. Further, the disturbance influence weakens. The disturbances do not practically affect the character of the dependence *p(t)* at the transition stages (pressure rise or drop). The increase in the program pressure level up to 9 MPa (curves 3) results in the growth of the disturbance influence on the gas generator operation quality.

Below (Figures 5.26 and 5.27) are the results of the calculations of gas generator pressure changes under the action of disturbances conforming to the periodic law. In the calculations, the oscillation amplitude was taken as 1% and 2% from the program pressure value. The oscillation frequency – 5 Hz, 10 Hz, 25 Hz and 50 Hz.

The calculations performed demonstrate that the influence of periodic disturbances on the pressure level inside the combustion chamber is mostly revealed at the initial stages of the operation. Further, the disturbance influence decreases which is conditioned by the free volume increase in the gas generator combustion chamber. The influence of the periodic disturbances is practically not observed at the transition stages of pressure changes. The analysis demonstrates that in the considered frequency intervals of the disturbing action, the disturbances with the frequency 10 Hz (Figure 5.26, oscillation amplitude $\delta_{max} = 1\%$ and Figure 5.27, oscillation amplitude $\delta_{max} = 2\%$) demonstrate the most influence on the gas generator operation. It should be noted that on the curves *p(t)* shown in these figures, the beats with the frequency ~2 Hz are observed. This fact is important for practical applications.

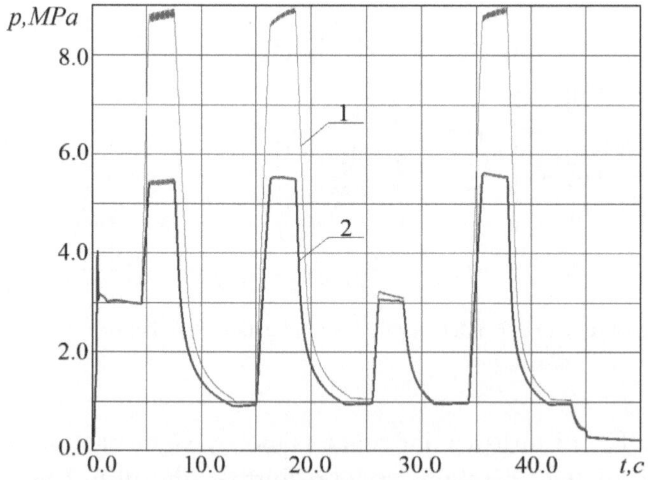

FIGURE 5.26 Pressure dependencies in the combustion chamber on time (oscillation amplitude $\delta_{max} = 1\%$).

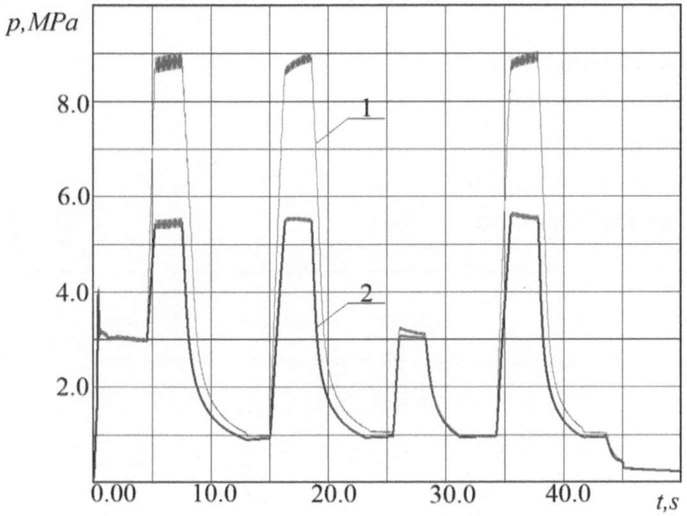

FIGURE 5.27 Pressure dependencies in the combustion chamber on time (oscillation amplitude $\delta_{max} = 2\%$).

It should be additionally pointed out that the regularities found correspond to the foregoing initial data. The algorithms and software developed allow analyzing the operation quality of the controllable gas generator at other initial data as well. Besides, it should be mentioned that when investigating the operation quality of dynamic systems it is advisable to apply the currently well-developed mathematical apparatuses of Furrier-analysis [112, 230] and wavelet-analysis [82, 112].

KEYWORDS

- **Brusselator**
- **Chebyshev theorem**
- **Dilatation method**
- **Dynamic Lorentz model**
- **Dynamic systems**
- **Euler method**
- **Gas generator**

- **Kutta-Merson method**
- **Le Verrier method**
- **Phase portraits**
- **Rosenbrock algorithm**
- **Runge-Kutta method**

CHAPTER 6

PHENOMENOLOGICAL MODELS IN ENGINEERING

CONTENTS

6.1 GENERAL IDEA OF PHENOMENOLOGICAL MODELS

We have been coming across phenomenological models since school time, though we have not been thinking about it. Such models are necessary to learn these or those complex regularities, peculiar for the object studied, with available simpler means. Let us give you a few examples.

Gas pressure is an important concept of thermodynamics. We use this concept to describe the gas laws (Boyle-Mariotte, Chapeyrone-Mendeleev, etc.). However, as the physical analysis is demonstrating (e.g., [133, 223]), the gas pressure is the manifestation of finer effects connected with the substance molecular and atomic structures. The same can be said about another thermodynamic concept – temperature of gas (or another material).

When formulating the concepts of gas pressure and temperature, the following assumptions are applied in statistic physics:

- pressure and temperature are the manifestation of the motion of the substance atoms and molecules and are proportional to the kinetic energy of their motion;
- translational motions of atoms and molecules are performed in three spatial directions, at the same time, the motion in each of the directions is equally probable;
- kinetic energy of atoms is the energy of translational motion of atoms in space;
- kinetic energy of molecules is the kinetic energy of a molecule as a whole, as well as the internal kinetic energy of rotations and oscillations of atoms in a molecule;
- pressure p – is the result of the molecule force action on the surface blocking the gas, proportional to the number of molecules N in the considered volume V and kinetic energy of the molecules $\frac{mv^2}{2}$ (m – molecule mass, v – molecule velocity);
- temperature T is the value proportional to the average kinetic energy of the molecule ($\frac{3}{2}kT \approx \frac{mv^2}{2}$; k=1.38×10^{-23} JK^{-1}).

Using the aforementioned assumptions we can write down the equation of state for the ideal gas in the form applicable in thermodynamics [129]

$$pV = RT$$

or

$$p = \rho(\gamma - 1)E$$

In the foregoing equations: R – gas constant, γ – adiabatic index (depends on the number of atoms in the molecule), E – internal energy whose value is proportional to the kinetic energy of all molecules contained in the volume considered.

In modern physics the equation of gas state is determined analytically. However, many of the gas laws used in engineering were established empirically and have been applied for over 150 years.

The phenomenon of liquid evaporation is also explained in physics based on the laws of statistic physics and kinetic theory. The analysis of molecule and atom behavior in the vicinity of the interface "liquid–gas" allows establishing the exponential law connecting the mass m of the liquid evaporated with the liquid temperature T

$$m \sim \exp(-\frac{W}{kT}) \tag{6.1}$$

Here W – work to be performed to convert the liquid mass into vapor. In this formula the value R can be used instead of Boltzmann constant k (k=1.380662×10^{-23} J/K) determined by the formula $R = N_0 k = 8.317 \frac{\text{J}}{\text{mol} \cdot \text{K}}$, in which N_0 – Avogadro number (the value R is called the universal gas constant).

Based on the physical messages, the empirical law can be arranged establishing the rate of evaporation of any liquid (the proportionality coefficient in the dependence (6.1) is found experimentally). Moreover, the empirical dependence determining the chemical substance decomposition rate, for example, explosive, or the rate of chemical reactions, or the rate of chemical reactions in the problems of chemical kinetics, etc. can be arranged in the similar way.

Other examples of phenomenological models are also referred to statistic physics. In particular, the diffusion processes are important in various engineering problems. In [223] it is discussed that the mixing of liquids and gases, heat transfer along the material, etc. can be described as the phenomenon of molecule collision. In this case, the diffusion coefficient D can be bound with the molecule kinetic energy by the following dependence

$$D = \mu kT$$

Here μ – mobility coefficient determined as the correlation of the average time τ between two collisions of the molecule to the mass m of this molecule – $\mu = \frac{\tau}{m}$.

Heat conductivity coefficient κ, determined as the coefficient of heat diffusion, can be found by the following correlation

$$\kappa = \frac{1}{\gamma - 1} \frac{kv}{\sigma_c} \tag{6.2}$$

In the formula (6.2) σ_c – effective section of the collision.

The foregoing examples show that the necessity to apply the considered class of mathematical models (phenomenological models) is conditioned, at least, by one of the following reasons:

- object or process studied are very complicated with the large number of internal bonds, about which the information is either unavailable or apocryphal;
- all bonds characterizing the investigated process are known, however, their recording in the actual models is practically impossible, for example, due to the limited computational resources.

Both reasons are determinative if the complex problem requires the obligatory solution. There are a lot of similar problems. In XX and early XXI century these are the problems of quantum and nuclear physics, problems of chemical kinematics and turbulence. In the nearest future we should expect the appearance of new phenomenological models in the problems of biophysics and biochemistry, microbiology and nanotechnologies, genetics, etc. This is connected with the fact that the solution of the aforementioned problems is very crucial for the humanity, however, in the nearest future the increasing computational power still does not allow solving these problems by the direct modeling.

Below are the examples of some phenomenological models applied in the problems of chemical kinetics and turbulence.

6.2 MODELS OF CHEMICAL KINETICS

In Chapter 5 we considered the problems on the processes in chemical reactors as the problems of dynamic systems. At the same time, we did not consider the issues connected with the selection of coefficients in the equation systems describing the processes in reactors. The missing information can be found experimentally that will require the considerable computational resources (by the volumes of the memory used, computer operational speed and problem solution time). In practice, the semi-empirical approach is applied to solve the problem within the acceptable time period and minimal consumption to determine the coefficients characterizing the considered chemical reactions. The sense of this approach applied when solving the problems on chemical reactions of materials is based on the application of the rate of chemical reactions of the equations of the type (6.1).

Let us consider the widely spread type of chemical reactions called the biomolecular [139].

Let us introduce the following designations:

- *A, B, C, D* – neutral atoms;
- *AB, AC, AD, BC, BD, CD* – diatomic molecules;
- *X, Y, Z, X,' Y,' Z'* – particles or their fragments consisting of the arbitrary number of atoms, of one atom, as well;
- *XY* – di- and more atomic molecule;
- *XYZ* – tri- and more atomic molecule;
- M – any component.

The reactions based on the following schemes are referred to biomolecular ones

$$X+Y\rightarrow Z+Z';\ X+Y\rightarrow XY;\ AB+C\rightarrow AC+B;$$
$$AB+CD\rightarrow AC+BD$$

The accurate solution of the problem of biomolecular chemical reaction in thermally equilibrium gas (the equilibrium between the internal and translational degrees of freedom of the molecules is not disturbed during the reaction) assumes the solution of the dynamic problem on the collision of two particles accompanied by the transformation. The phenomenological approach consists in the assumption that the biomolecular reaction rate k is the temperature function $k = k(T)$ and satisfies Arrenius law [43, 139]

$$k(T)=k_0\exp(-\frac{E_a}{RT}) \tag{6.3}$$

or the generalized Arrenius law

$$k(T)=k_0'T^n\exp(-\frac{E_a}{RT}) \tag{6.4}$$

In the Eqs. (6.3) and (6.4) E_a– activation energy, k_0, k_0' – pre-exponential coefficients in Arrenius formula, n – degree index in the temperature factor of pre-exponential multiplier, R – universal gas constant or Boltzmann constant. The coefficient values in the Eqs. (6.3) and (6.4) are found experimentally. The dependence $k(T)$ is such that even with the insignificant change in the reagent temperature, the reaction rate can change by orders. In the majority of typical chemical reactions the values of pre-exponential coefficients change in the interval from 10^4 s^{-1} to 10^{20} s^{-1}. We can find out about various methods for defining the coefficients in Arrenius formula, for example, in [43, 152, 204, 205, 212, 255].

Monomolecular reactions with multi-atom molecules are more complex. The reaction schemes can be as follows:

$$XYZ + M \rightarrow XY + Z + M;\ XYZ \rightarrow XY + Z;$$

$$XYZ + M \rightarrow YXZ + M;\ XYZ \rightarrow YXZ$$

The reaction group in question consists of two stages. At the first stage, the multi-atom molecule is activated when colliding with other particles. At the second stage, the molecule spontaneously decomposes or isomerizes. Depending on the temperature and pressure, the reaction can follow any foregoing scheme in the chemical reactor.

In case of monomolecular reactions, the equation of chemical reaction rate is also written down in the form of Arrenius formula, however, it looks more cumbersome. There are several techniques to calculate the reaction rate constants [139] and there is a special equation type for each technique $k(T)$. Let us point out that in these equations the number of atoms in the molecule is taken into account, as well as the reduced mass of the colliding particles and reduced frequency of collisions, energy of null oscillations of the molecule XYZ, number of the molecule oscillation freedoms XYZ and frequencies of their oscillations, moments of their oscillations, inertia moments for each of the internal rotations revealed, etc.

The same complex are the models for trimolecular reactions, in which three particles interact with the formation of two new ones

$$X + Y + Z \rightarrow X' + Y'$$

In particular, the models in which it is assumed that the triple collision passes three stages is applied in practice. At the first stage, the intermediary molecule with a very short life period (~10^{-12} sec) is formed as a result of interaction between X and Y. At the second stage, the intermediary molecule interacts with the particle Z with the formation of finite reaction products X' and Y'. The equation for the reaction constant written down in Arrenius form contains the components taking into account the masses of the particles

X, Y, Z, gas-kinetic radii of double and triple collisions, etc. In [139] there are other models of the variants of chemical reactions found in practice.

One more circumstance important for practice needs to be mentioned. The perfect models of chemical processes in reactors are those, in which all the initial reagents, all intermediary substances obtained during the chemical reactions, all finite products of the reaction are detailed, as well as all elementary chemical reactions in the reactor with the indication of their sequence and interconnection. For each stage, the values of kinetic parameters of all reactions, including direct and inverse ones, are established in the models. Such models are applied, for example, to improve the formulas of modern rocket fuels. Actually, the detailed analysis of intermediary reagents and finite products of all chemical reactions in the reactor allows arranging the processes in such a way as to suppress "parasite" energy-consuming intermediary reactions. When solving the ecological problems, the application of such detailed models gives the possibility to find out how toxic and other harmful substances can be excluded from the composition of finite products.

The models with detailed kinetic schemes of chemical reactions allow obtaining the most reliable results. However, the creation of such models requires considerable material resources connected with the conducting of original experiments, solution of theoretical problems aimed at the establishment of coefficient values in the equations for the rates of chemical reactions. Besides, the application of such models for the analysis of the processes, which really take place in chemical reactors, is also consumptive. The systems of differential equations used for solving problems in chemical reactors are stiff, as a rule (see Chapter 5). The rates of chemical reactions are such that the integration step of these equations needs to be very small. These circumstances require the application of highly productive computers.

The actual kinetic schemes of chemical reactions are simplified in practice. At the same time, the simplification is performed so as for the reduced (simplified) scheme of chemical reactions to adequately reflect the behavior of the main parameters of the processes in the reactor (pressure, temperature, component concentration, etc.) with time.

The kinetic schemes of chemical reactions can be maximally simplified. The complex kinetic scheme of chemical reactions can be represented in the form of one or several simple chemical reactions for which the initial substances and finite products are known, as well as all the parameters determining the chemical kinetics of these reactions. Such scheme of chemical reactions is called global or brutto-scheme.

The brutto-scheme of chemical reactions is the most convenient model when chemical processes constitute only the insignificant part of the complex physical-chemical process. In particular, these are the processes of chemically reacting products in the combustion chamber of the rocket engine, in the combustion chamber of the internal combustion engine, in the furnace of the boiler plant, etc.

In [139] you can find the experimentally obtained parameters of brutto-reactions of Arrenius type for the combustion reactions of hydrocarbons

$$C_nH_mO_r + (n + \frac{m}{4} - \frac{r}{2})O_2 \to nCO_2 + \frac{m}{2}H_2O$$

The total combustion rate of hydrocarbons is found by the following formula

$$k(T) = k_0 \exp(-\frac{E_a}{RT})\alpha_1^{n_1}\alpha_2^{n_2} \tag{6.5}$$

In the Eq. (6.5) the concentrations α_1, α_2 have the dimensionalities $[\alpha] = \frac{mol}{cm^3}$, pre-exponential factor dimensionality $[k_0] = \frac{1}{s} \cdot \left(\frac{cm^3}{mol}\right)^{n_1+n_2-1}$, activation energy dimensionality $[E_a] = \frac{J}{mol}$. The values of the reaction parameters for the combustion chemical reactions of some hydrocarbons comprised by the aforementioned Arrenius law are given in Table 6.1.

6.3 MODELING OF TURBULENT VISCOSITY

Turbulence phenomena affect various branches of science and engineering. In particular, these are aerodynamics and hydraulics, external and internal ballistics in rocket engineering, these are oil- and gas-producing, and chemical industries. The issues of turbulence are extremely important when solving the problems of metrology and oceanography, etc. Despite the extended study period of this phenomenon (nearly a century), the issues of turbulence appearing and modeling are not completely clear. The direct methods of turbulence modeling have been recently developed due to the growing capabilities of computers (e.g., [26, 30, 137]). However, the solution of problems

TABLE 6.1 Values of Constants in Arrenius Law

Fuel	K_0	E_a	n_1	n_2
CH_4, methane	$1.3 \cdot 10^8$	202,800	–0.3	1.3
C_2H_6, ethane	$1.1 \cdot 10^{12}$	125,700	0.1	1.65
C_2H_2, acetylene	$6.5 \cdot 10^{12}$	125,700	0.5	1.25
C_6H_6, benzene	$2.0 \cdot 10^{11}$	125,700	–0.1	1.85
CH_3OH, methanol	$3.2 \cdot 10^{12}$	125,700	0.25	1.5
C_2H_5OH, ethanol	$1.5 \cdot 10^{12}$	125,700	0.15	1.6
C_8H_{18}, isooctane	$7.2 \cdot 10^{11}$	167,600	0.25	1.5

of aerodynamics and hydrodynamics do not seem complete in these works. Moreover, as a result of these works, new questions arise which still remain unanswered.

Nevertheless, the necessity of solving the turbulence problems is topical. With the modern development of experimental means to investigate turbulence and the existing level of computational resources, semi-empirical phenomenological models are the most substantiated method for modeling the turbulence effects. The issues of turbulence and turbulence models are described in detail, for instance, in [50, 51, 59, 60, 100, 101, 138, 149]. Below you can find only a small part of models, the application of which is experimentally substantiated in the problems of hydro- and gas-dynamics, problems on gas flow in heat engines, etc. Moreover, only the modeling methods in the problems of liquid and gas mechanics of "turbulent" viscosity μ_δ coefficient, which is added to the dynamic viscosity μ in momentum and energy equations, will be considered below. You can find the details of the equations of turbulent motion of liquid and gas and analysis of the turbulence problems solved, for example, in [138, 249].

6.3.1 MODELS OF TURBULENT VISCOSITY OF ALGEBRAIC TYPE

Models of algebraic type are relatively simple, but the area of their application is limited by a small class of liquid and gas flows.

6.3.1.1 Flows in the Boundary Layer

The physical picture of solid body surface streamlining by the flow of viscous liquid (or gas) is demonstrated in [60, 61, 249]. In the little vicinity from the streamlined body a relatively small region is developed along the normal to the surface, in which the liquid rate changes from the rate value in the apprising flow (external boundary of the flow) to the zero value on the body surface (Figure 6.1). This region is called the boundary layer. When streamlining the solid body, the region occupied by the boundary layer (the boundary layer thickness) increases (as demonstrated in Figure 6.1). However, there are flows, in which the boundary layer thickness can decrease (e.g., in supersonic nozzle blocks).

Viscous effects in the boundary layer can be laminar (determined by the diffusion viscosity conditioned by the collision of liquid molecules) and turbulent (determined by macroeffects in the liquid flow).

To describe the turbulent viscosity in the boundary layer, the relatively simple models of algebraic type can be successfully applied.

Prandtl model is based on the assumption on the similarity of the liquid particle motion in the turbulent flow and molecule motion. The length scale l called the mixing length and on which the liquid particles preserve their momentum average values is introduced in this model. The value of the mixing length is found experimentally. In the problems on the flows in the

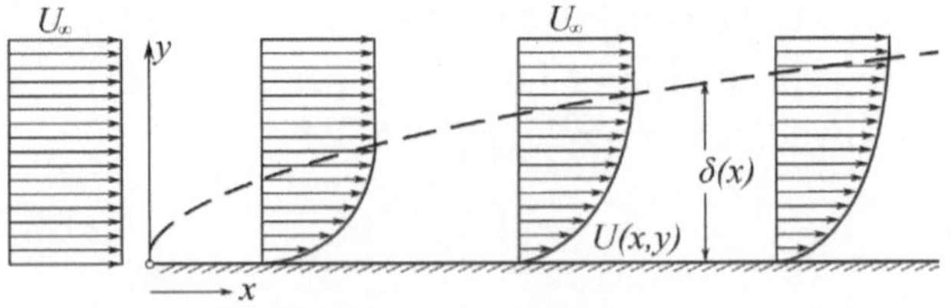

FIGURE 6.1 Scheme of the boundary layer development on the flat plate streamlined longitudinally.

boundary layers the linear dependence of the mixing length on the thickness δ of the boundary layer – $l = k\delta$. The value of the proportionality coefficient in this dependence is k ≈ 0.4.

The dependence of the turbulent viscosity in Prandtl model for the boundary layer is proportional to the liquid density ρ and rate gradient u in the direction y perpendicular to the streamlined surface is written down as follows:

$$\mu_m = \rho l^2 \left(\frac{\partial u}{\partial y}\right)^2 \tag{6.6}$$

In Karman model the value of turbulent viscosity in the boundary layer is determined by the dependence, in which beside the rate gradient $\frac{\partial u}{\partial y}$, the information about the second derivative $\frac{\partial^2 u}{\partial y^2}$ is also contained. This model is constructed using the assumption that the local models of the flows are statistically similar in any vicinity within the boundary layer.

The model type

$$\mu_m = \rho \kappa^2 \left(\frac{\partial u}{\partial y}\right)^4 \left(\frac{\partial^2 u}{\partial y^2}\right)^{-2} \tag{6.7}$$

Let us point out that if we designate $l = k\frac{\partial u}{\partial y}(\frac{\partial^2 u}{\partial y^2})^{-1}$, the formula (6.7) will coincide with the formula (6.6).

6.3.1.2 Flows in Pipes

When modeling the processes in pipes, the dependence (6.6) is successfully applied, in which the mixing length is established by the interpolation formula [249]

Is the

$$\frac{l}{R} = 0.14 - 0.08(1-\frac{y}{R})^2 - 0.06(1-\frac{y}{R})^4 \tag{6.8}$$

In the formula (6.8) R is the pipe radius. Besides, we can demonstrate that when approximating the wall ($y \to \infty$), the formula (6.8) is simplified to the type that is true for the boundary layers – $l \approx 0.4y$.

6.3.1.3 Jet Turbulent Flows

In jet flows the turbulent viscosity, in the majority of situations, can be calculated using Prandtl model

$$\mu_m = \rho c^2 x^2 \left(\frac{\partial u_x}{\partial y} \right)^2 \quad (6.9)$$

In the Eq. (6.9) c – coefficient found experimentally, x – longitudinal coordinate for which the turbulent viscosity is determined, u_x – longitudinal rate of the liquid on the jet external boundary.

6.3.1.4 Differential Models of Turbulent Viscosity

Differential models, the same as the previous model groups, refer to semi-empirical ones. However, they are constructed in the same way as when making the conservation laws (see Chapter 7). These equations can contain convective and diffusive components, source elements considering the formation of turbulent substances and their disappearance. The dimensionality of the equations, as a rule, coincides with the dimensionality of the main problem to be solved. There are much more differential models of turbulent viscosity than models represented by algebraic equations. Let us consider only two models applied in native investigations. The models are written down for the non-stationary case corresponding to two-dimensional motion of the liquid.

The first group of models is based on the solution of additional equation for the turbulent viscosity together with the equations of viscous liquid motion (or its analog). In general, the equation for turbulent viscosity is written down as follows:

$$\frac{\partial \rho v_m}{\partial t} + \rho u \frac{\partial v_m}{\partial x} + \rho v \frac{\partial v_m}{\partial y} = \frac{\partial}{\partial y}(\mu + k\mu_m)\frac{\partial v_m}{\partial y} + S_1 v_m - S_2 v_m^2 + S_3 \quad (6.10)$$

In the Eq. (6.10) the components $S_1 v_m$, $S_2 v_m^2$, S_3, respectively, take into account the turbulence production, its dissipation and transition from the laminarity to turbulence in the boundary layer.

Sekundov model refers to the type considered. In [41] this model is used to calculate the solid fuel erosive combustion (erosive combustion of the fuel is determined by the intensity of heat processes in the boundary layer in the vicinity of the combusting fuel surface).

In [41], Sekundov model is used as follows:

$$\frac{\partial \rho v_m}{\partial t} + \rho u \frac{\partial v_m}{\partial x} + \rho v \frac{\partial v_m}{\partial y} = \frac{\partial}{\partial y}(\mu + k\mu_m)\frac{\partial v_m}{\partial y} + \alpha \mu_m \left|\frac{\partial u}{\partial y}\right| - \frac{\gamma \mu_m (v + \beta v_m)}{y^2} + \xi v_m u \frac{\partial \rho}{\partial x}$$

$$\alpha = \begin{cases} 0,28(\frac{v_m}{7v})^{0,71} & at \ \frac{v_m}{7v} \le 1, \\ 0,28 & at \ \frac{v_m}{7v} > 1 \end{cases}$$

$$\mu_m = \rho v_m$$

$$\mu = \rho v$$

The coefficients k, γ, β, ξ in the equations are found experimentally. There are four empirical coefficients in this model, therefore the model (6.10) possesses greater universality in comparison with the foregoing models (6.6)–(6.9). For the boundary layer flows of liquid and gas the values of empirical coefficients are taken as follows:

$$k = 2.0; \quad \gamma = 12; \quad \beta = 0.34; \quad \xi = 0.7$$

Another group of models for determining the turbulent viscosity is based on the application of Kolmogorov-Prandtl formula [50]

$$\mu_m = c_\mu \frac{K^2 \rho}{\varepsilon} \tag{6.11}$$

In this formula K – turbulence kinetic energy, ε – turbulence dissipation rate, c_μ – coefficient accepted in the majority of models as follows – $c_\mu = 0.09$.

The turbulence kinetic energy and turbulence dissipation rate are found when solving additional equations written down in the form similar to the conservation equations [24, 50]

$$\frac{\partial \rho K}{\partial t} + \rho u \frac{\partial K}{\partial x} + \rho v \frac{\partial K}{\partial y} = \frac{\partial}{\partial y}\left(\mu + \frac{\mu_m}{\sigma_K}\right)\frac{\partial K}{\partial y} + S_1 + S_2 - \rho\varepsilon \quad (6.12)$$

$$\frac{\partial \rho \varepsilon}{\partial t} + \rho u \frac{\partial \varepsilon}{\partial x} + \rho v \frac{\partial \varepsilon}{\partial y} = \frac{\partial}{\partial y}\left(\mu + \frac{\mu_m}{\sigma_\varepsilon}\right)\frac{\partial \varepsilon}{\partial y} + \frac{\varepsilon}{K}(c_1(S_1 + S_2) - c_2\rho\varepsilon)$$

In the Eqs. (6.12) the component S_1 takes into account the turbulence production, and component S_2 takes into account the buoyant forces (proportional to the gradients of liquid density).

When solving the problem on erosive combustion of the solid fuel [41], the model (6.12) is used in the variant known as Jones-Launder model

$$\frac{\partial \rho K}{\partial t} + \rho u \frac{\partial K}{\partial x} + \rho v \frac{\partial K}{\partial y} = \frac{\partial}{\partial y}\left(\mu + \frac{\mu_m}{\sigma_K}\right)\frac{\partial K}{\partial y} + \mu_m\left(\frac{\partial u}{\partial y}\right)^2 - \rho\varepsilon - 2\frac{\mu K}{y^2},$$

$$\frac{\partial \rho \varepsilon}{\partial t} + \rho u \frac{\partial \varepsilon}{\partial x} + \rho v \frac{\partial \varepsilon}{\partial y} = \frac{\partial}{\partial y}\left(\mu + \frac{\mu_m}{\sigma_\varepsilon}\right)\frac{\partial \varepsilon}{\partial y} + c_1\frac{\varepsilon}{K}\mu_m\left(\frac{\partial u}{\partial y}\right)^2$$

$$-\rho\frac{\varepsilon}{K}\left[c_2\varepsilon\left(1 - 0{,}22\exp\left(-\frac{K^2\rho}{6\mu\varepsilon}\right)^2\right) + 2\frac{\mu K}{\rho y^2}\exp\left(-\frac{c_4 u_\tau y\rho}{\mu}\right)\right]$$

$$u_\tau = \sqrt{\frac{\tau_s}{\rho_s}}, \quad \tau_s = \frac{\Pr\lambda}{c_p}\cdot\frac{\partial u}{\partial y},$$

$$\mu_m = c_\mu\frac{K^2\rho}{\varepsilon}\left(1 - \exp\left(-\frac{c_3 u_\tau y\rho}{\mu}\right)\right)$$

The values of the coefficients comprised by the latter equations are accepted as follows in [41]

$$c_1 = 1.35; \quad c_2 = 1.8; \quad c_3 = 0.01;$$
$$c_4 = 0.5; \quad \sigma_K = 1; \quad \sigma_\varepsilon = 1.3$$

KEYWORDS

- **Boyle-Mariotte**
- **Chapeyrone-Mendeleev**
- **Gas pressure**
- **Jones-Launder model**
- **thermodynamics**
- **turbulence**

CHAPTER 7

MATHEMATICAL MODELS DEVELOPED WITH THE APPLICATION OF FUNDAMENTAL LAWS OF PHYSICS

CONTENTS

7.1 BASIC LAWS OF CONTINUUM MECHANICS

7.1.1 GENERAL IDEAS OF FUNDAMENTAL LAWS

To the fundamental laws of physics we refer conservation laws which are equally true for a substance and field. The issues of continuum mechanics hold a high position in the problems connected with power machines and heat engines. The mathematical models developed on the main conceptual issues of continuum mechanics comprise the equations of conservation of mass, energy, momentum, moment of momentum.

In continuum mechanics the models can be developed with the application of two possible points of view of continuum motion – Lagrange and Euler.

Lagrangian approach consists in the representation of the particle (continuum element) coordinates at any time point and its physical parameters as the function of its coordinates at the initial time point t_0

$$a = x(t_0); \quad b = y(t_0); \quad c = z(t_0);$$

$$x = x(a,b,c,t); \quad y = y(a,b,c,t); \quad z = z(a,d,c,t);$$

$$\rho = \rho(a,d,c,t); \quad \mathbf{v} = \mathbf{v}(a,b,c,t)$$

The efficiency of Lagrangian approach application is revealed to the full extent, for instance, when solving the problems on the flow of multiphase media in which it is necessary to determine the motion trajectories of solid or liquid particles in the investigated computational domain [235]. Lagrangian approach is applicable when it is required to separate the interfaces of two media during the computational process (e.g., wave crest when computing tsunami, filling of the rocket engine body with the polymerized viscoelastic liquid of fuel components, etc.).

The profiling of supersonic part of nozzle block is an important task when designing rocket engines. The solid fuel combustion products are the multiphase mixture containing the gaseous phase, liquid and solid particles of various sizes (from one micron to hundreds of microns). Generally speaking, the motion trajectories of different elements of the mixture can differ considerably. In Figure 7.1 you can see the motion trajectory *1* of the particle of the gaseous phase of combustion products and motion trajectory *2* of the solid particle. The trajectories of gaseous and solid phases will differ due to the motion inertness. The profile of the nozzle supersonic part should be selected from the condition of the maximal reactive force provision. If the combustion products contain only the gaseous phase, the nozzle profile in its supersonic part has to coincide with the trajectory of combustion products. If the combustion products contain solid particles, then, starting from some section, solid particles will intensively fall out onto the nozzle walls. This may cause the wall destruction. The design solution providing high power indexes of the nozzle block consists in applying the nozzle blocks with the angular point. In the vicinity of the beginning of intensive fall out of the particles onto the nozzle side wall its profile is selected as conical, and the cosine angle is selected in accordance with the trajectory of solid particles.

Lagrangian approach assumes the consideration of continuum as the finite aggregation of particles possessing various properties. The more particles are used in the computational domain, the more precise will be the problem

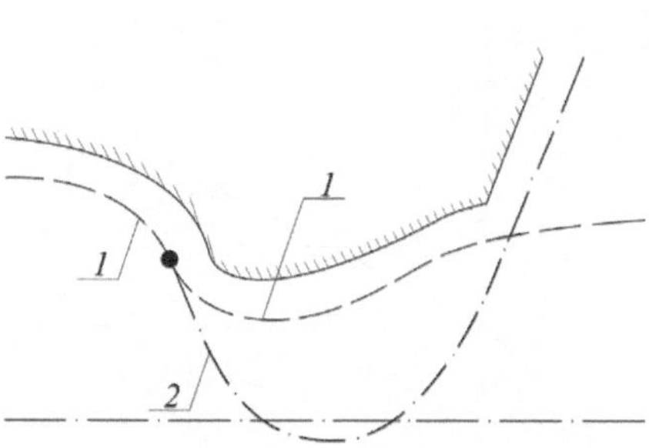

FIGURE 7.1 Scheme of the two-phase flow in the supersonic nozzle (*1* – current line of the gaseous phase; *2* – trajectory of the condensed particle).

solution. The models providing the problem computation are based on the solution of the systems of ordinary differential equations, the number of which is multiple of the number of particles considered. Lagrangian approach is rather demanding to the computer resources applied. Nevertheless, Lagrangian approach is used when solving practical problems, and it is the only one in the problems of nanotechnologies.

Euler approach assumes the setting up of all physical parameters as the function of certain spatial coordinates *x, y, z* and time *t*. The advantage of Euler approach is the visualization and convenience of analyzing the processes in the computational domain, both by the spatial coordinates and time of the process. The mathematical models based on the application of Euler approach are the systems of differential equations in partial derivatives. Nevertheless, Euler approach allows obtaining the high-quality results even with the help of the most widely spread computers with average productivity.

The systematic representation of the models of continuum mechanics with the application of Lagrangian and Euler approaches can be found, for instance, in [53, 201, 243]. Below we will restrict ourselves to considering the mathematical models developed with the application of Euler approach.

7.1.2 SELECTION OF THE COORDINATE SYSTEM

Cartesian coordinate system (Figure 7.2) is the simplest coordinate system used in the majority of applications (including the problems of internal ballistics, gas dynamics, heat conductivity, etc.). This coordinate system

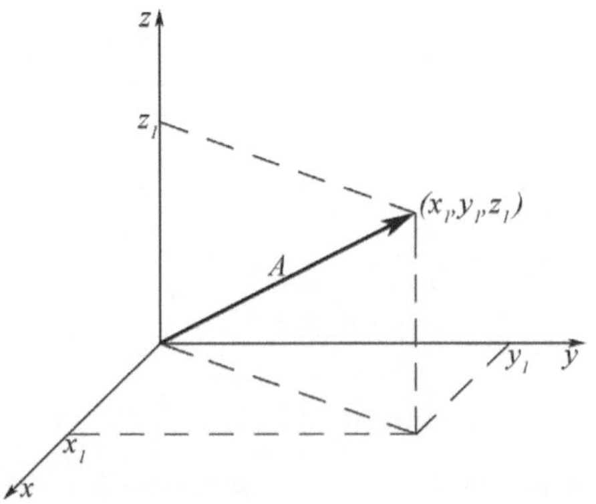

FIGURE 7.2 Cartesian coordinate system.

is formed by three families of planes orthogonal to each other $x = \text{const}$, $y = \text{const}$, $z = \text{const}$, the coordinate lines x, y, z are direct and mutually orthogonal. Cartesian coordinate system is relatively simple for understanding and appears to be quite acceptable when solving a lot of problems. Nevertheless, we can give some examples, in which the application of Cartesian coordinate system is inconvenient. Let us, for example, analyze the development of processes connected with the disaster (e.g., explosion) in the point with the coordinates x_1, y_1, z_1 of Cartesian coordinate system at the time moment $t = t_0$.

It follows from the problem setting up that the process development has to possess the spherical symmetry relating to the point with the coordinates x_1, y_1, z_1. Therefore, the application of the spherical coordinate system with the selection of the coordinate center in the point, in which the disaster occurs (corresponds to the coordinates x_1, y_1, z_1 of Cartesian coordinate system) looks the most favorable approach when solving the problem considered. Actually, when applying Cartesian coordinate system, the process development should be considered in the spatial setting, then the application of the spherical coordinate system allows obtaining the solution even in the one-dimensional setting. The problem on liquid or gas motion along the cylindrical pipe can be another example. The solution of this problem and processing of the solution results are much simplified if we refuse from Cartesian coordinate system in favor of the cylindrical one. Any number of similar examples can be given.

The coordinate system in three-dimensional space can be defined from different considerations. Let three numbers q_1, q_2, q_3 be put to any space point, which will be called the point coordinates. The lines along which $q_j = \text{const}$, $q_k = \text{const}$ (j, k =1.3; j, $k \neq i$) will be called the coordinate lines q_i(i = 1.3). In three-dimensional space the number of coordinate lines equals three. In the most general case, the coordinate lines q_1, q_2, q_3 are not straight and form the curvilinear coordinate system. If the tangents to coordinate lines are mutually perpendicular in each point, the curvilinear coordinate system is called orthogonal. Let us consider the cases when there are the direct (7.1) and inverse (7.2) dependencies linking Cartesian coordinates x, y, z with curvilinear ones q_1, q_2, q_3

$$x = x(q_1, q_2, q_3), \quad y = y(q_1, q_2, q_3), \quad z = z(q_1, q_2, q_3); \tag{7.1}$$

$$q_1 = q_1(x, y, z), \quad q_2 = q_2(x, y, z), \quad q_3 = q_3(x, y, z) \tag{7.2}$$

The existence of the bonds (7.1) and (7.2) allows linking the metrics of any coordinate system with the metrics of Cartesian coordinate system. The length of the element dS composed of the elementary unit axes dx, dy, dz of Cartesian coordinate system can be determined by the following equation

$$dS^2 = dx^2 + dy^2 + dz^2 = \sum_{ij} H_{ij}^2 dq_i dq_j \tag{7.3}$$

The coefficients H_{ij} are called Lame coefficients and are determined by the following correlations

$$H_{ij}^2 = \frac{\partial x}{\partial q_i}\frac{\partial x}{\partial q_j} + \frac{\partial y}{\partial q_i}\frac{\partial y}{\partial q_j} + \frac{\partial z}{\partial q_i}\frac{\partial z}{\partial q_j} \tag{7.4}$$

If the curvilinear coordinate system q_1, q_2, q_3 is orthogonal, the following correlation is true for Lame coefficients

$$H_{ij} = 0, \quad \text{if } i \neq j \tag{7.5}$$

Further, when applying the orthogonal coordinate systems, for the simplification we will designate $H_{ii} \equiv H_i$.

Taking into account (7.5), the Eqs. (7.3) and (7.4) will be simplified and written down as follows:

$$dS^2 = dS_1^2 + dS_2^2 + dS_3^2 = (H_1 dq_1)^2 + (H_2 dq_2)^2 + (H_3 dq_3)^2 \tag{7.6}$$

$$H_i = \sqrt{(\frac{\partial x}{\partial q_i})^2 + (\frac{\partial y}{\partial q_i})^2 + (\frac{\partial z}{\partial q_i})^2} \quad i = 1.3 \tag{7.7}$$

Values of the lengths of the elementary edges dS_i, area of the elementary faces $d\sigma_{ij}$ and elementary volume dW (Figure 7.3) can be established by the following correlations

$$dS_i = H_i dq_i$$

$$d\sigma_{ij} = dS_i dS_j = H_i H_j dq_1 dq_2$$

$$dW = dS_1 dS_2 dS_3 = H_1 H_2 H_3 dq_1 dq_2 dq_3 \tag{7.8}$$

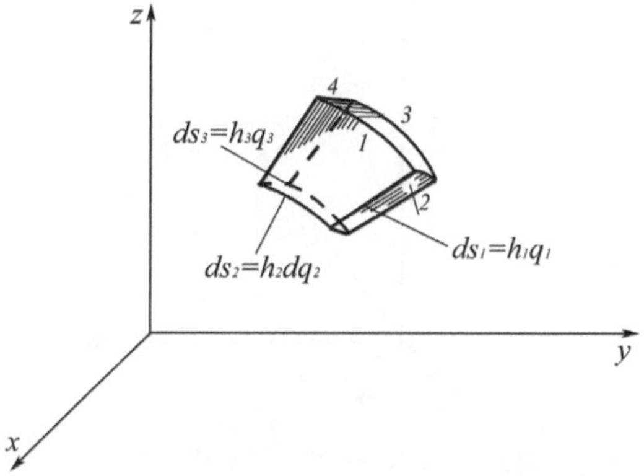

FIGURE 7.3 Metrics of the elementary volume in the arbitrary orthogonal coordinate system.

According to the definition, the values of Lame coefficients can be written down without the development for Cartesian coordinate system – $H_1 = 1$, $H_2 = 1$, $H_3 = 1$. In any other coordinate system the values of Lame coefficients should be determined using the Eqs. (7.4) or (7.7).

Let us consider some other coordinate systems [16], the application of which is spread in the problems of gas dynamics, internal ballistics, etc.

7.1.2.1 Cylindrical Coordinate System

The cylindrical coordinate system is formed by three families of coordinate surfaces – right circular cylinders with the axis z and radius $r = \sqrt{x^2 + y^2}$, semi-planes passing through the axis z – $\theta = \mathrm{arctg}\frac{y}{x}$ and planes parallel to the plane XY – $z = \mathrm{const}$ (Figure 7.4). In the cylindrical coordinate system the coordinate axes are designated as $q_1 = r$, $q_2 = \theta$, $q_3 = z$.

The connection of these coordinates with the coordinates of Cartesian coordinate system is established by the correlation $x = r\cos\theta$, $y = r\sin\theta$, $z = z$, and the values of Lame coefficients take on the values $H_1 = 1$, $H_2 = r$, $H_3 = 1$.

It should be pointed out that the axisymmetric coordinate system which can be considered as an individual case of the cylindrical coordinate system (when applying the axisymmetric coordinate system, the symmetry of the process studied by the angular coordinate $q_2 = \theta$ is assumed) is frequently applied in practice.

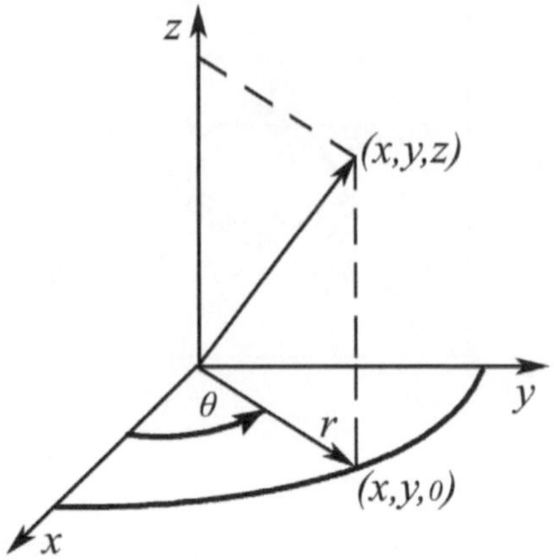

FIGURE 7.4 Cylindrical coordinate system.

7.1.2.2 Spherical Coordinate System

The spherical coordinate system is formed by the intersection of three families of coordinate surfaces of the concentric spheres with the center in the coordinate center and radius $r = \sqrt{x^2 + y^2 + z^2}$, concentric surfaces of right circular cones with the polar axis z and crests in the coordinate center $\theta = \arccos\frac{z}{\sqrt{x^2 + y^2 + z^2}}$, as well as the semi-planes passing through the axis $z - \varphi = \text{arctg}\frac{y}{x}$ (Figure 7.5). In the spherical coordinate system the axes are designated as $q_1 = r,\ q_2 = \theta,\ q_3 = \varphi$. The bond of these coordinates with the coordinates of Cartesian coordinate system are established by the following correlations $x = r\sin\theta\cos\varphi$, $y = r\sin\theta\sin\varphi$, $z = r\cos\theta$, and the values of Lame coefficients take on the values $H_1 = 1$, $H_2 = r$, $H_3 = r\sin\theta$.

7.1.2.3 Elliptic Cylindrical Coordinate System

The coordinate grid is formed by the intersection of elliptic cylinders, hyperbolic cylinders and planes (Figure 7.6). Designation of the coordinates – $q_1 = u,\ q_2 = v,\ q_3 = z$.

The bond with Cartesian coordinate system is established by the following correlations

$$x = a\,\text{ch}\,u\cos v, \qquad y = a\,\text{sh}\,u\sin v, \qquad z = z$$

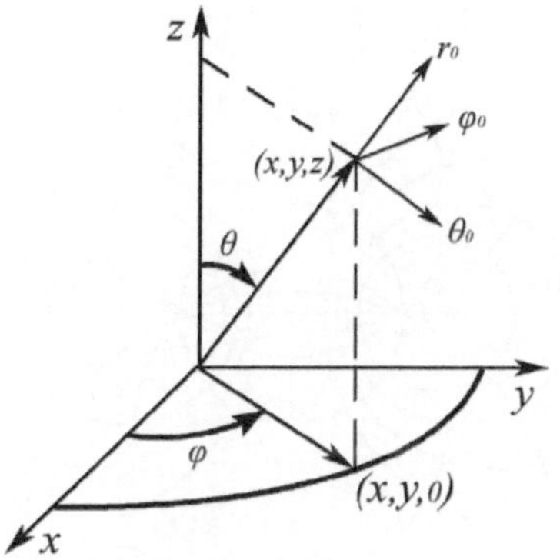

FIGURE 7.5 Spherical coordinate system.

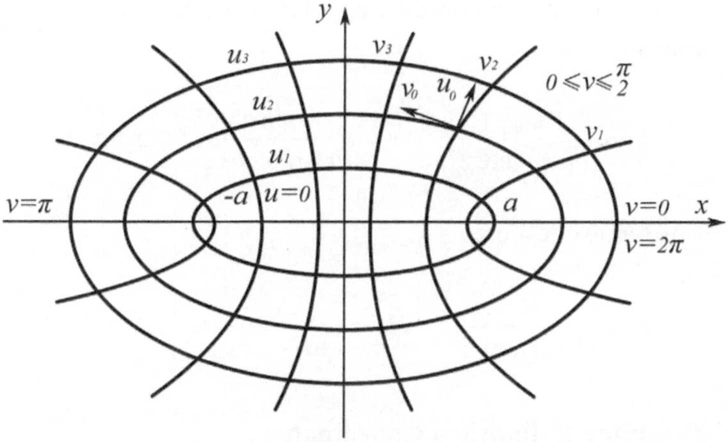

FIGURE 7.6 Elliptic cylindrical coordinate system.

The values of Lame coefficients

$$H_1 = a\sqrt{\text{sh}^2 u + \sin^2 v}\,,\; H_2 = a\sqrt{\text{sh}^2 u + \sin^2 v}\,,\; H_3 = 1$$

7.1.2.4 Bipolar Coordinates

The coordinate grid is formed by the intersection of two circular cylinders and plane (Figure 7.7). Designation of the coordinates – $q_1 = \xi$, $q_2 = \eta$, $q_3 = z$.

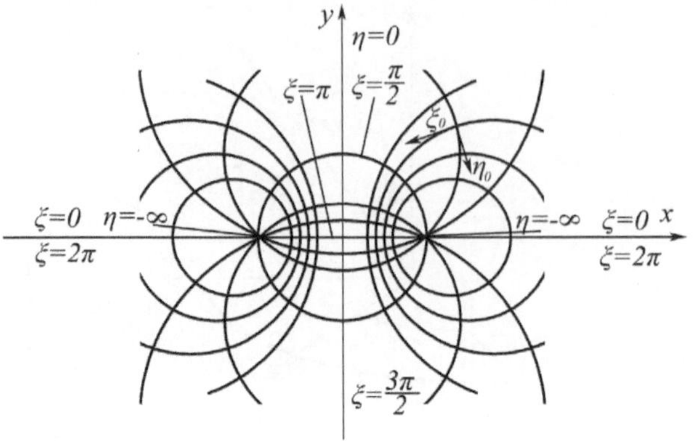

FIGURE 7.7 Bipolar coordinate system.

The bond with Cartesian coordinate system is established by the following correlations

$$x = a\frac{\operatorname{sh}\eta}{\operatorname{ch}\eta - \cos\xi}, \qquad y = a\frac{\sin\xi}{\operatorname{ch}\eta - \cos\xi}, \qquad z = z$$

The values of Lame coefficients

$$H_1 = \frac{a}{\operatorname{ch}\eta - \cos\xi}, \qquad H_2 = \frac{a}{\operatorname{ch}\eta - \cos\xi}, \qquad H_3 = 1$$

7.1.2.5 Parabolic Cylindrical Coordinates

The coordinate grid is formed by the intersection of two parabolic cylinders and plane (Figure 7.8). Designation of the coordinates – $q_1 = \xi$, $q_2 = \eta$, $q_3 = z$.

The bond with Cartesian coordinate system is established by the following correlations

$$x = \xi\eta, \quad y = \frac{1}{2}(\eta^2 - \xi^2), \quad z = z$$

The values of Lame coefficients

$$H_1 = \sqrt{\xi^2 + \eta^2}, \quad H_2 = \sqrt{\xi^2 + \eta^2}, \quad H_3 = 1$$

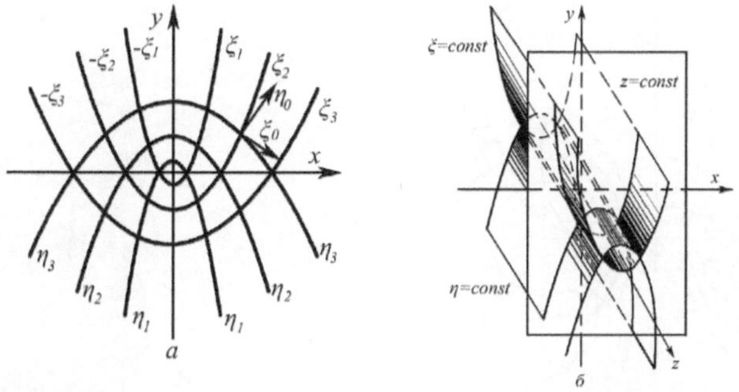

FIGURE 7.8 Parabolic cylindrical coordinate system (a – coordinate grid; b – isometry).

7.1.2.6 Toroidal Coordinates

The coordinate grid is formed by the intersection of spheres, toruses and semi-planes (Figure 7.9). Designation of the coordinates – $q_1 = \xi$, $q_2 = \eta$, $q_3 = \varphi$

The bond with Cartesian coordinate system is established by the following correlations

$$x = \frac{a\,\mathrm{sh}\,\eta\cos\varphi}{\mathrm{ch}\,\eta - \cos\varphi}, \qquad y = \frac{a\,\mathrm{sh}\,\eta\sin\varphi}{\mathrm{ch}\,\eta - \cos\xi}, \qquad y = \frac{a\sin\xi}{\mathrm{ch}\,\eta - \cos\xi}$$

The values of Lame coefficients

$$H_1 = \frac{a}{\mathrm{ch}\,\eta - \cos\xi}, \qquad H_2 = \frac{a}{\mathrm{ch}\,\eta - \cos\xi}, \qquad H_3 = \frac{a\,\mathrm{sh}\,\eta}{\mathrm{ch}\,\eta - \cos\xi}$$

7.1.3 CONSERVATION LAWS IN CONTINUUM MECHANICS

7.1.3.1 Law of Mass Conservation (Continuity Equation)

Let us write down the derivation of the continuity equation in the arbitrary orthogonal curvilinear coordinate system q_1, q_2, q_3 (Figure 7.10) which is connected with Cartesian coordinate system *x, y, z* by the following correlations

$$x = x(q_1, q_2, q_3); \quad y = y(q_1, q_2, q_3); \quad z = z(q_1, q_2, q_3)$$

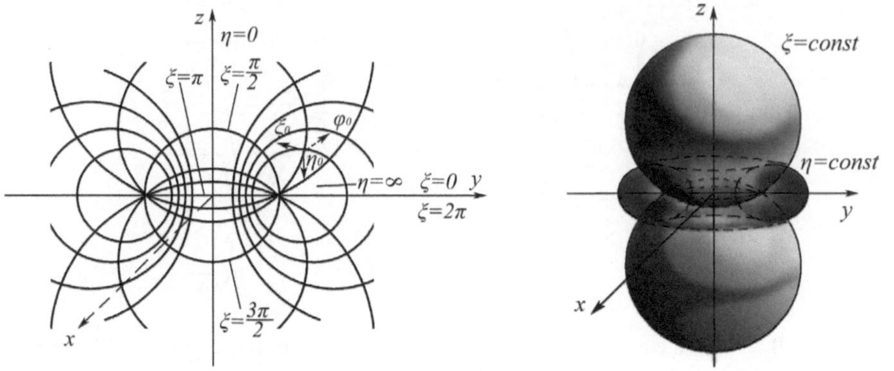

FIGURE 7.9 Toroidal coordinate system.

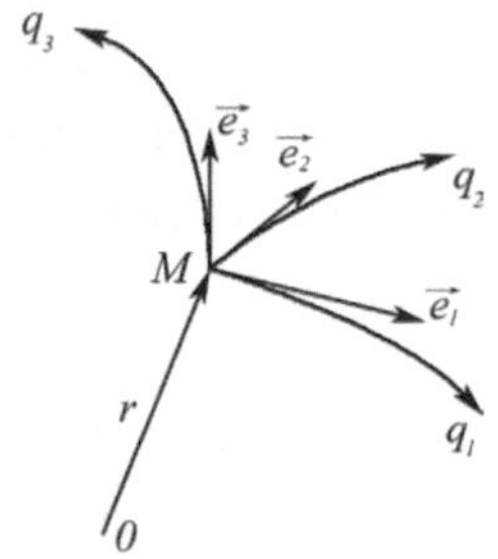

FIGURE 7.10 Curvilinear orthogonal coordinate system.

Let us consider the elementary volume (Figure 7.11) formed by the coordinate surfaces

$$q_1 = \text{const}\,,\quad q_2 = \text{const}\,,\quad q_3 = \text{const}$$

$$q_1 + dq_1 = \text{const}\,,\quad q_2 + dq_2 = \text{const}\,,\quad q_3 + dq_3 = \text{const}$$

The development of continuity equations is based on the assumption that the mass changes in the volume considered within any small time period Δt due to the functioning of the sources distributed within the volume (positive or negative), as well as due to the fact that the unequal amounts of the continuum considered (liquid, gas, deformed body, etc.) inlet and outlet through the opposite faces of the elementary volume. Let us designate the continuum mass ΔM, $\Delta M'$ in the volume dW at the time points t and $(t+\Delta t)$. Taking

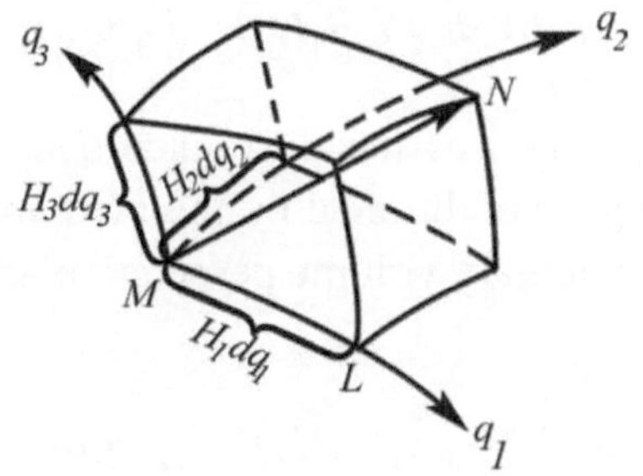

FIGURE 7.11 Elementary volume in the arbitrary orthogonal coordinate system.

into account the accepted assumption of the expression for ΔM, $\Delta M'$, the following can be written down

$$\Delta M = \rho\,(t) H_1 H_2 H_3 dq_1 dq_2 dq_3$$

$$\Delta M' = \rho\,(t+\Delta t) H_1 H_2 H_3 dq_1 dq_2 dq_3$$

In this case, the change of the mass δM will be as follows:

$$\delta M = \Delta M - \Delta M' = \frac{\rho(t+\Delta t) - \rho(t)}{\Delta t} H_1 H_2 H_3 dq_1 dq_2 dq_3 \Delta t$$

The time Δt is indefinitely small, therefore, it seems possible to accomplish the passage to the limit, taking $\Delta t \to 0$. Then

$$\delta M = \lim_{\Delta t \to 0} \delta M = \frac{\partial \rho}{\partial t} H_1 H_2 H_3 dq_1 dq_2 dq_3 dt$$

Let us designate the values of mass growth in volume as δm, δm_1, δm_2, δm_3, respectively, due to the sources functioning and difference of mass flows through the faces perpendicular to the axes q_1, q_2, q_3. Then

$$\delta M = \delta m + \delta m_1 + \delta m_2 + \delta m_3$$

Let the power of the source functioning in the volume dW be equal ε. In this case, the mass input into the elementary volume from this source can be determined by the following correlation

$$\delta m = \varepsilon dW\, dt = \varepsilon H_1 H_2 H_3 dq_1 dq_2 dq_3 dt$$

The value δm_1 is determined as the difference $\delta m_1 = \delta M_1 - \delta M_1'$, where δM_1, $\delta M_1'$– mass changes conditioned by the medium transfer through the opposite faces of the elementary volume perpendicular to the axis q_1. We can write down

$$\delta M_1 = (\rho v_1 H_2 H_3)_{q_1} dq_2 dq_3 dt$$

$$\delta M_1' = (\rho v_1 H_2 H_3)_{q_1 + dq_1} dq_2 dq_3 dt$$

In this case, to determine the value δm_1 we obtain

$$\delta m_1 = \frac{(\rho v_1 H_2 H_3)_{q_1} - (\rho v_1 H_2 H_3)_{q_1 + dq_1}}{dq_1} dq_1 dq_2 dq_3 dt$$
$$= -\frac{\partial\, \rho v_1 H_2 H_3}{\partial q_1} dq_1 dq_2 dq_3 dt$$

The values δm_2, δm_3 are found in a similar way:

$$\delta m_2 = -\frac{\partial\, \rho v_2 H_1 H_3}{\partial q_2} dq_1 dq_2 dq_3 dt$$

$$\delta m_3 = -\frac{\partial\, \rho v_3 H_1 H_2}{\partial q_3} dq_1 dq_2 dq_3 dt$$

Writing down the total balance in the differential form, we have:

$$H_1 H_2 H_3 \frac{\partial \rho}{\partial t} + \frac{\partial}{\partial q_1} \rho v_1 H_2 H_3 + \frac{\partial}{\partial q_2} \rho v_2 H_1 H_3 + \frac{\partial}{\partial q_3} \rho v_3 H_1 H_2 = \varepsilon H_1 H_2 H_3 \tag{7.9}$$

As defined in [42, 254], the expression

$$\frac{1}{H_1 H_2 H_3}\left(\frac{\partial}{\partial q_1} v_1 H_2 H_3 + \frac{\partial}{\partial q_2} v_2 H_1 H_3 + \frac{\partial}{\partial q_3} v_3 H_1 H_2\right) = \operatorname{div} \mathbf{v}$$

In this case, the continuity equation can be written down as follows:

$$\frac{\partial \rho}{\partial t}+\frac{v_1}{H_1}\frac{\partial \rho}{\partial q_1}+\frac{v_2}{H_2}\frac{\partial \rho}{\partial q_2}+\frac{v_3}{H_3}\frac{\partial \rho}{\partial q_3}+\rho\,\mathrm{div}\,\mathbf{v}=\varepsilon \tag{7.10}$$

or

$$\frac{d\rho}{dt}+\rho\,\mathrm{div}\,\mathbf{v}=\varepsilon \tag{7.11}$$

The latter record is invariant for any coordinate system. In the first variant of the record (Eq. (7.10)) the values H_1, H_2, H_3 can differ depending on the selected coordinate system.

Let us consider the special cases:

- *Cartesian coordinate system*

$$q_1 = x,\quad q_2 = y,\quad q_3 = z;\qquad H_1 = 1,\quad H_2 = 1,\quad H_3 = 1;$$

$$\frac{\partial \rho}{\partial t}+\frac{\partial}{\partial x}\rho v_1+\frac{\partial}{\partial y}\rho v_2+\frac{\partial}{\partial z}\rho v_3=\varepsilon\,; \tag{7.12}$$

- *Cylindrical coordinate system*

$$q_1 = r,\qquad q_2 = \theta,\qquad q_3 = z;$$

$$H_1 = 1,\qquad H_2 = r,\qquad H_3 = 1;$$

$$\frac{\partial \rho}{\partial t}+\frac{\partial}{\partial r}\rho v_r r+\frac{\partial}{\partial \theta}\rho v_\theta+\frac{\partial}{\partial z}\rho v_z r=\varepsilon\cdot r \tag{7.13}$$

- *Spherical coordinate system*

$$q_1 = r,\qquad q_2 = \theta,\qquad q_3 = \varphi;$$

$$H_1 = 1,\qquad H_2 = r,\qquad H_3 = r\cdot\sin\theta;$$

$$r^2\sin\theta\frac{\partial \rho}{\partial t}+\frac{\partial}{\partial r}\rho v_r r^2\sin\theta+\frac{\partial}{\partial \theta}\rho v_\theta r\sin\theta+\frac{\partial}{\partial \varphi}\rho v_\varphi r=\varepsilon r^2\sin\theta \tag{7.14}$$

In accordance with the original assumptions on the substance conservation in the volume **W**, the law of mass conservation can be written down in the integral form as follows:

$$\frac{d}{dt}\iiint_W \rho\, dW = \iiint_W \varepsilon\, dW \tag{7.15}$$

Further, the laws of conservation of energy, momentum and moment of momentum will be also written down in integral and differential forms.

7.1.3.2 Law of Energy Conservation

The law of mass conservation (equation of continuum continuity) has been introduced in differential form. Such approach is also possible when developing other fundamental conservation laws for continuum. However, it is also possible to apply the integral approach when developing equations. Thus, when formulating the law of continuum energy conservation in arbitrary limited volume, the assumption is true that the change in the total energy (sum of internal and kinetic energies) is determined by the action of energy sources ε_E located in the elementary volume *dW*, work of mass forces **F**, work of surface forces $(\tau_{\mathbf{n}} \cdot \mathbf{v})$ – pressure forces and friction forces, heat flows *Qn* through the volume surfaces.

In accordance with the aforementioned definition the equation of energy conservation is written down in the integral form as follows:

$$\frac{d}{dt}\iiint_W \rho E\; dW = \iiint_W \rho\,(\mathbf{F}\cdot\mathbf{v})dW + \iiint_W \varepsilon_E dW + \iint_S (\tau_{\mathbf{n}}\cdot\mathbf{v})dS + \iint_S Q_n dS \tag{7.16}$$

In the Eq. (7.16) $\tau_{\mathbf{n}}$ – value of stress tensor of surface forces (pressure and friction forces) acting by the normal to the elementary surface (to the faces of elementary volume *dW*).

To obtain the recording of the energy equation in differential form, it is necessary to change from integrals by the surface $\iint_S (\tau_{\mathbf{n}}\cdot\mathbf{v})dS$, $\iint_S Q_n dS$ to the integrals by volume in the Eq. (7.16). Such transition can be performed using Gauss theorem ([201], Section 7.2). In particular, for the recorded integrals by the surface according to the theorem conditions we have

$$\iint_S (\tau_{\mathbf{n}} \cdot \mathbf{v}) dS = \iiint_W \left(\begin{array}{l} \dfrac{1}{H_1 H_2 H_3} (\dfrac{\partial}{\partial q_1} H_2 H_3 (\tau_1 \cdot \mathbf{v}) \\ + \dfrac{\partial}{\partial q_2} H_1 H_3 (\tau_2 \cdot \mathbf{v}) + \dfrac{\partial}{\partial q_3} H_1 H_2 (\tau_3 \cdot \mathbf{v})) dW \end{array} \right)$$

$$\iint_S Q_n dS = \iiint_W \frac{1}{H_1 H_2 H_3} \left(\frac{\partial H_2 H_3 Q_1}{\partial q_1} + \frac{\partial H_1 H_3 Q_2}{\partial q_2} + \frac{\partial H_1 H_2 Q_3}{\partial q_3} \right) dW$$

Besides, it should be taken into account [42, 254] that

$$\frac{d}{dt} \iiint_W \rho E \, dW = \iiint_W (\frac{d}{dt} \rho E + \rho E \operatorname{div} \mathbf{v}) \, dW$$

$$= \iiint_W (\rho (\mathbf{v} \cdot \frac{d\mathbf{v}}{dt}) + \rho \frac{d(E - \frac{1}{2} v^2)}{dt}) \, dW$$

$$\rho (\mathbf{v} \cdot \frac{d\mathbf{v}}{dt}) - \rho (\mathbf{F} \cdot \mathbf{v}) - \frac{1}{H_1 H_2 H_3} ((\mathbf{v} \cdot \frac{\partial H_2 H_3 \tau_1}{\partial q_1}) - (\mathbf{v} \cdot \frac{\partial H_1 H_3 \tau_2}{\partial q_2}) - (\mathbf{v} \cdot \frac{\partial H_1 H_2 \tau_3}{\partial q_3})) = 0$$

Further, due to the arbitrariness of the selected elementary volume it seems possible to perform the differential transition and obtain the energy equation in the invariant form

$$\rho \frac{d}{dt} E = \frac{1}{H_1 H_2 H_3} \left(\frac{\partial H_2 H_3 Q_1}{\partial q_1} + \frac{\partial H_1 H_3 Q_2}{\partial q_2} + \frac{\partial H_1 H_2 Q_3}{\partial q_3} \right) + \varepsilon_E$$
$$+ \frac{1}{H_1 H_2 H_3} (\frac{\partial}{\partial q_1} H_2 H_3 (\tau_1 \cdot \mathbf{v}) + \frac{\partial}{\partial q_2} H_1 H_3 (\tau_2 \cdot \mathbf{v}) + \frac{\partial}{\partial q_3} H_1 H_2 (\tau_3 \cdot \mathbf{v})) \quad (7.17)$$

In the ideal gas the value of friction forces and value of the external heat flow are assumed to equal zero. Therefore, for the ideal gas in the arbitrary orthogonal coordinate system the equation of energy conservation in the differential form can be written down as follows:

$$H_1H_2H_3\frac{\partial}{\partial t}\rho E+\frac{\partial}{\partial q_1}\rho v_1(E+{}^{p}\!/_{\rho})H_2H_3+\frac{\partial}{\partial q_2}\rho v_2(E+{}^{p}\!/_{\rho})H_1H_3$$

$$+\frac{\partial}{\partial q_3}\rho v_3(E+{}^{p}\!/_{\rho})H_1H_2=\varepsilon_E E\,H_1H_2H_3 \tag{7.18}$$

It should be mentioned that the above law of energy conservation is sometimes called the first law of thermodynamics.

7.1.3.3 Law of Momentum Conservation

Let us write down the law of momentum conservation in the integral form. The law should be developed from the assumption that the change in the mass momentum in the elementary volume dW is determined by the action of mass $\mathbf{F}$ and surface $\boldsymbol{\tau}_{\mathbf{n}}$ forces on the elementary volume

$$\frac{d}{dt}\iiint_W \rho\cdot\mathbf{v}\cdot dW=\iiint_W \rho\cdot\mathbf{F}\cdot dW+\iint_S \tau_{\mathbf{n}}\,dS \tag{7.19}$$

It is known [42, 254] that

$$\frac{d}{dt}\iiint_W \rho\mathbf{v}\;dW=\iiint_W\left(\frac{d\rho\mathbf{v}}{dt}+\rho\mathbf{v}\,\mathrm{div}\,\mathbf{v}\right)dW$$

Besides, in accordance with Gauss theorem and Cauchy formula [42] the latter integral by the surface can be written down in the Eq. (7.19) as follows:

$$\iint_S \tau_{\mathbf{n}}\cdot dS=\iiint_W\frac{1}{H_1H_2H_3}\left(\frac{\partial H_2H_3\tau_1}{\partial q_1}+\frac{\partial H_1H_3\tau_2}{\partial q_2}+\frac{\partial H_1H_2\tau_3}{\partial q_3}\right)\cdot dW$$

Applying the latter two equations and performing the limiting transition ($dW\to 0$) in the Eq. (7.19) due to the arbitrariness of the selected elementary volume dW, we obtain the equation of momentum in the invariant differential form

$$\frac{d\rho\mathbf{v}}{dt}+\rho\mathbf{v}\,\mathrm{div}\,\mathbf{v}=\rho\mathbf{F}+\frac{1}{H_1H_2H_3}\left(\frac{\partial H_2H_3\tau_1}{\partial q_1}+\frac{\partial H_1H_3\tau_2}{\partial q_2}+\frac{\partial H_1H_2\tau_3}{\partial q_3}\right) \tag{7.20}$$

The equation of momentum conservation for the ideal gas in the arbitrary orthogonal coordinate system in the projections onto the coordinate axes can be written down as follows [113]

$$H_1H_2H_3\frac{\partial}{\partial t}\rho v_1+\frac{\partial}{\partial q_1}\rho v_1^2H_2H_3+\frac{\partial}{\partial q_2}\rho v_1v_2H_1H_3+\frac{\partial}{\partial q_3}\rho v_1v_3H_1H_2$$
$$=-H_2H_3\frac{\partial p}{\partial q_1}-\rho v_2H_3d_{12}-\rho v_3H_2d_{13}+H_1H_2H_3\rho F_1$$
;

$$H_1H_2H_3\frac{\partial}{\partial t}\rho v_2+\frac{\partial}{\partial q_1}\rho v_2v_1H_2H_3+\frac{\partial}{\partial q_2}\rho v_2^2H_1H_3+\frac{\partial}{\partial q_3}\rho v_2v_3H_1H_2$$
$$=-H_1H_3\frac{\partial p}{\partial q_2}-\rho v_1H_3d_{21}-\rho v_3H_1d_{23}+H_1H_2H_3\rho F_2$$
;

$$H_1H_2H_3\frac{\partial}{\partial t}\rho v_3+\frac{\partial}{\partial q_1}\rho v_3v_1H_2H_3+\frac{\partial}{\partial q_2}\rho v_3v_2H_1H_3+\frac{\partial}{\partial q_3}\rho v_3^2H_1H_2$$
$$=-H_1H_2\frac{\partial p}{\partial q_3}-\rho v_1H_2d_{31}-\rho v_2H_1d_{32}+H_1H_2H_3\rho F_3$$

Here it is designated:

$$d_{12}=v_1\frac{\partial H_1}{\partial q_2}-v_2\frac{\partial H_2}{\partial q_1};\quad d_{13}=v_1\frac{\partial H_1}{\partial q_3}-v_3\frac{\partial H_3}{\partial q_1}$$

$$d_{21}=v_2\frac{\partial H_2}{\partial q_1}-v_1\frac{\partial H_1}{\partial q_2};\quad d_{23}=v_2\frac{\partial H_2}{\partial q_3}-v_3\frac{\partial H_3}{\partial q_2}$$

$$d_{31}=v_3\frac{\partial H_3}{\partial q_1}-v_1\frac{\partial H_1}{\partial q_3};\quad d_{32}=v_3\frac{\partial H_3}{\partial q_2}-v_2\frac{\partial H_2}{\partial q_3}$$

7.1.3.4 Law of the Conservation of the Moment of Momentum

For some mechanical systems (in particular, in deformable solid mechanics) apart from the aforementioned fundamental laws, it is necessary to apply the law of the conservation of the moment of momentum. It should be pointed out that it is practically unnecessary to consider the law of the conservation of the moment of momentum in the problems of liquid and gas mechanics, since in the majority of problems this law is the result of the law of momentum conservation. The development of the law of the conservation of the moment of momentum is based on the assumption that the moment of momentum **M** in the closed system changes due to the action of the moments of external forces $\mathbf{M}_0$ (main orbital moment from the external force **F**) and $\mathbf{M}_s$ (main orbital moment of surface forces) acting upon this system, rate of the generation **Π** of the internal moment of momentum $\mathbf{M}_0^{in}$ and density π_n of the flow (penetration) of the internal moment of momentum $\mathbf{M}_s^{in}$ through the surface limiting the mechanical system.

Let us designate:

- $\mathbf{r}$ – radius-vector determining the spatial coordinate in the initially selected coordinate system for the mechanical system;
- $\mathbf{L} = \iiint\limits_{\mathbf{W}} ((\mathbf{r} \times \rho \mathbf{v}) + \rho \mathbf{M}) dW$ – total moment of the mass momentum;
- $\mathbf{M}_0 = \iiint\limits_{W} (\mathbf{r} \times \rho \mathbf{F}) dW$ – main orbital moment of the mass forces;
- $\mathbf{M}_s = \iint\limits_{S} (\mathbf{r} \times \tau_n) dS$ – main orbital moment of the surface forces;
- $\mathbf{M}_0^{in} = \iiint\limits_{W} \rho \Pi \, dW$ – volume gain of the moment of system momentum;
- $\mathbf{M}_s^{in} = \iint\limits_{S} \pi_{\mathrm{n}} dS$ – gain of the moment of system momentum due to the penetration through the boundaries.

In accordance with the definition of the law of the conservation of the moment of momentum we can write down as follows:

$$\frac{dL}{dt} = \mathbf{M}_0 + \mathbf{M}_s + \mathbf{M}_0^{in} + \mathbf{M}_s^{in} \tag{7.21}$$

Taking into account the foregoing designations and simplifications, the Eq. (7.21) can be written down in the following forms:

- integral form of the law

$$\iiint\limits_{W} \rho \frac{d\mathbf{M}}{dt} dW = \iiint\limits_{W} (\mathbf{e}_1 \times \tau_1 + \mathbf{e}_2 \times \tau_2 + \mathbf{e}_3 \times \tau_3) dW + \iiint\limits_{W} \rho \mathbf{\Pi} dW + \iint\limits_{S} \pi_n dS; \tag{7.22}$$

- differential form of the law

$$\rho\frac{d\mathbf{M}}{dt}=\rho\Pi+\frac{1}{H_1H_2H_3}\left(\frac{\partial H_2H_3\pi_1}{\partial q_1}+\frac{\partial H_1H_3\pi_2}{\partial q_2}+\frac{\partial H_1H_2\pi_3}{\partial q_3}\right)$$
$$+\mathbf{e}_1\times\tau_1+\mathbf{e}_2\times\tau_2+\mathbf{e}_3\times\tau_3 \qquad (7.23)$$

7.1.4 UNIQUENESS CONDITIONS OF SOLVING THE PROBLEMS OF CONTINUUM MECHANICS

The conservation laws discussed above are not the only fundamental laws. In particular, when investigating closed thermodynamic systems, the second law of thermodynamics is important postulating the idea that the system state can change with time in the direction not decreasing the entropy value of this system. In electrodynamics (field also related to continuum mechanics) Maxwell equations containing Coulomb law, equations of gravity field, equations establishing the bond between the field electric and magnetic strengths, etc. become principal. Nevertheless, the Eqs. (7.11)–(7.23) can be taken as the basis when recording closed mathematical models of continuum mechanics or mechanical systems.

The considered class of problems in the given setting cannot be solved yet. Actually, the most various flows of liquid, gas and deformed body can be described by these equations. The studied problem can be solved using additional conditions characterizing the certain problem setting. Additional conditions need to be written down in the form of mathematical equations and their number has to be sufficient to obtain the only solution of the problem. Additional conditions can specify the continuum rheological properties (viscous stress tensor, for example), thermodynamic properties of the medium (values of heat capacity coefficients, gas constant, etc.), reactive properties of the medium (laws describing the chemical kinetics of processes in the continuum, diffusion laws) and etc. When closing the equation system, the special role is played by so-called initial and boundary (edge) conditions.

Initial conditions are required when considering problems on the continuum non-stationary motion. The continuum state in the arbitrary point of the computational domain at the given time point $t = t_0$ needs to be described in the initial conditions. If further the problem is solved, in which $t > t_0$, such problem is called evolutionary. The problems with the time reverse direction can be also interesting in practice ($t < t_0$). Sometimes such problems are called inverse. Their setting is more complicated and can be incorrect (solution can be not the only one).

Boundary conditions assume that at the arbitrary time point t, for which it is necessary to find the medium state in some closed computational domain, the medium values on the boundaries of this computational domain are known or the mathematical dependencies determining the continuum interaction with the external space can be recorded. In some problems, when the computational domain boundaries cannot be determined in advance, additional correlations can be necessary. Such correlations can be, for instance, conditions determining the computational domain boundary location (e.g., free boundary of liquid and gas in the solid vessel at agitation). In a particular case, the boundary location can be selected at infinity. It should be pointed out that the selection of boundaries at infinity is a convenient idealization of the computational domain which is frequently used when solving practical problems.

The number of boundary conditions, which need to be written down for the system to be closed, depends on the type of differential equations describing the continuum state. The equations can be classified into elliptic, parabolic and hyperbolic types. For example, when considering the ideal gas motion, the class of equations solved can be linked with the gas motion velocity and dependence of the parameters on the process time. The equations of non-stationary processes are referred to the hyperbolic type at any motion velocities of the ideal gas. The equations of stationary motion of the ideal gas at subsonic velocities are referred to the elliptic type, at the motion with sound velocity – to the parabolic type and with supersonic velocity – to the hyperbolic type. The equations of the stationary flow of viscous gas in the thin boundary layer can be referred to the parabolic class of equations. The equations of elliptic type are most demanding to the boundary conditions. The least number of boundary conditions are required when solving the problems of hyperbolic type. Ideally, the value or functional dependence for each sought parameter of the continuum ($\rho, \mathbf{v}, E$ and etc.) should be described in the boundary conditions. However, such number of boundary conditions is excessive for parabolic and hyperbolic equations. Therefore, when correctly setting up the problem, some boundary correlations have to depend on others.

Let us enlist the principal variants of setting up the boundary conditions in continuum mechanics.

1) Conditions of the medium motion at solid fixed boundaries

- viscous gas – $\mathbf{v} = 0$ – no-slip conditions;
- ideal gas – $v_n = 0$; $\partial v_\tau / \partial q_n = 0$ – slip conditions.

If the boundary transfers, the motion velocity vector of the boundary $\mathbf{w}_{\text{гр}}$ needs to be taken into account in the boundary conditions. The sense of the boundary conditions remains the same.

2) Conditions at the flow free boundary

- viscous gas or elastic solid material – the density of surface forces is set up – $\mathbf{P}_n = \mathbf{P}_{\mathbf{nn}} + \mathbf{P}_{\mathbf{n}\tau} = f(q_1, q_2, q_3)$
- ideal gas – $p = p_0$; here p_0 – environment pressure.

In the same way as at free boundaries, the conditions can be recorded at "infinity." Here, the conditions taking into account the possible periodicity of the medium motion, its stochastic character, symmetry, etc. can be set. Let us point out another important circumstance. When practically solving problems (especially when applying finite-difference methods), the form of boundary conditions can differ from the one described above. However, the initial assumptions leading to recording these boundary conditions need to be similar to those indicated above.

7.2 METHODS OF SIMPLIFYING MATHEMATICAL MODELS DEVELOPED WITH THE APPLICATION OF FUNDAMENTAL LAWS OF PHYSICS

Mathematical models developed with the application of fundamental laws are sufficiently nonlinear and, as a rule, rather complicated. In general, there are no analytical solutions for such problems. Their solution with computer facilities is not always methodologically conditioned, and requires more computational resources if there are no methodological difficulties. Actually, in practice, it is rarely necessary to solve problems by complete models developed in accordance with fundamental laws, and, in some cases, these models can be considerably simplified. The simplifications are achieved by different methods, some of which are considered below.

7.2.1 APPLICATION OF ASSUMPTION ON THE PROCESS ESTABLISHED CHARACTER

The assumption on the process established character can be accepted when solving many problems from practical applications. Thus, when solving problems in rocket engines – it is the main stationary part of the engine performance, when solving the problems of external ballistics – it is the aircraft flight with constant speed and unchanged height. When solving hydrodynamic problems connected, for instance, with designing a dam, these are the modes in which the water rise level is unchanged, etc.

Actually, there are no stationary modes in any real physical process or phenomenon, since there are always disturbing factors influencing the process stationary picture. Nevertheless, the acceptance of the assumptions on the process established character means that we can neglect the influence of disturbing factors on the investigated process or phenomenon in certain applications. In the most general case, the modes similar to the established (stationary) ones are called equilibrium.

Sometimes when analyzing the equilibrium modes, we accept the assumptions on the process quasi-stationarity. The sense of such assumption consists in the fact that it is possible to neglect the change rate (with time) of the process parameters at any arbitrary time point, although in the long time interval the influence of the change rate of the parameters is significant. Such method also allows simplifying the initial equations, but forces to calculate the nearly stationary (slightly changing with time) picture for several time moments. As an example, we can use the operation period of the solid fuel rocket engine, in which the combustion surface value is a slightly changing function of the burnt dome. When solving the problem of the aircraft external ballistics – this is, for example, the time period when the flight height changes slowly.

In the aforementioned cases, the terms "weakly changing function" and "slow change" have to be substantiated. They can be substantiated by comparing the characteristic times – the time of establishing the gas-dynamic process in the engine chamber and time during which the pressure of the combustion products significantly changes in the intrachamber volume. The time of establishing the gas-dynamic processes in the engine chamber is determined by the time of spreading the weak disturbance occurring in some point in the intrachamber volume, by the overall volume and its fading time.

The assumption on the established character of the processes allows accepting the partial derivative value by time on the function investigated to equal zero. In particular, the closed equation system comprising the equations of continuity, momentum, energy and gas pressure can be obtained for the ideal gas in the arbitrary orthogonal coordinate system. In accordance with the assumption on the partial derivative by time to equal zero, the equation system of the ideal gas motion (see Section 7.1) is as follows:

$$\frac{\partial}{\partial q_1}\rho v_1 H_2 H_3 + \frac{\partial}{\partial q_2}\rho v_2 H_1 H_3 + \frac{\partial}{\partial q_3}\rho v_3 H_1 H_2 = \varepsilon H_1 H_2 H_3$$

$$\frac{\partial}{\partial q_1}\rho v_1^2 H_2 H_3 + \frac{\partial}{\partial q_2}\rho v_1 v_2 H_1 H_3 + \frac{\partial}{\partial q_3}\rho v_1 v_3 H_1 H_2$$

$$= -H_2 H_3 \frac{\partial p}{\partial q_1} - \rho v_2 H_3 d_{12} - \rho v_3 H_2 d_{13} + H_1 H_2 H_3 \rho F_1$$

$$\frac{\partial}{\partial q_1}\rho v_2 v_1 H_2 H_3 + \frac{\partial}{\partial q_2}\rho v_2^2 H_1 H_3 + \frac{\partial}{\partial q_3}\rho v_2 v_3 H_1 H_2$$

$$= -H_1 H_3 \frac{\partial p}{\partial q_2} - \rho v_1 H_3 d_{21} - \rho v_3 H_1 d_{23} + H_1 H_2 H_3 \rho F_2$$

$$\frac{\partial}{\partial q_1}\rho v_3 v_1 H_2 H_3 + \frac{\partial}{\partial q_2}\rho v_3 v_2 H_1 H_3 + \frac{\partial}{\partial q_3}\rho v_3^2 H_1 H_2$$

$$= -H_1 H_2 \frac{\partial p}{\partial q_3} - \rho v_1 H_2 d_{31} - \rho v_2 H_1 d_{32} + H_1 H_2 H_3 \rho F_3$$

$$\frac{\partial}{\partial q_1}\rho v_1 (E + p/\rho) H_2 H_3 + \frac{\partial}{\partial q_2}\rho v_2 (E + p/\rho) H_1 H_3$$

$$+ \frac{\partial}{\partial q_3}\rho v_3 (E + p/\rho) H_1 H_2 = \varepsilon_E E H_1 H_2 H_3, \quad p = \rho(k-1)\left(E - \frac{v_1^2 + v_2^2 + v_3^2}{2}\right);$$

$$d_{12} = v_1 \frac{\partial H_1}{\partial q_2} - v_2 \frac{\partial H_2}{\partial q_1}; \quad d_{13} = v_1 \frac{\partial H_1}{\partial q_3} - v_3 \frac{\partial H_3}{\partial q_1};$$

$$d_{21} = v_2 \frac{\partial H_2}{\partial q_1} - v_1 \frac{\partial H_1}{\partial q_2}, \quad d_{23} = v_2 \frac{\partial H_2}{\partial q_3} - v_3 \frac{\partial H_3}{\partial q_2}$$

$$d_{31} = v_3 \frac{\partial H_3}{\partial q_1} - v_1 \frac{\partial H_1}{\partial q_3}, \quad d_{32} = v_3 \frac{\partial H_3}{\partial q_2} - v_2 \frac{\partial H_2}{\partial q_3}$$

Another option of simplification consists in the assumption that the sought function $f(t, q_1, q_2, q_3)$ (e.g., the density ρ, velocity vector $\mathbf{v}$, energy E, pressure p, etc. can be considered under the function f) can be presented as the product of two functions $f_1(t)$ and $f_2(q_1, q_2, q_3)$

$$f(t, q_1, q_2, q_3) = f_1(t) \cdot f_2(q_1, q_2, q_3) \tag{7.24}$$

The successful selection of the functional dependencies $f_1(t)$ and $f_2(q_1, q_2, q_3)$ can considerably simplify the initial mathematical problem.

The additional simplifications can be achieved if the function $f_1(t)$ is periodic. It should be reminded that the function $f_1(t)$ is called periodic with the period T if the following correlation is fulfilled

$$f_1(t) = f_1(t + iT), \quad i = 1, \ldots, N \tag{7.25}$$

The examples of the processes, for which the similar assumptions are applied, can be the ones in internal combustion engines, compressors, etc.

For some periodic processes, the values of the parameters averaged for the period during which the parameters change are interesting. If T is the period during which the parameter $f(t)$ changes, the average value of the parameter f is determined by the following expression

$$\bar{f} = \frac{1}{T} \int_{t}^{t+T} f(t, q_1, q_2, q_3) \cdot dt \tag{7.26}$$

For the function $f(t, q_1, q_2, q_3)$ satisfying the condition (7.24), the value averaged during the period T can be established by the following expression

$$\bar{f} = \frac{1}{T} \int_{t}^{t+T} f_1(t) f_2(q_1, q_2, q_3) dt = \left(\frac{1}{T} \int_{t}^{t+T} f_1(t) dt \right) f_2(q_1, q_2, q_3)$$

7.2.2 USE OF THE ASSUMPTION ON THE FLOW DIMENSIONALITY

In some cases, we can definitely say that the process (object) parameters do not change in some spatial direction or they can be neglected. The gas or liquid flow in a cylindrical pipe can be used as the example. It can be demonstrated that in many practically significant cases the gas-dynamic (hydrodynamic) parameters change symmetrically along any angular direction. In the previous example there are no doubts about the possibility of simplifying the mathematical problem setting due to the symmetry of the problem being solved. We can give multiple examples from engineering applications in which the change in the parameters determining the process along any direction differs significantly (in five and more times) from the change in the parameters along other directions. In such cases, it is necessary to strive for simplifying the mathematical setting of the problem being solved. The simplification is achieved by accepting the assumption that there is no functional dependence of the parameter f on some spatial coordinate (q_1, q_2, q_3), and the value of the partial derivative from the parameter f along this spatial coordinate (e.g., along the coordinate q_3) equals zero

$$f(q_1, q_2, q_3) \equiv f(q_1, q_2), \quad \frac{\partial f(q_1, q_2, q_3)}{\partial q_3} = 0 \tag{7.27}$$

We can apply additional simplification methods along some spatial coordinate taking into account specific features of the problem being solved in the same way as when analyzing the possibility of simplifying fundamental equations by time. In particular, we can neglect the change of the derivatives along only some coordinate – $\frac{\partial f}{\partial q_i} \approx 0$ and solve the initial equations for several sections of q_i. The parameters can be integrated along any coordinate q_i, for instance

$$\bar{f}(q_1, q_2) = \frac{1}{(q_{3\max} - q_{3\min})} \int_{q_{3\min}}^{q_{3\max}} f(q_1, q_2, q_3) dq_3 \tag{7.28}$$

If from some ideas we can imagine the dependence $f(q_1, q_2, q_3)$, for instance, as the product $f(q_1, q_2, q_3) = f_1(q_1, q_2) \cdot f_2(q_3)$, then this also allows simplifying the initial equations.

Let us give some examples. The main motion equations (conservation of mass, momentum and energy) recorded for the ideal gas will be applied in the examples (see Section 7.1).

1. The equations of the ideal gas established flow in the curvilinear orthogonal coordinate system are given above. Let us accept the additional assumption on the fairness for this equation system (7.27). It should be pointed out

that the possibility of accepting the assumptions (7.27) depends, to some extent, on the metrics of the initial coordinate system. Actually the application of the assumptions (7.27) makes the initial equation as follows:

$$\frac{\partial}{\partial q_1}\rho v_1 H_2 H_3 + \frac{\partial}{\partial q_2}\rho v_2 H_1 H_3 = \varepsilon H_1 H_2 H_3 \tag{7.29}$$

$$\frac{\partial}{\partial q_1}\rho v_1^2 H_2 H_3 + \frac{\partial}{\partial q_2}\rho v_1 v_2 H_1 H_3 = -H_2 H_3 \frac{\partial p}{\partial q_1} - \rho v_2 H_3 d_{12} + H_1 H_2 H_3 \rho F_1 \tag{7.30}$$

$$\frac{\partial}{\partial q_1}\rho v_2 v_1 H_2 H_3 + \frac{\partial}{\partial q_2}\rho v_2^2 H_1 H_3 = -H_1 H_3 \frac{\partial p}{\partial q_2} - \rho v_1 H_3 d_{21} + H_1 H_2 H_3 \rho F_2 \tag{7.31}$$

$$\frac{\partial}{\partial q_1}\rho v_1 (E + p/\rho) H_2 H_3 + \frac{\partial}{\partial q_2}\rho v_2 (E + p/\rho) H_1 H_3 = \varepsilon_E E H_1 H_2 H_3 \tag{7.32}$$

$$p = \rho(k-1)\left(E - \frac{v_1^2 + v_2^2}{2}\right) \tag{7.33}$$

$$d_{12} = v_1 \frac{\partial H_1}{\partial q_2} - v_2 \frac{\partial H_2}{\partial q_1}; \quad d_{13} = v_1 \frac{\partial H_1}{\partial q_3}; \tag{7.34}$$

$$d_{21} = v_2 \frac{\partial H_2}{\partial q_1} - v_1 \frac{\partial H_1}{\partial q_2}, \quad d_{23} = v_2 \frac{\partial H_2}{\partial q_3} \tag{7.35}$$

$$d_{31} = -v_1 \frac{\partial H_1}{\partial q_3}, \quad d_{32} = -v_2 \frac{\partial H_2}{\partial q_3} \tag{7.36}$$

It is easy to demonstrate that applying the assumption (7.27) for the velocity component v_3 of the momentum equation leads to the following equation

$$v_1^2 H_2 \frac{\partial H_1}{\partial q_3} + v_2^2 H_1 \frac{\partial H_2}{\partial q_3} = 0 \tag{7.37}$$

binding the space metrics with the velocity vector components v_1, v_2. It is obvious that the latter equation can contradict the Eqs. (7.30), (7.31). However, the Eqs. (7.29)–(7.33) will be true if the following metric correlations are fulfilled

$$\frac{\partial H_1}{\partial q_3}=0,\quad \frac{\partial H_2}{\partial q_3}=0,\quad (d_{13}=0,\quad d_{23}=0)$$

If the following metric correlations are also true

$$\frac{\partial H_3}{\partial q_1}=0,\quad \frac{\partial H_3}{\partial q_2}=0,\quad (d_{31}=0,\quad d_{32}=0)$$

the Eq. (7.29)–(7.33) is additionally simplified and rewritten as follows:

$$\frac{\partial}{\partial q_1}\rho v_1 H_2+\frac{\partial}{\partial q_2}\rho v_2 H_1=\varepsilon H_1 H_2 \tag{7.38}$$

$$\frac{\partial}{\partial q_1}\rho v_1^2 H_2+\frac{\partial}{\partial q_2}\rho v_1 v_2 H_1=-H_2\frac{\partial p}{\partial q_1}-\rho v_2 d_{12}+H_1 H_2 \rho F_1 \tag{7.39}$$

$$\frac{\partial}{\partial q_1}\rho v_2 v_1 H_2+\frac{\partial}{\partial q_2}\rho v_2^2 H_1=-H_1\frac{\partial p}{\partial q_2}-\rho v_1 d_{21}+H_1 H_2 \rho F_2 \tag{7.40}$$

$$\frac{\partial}{\partial q_1}\rho v_1(E+p/\rho)H_2+\frac{\partial}{\partial q_2}\rho v_2(E+p/\rho)H_1=\varepsilon_E E H_1 H_2 \tag{7.41}$$

$$p=\rho(k-1)(E-\frac{v_1^2+v_2^2}{2}) \tag{7.42}$$

2. For the initial equation system of the ideal gas motion recorded for the arbitrary curvilinear orthogonal system we accept the assumption that the gas parameter values and their derivatives along the coordinates q_2, q_3 should be neglected. The gas flow will be considered no-stationary. The assumptions accepted allow writing down the unknown equation system as follows:

$$H_1 H_2 H_3\frac{\partial \rho}{\partial t}+\frac{\partial}{\partial q_1}\rho v_1 H_2 H_3=\varepsilon H_1 H_2 H_3 \tag{7.43}$$

$$H_1H_2H_3\frac{\partial}{\partial t}\rho v_1+\frac{\partial}{\partial q_1}\rho v_1^2H_2H_3=-H_2H_3\frac{\partial p}{\partial q_1}+H_1H_2H_3\rho F_1 \quad (7.44)$$

$$H_1H_2H_3\frac{\partial}{\partial t}\rho E+\frac{\partial}{\partial q_1}\rho v_1(E+{}^p\!/_\rho)H_2H_3=\varepsilon_E EH_1H_2H_3 \quad (7.45)$$

$$p=\rho(k-1)(E-\frac{v_1^2}{2}) \quad (7.46)$$

In the same way as in the previous case, the truthfulness of the assumptions accepted when recording the Eqs. (7.43)–(7.46) depends on the coordinate system metrics. The momentum equations recorded along the coordinate directions q_2, q_3 are identically satisfied if the conditions $\frac{\partial H_1}{\partial q_2}=0,\ \frac{\partial H_1}{\partial q_3}=0$ are fulfilled. Besides, if the conditions $\frac{\partial H_2}{\partial q_1}=0,\ \frac{\partial H_3}{\partial q_1}=0$ are fulfilled, the Eqs. (7.43)–(7.45) are additionally simplified and look as follows:

$$H_1\frac{\partial \rho}{\partial t}+\frac{\partial}{\partial q_1}\rho v_1=\varepsilon H_1 \quad (7.47)$$

$$H_1\frac{\partial}{\partial t}\rho v_1+\frac{\partial}{\partial q_1}\rho v_1^2=-\frac{\partial p}{\partial q_1}+H_1\rho F_1 \quad (7.48)$$

$$H_1\frac{\partial}{\partial t}\rho E+\frac{\partial}{\partial q_1}\rho v_1(E+{}^p\!/_\rho)=\varepsilon_E EH_1 \quad (7.49)$$

It should be pointed out that if the conditions $\frac{\partial H_1}{\partial q_2}=0,\ \frac{\partial H_1}{\partial q_3}=0$ are not fulfilled, the assumptions that $\frac{\partial p}{\partial q_2}=0,\ \frac{\partial p}{\partial q_3}=0$ are not true. When solving a number of problems connected with flowing around a body whose bluntness has a higher curvature, the equations of momentum along the coordinate directions q_2, q_3 are solved apart from the Eq. (7.44) to define the pressure distribution in the vicinity of the body flown around written down as follows:

$$-H_1H_3\frac{\partial p}{\partial q_2} - \rho v_1 H_3 d_{21} + H_1H_2H_3\rho F_2 = 0 \tag{7.50}$$

$$-H_1H_2\frac{\partial p}{\partial q_3} - \rho v_1 H_2 d_{31} + H_1H_2H_3\rho F_3 = 0 \tag{7.51}$$

3. When recording the Eqs. (7.29)–(7.51), the transformations connected with the boundary conditions for differential equations in partial derivatives are not considered. Formally, the recording is not difficult, especially, if the region boundaries, in which the gas flows, are impermeable, do not produce friction and are not heat-conductive. Nevertheless, the following example can be given. Let the gas move in the cylindrical pipe, along the side surface of which the gaseous products (e.g., the fuel combustion products) inflow. When solving the problem on gas motion in the cylindrical pipe in 2D setting, the product inflow along the pipe side surface will be taken into account in the boundary layers. At the same time, when solving the problem in one-dimensional setting (the longitudinal coordinate coincides with the pipe axis), the product inflow along the pipe surface will be taken into account as the source of mass and energy (Figure 7.12).

Based on the previous remark, the correct transition from the two-dimensional problem to one-dimensional one is possible when applying the averaging procedure similar to Eq. (7.28). To illustrate this method application, we will use the equation system of gas motion in the cylindrical coordinate system x, r [9]:

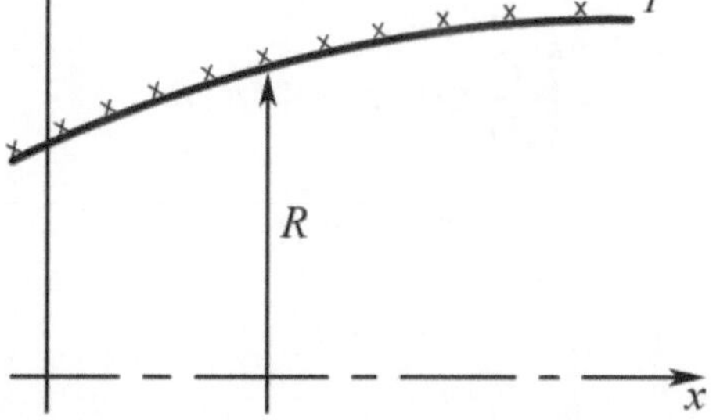

FIGURE 7.12 Coordinate system when calculating the gas motion in the cylindrical pipe.

$$\frac{\partial}{\partial t} r\rho + \frac{\partial}{\partial x} r\rho v_1 + \frac{\partial}{\partial r} r\rho v_2 = 0 \tag{7.52}$$

$$\frac{\partial}{\partial t} r\rho v_1 + \frac{\partial}{\partial x} r\rho v_1^2 + \frac{\partial}{\partial r} r\rho\, v_1 v_2 + r\frac{\partial p}{\partial x} = 0 \tag{7.53}$$

$$\frac{\partial}{\partial t} r\rho v_2 + \frac{\partial}{\partial x} r\rho\, v_1 v_2 + \frac{\partial}{\partial r} r\rho v_2^2 + r\frac{\partial p}{\partial r} = 0 \tag{7.54}$$

$$\frac{\partial}{\partial t} r\rho E + \frac{\partial}{\partial x} r\rho v_1 E + \frac{\partial}{\partial r} r\rho v_2 E + \frac{\partial}{\partial x} rpv_1 + \frac{\partial}{\partial r} rpv_2 = 0 \tag{7.55}$$

The values ρ, v_1, v_2, E will be averaged by the coordinate perpendicular to the channel axis (along the axis r). When averaging, the correlations true for the arbitrary function Φ [117, 244] are considered

$$\int_0^R \Phi(r)\,dr = \bar{\Phi} R$$

$$\int_0^R (\Phi(r) r)\,dr = \bar{\Phi}\frac{R^2}{2}$$

$$\int_0^R \frac{\partial}{\partial x}\Phi\,dr = \frac{\partial}{\partial x}\bar{\Phi} R$$

$$\int_0^R \frac{\partial}{\partial x}\Phi r\,dr = \frac{\partial}{\partial x}\left(\bar{\Phi}\frac{R^2}{2}\right)$$

$$\int_0^R \frac{\partial}{\partial r}\Phi\,dr = \Phi(R) - \Phi(0)$$

$$\int_0^R \frac{\partial}{\partial r}\Phi r\,dr = \Phi(R) R$$

$$\Pi = 2\pi R$$

$$F = \pi R^2 \tag{7.56}$$

Besides, the boundary conditions on the channel axis and fuel surface are taken into consideration

$$v_2\big|_{r=0} = 0$$

$$v_2\big|_{r=R} = \frac{\rho_m u_m}{\rho}$$

$$(\rho v_2 (E + p/\rho))\big|_{r=0} = 0$$

$$(\rho v_2 (E + p/\rho))\big|_{r=R} = -\rho_m u_m H_m \qquad (7.57)$$

When integrating the Eqs. (7.52)–(7.55) by the coordinate r on the interval from 0 to R considering the Eqs. (7.56), (7.57), it allows obtaining one-dimensional equations of gas (combustion products) motion in the following form

$$\frac{\partial}{\partial t}\bar{\rho}F + \frac{\partial}{\partial x}\bar{\rho}F\bar{v}_1 = \rho_m u_m \Pi$$

$$\frac{\partial}{\partial t}\bar{\rho}F\bar{v}_1 + \frac{\partial}{\partial x}\bar{\rho}F\bar{v}_1^{\,2} + F\frac{\partial \bar{p}}{\partial x} = 0$$

$$\frac{\partial}{\partial t}\bar{\rho}F\bar{E} + \frac{\partial}{\partial x}\bar{\rho}F\bar{v}_1(\bar{E} + \frac{\bar{p}}{\bar{\rho}}) = \rho_m u_m H_m \Pi$$

$$\bar{v}_2 \equiv 0 \qquad (7.58)$$

It should be pointed out that when recording the equations averaged by some coordinate, the averaging sign is omitted. It should be additionally noted that it is possible to refuse from performing the condition $\bar{v}_2 \equiv 0$ when averaging, given the transverse velocity profile $v_2 = v_2(x,r)$, for instance, based on the experimental results.

4. Let us consider the computational diagrams given in Figure 7.13. Such regions are found in rocket engines with front combustion charges (Figure 7.13a) and tubular charges non-symmetrically inserted into the engine chamber (Figure 7.13b). In the figures, the double hatching corresponds to the permeable boundaries (the combustion products inflow through these boundaries), the regular hatching – to the impermeable boundaries. The characteristic feature of the computational diagrams is the fact

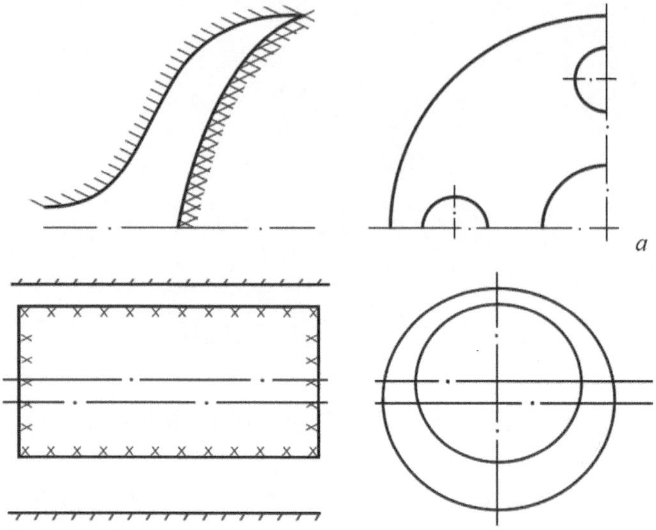

FIGURE 7.13 Computational diagrams applied when modeling the processes in rocket engines (a – front combustion charge; b – cylindrical charge non-symmetrically inserted into the chamber).

that the clearances between the permeable and impermeable boundaries are small and the gas motion equations can be integrated by the clearance width.

Let us derive the gas motion equation for the computational cases considered. The unknown equations will be obtained from the overall equation system (see Section 7.1) recorded for the coordinate system demonstrated in Figure 7.14 (here, the impermeable wall is designated with the index ***w***, index ***M*** – fuel, ***R*** – curvature radius). The equations are written down for the coordinate system ξ,η,θ ($q_1 \equiv \xi$, $q_2 \equiv \eta$, $q_3 \equiv \theta$) in the matrix form. We will designate the velocity vector components as u,v,w (correspond to the velocity vector components $v_1 \equiv u$, $v_2 \equiv v$, $v_3 \equiv w$ used in Section 7.1).

$$H_1H_2H_3\frac{\partial}{\partial t}\mathbf{A}+\frac{\partial}{\partial \xi}H_2H_3\,\mathbf{B}+\frac{\partial}{\partial \eta}H_1H_3\,\mathbf{C}+\frac{\partial}{\partial \theta}H_1H_2\,\mathbf{D}=\mathbf{F}\,; \quad (7.59)$$

$$\mathbf{A}=\begin{Bmatrix}\rho\\ \rho u\\ \rho v\\ \rho w\\ \rho E\end{Bmatrix}; \qquad \mathbf{B}=\begin{Bmatrix}\rho u\\ \rho u^2\\ \rho uv\\ \rho uw\\ \rho u(E+p/\rho)\end{Bmatrix};$$

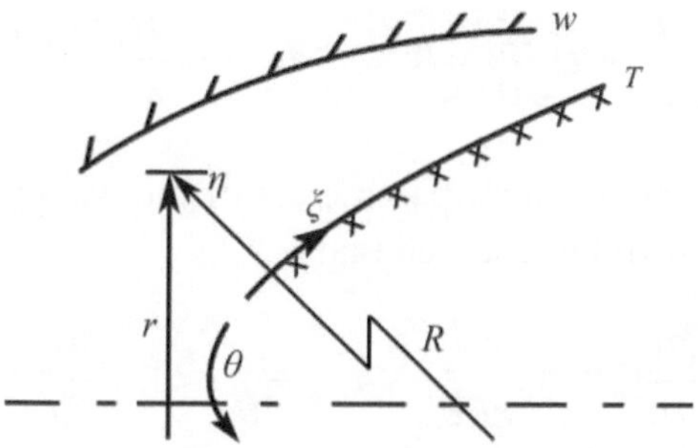

FIGURE 7.14 Computational diagram of modeling the gas flow in front cavity.

$$\mathbf{C}=\begin{Bmatrix}\rho v\\ \rho uv\\ \rho v^2\\ \rho vw\\ \rho v(E+p/\rho)\end{Bmatrix};\qquad \mathbf{D}=\begin{Bmatrix}\rho w\\ \rho wu\\ \rho wv\\ \rho w^2\\ \rho w(E+p/\rho)\end{Bmatrix};$$

$$\mathbf{F}=\begin{Bmatrix}H_1H_2H_3\varepsilon_\rho\\ H_1H_2H_3\varepsilon_u-H_2H_3\dfrac{\partial p}{\partial\xi}-\rho vH_3d_{12}-\rho wH_2d_{13}\\ H_1H_2H_3\varepsilon_v-H_1H_3\dfrac{\partial p}{\partial\eta}-\rho uH_3d_{21}-\rho wH_1d_{23}\\ H_1H_2H_3\varepsilon_w-H_1H_2\dfrac{\partial p}{\partial\theta}-\rho uH_2d_{31}-\rho vH_1d_{32}\\ H_1H_2H_3\varepsilon_E\end{Bmatrix};$$

Lame coefficients for the selected coordinate system equal $H_1=\dfrac{R+\eta}{R}$, $H_2=1$, $H_3=r$, respectively. In this case, the values of the coefficients $d_{12},d_{13},d_{21},d_{23},d_{31},d_{32}$ will be as follows:

$$d_{12}=u\frac{\partial H_1}{\partial\eta}-v\frac{\partial H_2}{\partial\xi}=\frac{u}{R};\qquad d_{13}=u\frac{\partial H_1}{\partial\theta}-w\frac{\partial H_3}{\partial\xi}=0;$$

$$d_{21}=v\frac{\partial H_2}{\partial\xi}-u\frac{\partial H_1}{\partial\eta}=-\frac{u}{R};\qquad d_{23}=v\frac{\partial H_2}{\partial\theta}-w\frac{\partial H_3}{\partial\eta}=0;$$

$$d_{31} = w\frac{\partial H_3}{\partial \xi} - u\frac{\partial H_1}{\partial \theta} = 0; \qquad d_{32} = w\frac{\partial H_3}{\partial \eta} - v\frac{\partial H_2}{\partial \theta} = 0$$

Let us average all gas-dynamic parameters of the gas by the coordinate η. When averaging, it will be assumed that

$$\frac{\partial \eta_w}{\partial \xi} << 1, \ \frac{\partial \eta_T}{\partial \xi} << 1, \ \frac{\partial \eta_w}{r\partial \theta} << 1, \ \frac{\partial \eta_T}{r\partial \theta} << 1; \ \frac{\partial \eta_w}{\partial t} = \frac{\partial \eta_T}{\partial t} \approx 0$$

Besides, the correlations true for the arbitrary function Φ will be used

$$\int_{\eta_T}^{\eta_w} \Phi(\eta) d\eta = \bar{\Phi}(\eta_w - \eta_T)$$

$$\int_{\eta_T}^{\eta_w} \Phi(\eta)\eta \, d\eta = (\bar{\Phi}\bar{\eta})(\eta_w - \eta_T) = \bar{\Phi}\frac{\eta_w^2 - \eta_T^2}{2}$$

$$\int_{\eta_T}^{\eta_w} \frac{\partial}{\partial S}\Phi d\eta = \frac{\partial}{\partial S}\bar{\Phi}(\eta_w - \eta_T), \ S \neq \eta$$

$$\int_{\eta_T}^{\eta_w} \frac{\partial}{\partial S}\Phi\eta \, d\eta = \frac{\partial}{\partial S}\left(\bar{\Phi}\frac{\eta_w^2 - \eta_T^2}{2}\right), \ S \neq \eta$$

$$\int_{\eta_T}^{\eta_w} \frac{\partial}{\partial \eta}\Phi d\eta = \Phi(\eta_w) - \Phi(\eta_T)$$

$$\int_{\eta_T}^{\eta_w} \frac{\partial}{\partial \eta}\Phi\eta \, d\eta = \Phi(\eta_w)\eta_w - \Phi(\eta_T)\eta_T \qquad (7.60)$$

For the affinity, we assume that combustion products flow into the computational area along both side boundaries $\eta = \eta_w$, $\eta = \eta_m$. Then the boundary conditions of the equations recorded will be as follows:

$$u\big|_{\eta=\eta_m} = 0, \quad w\big|_{\eta=\eta_m} = 0, \quad v\big|_{\eta=\eta_m} = \frac{\sum_{i=1}^{N} \rho_{mi} u_{mi}}{\rho}$$

$$u\big|_{\eta=\eta_w}=0,\quad w\big|_{\eta=\eta_w}=0,\quad v\big|_{\eta=\eta_w}=\frac{\sum_{j=1}^{M}\rho_{mj}u_{mj}}{\rho}$$

$$\left(\rho v(E+{}^{p}\!/_{\rho})\right)\Big|_{\eta=\eta_T}=\sum_{i=1}^{N}\rho_{mi}u_{mi}H_{mi}$$

$$\left(\rho v(E+{}^{p}\!/_{\rho})\right)\Big|_{\eta=\eta_w}=-\sum_{j=1}^{M}\rho_{mj}u_{mj}H_{mj} \tag{7.61}$$

Taking into account (7.60) and (7.61), the Eq. (7.59) is finally as follows:

$$a\frac{\partial \mathbf{k}_1}{\partial t}+\frac{\partial \mathbf{k}_2}{\partial \xi}+\frac{\partial \mathbf{k}_3}{\partial \theta}=\mathbf{k}_4 \tag{7.62}$$

Here: $a=\frac{r}{R}\left(R+\frac{\eta_m+\eta_w}{2}\right)$; $\mathbf{k}_1=\begin{Bmatrix}\overline{\rho}\\ \overline{\rho u}\\ \overline{\rho w}\\ \overline{\rho E}\end{Bmatrix}$; $\mathbf{k}_2=\begin{Bmatrix}r\overline{\rho u}\\ r\overline{\rho u^2}\\ r\overline{\rho uw}\\ r\overline{\rho E}\end{Bmatrix}$;

$$\mathbf{k}_3=\left(1+\frac{\eta_T+\eta_w}{2R}\right)\begin{Bmatrix}\overline{\rho}\,\overline{w}\\ \overline{\rho}\,\overline{uw}\\ \overline{\rho}\,\overline{w}^2\\ \overline{\rho wE}\end{Bmatrix};$$

$$\mathbf{k}_4=\begin{Bmatrix}a\varepsilon_\rho+a\left(\sum_{i=1}^{N}\rho_{mi}u_{mi}+\sum_{j=1}^{M}\rho_{mj}u_{mj}\right),\\ a\varepsilon_u-r\frac{\partial p}{\partial \xi},\\ a\varepsilon_w-a\frac{\partial p}{\partial \theta},\\ a\varepsilon_E+a\left(\sum_{i=1}^{N}\rho_{mi}u_{mi}H_{mi}+\sum_{j=1}^{M}\rho_{mj}u_{mj}H_{mj}\right).\end{Bmatrix}$$

Let us consider the possible simplifications of the equation systems (7.62). If the curvature influence in the fuel solid rocket engine given in Figure 7.13a is insignificant ($h<<R$; h – clearance value between the engine body and fuel charge, R – body radius), then after averaging the gas parameters by the clearance width, the equations of gas (combustion products) motion are written down as follows:

$$h\frac{\partial r\bar{\rho}}{\partial t}+\frac{\partial hr\overline{\rho v}}{\partial r}+\frac{1}{r}\frac{\partial hr\overline{\rho w}}{\partial \theta}=\rho_T u_T r+\varepsilon_\rho h$$

$$h\frac{\partial r\overline{\rho v}}{\partial t}+\frac{\partial hr\overline{\rho v}^2}{\partial r}+\frac{1}{r}\frac{\partial hr\overline{\rho vw}}{\partial \theta}+r\frac{\partial \overline{p}h}{\partial r}=\overline{\rho w}^2 h+\varepsilon_v h$$

$$h\frac{\partial r\overline{\rho w}}{\partial t}+\frac{\partial hr\overline{\rho vw}}{\partial r}+\frac{1}{r}\frac{\partial hr\overline{\rho w}^2}{\partial \theta}+h\frac{\partial \overline{p}}{\partial \theta}=-\overline{\rho vw}h+\varepsilon_w h$$

$$h\frac{\partial r\overline{\rho E}}{\partial t}+\frac{\partial hr\overline{\rho v}(\overline{E}+\overline{p}/\overline{\rho})}{\partial r}+\frac{1}{r}\frac{\partial hr\overline{\rho w}(\overline{E}+\overline{p}/\overline{\rho})}{\partial \theta}=\rho_T u_T H_T r+\varepsilon_E h$$

$$\overline{p}=\overline{\rho}(k-1)(\overline{E}-\frac{1}{2}(\overline{v}^2+\overline{w}^2))$$

In the recorded equations $h=h(r,\theta)$.

Similarly to the diagram given in Figure 7.13b, the equations of gas (combustion products) motion can be written down as follows:

$$\frac{\partial}{\partial t}\mathbf{A}+\frac{\partial}{\partial x}\mathbf{B}+\frac{1}{R_2}\frac{\partial}{\partial \theta}\mathbf{C}+\mathbf{D}=\mathbf{E} \tag{7.63}$$

$$R_1=r_w-r_T;\quad R_2=\frac{1}{2}(r_w+r_T);$$

$$\mathbf{A}=\begin{Bmatrix}\overline{\rho}R_1R_2,\\ \overline{\rho u}R_1R_2,\\ \overline{\rho w}R_1R_2,\\ \overline{\rho E}R_1R_2\end{Bmatrix};\qquad \mathbf{B}=\begin{Bmatrix}\overline{\rho u}R_1R_2,\\ \overline{\rho u^2}R_1R_2,\\ \overline{\rho uw}R_1R_2,\\ \overline{\rho u}(\overline{E}+\overline{p}/\overline{\rho})R_1R_2,\end{Bmatrix};$$

$$\mathbf{C}=\begin{Bmatrix}\bar{\rho}\bar{w}R_1R_2,\\ \bar{\rho}\,\overline{uw}R_1R_2,\\ \bar{\rho}\,\overline{w}^2R_1R_2,\\ \overline{\rho w}(\bar{E}+\bar{p}/\bar{\rho})R_1R_2,\end{Bmatrix};\qquad \mathbf{D}=\begin{Bmatrix}0\\ R_1R_2\dfrac{\partial}{\partial x}\bar{p}\\ R_1\dfrac{\partial}{\partial\theta}\bar{p}\\ 0\end{Bmatrix};$$

$$\mathbf{E}=\begin{Bmatrix}\varepsilon_\rho R_1+\sum_{i=1}^{N}\rho_{mi}u_{mi}r_0+\sum_{j=1}^{M}\rho_{mj}u_{mj}r_k\\ \varepsilon_u R_1\\ \varepsilon_w R_1\\ \varepsilon_E R_1+\sum_{i=1}^{N}\rho_{mi}u_{mi}H_{mi}r_0+\sum_{j=1}^{M}\rho_{mj}u_{mj}H_{mj}r_k.\end{Bmatrix}.$$

In practice, the case when the boundaries η_w, η_T move with time ($\eta_w=\eta_w(t), \eta_T=\eta_T(t)$) can be implemented. In this case, the partial derivatives of the arbitrary parameter Φ by time should be averaged based on the expression [117, 202]

$$\int_{\eta_T(t)}^{\eta_w(t)}\frac{\partial}{\partial t}\Phi(\eta,t)d\eta=\frac{\partial}{\partial t}\int_{\eta_T(t)}^{\eta_w(t)}\Phi(\eta,t)d\eta-\Phi(\eta_w,t)\frac{d\eta_w}{dt}+\Phi(\eta_T,t)\frac{d\eta_T}{dt}$$

If it is assumed that $\Phi(\eta_w,t)\approx\Phi(\eta_T,t)\approx\bar{\Phi}(t)$, the latter expression can be written down in the following way

$$\int_{\eta_T(t)}^{\eta_w(t)}\frac{\partial}{\partial t}\Phi(\eta,t)d\eta=\frac{\partial}{\partial t}\left[(\eta_w(t)-\eta_T(t))\bar{\Phi}(t)\right]-\bar{\Phi}(t)\frac{d(\eta_w(t)-\eta_T(t))}{dt}$$

$$=(\eta_w(t)-\eta_T(t))\frac{\partial}{\partial t}\bar{\Phi}(t)$$

The analysis demonstrates that the Eqs. (7.62) and (7.63) can be also applied in the situation when η_w, η_T are not functions of time. The accepted assumption that $\Phi(\eta_w,t)\approx\Phi(\eta_T,t)\approx\bar{\Phi}(t)$ does not decrease the problem affinity, moreover, its application is only necessary when averaging the partial derivative in the continuity and energy equations. In the momentum equations the values $\Phi(\eta_w,t) \approx \Phi(\eta_T,t) \approx 0$.

7.2.3 COORDINATE TRANSFORMATION

The possibility of recording the conservation laws in the arbitrary orthogonal coordinate system was considered before. Such problem can be given in a more general setting. Let us consider the equation which can be written down in the matrix form as follows:

$$\mathbf{A}\frac{\partial \mathbf{U}}{\partial t}+\mathbf{B}\frac{\partial \mathbf{U}}{\partial x}+\mathbf{C}\frac{\partial \mathbf{U}}{\partial y}+\mathbf{D}\frac{\partial \mathbf{U}}{\partial z}=\mathbf{f}(x,y,z,t) \tag{7.64}$$

It is designated in the Eq. (7.64):

$$\mathbf{U}=\begin{pmatrix} u_1 \\ u_2 \\ \cdots \\ u_n \end{pmatrix}; \ \mathbf{A}=\begin{pmatrix} a_{11} & a_{12} & \cdots & a_{1n} \\ a_{21} & a_{22} & \cdots & a_{2n} \\ \cdots & \cdots & \cdots & \cdots \\ a_{n1} & a_{n2} & \cdots & a_{nn} \end{pmatrix}; \ \mathbf{B}=\begin{pmatrix} b_{11} & b_{12} & \cdots & b_{1n} \\ b_{21} & b_{22} & \cdots & b_{2n} \\ \cdots & \cdots & \cdots & \cdots \\ b_{n1} & b_{n2} & \cdots & b_{nn} \end{pmatrix};$$

$$\mathbf{C}=\begin{pmatrix} c_{11} & c_{12} & \cdots & c_{1n} \\ c_{21} & c_{22} & \cdots & c_{2n} \\ \cdots & \cdots & \cdots & \cdots \\ c_{n1} & c_{n2} & \cdots & c_{nn} \end{pmatrix}; \ \mathbf{D}=\begin{pmatrix} d_{11} & d_{12} & \cdots & d_{1n} \\ d_{21} & d_{22} & \cdots & d_{2n} \\ \cdots & \cdots & \cdots & \cdots \\ d_{n1} & d_{n2} & \cdots & d_{nn} \end{pmatrix}; \ \mathbf{f}=\begin{pmatrix} f_1 \\ f_2 \\ \cdots \\ f_n \end{pmatrix}$$

The matrix equation written down is equivalent to the following equation system:

$$a_{11}\frac{\partial u_1}{\partial t}+a_{12}\frac{\partial u_2}{\partial t}+\ldots+a_{1n}\frac{\partial u_n}{\partial t}+b_{11}\frac{\partial u_1}{\partial x}+b_{12}\frac{\partial u_2}{\partial x}+\ldots+b_{1n}\frac{\partial u_n}{\partial x}$$

$$+c_{11}\frac{\partial u_1}{\partial y}+c_{12}\frac{\partial u_2}{\partial y}+\ldots+c_{1n}\frac{\partial u_n}{\partial y}+d_{11}\frac{\partial u_1}{\partial z}+d_{12}\frac{\partial u_2}{\partial z}+\ldots+d_{1n}\frac{\partial u_n}{\partial z}=f_1$$

$$a_{21}\frac{\partial u_1}{\partial t}+a_{22}\frac{\partial u_2}{\partial t}+\ldots+a_{2n}\frac{\partial u_n}{\partial t}+b_{21}\frac{\partial u_1}{\partial x}+b_{22}\frac{\partial u_2}{\partial x}+\ldots+b_{2n}\frac{\partial u_n}{\partial x}$$

$$+c_{21}\frac{\partial u_1}{\partial y}+c_{22}\frac{\partial u_2}{\partial y}+...+c_{2n}\frac{\partial u_n}{\partial y}+d_{21}\frac{\partial u_1}{\partial z}+d_{22}\frac{\partial u_2}{\partial z}+...+d_{2n}\frac{\partial u_n}{\partial z}=f_2$$

$$...$$

$$a_{n1}\frac{\partial u_1}{\partial t}+a_{n2}\frac{\partial u_2}{\partial t}+...+a_{nn}\frac{\partial u_n}{\partial t}+b_{n1}\frac{\partial u_1}{\partial x}+b_{n2}\frac{\partial u_2}{\partial x}+...+b_{nn}\frac{\partial u_n}{\partial x}$$

$$+c_{n1}\frac{\partial u_1}{\partial y}+c_{n2}\frac{\partial u_2}{\partial y}+...+c_{nn}\frac{\partial u_n}{\partial y}+d_{n1}\frac{\partial u_1}{\partial z}+d_{n2}\frac{\partial u_2}{\partial z}+...+d_{nn}\frac{\partial u_n}{\partial z}=f_n$$

Further, mainly the matrix form of recording equations will be used as a more compact one.

Let us mark out the surface described by the equation $\varphi(x,y,z,t)=0$ in the initial coordinate system and satisfying the condition that grad $\varphi\neq 0$. The following additional functions are introduced

$$\alpha=\alpha(x,y,z,t)$$

$$\beta=\beta(x,y,z,t)$$

$$\gamma=\gamma(x,y,z,t)$$

for all x, y, z, t satisfying the condition

$$\begin{vmatrix}\varphi_x & \varphi_y & \varphi_z & \varphi_t\\ \alpha_x & \alpha_y & \alpha_z & \alpha_t\\ \beta_x & \beta_y & \beta_z & \beta_t\\ \gamma_x & \gamma_y & \gamma_z & \gamma_t\end{vmatrix}\neq 0$$

Let us consider the functions $\varphi,\alpha,\beta,\gamma$ as the main coordinate system. The transition from the initial coordinate system (t,x,y,z) to the new one $(\varphi,\alpha,\beta,\gamma)$ can be written down in the matrix form [62]

$$(\frac{\partial\varphi}{\partial t}\mathbf{A}+\frac{\partial\varphi}{\partial x}\mathbf{B}+\frac{\partial\varphi}{\partial y}\mathbf{C}+\frac{\partial\varphi}{\partial z}\mathbf{D})\cdot\frac{\partial\mathbf{U}}{\partial\varphi}$$

$$+(\frac{\partial\alpha}{\partial t}\mathbf{A}+\frac{\partial\alpha}{\partial x}\mathbf{B}+\frac{\partial\alpha}{\partial y}\mathbf{C}+\frac{\partial\alpha}{\partial z}\mathbf{D})\cdot\frac{\partial\mathbf{U}}{\partial\alpha}$$

$$+(\frac{\partial \beta}{\partial t}\mathbf{A}+\frac{\partial \beta}{\partial x}\mathbf{B}+\frac{\partial \beta}{\partial y}\mathbf{C}+\frac{\partial \beta}{\partial z}\mathbf{D})\cdot\frac{\partial \mathbf{U}}{\partial \beta}$$

$$+(\frac{\partial \gamma}{\partial t}\mathbf{A}+\frac{\partial \gamma}{\partial x}\mathbf{B}+\frac{\partial \gamma}{\partial y}\mathbf{C}+\frac{\partial \gamma}{\partial z}\mathbf{D})\cdot\frac{\partial \mathbf{U}}{\partial \gamma}=\mathbf{f}(\alpha,\beta,\gamma,\varphi) \qquad (7.65)$$

The following examples will be considered to illustrate the aforementioned.

1. To write down the transfer equation in the coordinate system q, t

$$a_0\frac{\partial f}{\partial t}+a_1\frac{\partial f}{\partial q}=0 \qquad (7.66)$$

in the new coordinate system q_*, t_* connected with the initial coordinate system by the following correlations

$$t_*=t, \quad q_*=q-ut \qquad (7.67)$$

Here u – coefficient independent from q, t and f.

Let us find the partial derivatives $\frac{\partial q_*}{\partial q}, \frac{\partial t_*}{\partial q}, \frac{\partial q_*}{\partial t}, \frac{\partial t_*}{\partial t}$ in accordance with the Eqs. (7.67):

$$\frac{\partial t_*}{\partial t}=1, \ \frac{\partial t_*}{\partial q}=0, \ \frac{\partial q_*}{\partial t}=-u, \ \frac{\partial q_*}{\partial q}=1$$

According to the Eqs. (7.65), the Eq. (7.66) can be given as follows:

$$(a_0\frac{\partial t_*}{\partial t}+a_1\frac{\partial t_*}{\partial q})\frac{\partial f}{\partial t_*}+(a_0\frac{\partial q_*}{\partial t}+a_1\frac{\partial q_*}{\partial q})\frac{\partial f}{\partial q_*}=0$$

Substituting the values of partial equations in the latter equation, we obtain the desired transformation

$$a_0\frac{\partial f}{\partial t_*}+(a_1-ua_0)\frac{\partial f}{\partial q_*}=0 \qquad (7.68)$$

2 Let us complicate the previous problem. Let us consider the differential equation of the non-stationary process differing from the Eq. (7.66) by the availability of the second derivative along the coordinate q and source component Φ:

$$\frac{\partial\, a_0 f}{\partial\, t} = \frac{\partial a_1 f}{\partial q} + \frac{\partial}{\partial q} a_2 \frac{\partial\, a_3 f}{\partial\, q} + \Phi \tag{7.69}$$

Here a_0, a_1, a_2, a_3 – values independent from the function *f*, source Φ depends on time and spatial variable.

As in the previous example, we will fulfill the transformation from the initial independent variables *q*, *t* to the new ones q_*, t_* determined by the correlations (7.67). The expressions for the partial derivatives of the arbitrary function *f* are written down as follows:

$$\frac{\partial a_0\, f}{\partial\, t} = \frac{\partial\, a_0\, f}{\partial\, q_*} \cdot \frac{\partial q_*}{\partial\, t} + \frac{\partial\, a_0\, f}{\partial\, t_*} \cdot \frac{\partial\, t_*}{\partial\, t} \tag{7.70}$$

$$\frac{\partial a_1\, f}{\partial\, q} = \frac{\partial\, a_1 f}{\partial\, q_*} \cdot \frac{\partial q_*}{\partial\, q} + \frac{\partial a_1\, f}{\partial\, t_*} \cdot \frac{\partial\, t_*}{\partial\, q} \tag{7.71}$$

Taking into account the previously obtained values of the partial derivatives $\frac{\partial\, q_*}{\partial q}$, $\frac{\partial\, t_*}{\partial\, q}$, $\frac{\partial\, q_*}{\partial\, t}$, $\frac{\partial\, t_*}{\partial\, t}$, the Eqs. (7.70) and (7.71) are written down as follows:

$$\frac{\partial\, a_0 f}{\partial\, t} = \frac{\partial\, a_0 f}{\partial\, t_*} - u \frac{\partial a_0\, f}{\partial\, x}; \quad \frac{\partial\, a_1 f}{\partial q} = \frac{\partial a_1\, f}{\partial q_*}$$

Similarly, the expression for the second derivative from the function *f* can be obtained along the spatial coordinate

$$\frac{\partial}{\partial q} a_2 \frac{\partial\, a_3 f}{\partial\, q} = \frac{\partial}{\partial q_*} a_2 \frac{\partial\, a_3 f}{\partial\, q_*}$$

The Eq. (7.69) will eventually look as follows in the new coordinate system

$$\frac{\partial\, a_0 f}{\partial\, t} = \frac{\partial a_1 f}{\partial q} + u \frac{\partial a_0 f}{\partial q} + \frac{\partial}{\partial q} a_2 \frac{\partial\, a_3 f}{\partial\, q} + \Phi \tag{7.72}$$

By its structure, the Eq. (7.72) is very close to the heat-conductivity equation recorded for the non-stationary case in the movable one-dimensional coordinate system x, t linked with the fuel solid surface moving by the normal along the axis x (inside the fuel) with the velocity u_m. In the fixed coordinate system (for instance, if $u_m = 0$) the heat-conductivity equation can be represented by the following formula

$$\frac{\partial T}{\partial t} = \frac{1}{c\rho}\frac{\partial}{\partial x}\lambda\frac{\partial T}{\partial x} + \Phi \tag{7.73}$$

In accordance with the aforementioned, the heat-conductivity equation can be written down in the movable coordinate system linked with the fuel surface and moving with the velocity $u_m \neq 0$.

$$\frac{\partial T}{\partial t} = u_m\frac{\partial T}{\partial x} + \frac{1}{c\rho}\frac{\partial}{\partial x}\lambda\frac{\partial T}{\partial x} + \Phi \tag{7.74}$$

3. Let us perform the transformation of the differential equation in partial derivatives of the type

$$\frac{\partial\, a_0 f}{\partial t} = \frac{\partial a_1 f}{\partial x} + \frac{\partial\, a_2 f}{\partial r} + \frac{\partial}{\partial x}a_3\frac{\partial\, a_4 f}{\partial x} + \frac{\partial}{\partial r}a_5\frac{\partial\, a_6 f}{\partial r} + \Phi \tag{7.75}$$

recorded in the axisymmetric coordinate system (t – process time, x, r – longitudinal and transverse coordinates of the computational area) with the transition to the new coordinate system

$$t_* = t,\; x_* = x,\; \xi = \frac{r - r_0(x)}{R(x) - r_0(x)} \tag{7.76}$$

The transformation of the type (7.76) is interesting for the computational area demonstrated in Figure 7.15. This area is peculiar because its upper and lower boundaries are curvilinear.

The transformation in question makes the computational area rectangular, in which the coordinate grid changes from 0 to 1 in the transverse direction. The expressions for partial derivatives during the coordinate transformation look as follows:

$$\frac{\partial a_0 f}{\partial t} = \frac{\partial a_0 f}{\partial t_*}\cdot\frac{\partial t_*}{\partial t} + \frac{\partial a_0 f}{\partial x_*}\cdot\frac{\partial x_*}{\partial t} + \frac{\partial a_0 f}{\partial \xi}\cdot\frac{\partial \xi}{\partial t}$$

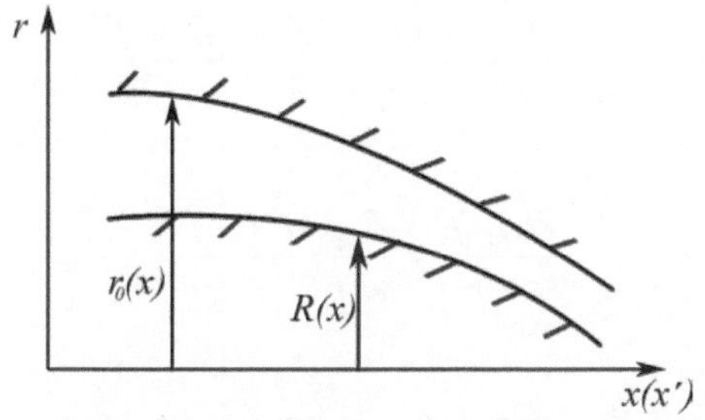

FIGURE 7.15 Computational diagram when modeling the gas flow in the circle region.

$$\frac{\partial a_1 f}{\partial x} = \frac{\partial a_1 f}{\partial t_*} \cdot \frac{\partial t_*}{\partial x} + \frac{\partial a_1 f}{\partial x_*} \cdot \frac{\partial x_*}{\partial x} + \frac{\partial a_1 f}{\partial \xi} \cdot \frac{\partial \xi}{\partial x}$$

$$\frac{\partial a_2 f}{\partial r} = \frac{\partial a_2 f}{\partial t_*} \cdot \frac{\partial t_*}{\partial r} + \frac{\partial a_2 f}{\partial x_*} \cdot \frac{\partial x_*}{\partial r} + \frac{\partial a_2 f}{\partial \xi} \cdot \frac{\partial \xi}{\partial r}$$

The second partial derivatives are found after simplifying the recorded expressions. Let us find the partial equations $\frac{\partial t_*}{\partial t}, \frac{\partial x_*}{\partial t}, \frac{\partial \xi}{\partial t}, \frac{\partial t_*}{\partial x}, \frac{\partial x_*}{\partial x}, \frac{\partial \xi}{\partial x}, \frac{\partial t_*}{\partial r}, \frac{\partial x_*}{\partial r}, \frac{\partial \xi}{\partial r}$ from the Eqs. (7.76):

$$\frac{\partial t_*}{\partial t} = 1, \ \frac{\partial x_*}{\partial t} = 0, \ \frac{\partial \xi}{\partial t} = 0$$

$$\frac{\partial t_*}{\partial x} = 0, \ \frac{\partial x_*}{\partial x} = 1, \ \frac{\partial \xi}{\partial x} = -\frac{\xi\left(\frac{dR(x)}{dx} - \frac{dr_0(x)}{dx}\right) + \frac{dr_0(x)}{dx}}{R(x) - r_0(x)}$$

$$\frac{\partial t_*}{\partial r} = 0, \ \frac{\partial x_*}{\partial r} = 0, \ \frac{\partial \xi}{\partial r} = \frac{1}{R(x) - r_0(x)}$$

After the simplification, the partial derivatives of the initial equation can be written down as follows:

$$\frac{\partial a_0 f}{\partial t} = \frac{\partial a_0 f}{\partial t_*}$$

$$\frac{\partial a_1 f}{\partial x} = \frac{\partial a_1 f}{\partial x_*} + \frac{\partial \xi}{\partial x} \cdot \frac{\partial a_1 f}{\partial \xi}$$

$$\frac{\partial a_2 f}{\partial r} = \frac{\partial \xi}{\partial r} \cdot \frac{\partial a_2 f}{\partial \xi}$$

Taking into account the previous expressions, the second derivatives $\frac{\partial}{\partial x} a_3 \frac{\partial a_4 f}{\partial x}$ and $\frac{\partial}{\partial r} a_5 \frac{\partial a_6 f}{\partial r}$ can be written down as follows:

$$\frac{\partial}{\partial x} a_3 \frac{\partial a_4 f}{\partial x} = \frac{\partial}{\partial x} a_3 (\frac{\partial a_4 f}{\partial x_*} + \frac{\partial \xi}{\partial x} \cdot \frac{\partial a_4 f}{\partial \xi}) = \frac{\partial}{\partial x_*} a_3 \frac{\partial a_4 f}{\partial x_*}$$
$$+ \frac{\partial \xi}{\partial x} \cdot \frac{\partial}{\partial \xi} a_3 \frac{\partial a_4 f}{\partial x_*} + \frac{\partial \xi}{\partial x} \cdot \frac{\partial}{\partial x_*} a_3 \frac{\partial a_4 f}{\partial \xi} + (\frac{\partial \xi}{\partial x})^2 \cdot \frac{\partial}{\partial \xi} a_3 \frac{\partial a_4 f}{\partial \xi}$$

$$\frac{\partial}{\partial r} a_5 \frac{\partial a_6 f}{\partial r} = \frac{\partial \xi}{\partial r} \cdot \frac{\partial}{\partial \xi} (a_5 \frac{\partial \xi}{\partial r} \cdot \frac{\partial a_6 f}{\partial \xi}) = (\frac{\partial \xi}{\partial r})^2 \cdot \frac{\partial}{\partial \xi} a_5 \frac{\partial a_6 f}{\partial \xi}$$

Eventually, the initial Eq. (7.75) in the new coordinate system (7.76) looks as follows:

$$\frac{\partial a_0 f}{\partial t_*} = \frac{\partial a_1 f}{\partial x_*} + \frac{\partial \xi}{\partial x} \cdot \frac{\partial a_1 f}{\partial \xi} + \frac{\partial \xi}{\partial r} \cdot \frac{\partial a_2 f}{\partial \xi} + \frac{\partial}{\partial x_*} a_3 \frac{\partial a_4 f}{\partial x_*}$$
$$+ \frac{\partial \xi}{\partial x} \cdot \frac{\partial}{\partial \xi} a_3 \frac{\partial a_4 f}{\partial x_*} + \frac{\partial \xi}{\partial x} \cdot \frac{\partial}{\partial x_*} a_3 \frac{\partial a_4 f}{\partial \xi} + (\frac{\partial \xi}{\partial x})^2 \cdot \frac{\partial}{\partial \xi} a_3 \frac{\partial a_4 f}{\partial \xi}$$
$$+ (\frac{\partial \xi}{\partial r})^2 \cdot \frac{\partial}{\partial \xi} a_5 \frac{\partial a_6 f}{\partial \xi} + \Phi \quad (7.77)$$

If the dependencies determining the change in the upper $R(x)$ and lower $r_0(x)$ boundaries of the computational area (Figure 7.15) $R(x)$, $r_0(x)$ satisfy the conditions $\frac{dR}{dx} \approx 0$, $\frac{dr_0}{dx} \approx 0$, the last equation is considerably simplified and can be transformed as follows:

$$\frac{\partial\, a_0 f}{\partial\, t_*} = \frac{\partial a_1 f}{\partial x_*} + \frac{\partial\, a_1 f}{\partial \xi} + \frac{\partial \xi}{\partial r} \cdot \frac{\partial\, a_2 f}{\partial \xi} + \frac{\partial}{\partial x_*} a_3 \frac{\partial a_4 f}{\partial x_*}$$

$$+ (\frac{\partial \xi}{\partial r})^2 \cdot \frac{\partial}{\partial \xi} a_5 \frac{\partial a_6 f}{\partial \xi} + \Phi \tag{7.78}$$

Let us come back to analyzing the Eq. (7.65). It can be assumed that there is such transition from the initial coordinate system to the new one when the determinant of any matrix composed of the coefficients at partial derivatives from the function **U** equals zero:

$$\det\left(\frac{\partial \varphi}{\partial t} \mathbf{A} + \frac{\partial \varphi}{\partial x} \mathbf{B} + \frac{\partial \varphi}{\partial y} \mathbf{C} + \frac{\partial \varphi}{\partial z} \mathbf{D} \right) = 0$$

or

$$\det\left(\frac{\partial \alpha}{\partial t} \mathbf{A} + \frac{\partial \alpha}{\partial x} \mathbf{B} + \frac{\partial \alpha}{\partial y} \mathbf{C} + \frac{\partial \alpha}{\partial z} \mathbf{D} \right) = 0$$

or

$$\det\left(\frac{\partial \beta}{\partial t} \mathbf{A} + \frac{\partial \beta}{\partial x} \mathbf{B} + \frac{\partial \beta}{\partial y} \mathbf{C} + \frac{\partial \beta}{\partial z} \mathbf{D} \right) = 0$$

or

$$\det\left(\frac{\partial \gamma}{\partial t} \mathbf{A} + \frac{\partial \gamma}{\partial x} \mathbf{B} + \frac{\partial \gamma}{\partial y} \mathbf{C} + \frac{\partial \gamma}{\partial z} \mathbf{D} \right) = 0 \tag{7.79}$$

The surfaces (lines), along which the aforementioned determinants equal zero, are called characteristic [62]. It is easy to see that the solution of the initial equation system in partial derivatives along the characteristic surface (line) is simpler than the solution of this system in the initial coordinate system. If at least one of the recorded determinants equals zero, the dimensionality of the equation being solved decreases by one. The foregoing method of coordinate transformation with the determination of characteristic directions is widely used in mathematical physics and calculus mathematics. It is the basis of the powerful computational method called "the method of characteristics" [184]. It is demonstrated in mathematical physics that for the equations in partial derivatives of hyperbolic or parabolic types there are characteristic directions (rules of determining the type of differential equations in partial derivatives are given, for example, in [62, 184]) and they can be determined analytically or applying numerical methods. For instance, the equations of sound wave propagation are as follows:

$$\frac{\partial u}{\partial t}+\frac{1}{\rho}\frac{\partial p}{\partial x}=0$$

$$\frac{\partial p}{\partial t}+\rho c^2\frac{\partial u}{\partial x}=0 \qquad (7.80)$$

Here: u – gas velocity, p – pressure, ρ – density, c – sound speed. The equation system (7.80) corresponds to the system (7.65), if we accept

$$\mathbf{U}=\begin{pmatrix}u\\p\end{pmatrix};\quad \mathbf{A}=\begin{pmatrix}1&0\\0&1\end{pmatrix};\quad \mathbf{B}=\begin{pmatrix}0&\frac{1}{\rho}\\\rho c^2&0\end{pmatrix};\quad \mathbf{f}=\begin{pmatrix}0\\0\end{pmatrix}$$

Let us consider the possibility of transiting to the new coordinate system linked with the initial one (x, t) by the linear correlations:

$$\varphi=ax+bt,\qquad \alpha=ax-bt \qquad (7.81)$$

In this case, we have

$$\frac{\partial\varphi}{\partial t}=b,\quad \frac{\partial\varphi}{\partial x}=a,\quad \frac{\partial\alpha}{\partial t}=-b,\quad \frac{\partial\alpha}{\partial x}=a$$

We can accept $a=1$ without violating the affinity. In this case, the further computations are insignificantly simplified.

Taking into account (7.81) and accepted value $a=1$, the first and second equations of the system (7.79) are as follows:

- first equation

$$b\begin{vmatrix}1&0\\0&1\end{vmatrix}+1\cdot\begin{vmatrix}0&\frac{1}{\rho}\\\rho c^2&0\end{vmatrix}=\begin{vmatrix}b&\frac{1}{\rho}\\\rho c^2&b\end{vmatrix}=b^2-c^2=0$$

- second equation

$$-b\begin{vmatrix}1 & 0\\ 0 & 1\end{vmatrix}+1\cdot\begin{vmatrix}0 & \dfrac{1}{\rho}\\ \rho c^2 & 0\end{vmatrix}=\begin{vmatrix}-b & \dfrac{1}{\rho}\\ \rho c^2 & -b\end{vmatrix}=b^2-c^2=0$$

Both determinants (7.79) take zero values if the condition $b=\pm c$ is fulfilled and the transformation (7.81) is as follows:

$$\varphi = x+ct\,, \qquad \alpha = x-ct$$

The first of these equations determines the first characteristic direction, and the second one – the direction of the second family characteristics.

The Eq. (7.65) is true along the characteristics of the first family ($\varphi = x+ct$) written down as follows:

$$\left(-c\begin{pmatrix}1 & 0\\ 0 & 1\end{pmatrix}+1\cdot\begin{pmatrix}0 & \dfrac{1}{\rho}\\ \rho c^2 & 0\end{pmatrix}\right)\frac{\partial \mathbf{U}}{\partial \alpha}=0$$

or

$$-c\frac{\partial u}{\partial \alpha}+\frac{1}{\rho}\frac{\partial p}{\partial \alpha}=0$$

or

$$\frac{\partial (u-\dfrac{p}{\rho c})}{\partial \alpha}=0$$

The Eq. (7.65) is true along the characteristics of the second family ($\alpha = x-ct$) written down as follows:

$$\left(c\begin{pmatrix}1 & 0\\ 0 & 1\end{pmatrix}+1\cdot\begin{pmatrix}0 & \dfrac{1}{\rho}\\ \rho c^2 & 0\end{pmatrix}\right)\frac{\partial \mathbf{U}}{\partial \varphi}=0$$

or

$$c\frac{\partial u}{\partial \varphi}+\frac{1}{\rho}\frac{\partial p}{\partial \varphi}=0$$

or

$$\frac{\partial (u+\frac{p}{\rho c})}{\partial \varphi}=0$$

It is demonstrated in mathematical physics [62, 184] that the existence of actual (or real-valued) characteristics in partial derivatives is defined by the class of these equations. Thus, there are always the actual characteristics in hyperbolic and parabolic equation systems, but they are not available, if the equation systems refer to the elliptic class. In particular, it is proved in mathematical physics that the equations describing the stationary modes of sound and supersonic flows of the ideal gas have additional characteristics. Actually, these characteristics can be established when solving the non-stationary problems of gas dynamics, etc.

7.2.4 APPLICATION OF THE THEOREMS OF MATHEMATICAL PHYSICS AND FIELD THEORY

1. Gauss theorem establishes the integral bond along the surface with the integral by volume.

Let us designate: W – volume limited by the surface S; $\boldsymbol{n}$ – external normal reduced to the surface S; $\mathbf{A}(t)$– vector existing inside the volume W and on its surface: $A_1(t,q_1,q_2,q_3), A_2(t,q_1,q_2,q_3), A_3(t,q_1,q_2,q_3)$ – vector projections $\mathbf{A}(t)$ on the coordinate axis q_1,q_2,q_3 of the arbitrary orthogonal coordinate system. In this case, the following correlation is true

$$\iint_S (\mathbf{A}\cdot\mathbf{n})dS=\iiint_W \frac{1}{H_1H_2H_3}\left(\frac{\partial A_1H_2H_3}{\partial q_1}+\frac{\partial A_2H_1H_3}{\partial q_2}+\frac{\partial A_3H_1H_2}{\partial q_3}\right)dW$$

or

$$\iint_S A_n dS=\iiint_W \operatorname{div}\mathbf{A}\, dW$$

Here A_n – projection of the vector $\mathbf{A}$ on the external normal, and div$\mathbf{A}$ – divergence of the vector $\mathbf{A}$.

The application of Gauss theorem for the problems of continuum mechanics was discussed before in Section 7.1.

2. Assumption on the potential existence

The field is considered potential for the vector $\mathbf{A}(x,y,z,t)$, if there is such function Φ (x, y, z, t) that $\mathbf{A}=\text{grad}\,\Phi(x, y, z, t)$. As an example, we can point out that the velocity field can have the potential in gas dynamics and hydrodynamics. In this case, the latter condition can be written down as follows:

$$d\Phi = \frac{\partial \Phi}{\partial x}dx + \frac{\partial \Phi}{\partial y}dy + \frac{\partial \Phi}{\partial z}dz = udx + vdy + wdz$$

In mathematical physics it is proved that the potential Φ is available if the following correlations are fulfilled

$$\frac{\partial u}{\partial y} = \frac{\partial v}{\partial x}; \quad \frac{\partial u}{\partial z} = \frac{\partial w}{\partial x}; \quad \frac{\partial v}{\partial z} = \frac{\partial w}{\partial y}$$

The assumption that the developing process has the potential allows shortening the number of equations being solved at the same time. Actually, in this case, when solving the problems of hydro- and gas-dynamics, only one equation for the potential can be solved instead of three momentum equations (or two – in case of two-dimensional flow).

The main values determining the medium thermodynamic state are: the pressure of gas (liquid, system) p, density ρ, temperature T, entropy S (entropy of the thermodynamic system is determined by the following expression $S = c_p \ln T\big/ p^{\frac{k-1}{k}} + \text{const}$).

The following potentials are established in thermodynamics:

- internal energy $U = U(S,\rho)$ determined by the expression

$$dU = \left(\frac{\partial U}{\partial S}\right)_\rho dS + \left(\frac{\partial U}{\partial \rho}\right)_S d\rho = TdS - p\,d\frac{1}{\rho};$$

- free energy (state function) $F = F(\rho,T)$ (by definition, $F = U - TS$) in the differential form is determined by the following dependence

$$dF = \frac{p}{\rho^2}d\rho - SdT$$

- enthalpy (heat storage) *i(p, S)* (by definition, the enthalpy $i = U + p/\rho$) whose differential equation is as follows:

$$di = \frac{1}{\rho} dp + TdS$$

- thermodynamic Gibbs potential $\Psi = \Psi(p,T)$ (by definition, Gibbs potential – $\Psi = U - TS - p/\rho = F - p/\rho$) determined in the differential form as follows:

$$d\Psi = \frac{1}{\rho} dp - SdT$$

The introduction of potentials in thermodynamics, as well as in other sections of physics, allows shortening the number of simultaneously solved equations (the number of the equations being solved shortens by one).

3. Application of the conditions on the existence of current lines (surfaces)

By the definition known from mathematical physics [62, 201], the line $f(x, y, z) = const$ is the current line if it satisfies the following condition

$$(\text{grad}\, f \cdot \mathbf{v}) = 0$$

This condition can be written down in differential form

$$u\frac{\partial f}{\partial x} + v\frac{\partial f}{\partial y} + w\frac{\partial f}{\partial z} = 0$$

It is demonstrated in continuum mechanics that the current lines exist only in the liquid and gas flow if the following conditions are fulfilled

$$\frac{dx}{u} = \frac{dy}{v} = \frac{dz}{w} = d\lambda$$

Here: $d\lambda$ – scalar parameter.

The equation for current lines can be applied as additional when analyzing the gas or fluid structure.

4. Application of the condition on the existence of vortex lines

By the definition known from mathematical physics [62, 201], the line $f_\omega(x, y, z) = \text{const}$ is a vortex line if it satisfies the following condition

$$(\text{grad}\, f_\omega \cdot \omega) = 0$$

This condition can be written down in differential form

$$\omega_x \frac{\partial f_\omega}{\partial x} + \omega_y \frac{\partial f_\omega}{\partial y} + \omega_z \frac{\partial f_\omega}{\partial z} = 0$$

The following designations are accepted here:

$$\omega_x = \frac{1}{2}(\frac{\partial w}{\partial y} - \frac{\partial v}{\partial z}); \quad \omega_y = \frac{1}{2}(\frac{\partial u}{\partial z} - \frac{\partial w}{\partial x}); \quad \omega_z = \frac{1}{2}(\frac{\partial v}{\partial x} - \frac{\partial u}{\partial y})$$

The equations of the current vortex lines can be applied when analyzing the gas or fluid flow structure.

5. Stokes theorem

The theorem establishes the integral bond along the closed contour L with the integral by the surface S pulled on the closed contour:

$$\oint v_L dL = 2\iint_S \omega_n dS$$

Here v_L – projection of the velocity vector on the contour L, ω_n – projection of the turbulence vector ωon the normal to the surface S.

Stokes theorem is applied in continuum mechanics in the regions, in the vicinity of which the parameter gaps can be available (e.g., in the vicinity of compression shocks in the gas supersonic flow).

The use of the aforementioned simplifications by introducing the turbulence functions of current lines is widely applied in practice. It should be pointed out that in the period of poor development of computer engineering (application of the computers of the second and third generations) the problem on the incompressible gas flow was actively solved in the problems of gas dynamics recorded in $\omega - \psi$ – derivatives [90, 170]

$$\frac{\partial \omega}{\partial t}+\frac{\partial u\omega}{\partial x}+\frac{\partial v\omega}{\partial y}-\frac{1}{\mathrm{Re}}\left(\frac{\partial^2 \omega}{\partial x^2}+\frac{\partial^2 \omega}{\partial y^2}\right)=0$$

$$\frac{\partial^2 \psi}{\partial x^2}+\frac{\partial^2 \psi}{\partial y^2}=\omega$$

$$\omega=\frac{\partial u}{\partial y}-\frac{\partial v}{\partial x}$$

$$u=\frac{\partial \psi}{\partial y}$$

$$v=-\frac{\partial \psi}{\partial x} \tag{7.82}$$

The practical shortcomings of solving the equations of gas dynamics in the foregoing form are, for instance, the impossibility of calculating the pressure from the equation systems (7.82), impossibility of solving the problem in spatial-3D setting, etc.

The above methods do not exhaust all the possibilities of simplifying the fundamental mathematical models. In particular, the original methods and approaches applied in physics and engineering can be found, for instance, in [15, 16, 77, 131–133].

KEYWORDS

- **bipolar coordinates**
- **Cartesian coordinate system**
- **conservation laws**
- **cylindrical coordinate system**
- **lame coefficients**
- **spherical coordinate system**

CHAPTER 8

METHODS OF COMPUTATIONAL ANALYSIS OF MODELS WITH DIFFERENTIAL EQUATIONS IN PARTIAL DERIVATIVES

CONTENTS

8.1 CLASSIFICATION OF THE EQUATIONS OF MATHEMATICAL PHYSICS

Differential equations in partial derivatives are one of the most difficult to solve in the problems of calculus mathematics. Nevertheless, such problems have a large share in engineering applications. The problems of continuum mechanics (hydro- and gas-dynamics, mechanics of deformable bodies), thermodynamics and heat transfer come down to differential equations in partial derivatives. The artificially developed mathematical models (phenomenological models, models of mathematical programming problems, etc.) can be demonstrated in partial derivatives.

Equations and equation systems in partial derivatives are much diversified, and their properties can eminently change even with seemingly insignificant change in recording the equations. The most principal properties of the equations in partial derivatives are established by the type of these equations.

Let us consider the equation in which the function *f* depends on two arguments – x_1, x_2

$$a_{11}\frac{\partial^2 f}{\partial x_1^2} + a_{12}\frac{\partial^2 f}{\partial x_1 \partial x_2} + a_{22}\frac{\partial^2 f}{\partial x_2^2} + b_1\frac{\partial f}{\partial x_1} + b_2\frac{\partial f}{\partial x_2} + cf = F(x_1, x_2) \quad (8.1)$$

In the Eq. (8.1) the coefficients $a_{11}, a_{12}, a_{22}, b_1, b_2, c$ can take arbitrary values, and $F(x_1, x_2)$ – arbitrary algebraic function of the arguments x_1, x_2. In special case, the value of this function can equal zero at any values of x_1, x_2 ($F(x_1, x_2) = 0$). In accordance with the theory of equations of mathematical physics (for instance, [62, 117]), the type of the Eq. (8.1) can be determined by the value of the parameter $D = a_{12}^2 - 4a_{11}a_{22}$. If the value D is positive ($D > 0$), the Eq. (8.1) refers to hyperbolic type. The equation refers to parabolic type, if $D = 0$. At $D < 0$ the Eq. (8.1) refers to elliptic type. By their properties, more complex equations in partial derivatives (over two arguments) can refer to the foregoing three group types.

Let us consider the special cases of equations in partial derivatives used in continuum mechanics.

8.1.1 TRANSFER EQUATION

$$\frac{\partial f}{\partial t} + a_1\frac{\partial f}{\partial x} + a_2\frac{\partial f}{\partial y} + a_3\frac{\partial f}{\partial z} = 0 \quad (8.2)$$

The transfer equation refers to hyperbolic type. Such equations can be obtained, for example, in the problems of gas dynamics. In this case, the coefficients a_1, a_2, a_3 – velocities of substance transfer (mass, momentum, energy) along the coordinates x, y, z.

8.1.2 WAVE EQUATION

$$\frac{1}{c^2}\frac{\partial^2 f}{\partial t^2}=\frac{\partial^2 f}{\partial x^2}+\frac{\partial^2 f}{\partial y^2}+\frac{\partial^2 f}{\partial z^2} \tag{8.3}$$

The wave equation refers to hyperbolic type. The equation can be used when solving the problems of acoustics to investigate the processes of sound wave propagation. In the problems of acoustics c – sound velocity.

8.1.3 DIFFUSION EQUATION

$$\frac{\partial f}{\partial t}+a_1\frac{\partial^2 f}{\partial x^2}+a_2\frac{\partial^2 f}{\partial y^2}+a_3\frac{\partial^2 f}{\partial z^2}=0 \tag{8.4}$$

The diffusion equation refers to parabolic type. The equation is found in the problems on gas mixture motion. The equation of heat diffusion, called the heat conductivity equation, is of the same type.

8.1.4 LAPLACE EQUATION AND POISSON EQUATION

$$\frac{\partial^2 f}{\partial x^2}+\frac{\partial^2 f}{\partial y^2}+\frac{\partial^2 f}{\partial z^2}=0$$

$$\frac{\partial^2 f}{\partial x^2}+\frac{\partial^2 f}{\partial y^2}+\frac{\partial^2 f}{\partial z^2}=F(x,y) \tag{8.5}$$

Laplace and Poisson equations refer to elliptic type. The peculiarities of this type of equations allow constructing the problem solution at any point inside the computational area by the function f values set on its boundaries.

8.1.5 BIHARMONIC EQUATION

$$\frac{\partial^4 f}{\partial x^4}+2\frac{\partial^4 f}{\partial x^2\partial y^2}+\frac{\partial^4 f}{\partial y^4}=F(x,y) \tag{8.6}$$

Biharmonic equation refers to elliptic type of equations. This equation is found when solving the problems on the mechanics of deformable body.

In particular, a similar equation is used to determine the plate deformation under the force action $F(x,y)$.

8.1.6 EULER EQUATION FOR THE PROBLEM OF GAS DYNAMICS

$$\frac{\partial \rho}{\partial t}+\frac{\partial \rho u}{\partial x}=\varepsilon_{\rho}$$

$$\frac{\partial \rho u}{\partial t}+\frac{\partial \rho(u^2+{p}/{\rho})}{\partial x}=\varepsilon_{u}$$

$$\frac{\partial \rho E}{\partial t}+\frac{\partial \rho u(E+{p}/{\rho})}{\partial x}=\varepsilon_{E}$$

$$p=\rho(k-1)(E-\frac{u^2}{2}) \tag{8.7}$$

The aforementioned equation system of the non-stationary flow of an ideal gas in one-dimensional spatial setting refers to hyperbolic type. It can be demonstrated that the system of these equations can come down to transfer equations [Eqs. (8.2)].

Equations and equation systems in partial derivatives can be solved only after setting up initial and boundary conditions. When setting up the initial conditions, the distribution of the function $f(t,x,y,z)$ at the time point $t=t_0$ corresponding to the beginning of the problem integration is considered as known. In special case, it can be accepted that $t_0=0$. The setting up of boundary (or edge) conditions assumes the setting up of the values of the function $f(t,x,y,z)$ on the boundaries of the computational area for any time point t. It should be noted that even at setting up initial and boundary conditions the systems of differential equations in partial derivatives do not always have a solution or they have several solutions. Generally speaking, the setting up of boundary conditions for the equations related to various types (hyperbolic, parabolic or elliptic) differs significantly [62, 184].

In some situations, equations in partial derivatives have analytical solutions. Let us take the problem on the non-stationary heating of the material considered in one-dimensional setting as an example (semi-infinite material, $0 \le x < \infty$):

- *heat conductivity equation*

$$\frac{\partial T}{\partial t} = a^2 \frac{\partial^2 T}{\partial x^2}; \tag{8.8}$$

- *initial conditions (correspond to the time moment* $t = 0$*)*

$$T = f(x);$$

- *boundary conditions (correspond to the coordinate* $x = 0$*)*

$T = g(t)$ corresponds to the coordinate $x = 0$;

$$\frac{\partial T}{\partial x} = 0, \quad \text{at } x = 0 \to \infty$$

The general solution of the heat conductivity equation at the set up initial and boundary conditions can be written down as follows [62, 124, 207]

$$T = \frac{1}{2a\sqrt{\pi t}} \cdot \int_0^{\infty} \left(\exp(-\frac{(x-\xi)^2}{4a^2 t}) - \exp(-\frac{(x+\xi)^2}{4a^2 t}) \right) \cdot f(\xi) \cdot d\xi$$

$$+ \frac{x}{2a\sqrt{\pi}} \cdot \int_0^{t} \exp(-\frac{x^2}{4a^2 (t-\tau)}) \cdot \frac{g(\tau)}{(t-\tau)^{1,5}} \cdot d\tau \tag{8.9}$$

For practical aims, the solution of the heat conductivity equation in the recorded analytical form is inconvenient, and, therefore, is used very seldom. The known analytical solutions of other differential equations and systems of differential equations in partial derivatives have the same shortcoming. In this connection, the numerical methods of solving equations in partial derivatives orientated on computer application have been rapidly developed recently.

8.2 METHODS OF CONSTRICTING THE COMPUTATIONAL AREA

For definiteness, we will talk about the solution of equation of transfer type (8.2). The transfer equation will be written down in one-dimensional setting as follows:

$$\frac{\partial f}{\partial t} + a_1 \frac{\partial f}{\partial x} = 0 \tag{8.10}$$

Let the problem on finding the function values $f(x,t)$ set up by the Eq. (8.10) be solved in the time period from $t = t_0$ to $t = t_k$ and for the values of the spatial coordinate x changing from $x = x_0$ to $x = x_k$ (Figure 8.1). The singularity of the problem solution (8.10) will be provided when setting up the function f values for the arbitrary value of the spatial coordinate x at the initial time point – $f(x,t_0)$, and when setting up the function f values for the arbitrary value of time t on the left and right boundaries of the computational area – $f(x_0,t)$ and $f(x_k,t)$. The value of the coefficient a_1 is the function of time and spatial coordinate – $a_1(t,x)$. In particular case, we can accept that $a_1 = const$.

The numerical solution of differential equations in partial derivatives is based on bringing these problems to the problems on the solution of the systems of linear and or (and) non-linear algebraic equations. For this, the function f satisfying the differential equation is replaced by its grid analog f_h whose values are established either in the grid nodes f_i^n (Figure 8.1a) or in the cell centers f_I^n (Figure 8.1b). The grid steps by time t and by spatial coordinate x can be constant or variable.

For the grid function f_h, the process current time will be recorded as t^n and determine it by the summation

$$t^n = \sum_{l=1}^{n} \Delta t_l \tag{8.11}$$

If when constructing the grid the step by time is accepted as constant – $\Delta t_l = const \ \forall l$, the correlation (8.11) will be as follows:

$$t^n = n\Delta t \tag{8.12}$$

Both variants of the formula for the time value t^n can be characterized by the value of the integral-valued parameter *n,* and this will be used further.

For the grid function f_h, the spatial coordinate will be recorded in the form x_i and will determine it by the total

$$x_i = \sum_{l=1}^{i} \Delta x_l \tag{8.13}$$

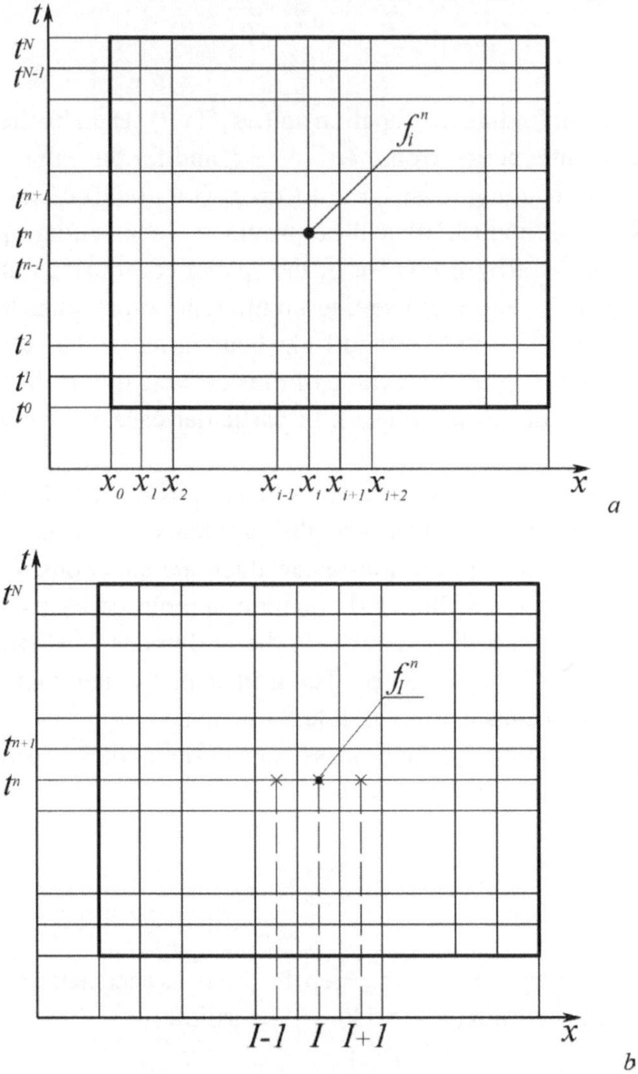

FIGURE 8.1 Splitting of the computational area into elementary cells (*a* – values of the function are established on the cell boundaries; *b* – values of the function are established in the cell centers).

If when constructing the grid the step by spatial coordinate is accepted as constant – $\Delta x_l = const \;\; \forall l$ (such grids are called regular), the correlation (8.13) will be as follows:

$$x_i = i\,\Delta x \tag{8.14}$$

Both variants of the formula for the value of the spatial coordinate x_i can be characterized by the value of the integral-valued parameter i (we will call this parameter as the integral-valued coordinate value), and this will be also used further.

When solving the equations recorded in multi-dimensional setting, the grid field is characterized by three coordinates. For instance, in Cartesian coordinate system this is x_i, y_j, z_k determined by the following correlation

$$x_i = \sum_{l=1}^{i} \Delta x_l$$

$$y_j = \sum_{l=1}^{j} \Delta y_l$$

$$z_k = \sum_{l=1}^{k} \Delta z_l \tag{8.15}$$

If when constructing the grid the steps are constant along each of the spatial coordinates, the correlations (8.15) are written down as follows:

$$x_i = i\,\Delta x, \qquad y_j = j\,\Delta y, \qquad z_k = k\,\Delta z \tag{8.16}$$

In accordance with (8.15) and (8.16), the location of the specific node of the computational grid is established by the coordinates x_i, y_j, z_k or values of their integral-valued coordinates *i, j, k*.

The examples of constructing spatial grids in two-dimensional situation are demonstrated in Figures 8.2 (*a* – Cartesian coordinate system, *b* – cylindrical coordinate system).

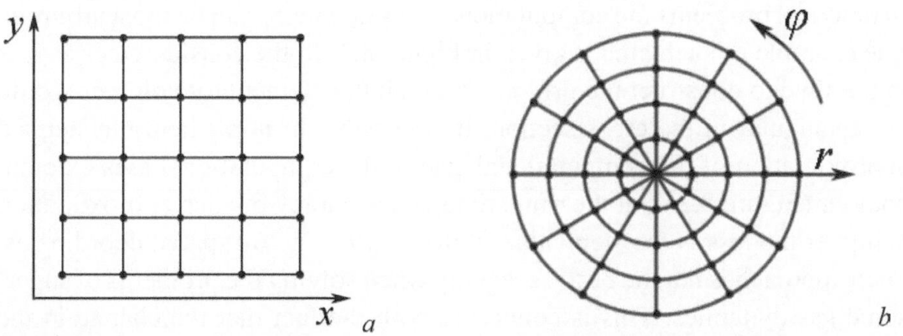

FIGURE 8.2 Grid construction for the situation of two-dimensional coordinates (*a* – Cartesian coordinate system; *b* – cylindrical coordinate system).

In finite-volumetric methods of solving differential equations in partial derivatives the elementary volumes (cells) located between the nodes *i, j, k* and *i*+1, *j*+1, *k*+1 (volume diagonal) are considered together with the grid nodes *i, j, k*. To indicate the coordinates of the center of such volume, the integral coordinates *I, J, K* are applied. For each elementary volume (cell) it is necessary to calculate its geometrical sizes – length of edges along each spatial coordinate, areas of flat sides, volume geometrical value. For arbitrary orthogonal coordinate system (Figure 7.4) these metric correlations are written down in the form (7.8).

In Cartesian coordinate system these correlations are written down as follows:

$$\Delta s_1 = \Delta x, \quad \Delta s_2 = \Delta y, \quad \Delta s_3 = \Delta z\,;$$

$$\sigma_1 = \Delta y \Delta z, \quad \sigma_2 = \Delta x \Delta z, \quad \sigma_3 = \Delta x \Delta y\,;$$

$$w = \Delta x \Delta y \Delta z \tag{8.17}$$

In cylindrical coordinate system the geometrical sizes of the elementary volume (cell) are calculated by the following correlations

$$\Delta s_1 = \Delta r, \quad \Delta s_2 = r\Delta\theta, \quad \Delta s_3 = \Delta z\,;$$

$$\sigma_1 = (r + \frac{1}{2}\Delta r)\,\Delta z\,\Delta\theta, \quad \sigma_2 = \Delta r\,\Delta z, \quad \sigma_3 = (r + \frac{1}{2}\Delta r)\,\Delta r\,\Delta\theta\,;$$

$$w = (r + \frac{1}{2}\Delta r)\Delta\theta\,\Delta r\,\Delta z \tag{8.18}$$

In practical problems the computational area geometry can be most arbitrary. The example of such area is given in Figure 8.3. In the considered case, it is impossible to construct the grid in which all the elementary volumes would be rectangular in the cross-section. It is possible to implement the stepped approximation of the computational grid to the computational area external contour (Figure 8.3a), at the same time, the accuracy of such approximation is higher the less is the step value of the grid Δx, Δy by spatial coordinates. Such approach is not the best, especially when solving the problems of supersonic gas dynamics. This is connected with the fact that the change in the external contour boundary of the computational area significantly changes the gas flow structure inside the area. Another approach, which can be used

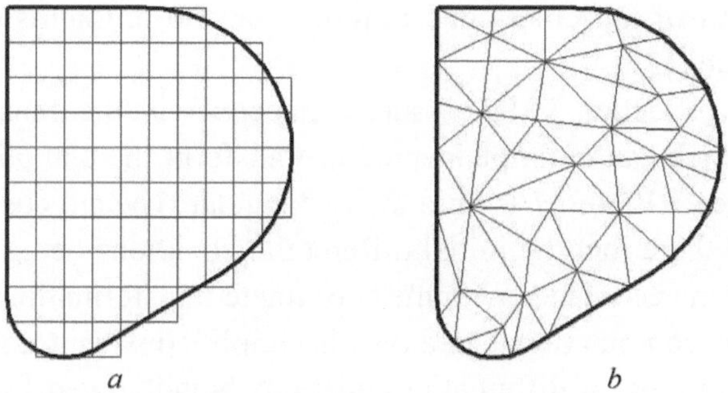

FIGURE 8.3 Regular (a) and irregular (b) computational areas.

when solving the problem, consists in applying the elementary volumes nonrectangular in the cross-section. These can be the volumes with trapezoidal or triangular cross-sections (Figure 8.3b). In most general case, the types of nonrectangular or fractional cells can be as those demonstrated in Figure 8.4.

It is impossible to apply the functions of (8.17) and (8.18) types to calculate the geometrical sizes of nonrectangular volumes. Besides, for these volumes it is necessary to modify the computational algorithms to solve differential equations. Finite-volumetric approaches considered further are the most convenient to apply to the problems on gas dynamics with fractional cells. The considered approach is applied, for instance, in [28].

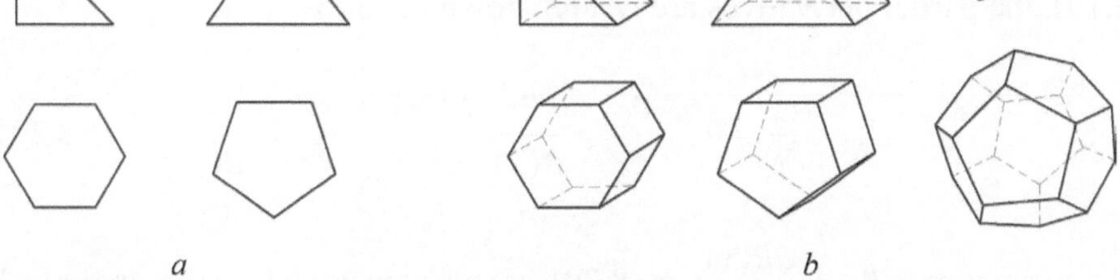

FIGURE 8.4 Types of fractional cells (volumes) (a – cells used when solving two-dimensional problems; *b* – cells used when solving three-dimensional problems).

The application of Dirichlet cells is topical in the problems of computational mechanics [37].

In Chapter 7 (Section 7.3) the issues connected with the transformation of coordinates which can be implemented to transform the computational area given in Figures 7.12 and 7.14 into the rectangular type are considered. It is also discussed there that the initial differential equations being solved need to be recorded in accordance with the coordinate transformations performed. As a rule, if the computational area type is simplified when transforming the coordinates, the type of differential equations being solved becomes more complicated. Generally speaking, such complication eventually results in decreasing the accuracy of the initial problem solution. Nevertheless, the coordinate transformation described in Chapter 7 is successfully applied in practice.

The computational area demonstrated in Figure 7.12 is of interest when solving the problems on the flow of the chemical fuel combustion products in the supersonic nozzle. Using the coordinate transformation, this computational area can be easily transformed into the rectangular form if the dependence *R(x)* defining the location of the nozzle external contour is known. In accordance with Section 7.2.3, let us apply the coordinate system transformation – we will change to the new coordinates instead of the initial coordinates *x, r*

$$x_* = x; \quad \xi = \frac{r}{R(x)} \tag{8.19}$$

The written down transformation converts the initial computational area into the rectangular one, in which $0 \le x \le x_*$ and $0 \le \xi \le 1$. To transform the equation of gas dynamics into the new coordinate system, it is necessary to establish the values of partial derivatives $\frac{\partial x_*}{\partial x}, \frac{\partial x_*}{\partial r}, \frac{\partial \xi}{\partial x}, \frac{\partial \xi}{\partial r}$. Taking into account (8.19), the partial derivatives are written down as follows:

$$\frac{\partial x_*}{\partial x} = 1, \quad \frac{\partial x_*}{\partial r} = 0, \quad \frac{\partial \xi}{\partial x} = \frac{\xi}{R(x)} \frac{dR(x)}{dx}, \quad \frac{\partial \xi}{\partial r} = \frac{1}{R(x)} \tag{8.20}$$

If the functions $R(x), \frac{dR(x)}{dx}$ in (8.20) are written down in the analytical form, it is not complicated to record the equations of gas dynamics in the new coordinate system. If the function $R(x)$ is set up tabularly, the value of the derivative $\frac{dR(x)}{dx}$ should be established by numerical differentiation.

From all the coordinate transformations those, at which the coordinate grid is orthogonal as the initial one, are of interest. In this case, it is possible to solve the initial problem with high accuracy. The methods of constructing the orthogonal grids are considered, for example, in [14, 226]. Among the known methods, Laplace and Poisson equations, as well as the variation methods, should be called the most effective. Let us point out that in [226] there are software products realizing the algorithms of constructing the orthogonal grids.

8.3 FINITE-DIFFERENCE METHODS OF SOLVING EQUATIONS IN PARTIAL DERIVATIVES

8.3.1 DEVELOPMENT OF FINITE-DIFFERENCE SCHEMES

In Chapter 2 it was discussed that the representation of the solutions of differential equations in the form of grid functions allows considering their tabular functions from one or several arguments. This gives the possibility to apply the apparatus of numerical differentiation consisting in the fact that all partial derivatives in the equation being solved are replaced by their difference analogs, when solving differential equations. The difference analogs can be recorded in different computational templates. Under the computational template we understand the nodes taken into account when approximating the derivatives.

The main approximation methods of the derivatives of grid (or tabular) functions are discussed in Section 2.3. In contrast to the problems connected with the solution of ordinary differential equations, it is necessary to approximate the derivatives by time and by each spatial coordinate when solving the equations in partial derivatives. Thus, for instance, the value of the partial derivative $\frac{\partial f(t,x,y,z)}{\partial t}$ and the computational error δ_t arising during its approximation can be established by the following formulas

$$\frac{\partial f(t,x,y,z)}{\partial t} = \frac{f(t_{n+1},x_i,y_j,z_k) - f(t_n,x_i,y_j,z_k)}{t_{n+1} - t_n} + \delta_t^{(1)}(t_n,x_i,y_j,z_k)$$

$$|\delta_t^{(1)}(t_n,x_i,y_j,z_k)| \le \frac{1}{2!}|t_{n+1}-t_n|\cdot\sup|\frac{\partial^2 f(\tau,x_i,y_j,z_k)}{\partial t^2}|,\ t_n \le \tau \le t_{n+1}$$

The values of the partial derivative $\frac{\partial f(t,x,y,z)}{\partial x}$ and computational error δ_x can be calculated, for example, as follows:

$$\frac{\partial f(t,x,y,z)}{\partial t} = \frac{f(t_n,x_{i+1},y_j,z_k)-f(t_n,x_i,y_j,z_k)}{x_{i+1}-x_i} + \delta_x^{(1)}(t_n,x_i,y_j,z_k);$$

$$|\delta_x^{(1)}(t_n,x_i,y_j,z_k)| \le \frac{1}{2!}|x_{i+1}-x_i|\cdot\sup|\frac{\partial^2 f(t,\xi,y_j,z_k)}{\partial x^2}|,\ x_i \le \xi \le x_{i+1};$$

or

$$\frac{\partial f(t,x,y,z)}{\partial t} = \frac{f(t_{n+1},x_{i+1},y_j,z_k)-f(t_{n+1},x_i,y_j,z_k)}{x_{i+1}-x_i} + \delta_x^{(1)}(t_{n+1},x_i,y_j,z_k);$$

$$|\delta_x^{(1)}(t_{n+1},x_i,y_j,z_k)| \le \frac{1}{2!}|x_{i+1}-x_i|\cdot\sup|\frac{\partial^2 f(t,\xi,y_j,z_k)}{\partial x^2}|,\ x_i \le \xi \le x_{i+1}$$

The equations can be approximated by direct or indirect finite-difference schemes. When writing down finite-difference schemes, it is assumed that when defining the parameters of the function f at the time moment t_{n+1}, the information on the value of the grid function f_i^n (corresponds to the time moment t_n) for all values of the spatial coordinate ($0 \le i \le i_{\max}$) is known.

Such finite-difference schemes, in which the definition of the grid function value of f_i^{n+1} function at the new time moment t_{n+1} is found by the simple transformation of the finite-difference equation, refer to direct ones. As a rule, the approximation of partial derivatives by the spatial coordinate in direct schemes contains the known values of the grid function corresponding to the time moments $t \le t_n$.

Such finite-difference schemes in which the values of the grid function f_i^{n+1} at the time moment t_{n+1} cannot be found directly but can be established by solving the system of algebraic equations, can refer to indirect ones. As a rule, in indirect schemes the approximation of partial derivatives by spatial coordinate contains the unknown values of the grid function corresponding to the time moments $t > t_n$.

Let us consider the most well-known variants of finite-difference schemes applied when solving the problems of mathematical physics (for instance, [14, 63, 87, 90, 185, 226]). For simplification, we will consider equations in partial derivatives recorded in non-stationary one-dimensional setting.

8.3.1.1 Transfer Equation

$$\frac{\partial f}{\partial t} + a\frac{\partial f}{\partial x} = 0 \tag{8.21}$$

This equation can be approximated by the templates given in Figure 8.5. Hereinafter the nodes used during the approximation of the derivatives by time are marked with circles, and the nodes used during the approximation of derivatives by spatial coordinate – with crosses. It is assumed during the approximation that the grid step by spatial coordinate $\Delta x = const.$

1. Direct schemes of the first order of accuracy by time and spatial coordinate (templates given in Figures 8.5a and 8.5b):

$$\frac{f_i^{n+1} - f_i^n}{\Delta t} + a\frac{f_i^n - f_{i-1}^n}{\Delta x} \approx 0; \tag{8.22}$$

$$\frac{f_i^{n+1} - f_i^n}{\Delta t} + a\frac{f_{i+1}^n - f_i^n}{\Delta x} \approx 0 \tag{8.23}$$

   The approximated equality of the right and left sides of the equation is written down in the Eqs. (8.22) and (8.23). In these equations the approximation error values $\delta_t^{(1)}$ and $\delta_x^{(1)}$ are discarded. In the following equations the approximated equality is written down as accurate for simplification.

2. "Tripod" scheme is the direct scheme of the first order of accuracy by time and second order – by spatial coordinate (template is given in Figure 8.5c):

$$\frac{f_i^{n+1} - f_i^n}{\Delta t} + a\frac{f_{i+1}^n - f_{i-1}^n}{2\Delta x} = 0; \tag{8.24}$$

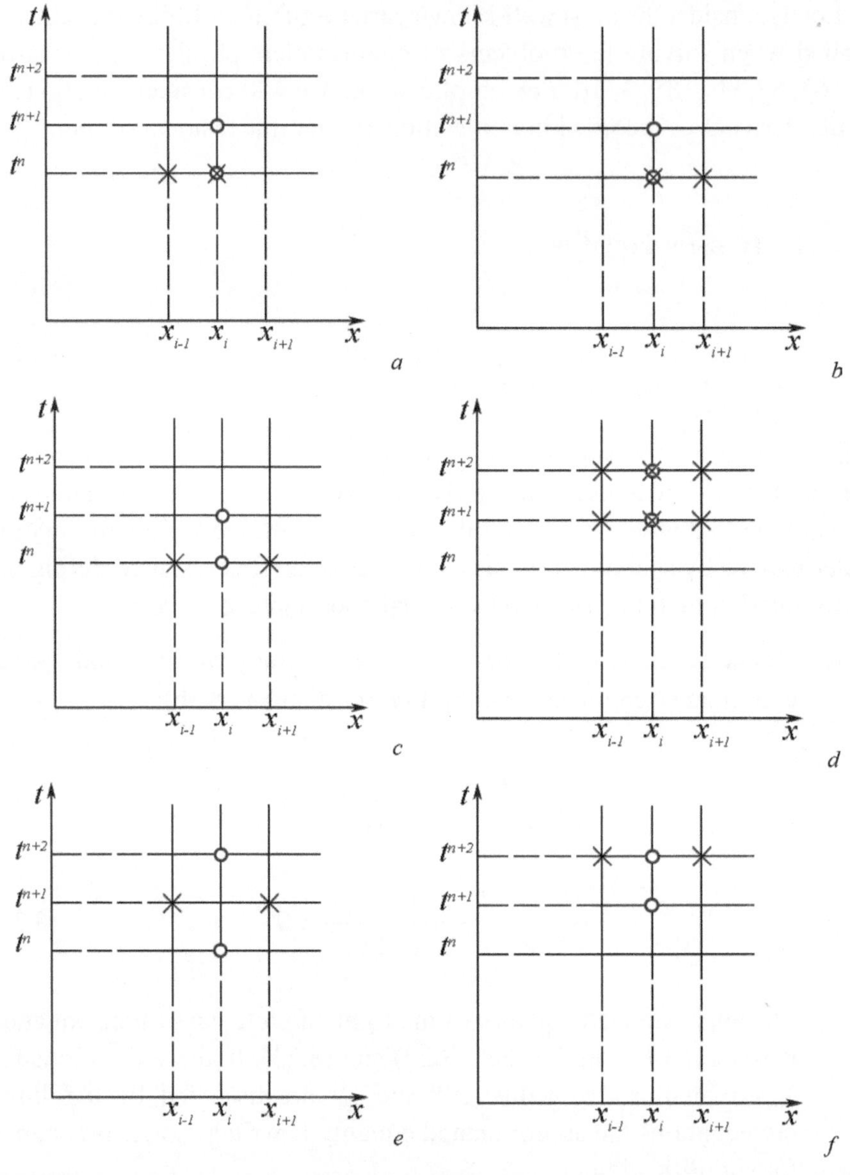

FIGURE 8.5 Computational templates used when solving transfer equations (a, b – schemes of the first order of accuracy by the spatial coordinate; c – "tripod" scheme; d – Lax scheme; e – "leapfrog" scheme; f – indirect scheme).

3. Lax scheme is the direct scheme of the first order of accuracy by time and second order – by spatial coordinate (template is given in Figure 8.5d):

$$\frac{f_i^{n+1} - 0.5(f_{i+1}^n + f_{i-1}^n)}{\Delta t} + a\frac{f_{i+1}^n - f_{i-1}^n}{2\Delta x} = 0\,; \tag{8.25}$$

4. "Leapfrog" scheme is the direct scheme of the second order of accuracy by time and second order – by spatial coordinate (template is given in Figure 8.5e):

$$\frac{f_i^{n+1} - f_i^{n-1}}{2\Delta t} + a\frac{f_{i+1}^n - f_{i-1}^n}{2\Delta x} = 0\,; \tag{8.26}$$

5. Lax-Werndroff scheme is the direct scheme of the type "predictor-corrector" of the second order of accuracy by time and second order – by spatial coordinate:
   - first step (predictor) is performed by Lax formula (8.25);
   - second step (corrector) is performed by the following formula

$$\frac{f_i^{n+2} - f_i^{n}}{2\Delta t} + a\frac{f_{i+1}^{n+1} - f_{i-1}^{n+1}}{2\Delta x} = 0\,; \tag{8.27}$$

6. MacCormak scheme is the direct scheme of the type "predictor-corrector" of the second order of accuracy by time and second order – by spatial coordinate:

   - at the first step (predictor) the intermediary value of the grid function $\tilde{f}_i^{n+1}$ is calculated by the formula of (8.22) type

$$\frac{\tilde{f}_i^{n+1} - f_i^{n}}{\Delta t} + a\frac{f_{i+1}^{n} - f_{i}^{n}}{\Delta x} = 0\,;$$

   - at the second step (corrector) the final value of the grid function f_i^{n+1} is calculated by the formula of (8.23) type

$$\frac{f_i^{n+1} - 0.5(f_i^n + \tilde{f}_i^{n+1})}{\Delta t} + a\frac{\tilde{f}_i^{n+1} - \tilde{f}_{i-1}^{n+1}}{\Delta x} = 0 \tag{8.28}$$

7. Indirect scheme of the first order of accuracy by time and second order – by spatial coordinate (template is given in Figure 8.5f):

$$\frac{f_i^{n+1} - f_i^{n}}{\Delta t} + a\frac{f_{i+1}^{n+1} - f_{i-1}^{n+1}}{2\Delta x} = 0 \tag{8.29}$$

8.3.1.2 Diffusion (Heat Conductivity) Equation

$$\frac{\partial f}{\partial t} - a\frac{\partial^2 f}{\partial x^2} = 0 \tag{8.30}$$

The diffusion equation can be approximated by the templates given in Figure 8.6. During the approximation we will consider the grid step by spatial coordinate $\Delta x = const$.

1. Direct scheme of the first order of accuracy by time and second order – by spatial coordinate (template is given in Figure 8.6a):

$$\frac{f_i^{n+1} - f_i^n}{\Delta t} - a\frac{f_{i+1}^n - 2f_i^n + f_{i-1}^n}{\Delta x^2} = 0; \tag{8.31}$$

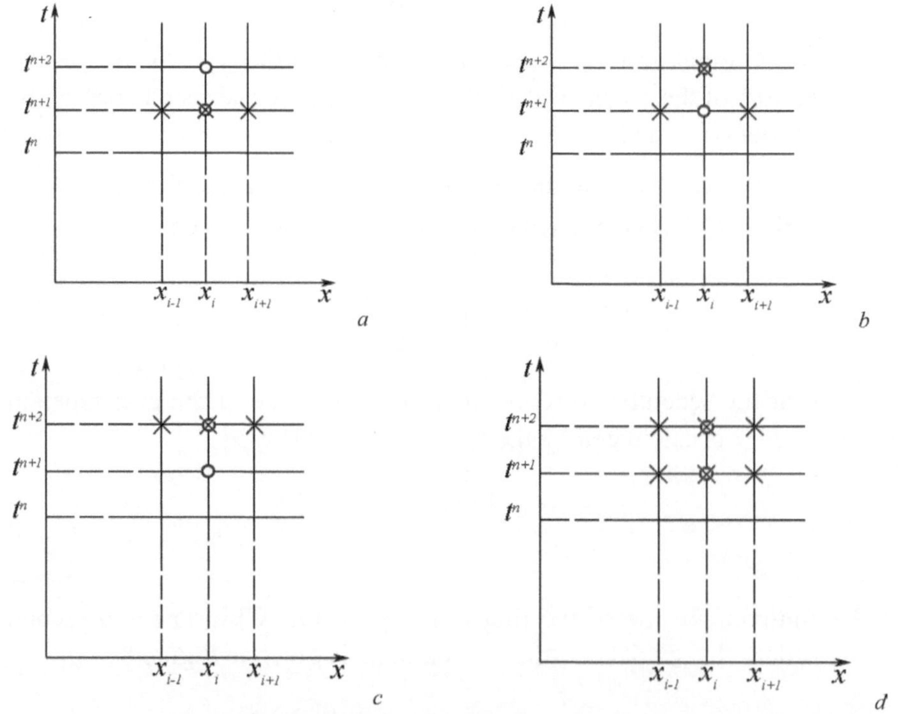

FIGURE 8.6 Computational templates used when solving diffusion equations (*a* – direct scheme; *b* – Dufort-Frankel scheme; *c* – indirect scheme; *d* – Crank-Nicolson scheme).

2. Direct Dufort-Frankel scheme of the first order of accuracy by time and second order – by spatial coordinate (template is given in Figure 8.6b):

$$\frac{f_i^{n+1} - f_i^n}{\Delta t} - a\frac{f_{i+1}^n - 2f_i^{n+1} + f_{i-1}^n}{\Delta x^2} = 0; \quad (8.32)$$

3. Indirect finite-difference scheme is written down as follows (template is given in Figure 8.6c):

$$\frac{f_i^{n+1} - f_i^n}{\Delta t} - a\frac{f_{i+1}^{n+1} - 2f_i^{n+1} + f_{i-1}^{n+1}}{\Delta x^2} = 0; \quad (8.33)$$

4. Indirect finite-difference Crank-Nicolson scheme (template is given in Figure 8.6d):

$$\frac{f_i^{n+1} - f_i^n}{\Delta t} - a(\sigma\frac{f_{i+1}^n - 2f_i^n + f_{i-1}^n}{\Delta x^2} + (1-\sigma)\frac{f_{i+1}^{n+1} - 2f_i^{n+1} + f_{i-1}^{n+1}}{\Delta x^2}) = 0 \quad (8.34)$$

In the Eq. (8.34) – $0 < \sigma < 1$. In particular, we can take $\sigma = 0.5$.

Let us consider the features of solving equations for grid functions f_i^{n+1} obtained during the finite-difference approximation of differential equations in partial derivatives. At the same time, it should be taken into account that according to setting up the problem on solving the equation in partial derivatives, the values of the grid function f_i^0 are known for all values of $i = 1, i_{\max}$ at the initial time point $t = t_0$. Besides, the grid functions f_1^{n+1}, $f_{i\max}^{n+1}$ are known for any time point $t = t_{n+1}$.

Let us consider the solution of the transfer equation fulfilled by the direct scheme (8.22). The equation is as follows:

$$f_i^{n+1} = f_i^n - a\frac{\Delta t}{\Delta x}(f_i^n - f_{i-1}^n) \quad (8.35)$$

The determination of the grid function f_i^n by the Eq. (8.35) is established by the following algorithm:

- let us take the value n=0 and set up the integration step Δt providing the stability of further calculations (order of selecting the integration step by time providing the stable computation of finite-difference equations is discussed in Section 8.3.3);

- we determine the values of the grid function f_i^0 for all values of $i = 1, i_{max}$ in accordance with the initial conditions;
- we establish the value of the grid function f_1^1 in accordance with the boundary conditions;
- we calculate the values of f_i^1 from $i = 2$ to $i = i_{max}$ in accordance with the Eq. (8.35);
- we increase the value n by one and return to the beginning of the algorithm.

The computational cycle is performed until $t < t_{max}$ that corresponds to the value $n < n_{max}$. Further it will be demonstrated that if the value of the coefficient a in the Eq. (8.22) is positive ($a > 0$) and the integration step Δt is selected in accordance with the condition of the numerical calculation stability, the recorded algorithm provides the solution of the initial problem.

Similarly, the calculations by other direct finite-difference schemes can be performed. The calculation in the direct form is also performed by Dufort-Frankel scheme (Eq. (8.32)), despite that the partial derivative approximation by the spatial coordinate is performed by the template in which the value of the grid function f_i^{n+1} on the non-calculated time layer t_{n+1} is available. The computational formula for the values of f_i^{n+1} in Dufort-Frankel method comes down to the following

$$f_i^{n+1} = \frac{1}{1 + a\frac{2\Delta t}{\Delta x^2}} \cdot \left(f_i^n + a\frac{\Delta t}{\Delta x^2}(f_{i-1}^n + f_{i+1}^n)\right) \tag{8.36}$$

Let us consider the solution of finite-difference equations for the grid functions f_i^{n+1} when applying the indirect scheme on the example of the diffusion equation (formula (8.33)). The Eq. (8.33) simultaneously contains three unknown values – f_{i-1}^{n+1}, f_i^{n+1}, f_{i+1}^{n+1}, corresponding to the time point t_{n+1}, and this does not allow solving the equation in the direct form. The determination of the grid function f_i^{n+1} at the new time step comes down to solving the following system of linear equations:

$$f_1^{n+1} = f_-^{n+1};$$

$$-\frac{a\Delta t}{\Delta x^2} f_{i-1}^{n+1} + \left(1 + 2\frac{a\Delta t}{\Delta x^2}\right) f_i^{n+1} - \frac{a\Delta t}{\Delta x^2} f_{i+1}^{n+1} = f_i^n, \qquad i = 2, i_{max-1};$$

$$f_{i_{\max}}^{n+1} = f_{+}^{n+1} \tag{8.37}$$

In the recorded equation system f_{-}^{n+1}, f_{+}^{n+1} are the boundary conditions on the left and right boundaries of the computational area, respectively. The number of equations $i_{\max}$ coincides with the number of the unknown values of the grid function f_i^{n+1}, therefore, the equation system has the only solution, if the values of the grid function f_i^n corresponding to the time point t_n are known. The equation system is multiply solved starting from the value $n=1$ (corresponds to the time point t_1) till the value $n=n_{\max}$ (corresponds to the time point $t_{n_{\max}}$). When $n=1$, the Eq. (8.37) is solved by the set up initial conditions (correspond to the grid values of the function $f_i^0, i=2, i_{\max-1}$). The system of linear equation (8.37) can be solved at any regular integration step with the methods discussed in Chapter 2 (Section 2.2.2). However, there are more effective algorithms for similar systems, and some of these algorithms are considered below.

The following should be additionally pointed out:

1. Before we discussed the variants of finite-difference approximation of the equations, in which the partial derivatives by time and spatial coordinates from one and the same function are available. However, in practice, it is necessary to solve the equation of (8.7) type, in which each of the partial derivatives in the equation is determined from its function. These functions, as a rule, are interconnected. The aforementioned methods of finite-difference approximation are applied for similar equations. For instance, when solving the following equation

$$\frac{\partial f}{\partial t} + \frac{\partial g}{\partial x} = 0 \tag{8.38}$$

   Lax scheme can be applied in the following finite-difference form

$$\frac{f_i^{n+1} - 0.5(f_{i+1}^n + f_{i-1}^n)}{\Delta t} + \frac{g_{i+1}^n - g_{i-1}^n}{2\Delta x} = 0 \tag{8.39}$$

   In the formulas (8.38), (8.39) g can be the function of the variables t, x ($g(t,x)$) or t, x, f ($g(t,x,f)$).
2. Direct finite-difference schemes for solving the multidimensional problems described by the equations in partial derivatives can be

constructed similar to the aforementioned ones. In this case, the grid function is designated, for instance, as $f^n_{i,j,k}$. Here *i, j, k* – integral coordinates of the grid function by three coordinate directions.

Let us consider the three-dimensional variant of the transfer equation

$$\frac{\partial f}{\partial t} + a_1 \frac{\partial f}{\partial x} + a_2 \frac{\partial f}{\partial y} + a_3 \frac{\partial f}{\partial z} = 0$$

and write down the direct finite-difference Lax scheme for its solution

$$\frac{f^{n+1}_{i,j,k} - \frac{1}{6}(f^n_{i+1,j,k} + f^n_{i-1,j,k} + f^n_{i,j+1,k} + f^n_{i,j-1,k} + f^n_{i,j,k+1} + f^n_{i,j,k-1})}{\Delta t}$$
$$+ a_1 \frac{f^n_{i+1,j,k} - f^n_{i-1,j,k}}{2\Delta x} + a_2 \frac{f^n_{i,j+1,k} - f^n_{i,j-1,k}}{2\Delta y} + a_3 \frac{f^n_{i,j,k+1} - f^n_{i,j,k-1}}{2\Delta z} = 0$$

The solution of the last equation is as simple as the solution of the Eq. (8.25).

3. Application of the indirect schemes for multidimensional problems significantly complicates their solution. Let us consider, for instance, the multidimensional variant of the diffusion equation

$$\frac{\partial f}{\partial t} - a\left(\frac{\partial^2 f}{\partial x^2} + \frac{\partial^2 f}{\partial y^2} + \frac{\partial^2 f}{\partial z^2}\right) = 0 \quad (8.40)$$

For this equation we can write down the following finite-difference scheme

$$\frac{f^{n+1}_{i,j,k} - f^n_{i,j,k}}{\Delta t} - a\left(\frac{f^{n+1}_{i+1,j,k} - 2f^{n+1}_{i,j,k} + f^{n+1}_{i-1,j,k}}{\Delta x^2} + \frac{f^{n+1}_{i,j+1,k} - 2f^{n+1}_{i,j,k} + f^{n+1}_{i,j-1,k}}{\Delta y^2}\right.$$
$$\left. + \frac{f^{n+1}_{i,j,k+1} - 2f^{n+1}_{i,j,k} + f^{n+1}_{i,j,k-1}}{\Delta z^2}\right) = 0 \quad (8.41)$$

In each of the Eq. (8.33) there are only three unknown values – $f^{n+1}_{i+1}, f^{n+1}_{i}, f^{n+1}_{i-1}$. Therefore, the Eq. (8.33) solution at each step by time comes down to the solution of the system of linear algebraic Eqs. (2.15), in which the matrix of the coefficients A has a three-diagonal form

$$\mathbf{A} = \begin{pmatrix} * & * & 0 & 0 & \dots & 0 & 0 & 0 & 0 \\ * & * & * & 0 & \dots & 0 & 0 & 0 & 0 \\ 0 & * & * & * & \dots & 0 & 0 & 0 & 0 \\ 0 & 0 & * & * & \dots & 0 & 0 & 0 & 0 \\ \dots & \dots & \dots & \dots & \dots & \dots & \dots & \dots & \dots \\ 0 & 0 & 0 & 0 & \dots & * & * & 0 & 0 \\ 0 & 0 & 0 & 0 & \dots & * & * & * & 0 \\ 0 & 0 & 0 & 0 & \dots & 0 & * & * & * \\ 0 & 0 & 0 & 0 & \dots & 0 & 0 & * & * \end{pmatrix}$$

There are effective computational methods, called the sweep methods, to solve such equation systems (these methods will be considered further). But it is impossible to apply these methods when solving the Eq. (8.41), since there are seven unknown values in each Eq. (8.41)

$$(f_{i+1,j,k}^{n+1}, f_{i,j,k}^{n+1}, f_{i-1,j,k}^{n+1}, f_{i,j+1,k}^{n+1}, f_{i,j-1,k}^{n+1}, f_{i,j,k+1}^{n+1}, f_{i,j,k-1}^{n+1})$$

There are effective computational methods to solve multidimensional problems called the methods of spatial separation, fractional step methods, whose essence consists in bringing the multidimensional problem solution to the aggregation of one-dimensional problems. The application of this method for the Eq. (8.40) requires the solution of three equation systems – (8.42)–(8.44).

$$\frac{\tilde{f}_{i,j,k}^{n+1} - f_{i,j,k}^{n}}{\Delta t} - a\frac{\tilde{f}_{i+1,j,k}^{n+1} - 2\tilde{f}_{i,j,k}^{n+1} + \tilde{f}_{i-1,j,k}^{n+1}}{\Delta x^2} = 0 \qquad (8.42)$$

$$\frac{\hat{f}_{i,j,k}^{n+1} - \tilde{f}_{i,j,k}^{n}}{\Delta t} - a\frac{\hat{f}_{i,j+1,k}^{n+1} - 2\hat{f}_{i,j,k}^{n+1} + \hat{f}_{i,j-1,k}^{n+1}}{\Delta y^2} = 0 \qquad (8.43)$$

$$\frac{f_{i,j,k}^{n+1} - \hat{f}_{i,j,k}^{n}}{\Delta t} - a\frac{f_{i,j,k+1}^{n+1} - 2f_{i,j,k}^{n+1} + f_{i,j,k-1}^{n+1}}{\Delta z^2} = 0 \qquad (8.44)$$

Each of the equation systems can be brought down to the type (8.33) together with the initial or boundary conditions with the three-diagonal matrix A.

4. Initial differential equation in partial derivatives can be brought to the type of the system of ordinary differential equations. For instance, for the case of diffusion Eq. (8.30) at the direct approximation of the second derivative by the spatial coordinate such system can be written down as follows:

$$\frac{df_i^n}{dt} = \frac{a}{\Delta x^2}(f_{i+1}^n - 2f_i^n + f_{i-1}^n), \quad i = 2, i_{\max-1} \tag{8.45}$$

When setting up the initial conditions (for all values $i = 2, i_{\max-1}$ the values of f_i^0 are known) and boundary conditions (at any time values the values of the function f_1^n, $f_{i_{\max}}^n$ are known), the Eq. (8.45) can be integrated by any known method (see Chapter 5), including the methods of high order of accuracy by time.
5. In recent years, the finite-difference schemes of high order of accuracy by time and spatial coordinates have been acquiring the special interest. The construction of such schemes is considered, for instance, in [137, 144, 146].

8.3.2 ANALYSIS OF THE CONVERGENCE OF FINITE-DIFFERENCE SCHEMES

The solution of the algebraic equations f_h approximating the differential equation in partial derivatives $f(t, q_1, q_2, q_3)$ and obtained when applying some finite-difference algorithm is called the convergent to the precise solution of the differential equation *f* in case, when the decrease in the sizes of the finite-difference grid ($\Delta t, \Delta q_1, \Delta q_2, \Delta q_3 \to 0$) results in the decrease in the difference between precise and approximated solutions ($|f_h - f| \to 0$). In practice, Lax theorem is applied to evaluate the convergence [226] which is true for linear problems with initial and boundary conditions. Lax theorem states that if the approximation of finite-difference equations to the initial equations is provided, the stability of the computational algorithm implemented by the applied finite-difference algorithm is the convergence necessary and sufficient condition.

The issues of the derivative approximation in differential equations, their finite-difference analogs were discussed in Chapters 2 and 5. The accuracy of the equation approximation is established by the approximation of the derivatives in these equations.

Let us consider the examples in which we will use the grid uniform by time and spatial coordinate ($\Delta t = const$, $\Delta x = const$).

The following correlations can be written down in accordance with the Eqs. (2.30), (2.31) and [250]

$$f_i^{n+1} \approx f_i^n + \Delta t \cdot (\frac{\partial f}{\partial t})_i^n + \frac{\Delta t^2}{2} \cdot (\frac{\partial^2 f}{\partial t^2})_i^n + \frac{\Delta t^3}{6} \cdot (\frac{\partial^3 f}{\partial t^3})_i^n + \frac{\Delta t^4}{24} \cdot (\frac{\partial^4 f}{\partial t^4})_i^n + \ldots;$$

$$f_{i\pm1}^n \approx f_i^n \pm \Delta x \cdot (\frac{\partial f}{\partial x})_i^n + \frac{\Delta x^2}{2} \cdot (\frac{\partial^2 f}{\partial x^2})_i^n \pm \frac{\Delta x^3}{6} \cdot (\frac{\partial^3 f}{\partial x^3})_i^n + \frac{\Delta x^4}{24} \cdot (\frac{\partial^4 f}{\partial x^4})_k^n + \ldots$$

The foregoing correlations allow establishing the values of the derivatives recorded in the finite-difference form, taking into account the residual elements obtained from Taylor decompositions:

- approximation of the derivative $\frac{\partial f}{\partial t}$ by the finite difference of the first order of accuracy by time

$$\frac{f_i^{n+1} - f_i^n}{\Delta t} = (\frac{\partial f}{\partial t})_i^n + \frac{\Delta t}{2} \cdot (\frac{\partial^2 f}{\partial t^2})_i^n + \frac{\Delta t^2}{6} \cdot (\frac{\partial^3 f}{\partial t^3})_i^n + \frac{\Delta t^3}{24} \cdot (\frac{\partial^4 f}{\partial t^4})_i^n + \ldots; \quad (8.46)$$

- approximation of the derivative $\frac{\partial f}{\partial t}$ by the finite difference by Lax scheme

$$\frac{f_i^{n+1} - 0,5(f_{i+1}^n + f_{i-1}^n)}{\Delta t} = (\frac{\partial f}{\partial t})_i^n + \frac{\Delta t}{2} \cdot (\frac{\partial^2 f}{\partial t^2})_i^n + \frac{\Delta t^2}{6} \cdot (\frac{\partial^3 f}{\partial t^3})_i^n + \frac{\Delta t^3}{24} \cdot (\frac{\partial^4 f}{\partial t^4})_i^n - \frac{\Delta x}{\Delta t} \cdot \frac{\Delta x}{2} \cdot (\frac{\partial^2 f}{\partial x^2})_i^n - \frac{\Delta x}{\Delta t} \cdot \frac{\Delta x^3}{24} \cdot (\frac{\partial^2 f}{\partial x^2})_i^n - \ldots; \quad (8.47)$$

- approximation of the derivative $\frac{\partial f}{\partial x}$ by the finite differences of the first and second orders of accuracy by spatial coordinate

$$\frac{f_i^n - f_{i-1}^n}{\Delta x} = (\frac{\partial f}{\partial x})_i^n - \frac{\Delta x}{2} \cdot (\frac{\partial^2 f}{\partial x^2})_i^n + \frac{\Delta x^2}{6} \cdot (\frac{\partial^3 f}{\partial x^3})_i^n - \frac{\Delta x^3}{24} \cdot (\frac{\partial^4 f}{\partial x^4})_i^n + \ldots; \quad (8.48)$$

$$\frac{f_{i+1}^n - f_i^n}{\Delta x} = (\frac{\partial f}{\partial x})_i^n + \frac{\Delta x}{2} \cdot (\frac{\partial^2 f}{\partial x^2})_i^n + \frac{\Delta x^2}{6} \cdot (\frac{\partial^3 f}{\partial x^3})_i^n + \frac{\Delta x^3}{24} \cdot (\frac{\partial^4 f}{\partial x^4})_i^n + \ldots; \quad (8.49)$$

$$\frac{f_{i+1}^n - f_{i-1}^n}{2\Delta x} = (\frac{\partial f}{\partial x})_i^n + \frac{\Delta x^2}{6} \cdot (\frac{\partial^3 f}{\partial x^3})_i^n + \ldots; \tag{8.50}$$

- approximation of the derivative $\frac{\partial^2 f}{\partial x^2}$ by the finite difference of the second order of accuracy by spatial coordinate

$$\frac{f_{i+1}^n - 2f_i^n + f_{i-1}^n}{\Delta x^2} = (\frac{\partial^2 f}{\partial x^2})_i^n + \frac{\Delta x^2}{12} \cdot (\frac{\partial^2 f}{\partial x^2})_i^n + \frac{\Delta x^4}{360} \cdot (\frac{\partial^4 f}{\partial x^4})_i^n + \ldots; \tag{8.51}$$

- approximation of the derivative $\frac{\partial^2 f}{\partial x^2}$ by the finite difference by Dufort-Frankel scheme

$$\begin{aligned} \frac{f_{i+1}^n - 2f_i^{n+1} + f_{i-1}^n}{\Delta x^2} &= (\frac{\partial^2 f}{\partial x^2})_i^n - \frac{2\Delta t}{\Delta x^2} \cdot (\frac{\partial f}{\partial t})_i^n - \frac{\Delta t^2}{\Delta x^2} \cdot (\frac{\partial^2 f}{\partial t^2})_i^n - \\ &- \frac{\Delta t^3}{3\Delta x^2} \cdot (\frac{\partial^3 f}{\partial t^3})_i^n - \frac{\Delta t^4}{12\Delta x^2} \cdot (\frac{\partial^4 f}{\partial t^4})_i^n + \frac{\Delta x^2}{12} (\frac{\partial^4 f}{\partial x^4})_i^n + \ldots \end{aligned} \tag{8.52}$$

The correlations (8.46)–(8.52) can be applied to evaluate the approximation accuracy of the equations by different finite-difference schemes. In the equations given below, the indexes i and n are omitted due to the arbitrariness of time and spatial coordinates:

- scheme (8.22) for the transfer Eq. (8.21)

$$\begin{aligned} &\frac{\partial f}{\partial t} + a\frac{\partial f}{\partial x} + \frac{\Delta t}{2}\frac{\partial^2 f}{\partial t^2} + a\frac{\Delta x}{2}\frac{\partial^2 f}{\partial x^2} + \frac{\Delta t^2}{6}\frac{\partial^3 f}{\partial t^3} \\ &+ a\frac{\Delta x^2}{6}\frac{\partial^3 f}{\partial x^3} + \frac{\Delta t^3}{24}\frac{\partial^4 f}{\partial t^4} + a\frac{\Delta x^3}{24}\frac{\partial^4 f}{\partial x^4} + \ldots = 0; \end{aligned}$$

- scheme (8.23) for the transfer Eq. (8.21)

$$\begin{aligned} &\frac{\partial f}{\partial t} + a\frac{\partial f}{\partial x} + \frac{\Delta t}{2}\frac{\partial^2 f}{\partial t^2} - a\frac{\Delta x}{2}\frac{\partial^2 f}{\partial x^2} + \frac{\Delta t^2}{6}\frac{\partial^3 f}{\partial t^3} \\ &+ a\frac{\Delta x^2}{6}\frac{\partial^3 f}{\partial x^3} + \frac{\Delta t^3}{24}\frac{\partial^4 f}{\partial t^4} - a\frac{\Delta x^3}{24}\frac{\partial^4 f}{\partial x^4} + \ldots = 0 \end{aligned}$$

In this equation (and further, as well) the attention should be paid to the multiplier value at the second derivative by the spatial coordinate $\frac{\partial^2 f}{\partial x^2}$

(multiplier $-a\frac{\Delta x}{2}$). The problems of hydro- and gas-dynamics in the equations of momentum and energy conservation comprise the component proportional to $-\mu\frac{\partial^2 u}{\partial x^2}$, which characterizes the liquid or gas viscosity influence on their flow. In this regard, the value $-a\frac{\Delta x}{2}\frac{\partial^2 f}{\partial x^2}$ is called the approximating viscosity. Looking ahead, it should be pointed out that the stability of the finite-difference problem is greater, the higher is the approximation viscosity.

- Lax scheme (8.25) for the transfer Eq. (8.21)

$$\frac{\partial f}{\partial t}+a\frac{\partial f}{\partial x}+\frac{\Delta t}{2}\frac{\partial^2 f}{\partial t^2}-\frac{\Delta x}{\Delta t}(\frac{\Delta x}{2}+\frac{\Delta x^3}{24})\frac{\partial^2 f}{\partial x^2}+\frac{\Delta t^2}{6}\frac{\partial^3 f}{\partial t^3}$$
$$+a\frac{\Delta x^2}{6}\frac{\partial^3 f}{\partial x^3}+\frac{\Delta t^3}{24}\frac{\partial^4 f}{\partial t^4}+\ldots=0 \qquad ;$$

- scheme (8.31) for the diffusion Eq. (8.30)

$$\frac{\partial f}{\partial t}-a\frac{\partial^2 f}{\partial x^2}+\frac{\Delta t}{2}\cdot\frac{\partial^2 f}{\partial t^2}+\frac{\Delta t^2}{6}\frac{\partial^3 f}{\partial t^3}-a\frac{\Delta x^2}{12}\frac{\partial^2 f}{\partial x^2}$$
$$+\frac{\Delta t^3}{24}\frac{\partial^4 f}{\partial t^4}-a\frac{\Delta x^4}{360}\cdot\frac{\partial^4 f}{\partial x^4}+\ldots=0 \qquad ;$$

- Dufort-Frankel schemes (8.32) for the diffusion Eq. (8.30)

$$\frac{\partial f}{\partial t}-a\frac{\partial^2 f}{\partial x^2}+a\frac{2\Delta t}{\Delta x^2}\cdot\frac{\partial f}{\partial t}+\Delta t(\frac{1}{2}+a\frac{\Delta t}{\Delta x^2})\frac{\partial^2 f}{\partial t^2}+\frac{\Delta t^2}{6}\frac{\partial^3 f}{\partial t^3}+a\Delta t^2\frac{\Delta t}{3\Delta x^2}\frac{\partial^3 f}{\partial t^3}+$$
$$+a\Delta t^3\frac{\Delta t}{12\Delta x^2}\frac{\partial^4 f}{\partial t^4}-a\frac{\Delta x^2}{12}\frac{\partial^4 f}{\partial x^4}+\ldots\ldots=0 \qquad (8.53)$$

It follows from the aforementioned formulas, in which the residual elements obtained by the decomposition into Taylor rows are preserved, that the finite-difference approximations of differential equations significantly change their initial properties. However, if the values of the steps Δt, Δx used during the discretization of the problems are indefinitely small, the equation approximation error decreases. The elements containing the additional derivatives with the multipliers proportional to Δt^{k_t}, Δx^{k_δ} are called residual and are designated as $O(\Delta t^{k_t})$, $O(\Delta x^{k_\delta})$. The coefficients k_t, k_x are called the approximation order. Any schemes for which the coefficients $k_t, k_x \geq 1$ are of interest for practice.

From the foregoing finite-difference approximations the approximation by Lax scheme (it contains the residual elements comprising the correlation $\left.\frac{\Delta x}{\Delta t}\right)$ and approximation by Dufort-Frankel scheme (it contains the residual elements comprising the correlation $\left.\frac{\Delta t}{\Delta x^2}\right)$ are the most interesting.

The approximation of differential equations with the order $k_t, k_x \geq 1$ is one of the conditions of the finite-difference equation convergence to the differential one. The second condition – the stability of the calculated algorithm in the finite-difference scheme base. The stability issues are considered, for instance, in Refs. [190–193]. Below certain methods for assessing the stability as applicable to the model transfer (8.21) and diffusion Eqs. (8.30) are discussed.

The stability evaluation assumes the determination of such values of elementary steps of the computational grid asΔt and Δx at which there is no catastrophic accumulation of computational errors. If for any values of Δt and Δx there is no catastrophic accumulation of computational errors, the finite-difference scheme applied is called absolutely stable. Otherwise the scheme is called conditionally stable, and the condition under which the stability is preserved should be established.

When analyzing the stability, the following method is often used. If the finite-difference scheme is stable (absolutely or under some condition), this scheme is stable at initially zero values of the function *f*, at the boundaries of the computational area and at the initial integration moment $t = t_0$, as well. At the same time, the stability is easier to analyze under such initially zero values of the unknown function *f*. If when calculating the function *f* any computational error δ occurs (e.g., due to the round-off errors), then for the stable finite-difference scheme this error needs to be further decreased.

8.3.2.1 Stability Analysis Using Power Function

The stability of the finite-difference algorithm to calculate the function *f* is investigated under the assumption that the error δ of the grid function f_k^n value computation (corresponds to the time point t_n and axial coordinate x_k)

can be represented by the formula $f_k^n = \delta = \lambda(-1)^k$. The stability condition of the computational algorithm can be derived from the condition that the values of the grid function f_k^{n+1}, f_k^{n+2} corresponding to the process development (time moments t_{n+1}, t_{n+2}, etc.), will be less than the values of the function f_k^n.

Taking the foregoing into account, the stability condition can be given as follows:

$$\left|\frac{f_k^{n+1}}{f_k^n}\right| \leq 1 \tag{8.54}$$

Let us consider the application of the aforementioned approach to solve the transfer Eq. (8.21) recorded as (8.22). For this equation the value f_k^{n+1} is established by the formula

$$f_k^{n+1} = (1 - a\frac{\Delta t}{\Delta x})f_k^n + a\frac{\Delta t}{\Delta x}f_{k-1}^n$$

Let us put the values $f_{k-1}^n = \lambda(-1)^{k-1}$, $f_k^n = \lambda(-1)^k$ into the equation and find the value of the grid function f_k^{n+1}

$$f_k^{n+1} = (1 - a\frac{\Delta t}{\Delta x})\lambda(-1)^k + a\frac{\Delta t}{\Delta x}\lambda(-1)^{k-1} = (1 - 2a\frac{\Delta t}{\Delta x})\lambda(-1)^k = (1 - 2a\frac{\Delta t}{\Delta x})f_k^n$$

The condition (8.47) will be fulfilled only if

$$\left|1 - 2a\frac{\Delta t}{\Delta x}\right| \leq 1$$

or

$$-1 \leq 1 - 2a\frac{\Delta t}{\Delta x} \leq 1$$

The stability condition is eventually written down as follows:

$$0 \leq a\frac{\Delta t}{\Delta x} \leq 1 \tag{8.55}$$

We draw a very important conclusion from the condition (8.55) – the finite-difference scheme (8.22) is stable only if the coefficient a value in the transfer Eq. (8.21) is positive ($a > 0$). It can be demonstrated that the finite-difference scheme (8.23) should be applied with the negative values of the coefficient a ($a < 0$), and, in this case, the stability condition is written down as follows:

$$0 \le -a\frac{\Delta t}{\Delta x} \le 1 \tag{8.56}$$

The foregoing method is relatively simple, however, it has limited capabilities. In particular, the application of this method for the schemes (8.24) and (8.25) does not allow obtaining the practically useful result.

8.3.2.2 Spectral Method (Neumann Method) for Stability Evaluation

The spectral method or Neumann method possesses greater generality than the previous one, relatively easy to use, and, therefore, the most widely spread. The method can be applied for linear problems with initial conditions and constant coefficients. The necessary and sufficient stability conditions are established for this class of problems. For a wider class of problems (nonlinear, at the availability of boundary conditions, etc.) only the necessary conditions are provided. Nevertheless, the conclusions obtained with the application of Neumann method when solving nonlinear problems are useful when analyzing the properties of finite-difference schemes.

Neumann method means that the errors δ arising at any computation stage can be represented in the form of the finite complex Fourier row $\delta = \sum_l q_l(t)\exp(i\varphi_l k)$. Here $q_l(t)$ – coefficient being the time functions (in particular, we can accept that $q_l(t_n) = q^n$), $i = \sqrt{-1}$, $\varphi_l = l\pi\Delta x$, $\exp(i\varphi_l k) = \cos(i\varphi_l k) + i\sin(i\varphi_l k)$. In accordance with the stability requirements, the initial computational error needs to diminish at further time layers. For Fourier row, this condition is true for every harmonics (any row element in the sum). This allows using the partial solution in which the error δ of the grid function f_k^n value computation can be represented by the formula $f_k^n = \delta = q^n \exp(i\varphi k)$.

The substitution of this expression into the transfer Eq. (8.21) recorded as (8.22) allows writing down the following

$$\frac{q^{n+1}\exp(i\varphi k)-q^{n}\exp(i\varphi k)}{\Delta t}+a\frac{q^{n}\exp(i\varphi k)-q^{n}\exp(i\varphi (k-1))}{\Delta x}=0$$

After simplifications, the previous equation can be written down as follows:

$$\frac{q-1}{\Delta t}+a\frac{1-\exp(-i\varphi)}{\Delta x}=0$$

From this follows the dependence for q

$$q=(1-a\frac{\Delta t}{\Delta x})+a\frac{\Delta t}{\Delta x}\cdot\exp(-i\varphi) \quad (8.57)$$

The solution for f_k^{n+1} will not increase catastrophically, if the value of q satisfies the condition $|q|\leq 1$. The solution (8.57) for q is the spectra of values located inside the circumference with the radius $\left|a\frac{\Delta t}{\Delta x}\right|$ and center in the point with the coordinate $(1-a\frac{\Delta t}{\Delta x})$ (Figure 8.7). It follows from (8.57) that the condition $|q|\leq 1$ is fulfilled at $0\leq a\frac{\Delta t}{\Delta x}\leq 1$, and this coincides with the result (8.55) obtained by a simpler method of stability evaluation.

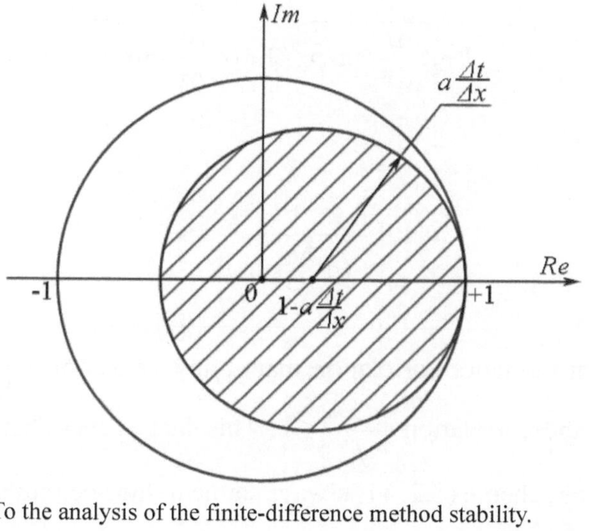

FIGURE 8.7 To the analysis of the finite-difference method stability.

Let us consider other examples:

- approximation of the transfer Eq. (8.21) by the scheme "tripod" (8.24)

$$\frac{q^{n+1}\exp(i\varphi k)-q^{n}\exp(i\varphi k)}{\Delta t}+a\frac{q^{n}\exp(i\varphi(k+1))-q^{n}\exp(i\varphi(k-1))}{2\Delta x}=0$$

After the simplification, the formula for the value q has the following form

$$(q-1)^2=(-a\frac{\Delta t}{\Delta x}i\sin\varphi)^2=(a\frac{\Delta t}{\Delta x}\sin\varphi)^2$$

and one of the solutions of this equation – $q=1+\left|a\frac{\Delta t}{\Delta x}\right|>1$ means that the scheme "tripod" is absolutely unstable;

- approximation of the transfer Eq. (8.21) by the implicit scheme (8.29)

$$\frac{q^{n+1}\exp(i\varphi k)-q^{n}\exp(i\varphi k)}{\Delta t}+a\frac{q^{n+1}\exp(i\varphi k)-q^{n+1}\exp(i\varphi(k-1))}{2\Delta x}=0$$

After the simplification, we have

$$q=\frac{1}{1+a\frac{\Delta t}{\Delta x}i\sin\varphi}=\frac{1-a\frac{\Delta t}{\Delta x}i\sin\varphi}{1+(a\frac{\Delta t}{\Delta x}\sin\varphi)^2}$$

or

$$|q|=\frac{1}{1+(a\frac{\Delta t}{\Delta x}\sin\varphi)^2}\leq 1$$

It follows from the latter correlation that at any values of $a\frac{\Delta t}{\Delta x}$ and at any values of $\frac{\Delta t}{\Delta x}$ the correlation $\left|\frac{f_k^{n+1}}{f_k^{n}}\right|\leq 1$. This fact proves that the implicit finite-difference scheme (8.29) is always stable to the accumulation of computational errors.

Finite-difference schemes stable at any values of $\frac{\Delta t}{\Delta x}$ are called absolutely stable. Implicit finite-difference schemes are usually absolutely stable. However, this property has to be compensated by the application (in comparison with explicit finite-difference schemes) of more complex computational algorithms;

- approximation of the diffusion Eq. (8.30) by the implicit scheme (8.31)

$$\frac{q^{n+1}\exp(i\varphi k) - q^{n}\exp(i\varphi k)}{\Delta t} -$$
$$a\frac{q^{n}\exp(i\varphi(k+1)) - 2q^{n}\exp(i\varphi k) + q^{n}\exp(i\varphi(k-1))}{\Delta x^2} = 0$$

After the simplifications, the expression for q is as follows:

$$q = 1 - 2a\frac{\Delta t}{\Delta x^2}(1-\cos\varphi) = 1 - 4a\frac{\Delta t}{\Delta x^2}\cdot\sin^2\frac{\varphi}{2},$$

and the stability condition of the computational algorithm ($|q| \le 1$) is written down as follows:

$$-1 \le 1 - 4a\frac{\Delta t}{\Delta x^2}\cdot\sin^2\frac{\varphi}{2} \le 1$$

It follows from the latter equation that the finite-difference (8.31) is conditionally stable, and the stability condition is written down as follows:

$$\frac{\Delta t}{\Delta x^2} \le \frac{1}{2a};$$

- approximation of the diffusion Eq. (8.30) by the implicit scheme (8.33)

$$\frac{q^{n+1}\exp(i\varphi k) - q^{n}\exp(i\varphi k)}{\Delta t}$$
$$-a\frac{q^{n+1}\exp(i\varphi(k+1)) - 2q^{n+1}\exp(i\varphi k) + q^{n+1}\exp(i\varphi(k-1))}{\Delta x^2} = 0$$

After the simplifications, the expression for q is as follows:

$$q = \frac{1}{1 + 2a\frac{\Delta t}{\Delta x^2}(1-\cos\varphi)} \le 1$$

- approximation of the diffusion Eq. (8.30) by Dufort-Frankel scheme (8.32)

$$\frac{q^{n+1}\exp(i\varphi k)-q^{n}\exp(i\varphi k)}{\Delta t}$$
$$-a\frac{q^{n}\exp(i\varphi(k+1))-2q^{n+1}\exp(i\varphi k)+q^{n}\exp(i\varphi(k-1))}{\Delta x^{2}}=0$$

After the simplifications and taking into account that $a>0$ and $|\cos\varphi|\leq 1$, the expression for q can be written down as follows:

$$q=\frac{1+2a\dfrac{\Delta t}{\Delta x^{2}}\cos\varphi}{1+2a\dfrac{\Delta t}{\Delta x^{2}}}\leq 1$$

Explicit Dufort-Frankel scheme is absolutely stable, and this is its clear advantage. However, special attention should be paid to its approximation properties. The approximation high accuracy in Dufort-Frankel scheme can be provided by very small values of the integration step by time (formula (8.53)).

The stability conditions for the other finite-difference schemes can be obtained similarly with the foregoing methods. However, it should be pointed out that when using more complicated variants of finite-difference schemes and for more complex differential equations, it is not always possible to explicitly obtain the stability conditions more convenient for practice.

8.3.2.3 Stability Analysis Method Using Eigen Values

The stability analysis method based on the analysis of eigen values is similar, by its ideology, to the spectral Neumann method. On the contrary to this method, the error δ value arising at the regular computation stage is represented in the form of the finite complex row of harmonics containing eigen values and eigen functions for each harmonics $\delta=\sum_{l}\lambda_{l}\exp(i\varphi_{l}k)$. In this expression, for the computational error λ_{l} and $\exp(i\varphi_{l}k)$, respectively, the eigen value and eigen function of the harmonics with the number l are applied.

The considered method can be substantiated as follows. In accordance with the theory of linear systems with constant coefficients, the general solution of the system

$$\frac{d\mathbf{y}}{dt} = \mathbf{A}\,\mathbf{y} \tag{8.58}$$

can be given in the form of the sum [108, 173]

$$y_l = \sum_{l=1}^{L} C_l \exp(\lambda_l t) \tag{8.59}$$

The following designations are used in the Eq. (8.58):

$$\mathbf{y} = \begin{pmatrix} y_1 \\ y_2 \\ \dots \\ y_l \\ \dots \\ y_{L-1} \\ y_L \end{pmatrix}; \quad \mathbf{A} = \begin{pmatrix} a_{11} & a_{12} & \dots & a_{1l} & \dots & a_{1,L-1} & a_{1L} \\ a_{21} & a_{22} & \dots & a_{2l} & \dots & a_{2,L-1} & a_{2L} \\ \dots & \dots & \dots & \dots & \dots & \dots & \dots \\ a_{l1} & a_{l2} & \dots & a_{ll} & \dots & a_{l,L-1} & a_{lL} \\ \dots & \dots & \dots & \dots & \dots & \dots & \dots \\ a_{L-1,1} & a_{L-1,2} & \dots & a_{L-1,l} & \dots & a_{L-1,L-1} & a_{L-1,L} \\ a_{L,1} & a_{L,2} & \dots & a_{L,l} & \dots & a_{L,L-1} & a_{L,L} \end{pmatrix}$$

In the Eq. (8.59), λ_l eigen matrix value can be found by the solution of the algebraic equation

$$\begin{vmatrix} a_{11}-\lambda & a_{12} & \dots & a_{1l} & \dots & a_{1,L-1} & a_{1L} \\ a_{21} & a_{22}-\lambda & \dots & a_{2l} & \dots & a_{2,L-1} & a_{2L} \\ \dots & \dots & \dots & \dots & \dots & \dots & \dots \\ a_{l1} & a_{l2} & \dots & a_{ll}-\lambda & \dots & a_{l,L-1} & a_{lL} \\ \dots & \dots & \dots & \dots & \dots & \dots & \dots \\ a_{L-1,1} & a_{L-1,2} & \dots & a_{L-1,l} & \dots & a_{L-1,L-1}-\lambda & a_{L-1,L} \\ a_{L,1} & a_{L,2} & \dots & a_{L,l} & \dots & a_{L,L-1} & a_{L,L}-\lambda \end{vmatrix} = 0$$

From (8.59) it follows that the solution of the Eq. (8.58) will be decreasing, when all the eigen values $\lambda_l, (l = 1, L)$ are less than one $|\lambda_l| \le 1$ by the absolute value.

In the Eq. (8.58) by vector **y** we can understand how the solutions for the function f determined by the solution of finite-difference equations and the computational error $\delta = f - \bar{f}$ (here: $f, \bar{f}$ – numerical and precise solutions of the equations, respectively). In both cases the matrix **A** will be the same for these equations.

In the same way as in the spectral method, when finding the computational error δ in the considered case, it is possible to use the partial solution in which the error of the grid function f_k^n computational value can be represented by the formula $f_k^n = \delta = \lambda^n \exp(i\varphi k)$. The substitution of this formula into finite-difference schemes allows obtaining exactly the same stability conditions as in the spectral method. This is clear because, in this case, the difference of both approaches consists in the values used in the decomposition of harmonics, which is not principal.

The stability analysis method using eigen values is interesting in those cases when it is possible to obtain the results differing from the aforementioned. Unfortunately, in the most general case, it is impossible to obtain all the matrix **A** eigen values constructed by the coefficients comprised by the finite-difference scheme. However, in this case, the corresponding computational method can be applied to obtain the numerical values of eigen values. It should be additionally mentioned that the stability method with the use of eigen values allows analyzing the stability both inside the computational area and on its boundaries.

Let us consider the practical example [13, 136].

The design of solid fuel rocket engines requires the solution of a number of problems connected with solid fuel heating and combustion, for which the application of conventional methods of solving heat conductivity problems is not possible or their application is linked with low accuracy. In particular, these are the problems of heating, ignition and combustion of the solid mixed fuel which can comprise, for instance, the organic binder and aluminum agglomerates and ammonium perchlorate contained in it. Another similar problem is the problem on the adjustment of the solid fuel combustion surface using thermal knives or non-recoverable heat conducting elements placed in the solid fuel charge body (Figure 8.8). The peculiarity of the aforementioned problems is the heterogeneity of the materials being heated. The thermophysical characteristics inside the structure element being heated can change by leaps and bounds, at the same time, the coefficient values of heat conductivity and specific heat capacities can change by orders. Besides, phase transitions and chemical reactions of exo- or endothermic types can take place in the heated fuel layer.

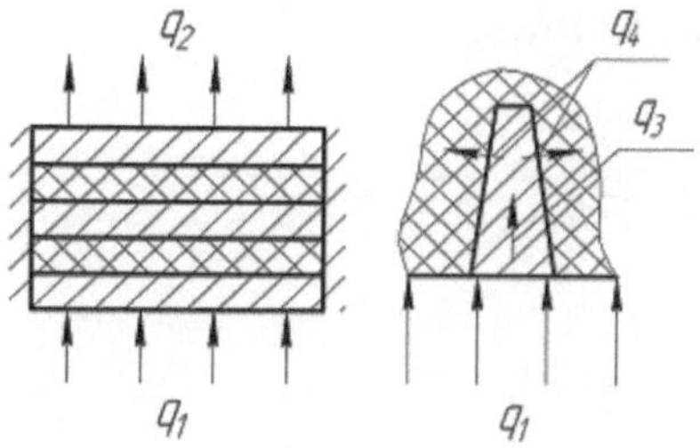

FIGURE 8.8 Schemes of materials with metal inclusions (q_1, q_2, q_3, q_4 – heat flows acting upon the materials being heated).

The problem on heating and igniting the solid fuel taking into account the heterogeneity of its thermophysical properties can be formulated as the problem on heat conductivity in the cylindrical coordinate system in axisymmetric setting. In this problem the values of the densities of the coefficients of heat conductivity and heat capacities are the functions of the computational area coordinates, and there are resource components taking into account the heat effect from the phase transitions and chemical reactions in the right side of the heat conductivity equation

$$\left.\begin{aligned} &\rho c\frac{\partial T}{\partial \tau}=\frac{1}{r}\frac{\partial}{\partial r}\left(kr\frac{\partial T}{\partial r}\right)+\frac{\partial}{\partial z}\left(k\frac{\partial T}{\partial z}\right)+\rho Q(1-\varphi)K_0\exp(-E_a/RT)\\ &\frac{\partial \varphi}{\partial \tau}=(1-\varphi)K_0\exp(-E_a/RT)\end{aligned}\right\} \tag{8.60}$$

Here, in the Eq. (8.60) and further, the following designations are applied:

ρ, c, k – density, specific heat capacity, heat conductivity coefficient; T – temperature; τ – time; r – radial coordinate; z – axial coordinate; Q – reaction thermal effect; φ – share of the substance reacted; K_0 – pre-exponential factor; E_a – effective energy of the decay process activation; R – gas constant.

For simplification, the design case is further considered, in which phase transitions or solid-phase chemical reactions are not discussed.

To solve the heat conductivity problem, the entire computational area is split into the family of elementary volumes (enumerated with two indexes $i=\overline{1,I}$ and $j=\overline{1,J}$ below), for each of them the finite-difference recording of the heat conductivity equation is performed.

The equations are as follows:

$$\frac{\partial T_{i,j}^{n+1}}{\partial \tau} = a_{i,j-\frac{1}{2}} T_{i,j-1}^{n} + a_{i-\frac{1}{2},j} T_{i-1,j}^{n} + a_{i,j} T_{i,j}^{n} - a_{i+\frac{1}{2},j} T_{i+1,j}^{n} - a_{i,j+\frac{1}{2}} T_{i,j+1}^{n} \tag{8.61}$$

Here it is designated:

$$a_{i-\frac{1}{2},j} = \frac{1}{\Delta z_{i-1,j}} \cdot \frac{S_{i-\frac{1}{2},j}}{\Delta v_{ij}} \cdot \frac{k_{i-\frac{1}{2},j}}{\rho_{ij} \cdot c_{ij}}; \quad a_{i+\frac{1}{2},j} = \frac{1}{\Delta z_{i+1,j}} \cdot \frac{S_{i+\frac{1}{2},j}}{\Delta v_{ij}} \cdot \frac{k_{i+\frac{1}{2},j}}{\rho_{ij} \cdot c_{ij}};$$

$$a_{i,j-\frac{1}{2}} = \frac{1}{\Delta r_{i,j-1}} \cdot \frac{S_{i,j-\frac{1}{2}}}{\Delta v_{ij}} \cdot \frac{k_{i,j-\frac{1}{2}}}{\rho_{ij} \cdot c_{ij}}; \quad a_{i,j+\frac{1}{2}} = \frac{1}{\Delta r_{i,j+1}} \cdot \frac{S_{i,j+\frac{1}{2}}}{\Delta v_{ij}} \cdot \frac{k_{i,j+\frac{1}{2}}}{\rho_{ij} \cdot c_{ij}};$$

$$a_{i,j} = a_{i+\frac{1}{2},j} + a_{i,j+\frac{1}{2}} - a_{i,j-\frac{1}{2}} - a_{i-\frac{1}{2},j}$$

The finite-difference approximation of the Eqs. (8.61) is fulfilled by the explicit scheme which does not provide the absolute stability of the computational algorithm. Due to the heterogeneity of the material thermophysical properties, the stability analysis using, for example, the spectral method is cumbersome. To analyze the stability of the computational algorithm applied together with the problem (8.61), we will consider the problem for calculating the error $\xi_m = \dfrac{T_m - \overline{T}_m}{\overline{T}_m}$ determined by the difference of its numerical solution for the temperature T_m and precise value of $\overline{T}_m$ at the partial time moment $t = \dfrac{\tau}{\tau_*}$ (τ_*– characteristic time in the problem being solved).

Taking into account (8.61), the equation system for the error of the numerical solution ξ_m $(m = 1 ... I \times J)$ is written down as follows:

$$\frac{d\xi_m}{dt} = \alpha_{m-1}\xi_{m-1} + \alpha_{m-J}\xi_{m-J} + \alpha_m \xi_m - \alpha_{m+1}\xi_{m+1} - \alpha_{m+J}\xi_{m+J} \tag{8.62}$$

In the Eqs. (8.62) the index m is connected with the indexes i and j by the formula $m = (i-1)J + j$, and the values of the coefficients α_m – by the formula $\alpha_m = a_{ij}\tau_*$.

The equation system (8.62) is the system of ordinary linear differential equations. It can be claimed that the selected finite-difference approximation of the heat conductivity equation will provide its solution stability in the time interval $\tau = 0 \ldots \tau_*$ (or $t = 0 \ldots 1$) when the modules of all eigen values of the matrix A composed of the coefficients α_m $(m = 1, I \times J)$ are less than one. Such conclusion allows bringing the problem on the analysis of the numerical solution stability of the heat conductivity equation to the problem on finding such time value τ_*, at which the maximal by module eigen values of the matrix A will be less than one. In the problem considered, it is advisable to use the value of the equation integration step by the process time $\Delta\tau$ as the characteristic time τ_*. To calculate the eigen values of the matrix A, any known computational methods can be applied, in particular, the orthogonal methods of equivalent transformations [23, 46].

The developed computational algorithms are checked when solving several problems. In particular, the heating of the homogeneous material and multilayer plates are calculated (Figure 8.8). The material is heated by the external heat flow with the lack of internal heat sources. The solution is performed on the irregular grid with the elementary volumes which are demonstrated in Figure 8.9. In the first variant (Figure 8.9a), the elementary volumes of square section are used, in the second (Figure 8.9b) and third (Figure 8.9c) variants – of triangular section. The centers of the elementary volumes (marked with dots) are located in the gravitation centers of the sections. From 60 up to 20,000 finite volumes are used in the calculations.

The two-layer composition is used in the calculations of the heterogeneous material. The first material in the composition is aluminum (thermophysical properties – k = 237 (J/m·s·K), c = 900 (J/kg·K), ρ = 2699 (kg/m³)). The second material – powder "N" (thermophysical properties – k = 0.234 (J/m·s·K), c = 1465 (J/kg·K), ρ = 1600 (kg/m³), ignition temperature – 618 K). The solution is performed in the assumption of the lack of solid-phase chemical reactions and phase transitions. At the initial time moment, the complex material temperature is taken as the same in all the computational area – $T_0 = 293$ K. The material is heated on the material external boundary by the convective thermal flow as a result of the high-temperature gas flow action

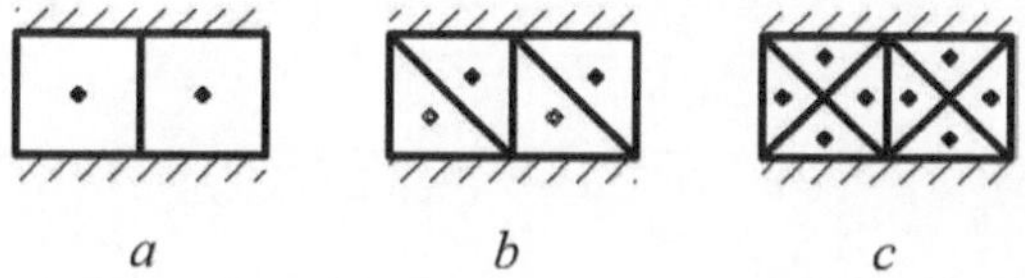

FIGURE 8.9 Types of elementary volumes (*a* – cells of square section; *b*, *c* – sections of triangular section).

(gas temperature – 2500 K), the value of heat transfer coefficient α is taken as constant $\alpha \approx 1000$ W/m²K.

The results of test calculations performed are demonstrated in Figures (8.10)–(8.12).

The results of the matrix A maximal eigen values as the parameter $\lg\left(a\frac{\Delta\tau}{\Delta z^2}\right)$ function are given in Figure 8.10 (here a – material temperature conductivity coefficient). The analysis is carried out for the homogeneous material. The results shown in Figure 8.10 demonstrate that the calculated values of $|\lambda_{\max}| < 1$ correspond to the values of $a\frac{\Delta\tau}{\Delta z^2} < \frac{1}{2}$, and this complies with the known data by the stability of explicit schemes for solving the heat conductivity problem.

The computation results of the multilayer plate heat-up when using the elementary volumes of different types are given in Figure 8.11. In the figures, curve 1 corresponds to the calculations in which the elementary volumes have the rectangular section (Figure 8.9a). Curve 2 corresponds to the calculations in which the elementary volumes have the triangular section of the type shown in Figure 8.9b. Curve 3 corresponds to the calculations of the triangular elementary volumes of the type shown in Figure 8.9c. Curve 4 corresponds to the analytical calculation of the multilayer plate in stationary mode. The results corresponding to the time moment $t \approx 0.001$ sec are shown in Figure 8.11a. In Figure 8.11b, time moment $t \approx 30.0$ sec. The analysis demonstrates that the computational error in all the considered cases is provided on the level allowed for practical calculations (the error does not exceed 2%).

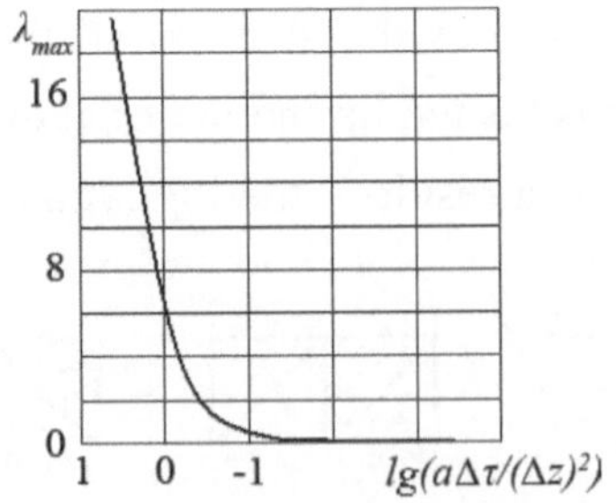

FIGURE 8.10 Dependence of the matrix A maximal eigen value on the complex $a\frac{\Delta\tau}{\Delta z^2}$ value.

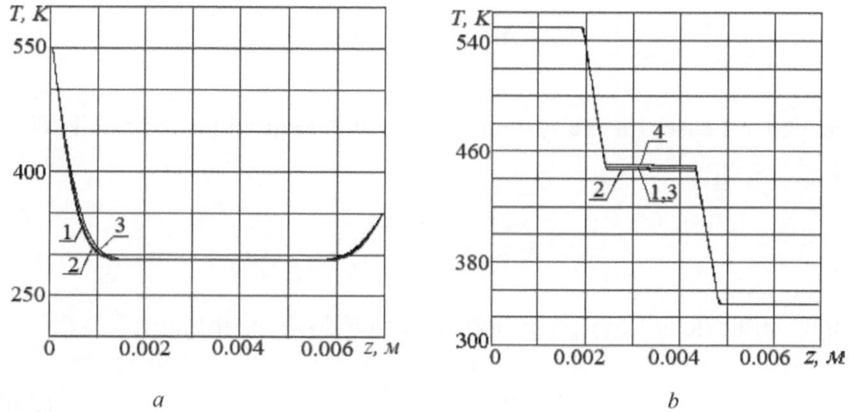

FIGURE 8.11 Multilayer plate heating.

8.3.2.4 Courant-Friedrichs-Levy Condition

The stability evaluation of finite-difference schemes with the help of Courant-Friedrichs-Levy condition [185, 226] is frequently applied in practice despite that it is the necessary stability condition. The sense of the condition is very simple – real and modeled physical processes need to be correlated. In particular, if any disturbing action occurred in any point of the computational area, its propagation velocity in the finite-difference scheme cannot exceed the physical velocity. Let, for example, the medium velocity in some cell be u, the cell length – Δx. In this case, the disturbance propagation time from one boundary to another one is $\Delta t_{max} = \dfrac{\Delta x}{u}$. Courant-Friedrichs-Levy condition requires for the integration step by time Δt in the actually selected calculations of finite-difference problems not to exceed the value Δt_{max}. This condition is written down as follows:

$$\Delta t \leq Ku\,\Delta t_{max} \tag{8.63}$$

In the formula (8.63), Ku – Courant number. In the calculations, $ku \leq 1$ is usually accepted.

In the problems of gas-dynamics the disturbance propagates with the velocity $(|u| + c)$. Here c – sound speed in gas. As a rule, for the problems of gas-dynamics Courant-Friedrichs-Levy stability condition is written down as follows:

$$\Delta t \le Ku \frac{\Delta x}{|u| + c} \tag{8.64}$$

In the general case, for the system of differential equations in partial derivatives of the type

$$\frac{\partial \mathbf{f}}{\partial t} + \mathbf{A}\frac{\partial \mathbf{f}}{\partial x} = 0$$

Courant-Friedrichs-Levy condition is written down as follows:

$$\Delta t \le Ku \frac{\Delta x}{|\lambda|_{\max}} \tag{8.65}$$

In the formula (8.65), $|\lambda|_{\max}$ – matrix A eigen value maximal by the absolute value.

8.3.2.5 Artificial Viscosity Application

It was already mentioned that the recording of differential equations in finite-difference form generates the approximation viscosity. Generally speaking, the availability of approximation viscosity improves the computation scheme stability. In this regard, in some practical situations the additional artificial viscosity whose exact formula is selected as to additionally increase "viscous" properties of the finite-difference scheme is applied.

The variant of the simplest finite-difference scheme of the first order of accuracy by V. V. Rusanov [185] applied at the early stages of hydromechanics development can be given as an example. In accordance with V. V. Rusanov method, instead of the Eq. (8.38) the equation of the following type is solved

$$\frac{\partial f}{\partial t} + \frac{\partial (g + \alpha \cdot f)}{\partial x} = 0$$

Here, the function $g(t,x,f) + \alpha f(t,x)$ is used in the partial derivative by spatial coordinate instead of the function $g(t,x,f)$. The coefficient α value is selected as follows:

$$\alpha = \omega \cdot Ku; \quad Ku = \frac{u + c}{\Delta x} \cdot \Delta t \le 0.9; \quad 0{,}75 > \omega > 0.25$$

The variants of finite-difference schemes with artificial viscosity are applied in different problems of mathematical physics. However, when constructing artificial viscosity, its qualitative influence on the accuracy of the problem quantitative solution and accuracy of the forecast of physical effects should be evaluated. For instance, the viscosity results in "smearing" of shock waves by space in problems of gas-dynamics.

8.3.3 METHODS OF SOLVING FINITE-DIFFERENCE EQUATIONS

When discussing the issues on constructing finite-difference schemes (Section 8.3.1), it was mentioned that the application of implicit schemes results in the necessity of solving the systems of linear algebraic equations. Some popular methods for solving the systems of linear algebraic equations which can be also applied when solving the equations in partial derivatives with implicit finite-difference schemes are considered in Chapter 2. However, in some cases, their application can be ineffective, for instance, due to the following reasons:

- coefficient matrix A is poorly conditioned in the system of linear equations $\mathbf{AX} = \mathbf{B}$. This means that with a large number of equations (number of equations can be over several millions), the computational error can reach the unacceptable level;
- when using the majority of finite-difference schemes, the coefficient matrix A is greatly sparse in the equation system (majority of coefficients equals zero), and this allows simplifying the equation system solution;
- in a number of schemes in the coefficient matrix A only the central diagonal and several diagonals adjoining it are non-zero, and this also allows simplifying the solution of the equation system.

There are original computational methods with high computational stability to solve the system of algebraic equations with poorly conditioned matrixes [79, 102, 167]. In these methods the computational error slightly depends on the number of equations and number of computational operations. As a rule, the final result is calculated with the error comparable with the error of the coefficients in the matrix A. The groups of methods are called orthogonal because they are based on the transformation of the matrix A.

The square matrix $\mathbf{Q}$ is called orthogonal when the expression $\mathbf{Q}^T = \mathbf{Q}^{-1}$ is true (transposed matrix coincides with the reverse matrix).

8.3.3.1 Rotation Method (Givens Method)

Let us consider the unity matrix in which in the lines with the numbers i, j on the diagonal the value of the diagonal elements a_{ii}, a_{jj} differ from one $a_{ii} = a_{jj} = c$, $|c| \le 1$. Besides, the values of the elements a_{ij}, a_{ji} differ from zero and satisfy the conditions $a_{ij} = -s$; $a_{ji} = s$; $c^2 + s^2 = 1$. Such matrix is called ***the rotation matrix***. Let us mark it as $\mathbf{R}_{ij}$. The matrix $\mathbf{R}_{ij}$ and transposed matrix $\mathbf{R}_{ij}^{-1}$ are written down below

$$\mathbf{R}_{\text{ij}} = \begin{pmatrix} 1 & 0 & 0 \ldots\ 0 \ldots 0 \ldots 0 \\ 0 & 1 & 0 \ldots\ 0 \ldots 0 \ldots 0 \\ \ldots\ldots\ldots & & \\ 0 & 0 & 0 \ldots\ c \ldots -s \ldots 0 \\ \ldots\ldots\ldots & & \\ 0 & 0 & 0 \ldots\ s \ldots\ c \ldots 0 \\ 0 & 0 & 0 \ldots\ 0 \ldots\ 0 \ldots 1 \end{pmatrix}, \quad \mathbf{R}_{ij}^{t} = \begin{pmatrix} 1 & 0 & 0 \ldots\ 0 \ldots 0 \ldots 0 \\ 0 & 1 & 0 \ldots\ 0 \ldots 0 \ldots 0 \\ \ldots\ldots\ldots & & \\ 0 & 0 & 0 \ldots\ c \ldots s \ldots 0 \\ \ldots\ldots\ldots & & \\ 0 & 0 & 0 \ldots -s \ldots\ c \ldots 0 \\ 0 & 0 & 0 \ldots\ 0 \ldots\ 0 \ldots 1 \end{pmatrix}$$

The matrix $\mathbf{R}_{ij}$ got its name due to the physical sense which it acquires in some application problems. Thus, in particular, the multiplication of such matrix from the left by the arbitrary matrix $\mathbf{A}$ (matrix size 2×2) sets up the turn of the coordinate axis i, j by some angle φ, $0 \le \varphi < 2\pi$ satisfying the conditions $\cos\varphi = c$, $\sin\varphi = s$ (Figure 8.12a).

The rotation matrix can be used when cancelling some elements of the matrix $\mathbf{A}$. Thus, if it is necessary to cancel the matrix element multiplying

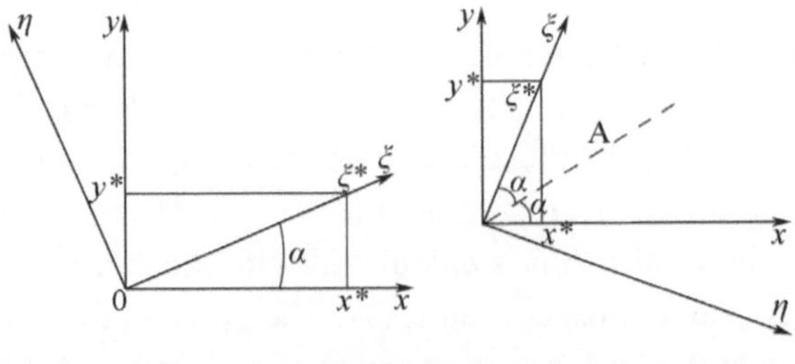

FIGURE 8.12 Geometric interpretation of orthogonal transformations (a – to the rotation methods; b – to the reflection method).

the matrix $\mathbf{R}_{ij}$ on the left by the matrix A, let us change the elements of the matrix a_{ik}, a_{jk} (parameter k changes from *1* to n) for the new $\tilde{a}_{ik}$, $\tilde{a}_{jk}$, calculated by the following correlations

$$\begin{aligned} \tilde{a}_{ik} &= c \cdot a_{ik} - s \cdot a_{jk}, \\ \tilde{a}_{jk} &= s \cdot a_{ik} + c \cdot a_{jk} \end{aligned} \tag{8.66}$$

To provide the equality $\tilde{a}_{ji} = 0$, the following should be accepted

$$c = \frac{a_{ii}}{\sqrt{a_{ii}^2 + a_{ji}^2}}, \qquad s = -\frac{a_{ji}}{\sqrt{a_{ii}^2 + a_{ji}^2}} \tag{8.67}$$

In this case, $\tilde{a}_{ii} = \sqrt{a_{ii}^2 + a_{ji}^2}$.

$$\begin{pmatrix} 1 & 0 & \dots & 0 & \dots & 0 & \dots & 0 & 0 \\ 0 & 1 & \dots & 0 & \dots & 0 & \dots & 0 & 0 \\ \dots & \dots & \dots & \dots & \dots & \dots & \dots & \dots & \dots \\ 0 & 0 & \dots & c & \dots & -s & \dots & 0 & 0 \\ \dots & \dots & \dots & \dots & \dots & \dots & \dots & \dots & \dots \\ 0 & 0 & \dots & s & \dots & c & \dots & 0 & 0 \\ \dots & \dots & \dots & \dots & \dots & \dots & \dots & \dots & \dots \\ 0 & 0 & \dots & 0 & \dots & 0 & \dots & 1 & 0 \\ 0 & 0 & \dots & 0 & \dots & 0 & \dots & 0 & 1 \end{pmatrix} \times$$

$$\begin{pmatrix} a_{11} & a_{12} & \dots & a_{1i} & \dots & a_{1j} & \dots & a_{1,n-1} & a_{1n} \\ a_{21} & a_{22} & \dots & a_{2i} & \dots & a_{2j} & \dots & a_{2,n-1} & a_{2n} \\ \dots & \dots & \dots & \dots & \dots & \dots & \dots & \dots & \dots \\ a_{i1} & a_{i2} & \dots & a_{ii} & \dots & a_{ij} & \dots & a_{i,n-1} & a_{in} \\ \dots & \dots & \dots & \dots & \dots & \dots & \dots & \dots & \dots \\ a_{j1} & a_{j2} & \dots & a_{ji} & \dots & a_{jj} & \dots & a_{j,n-1} & a_{jn} \\ \dots & \dots & \dots & \dots & \dots & \dots & \dots & \dots & \dots \\ a_{n-1,1} & a_{n-1,2} & \dots & a_{n-1,i} & \dots & a_{n-1,j} & \dots & a_{n-1,n-1} & a_{n-1,1} \\ a_{n1} & a_{n2} & \dots & a_{ni} & \dots & a_{nj} & \dots & a_{n-1,n} & a_{nn} \end{pmatrix}$$

$$
=\begin{pmatrix}
a_{11} & a_{12} & \cdots & a_{1i} & \cdots & a_{1j} & \cdots & a_{1,n-1} & a_{1n} \\
a_{21} & a_{22} & \cdots & a_{2i} & \cdots & a_{2j} & \cdots & a_{2,n-1} & a_{2n} \\
\cdots & \cdots & \cdots & \cdots & \cdots & \cdots & \cdots & \cdots & \cdots \\
\tilde{a}_{i1} & \tilde{a}_{i2} & \cdots & \tilde{a}_{ii} & \cdots & \tilde{a}_{ij} & \cdots & \tilde{a}_{i,n-1} & \tilde{a}_{in} \\
\cdots & \cdots & \cdots & \cdots & \cdots & \cdots & \cdots & \cdots & \cdots \\
\tilde{a}_{j1} & \tilde{a}_{j2} & \cdots & \mathbf{0} & \cdots & \tilde{a}_{jj} & \cdots & \tilde{a}_{j,n-1} & \tilde{a}_{jn} \\
\cdots & \cdots & \cdots & \cdots & \cdots & \cdots & \cdots & \cdots & \cdots \\
a_{n-1,1} & a_{n-1,2} & \cdots & a_{n-1,i} & \cdots & a_{n-1,j} & \cdots & a_{n-1,n-1} & a_{n-1,1} \\
a_{n1} & a_{n2} & \cdots & a_{ni} & \cdots & a_{nj} & \cdots & a_{n-1,n} & a_{nn}
\end{pmatrix}
$$

Similarly, the rotation matrix $\mathbf{R}_{ij}$ allows cancelling any element a_{jl}, $l \neq i, j$. For this, the following needs to be accepted

$$
c = \frac{a_{il}}{\sqrt{a_{il}^2 + a_{jl}^2}}, \qquad s = -\frac{a_{jl}}{\sqrt{a_{il}^2 + a_{jl}^2}}
$$

Let us point out the important property of the rotation matrixes. Let us designate $\mathbf{R}_{ij}^T$ – transposed matrix and consider the multiplication result $\mathbf{R}_{ij} \times \mathbf{R}_{ij}^T$:

$$
\begin{pmatrix}
1 \;\; 0 \;\; 0 \ldots 0 \ldots 0 \ldots 0 \\
0 \;\; 1 \;\; 0 \ldots 0 \ldots 0 \ldots 0 \\
\ldots\ldots\ldots\ldots\ldots\ldots \\
0 \;\; 0 \;\; 0 \ldots c \,..\, -s \ldots 0 \\
\ldots\ldots\ldots\ldots\ldots\ldots \\
0 \;\; 0 \;\; 0 \ldots s \ldots c \ldots 0 \\
\ldots\ldots\ldots\ldots\ldots\ldots \\
0 \;\; 0 \;\; 0 \ldots 0 \ldots 0 \ldots 1
\end{pmatrix}
\times
\begin{pmatrix}
1 \;\; 0 \;\; 0 \ldots 0 \ldots 0 \ldots 0 \\
0 \;\; 1 \;\; 0 \ldots 0 \ldots 0 \ldots 0 \\
\ldots\ldots\ldots\ldots\ldots\ldots \\
0 \;\; 0 \;\; 0 \ldots c \ldots s \ldots 0 \\
\ldots\ldots\ldots\ldots\ldots\ldots \\
0 \;\; 0 \;\; 0 \ldots -s \ldots c \ldots 0 \\
\ldots\ldots\ldots\ldots\ldots\ldots \\
0 \;\; 0 \;\; 0 \ldots 0 \ldots 0 \ldots 1
\end{pmatrix}
=
\begin{pmatrix}
1 \;\; 0 \;\; 0 \ldots 0 \ldots 0 \ldots 0 \\
0 \;\; 1 \;\; 0 \ldots 0 \ldots 0 \ldots 0 \\
\ldots\ldots\ldots\ldots\ldots\ldots \\
0 \;\; 0 \;\; 0 \ldots 1 \ldots 0 \ldots 0 \\
\ldots\ldots\ldots\ldots\ldots\ldots \\
0 \;\; 0 \;\; 0 \ldots 0 \ldots 1 \ldots 0 \\
\ldots\ldots\ldots\ldots\ldots\ldots \\
0 \;\; 0 \;\; 0 \ldots 0 \ldots 0 \ldots 1
\end{pmatrix}
$$

In the matrix form, the matrix multiplication result is written down as follows: $\mathbf{R}_{ij} \cdot \mathbf{R}_{ij}^T = \mathbf{E}$. The property mentioned indicates that the rotation matrix is orthogonal. This property is very important and allows using this matrix when creating the orthogonal methods for solving the systems of linear equations.

Let us consider the algorithm of solving the system of linear equations using rotation matrixes:

1. At the first stage, let us multiply the initial equation $\mathbf{AX}=\mathbf{B}$ by the rotation matrix $\mathbf{R}_{n,1}$ (values of the coefficients $c^{(1)}$ and $s^{(1)}$ are calculated by the Eq. (8.67)). When multiplying, all the elements located in the first and last lines of the matrix will change their meanings in the matrix **A** (new values $a_{1,k}$, $a_{n,k}$ for *k=1, n* are calculated by the Eq. (8.66)), at the same time, the element $a_{n,1}$ becomes null – $a_{n,1}=0$;
2. At the second stage, let us multiply the changed matrix $\mathbf{A}^{(1)}=\mathbf{R}_{n,1}\mathbf{A}$ by the rotation matrix $\mathbf{R}_{n-1,1}$ changing the elements in the first and second to last lines of the matrix **A** and cancelling the elements of the matrix $a_{n-1,1}=0$. Repeating this procedure *(n-1)* times, we will cancel the matrix elements located in the first line of the matrix **A** below the central diagonal.

In the matrix form, the transformations indicated are written down as follows:

$$\mathbf{R}_{21}\mathbf{R}_{31}...\mathbf{R}_{i,1}...\mathbf{R}_{n-1,1}\mathbf{R}_{n,1}\mathbf{A}\,\mathbf{X}=\mathbf{R}_{21}\mathbf{R}_{31}...\mathbf{R}_{i,1}...\mathbf{R}_{n-1,1}\mathbf{R}_{n,1}\mathbf{B} \quad (8.68)$$

Let us designate $\mathbf{R}_1=\mathbf{R}_{21}\mathbf{R}_{31}...\mathbf{R}_{i,1}...\mathbf{R}_{n-1,1}\mathbf{R}_{n,1}$; $\mathbf{A}_1=\mathbf{R}_1\mathbf{A}$. Taking these designations into account, the matrix Eq. (8.68) is written down as follows:

$$\mathbf{A}_1\mathbf{X}=\mathbf{R}_1\mathbf{B}$$

Here, all the elements of the matrix $\mathbf{A}_1$ do not coincide with the elements of the matrix **A**

$$\mathbf{A}_1=\begin{pmatrix}
a_{11}^{(2,1)} & a_{12}^{(2,1)} & a_{13}^{(2,1)} & \dots & a_{1,i}^{(2,1)} & \dots & a_{1,n-2}^{(2,1)} & a_{1,n-1}^{(2,1)} & a_{1,n}^{(2,1)} \\
0 & a_{22}^{(2,1)} & a_{23}^{(2,1)} & \dots & a_{2,i}^{(2,1)} & \dots & a_{2,n-2}^{(2,1)} & a_{2,n-1}^{(2,1)} & a_{2,n}^{(2,1)} \\
0 & a_{32}^{(3,1)} & a_{33}^{(3,1)} & \dots & a_{3,i}^{(3,1)} & \dots & a_{3,n-2}^{(3,1)} & a_{3,n-1}^{(3,1)} & a_{3,n}^{(3,1)} \\
\dots & \dots & \dots & \dots & \dots & \dots & \dots & \dots & \dots \\
0 & a_{i,2}^{(i,1)} & a_{i,3}^{(i,1)} & \dots & a_{i,i}^{(i,1)} & \dots & a_{i,n-2}^{(i,1)} & a_{i,n-1}^{(i,1)} & a_{i,n}^{(i,1)} \\
\dots & \dots & \dots & \dots & \dots & \dots & \dots & \dots & \dots \\
0 & a_{n-2,2}^{(n-2,1)} & a_{n-2,3}^{(n-2,1)} & \dots & a_{n-2,i}^{(n-2,1)} & \dots & a_{n-2,n-2}^{(n-2,1)} & a_{n-2,n-1}^{(n-2,1)} & a_{n-2,n}^{(n-2,1)} \\
0 & a_{n-1,2}^{(n-1,1)} & a_{n-1,3}^{(n-1,1)} & \dots & a_{n-1,i}^{(n-1,1)} & \dots & a_{n-1,n-2}^{(n-1,1)} & a_{n-1,n-1}^{(n-1,1)} & a_{n-1,n}^{(n-1,1)} \\
0 & a_{n,2}^{(n,1)} & a_{n,3}^{(n,1)} & \dots & a_{n,i}^{(n,1)} & \dots & a_{n,n-2}^{(n,1)} & a_{n,n-1}^{(n,1)} & a_{n,n}^{(n,1)}
\end{pmatrix};$$

3. Similarly, the matrix $\mathbf{R}_2$ can be constructed

$$\mathbf{R}_2 = \mathbf{R}_{32}\mathbf{R}_{42}...\mathbf{R}_{i,2}...\mathbf{R}_{n-1,2}\mathbf{R}_{n,2}$$

which allows cancelling the elements of the matrix $\mathbf{A}_1$ located in the second column below the central diagonal by the matrix multiplication $\mathbf{R}_2\mathbf{A}_1$.

Applying this procedure $(n-1)$ times, we eventually obtain the matrix $\mathbf{A}_{n-1} = \mathbf{R}$ (here $\mathbf{R} = \mathbf{R}_{n-1}\mathbf{R}_{n-2}...\mathbf{R}_i...\mathbf{R}_2\mathbf{R}_1\mathbf{A}$), in which all the elements below the central diagonal equal zero

$$\mathbf{R} = \mathbf{A}_{n-1} = \begin{pmatrix} a_{11}^{(2,1)} & a_{12}^{(2,1)} & a_{13}^{(2,1)} & ... & a_{1,i}^{(2,1)} & ... & a_{1,n-2}^{(2,1)} & a_{1,n-1}^{(2,1)} & a_{1,n}^{(2,1)} \\ 0 & a_{22}^{(3,2)} & a_{23}^{(3,2)} & ... & a_{2,i}^{(3,2)} & ... & a_{2,n-2}^{(3,2)} & a_{2,n-1}^{(3,2)} & a_{2,n}^{(3,2)} \\ 0 & 0 & a_{33}^{(4,3)} & ... & a_{3,i}^{(4,3)} & ... & a_{3,n-2}^{(4,3)} & a_{3,n-1}^{(4,3)} & a_{3,n}^{(4,3)} \\ ... & ... & ... & ... & ... & ... & ... & ... & ... \\ 0 & 0 & 0 & ... & a_{i,i}^{(i+1,i)} & ... & a_{i,n-2}^{(i+1,i)} & a_{i,n-1}^{(i+1,i)} & a_{i,n}^{(i+1,i)} \\ ... & ... & ... & ... & ... & ... & ... & ... & ... \\ 0 & 0 & 0 & ... & 0 & ... & a_{n-2,n-2}^{(n-1,n-2)} & a_{n-2,n-1}^{(n-1,n-2)} & a_{n-2,n}^{(n-1,n-2)} \\ 0 & 0 & 0 & ... & 0 & ... & 0 & a_{n-1,n-1}^{(n,n-1)} & a_{n-1,n}^{(n,n-1)} \\ 0 & 0 & 0 & ... & 0 & ... & 0 & 0 & a_{n,n}^{(n,n-1)} \end{pmatrix}$$

After the recorded transformations, the initial equation $\mathbf{AX} = \mathbf{B}$ is written down as follows:

$$\mathbf{RX} = \mathbf{R}_{n-1}\mathbf{R}_{n-2}...\mathbf{R}_i...\mathbf{R}_2\mathbf{R}_1\mathbf{B} \tag{8.69}$$

4. Let us consider the multiplication of the matrixes $\mathbf{Q} = \mathbf{R}_1^T\mathbf{R}_2^T...\mathbf{R}_i^T...$ $\mathbf{R}_{n-2}^T\mathbf{R}_{n-1}^T$. Since the matrixes $\mathbf{R}_i$ are orthogonal, the matrix $\mathbf{Q}$ will be also orthogonal. Let us multiply both sides in the Eq. (8.69) by the matrix $\mathbf{Q}$ on the left. Let us consider that

$$\mathbf{R}_1^T\mathbf{R}_2^T...\mathbf{R}_i^T...\mathbf{R}_{n-2}^T\mathbf{R}_{n-1}^T\mathbf{R}_{n-1}\mathbf{R}_{n-2}...\mathbf{R}_i...\mathbf{R}_2\mathbf{R}_1 = \mathbf{E}$$

The latter matrix multiplication allows writing down the initial equation $\mathbf{AX} = \mathbf{B}$ in the eventually transformed form

$$\mathbf{QRX} = \mathbf{B}; \tag{8.70}$$

5. Comparing the Eq. (8.70) with the initial one, we can make the conclusion that the initial coefficient matrix **A** is reduced to the product of two matrixes by the rotation method

$$\mathbf{A} = \mathbf{QR}$$

in which the matrix **R** – upper triangular matrix, and the matrix **Q** – orthogonal.

The solution of the Eq. (8.70) does not present difficulties. Actually, let us designate **Y** = **RX**. Then the Eq. (8.70) is written down as **QY** = **B**. The matrix **Y** is established from the last equation by the ordinary matrix multiplication

$$\mathbf{Y} = \mathbf{Q}^T\mathbf{B} \tag{8.71}$$

and the vector of the unknown of **X** is established by the found value of the matrix **Y** solving the following equation

$$\mathbf{RX} = \mathbf{Y} \tag{8.72}$$

8.3.3.2 Reflection Method (Householder Method)

The shortcoming of applying the rotation method for solving the system of linear algebraic equations is that at one multiplication of the rotation matrix $\mathbf{R}_{ij}$ by matrix **A** only one element of the matrix a_{ij} can be cancelled. Another orthogonal element, in which the reflection matrixes **H** are applied, does not have this shortcoming. In this method, when multiplying the reflection matrix **H** by the matrix **A**, we manage to simultaneously cancel all the elements of the matrix **A** located in one column below the main diagonal. However, the development of the computational algorithm HH using the orthogonal method of reflections is more cumbersome than when applying rotation matrixes. The matrix **H** is called the reflection matrix due to the physical sense, which it obtains in some applied problems (multiplication of the matrix **H** by the matrix with the dimensions 2 × 2 is interpreted as the reflection geometric transformation shown in Figure 8.12b).

The system of linear algebraic equations **AX** = **B** when using the reflection matrixes is solved by the following algorithm:

1. Let us consider the reflection matrix $\mathbf{H}_{(1)}$ with the dimensions $n \times n$ providing the cancellation of the elements of the matrix **A** located in the first column
   - at the first stage, we construct the matrix $\mathbf{W}_{(1)}$ comprising one column and n lines, and we calculate the elements of this matrix by the known values of the elements of the matrix **A** located in the first column

$$\mathbf{W}_{(1)} = \beta_{(1)} \cdot \begin{Bmatrix} a_{11} \mp s_{(1)} \\ a_{21} \\ a_{31} \\ \dots\dots\dots \\ a_{n1} \end{Bmatrix}$$

   Here $s_{(1)} = \sqrt{\sum_{i=1}^{n} a_{i1}^2}$; $\beta_{(1)} = \dfrac{1}{\sqrt{2 \cdot s_{(1)}(s_{(1)} \mp a_{11})}}$. $\beta_{(1)}$ – normalizing coefficient whose value provides the execution of the condition $\mathbf{W}_{(1)}^T \mathbf{W}_{(1)} = 1$ obligatory for this method. The similar condition is also obligatory at the following stages of the method. The availability of the sign $\mp$ in the equations for determining $\mathbf{W}_{(1)}$ and $\beta_{(1)}$ demonstrates that the problem on constructing the reflection matrix has two independent solutions;
   - matrix product $\mathbf{W}_{(1)} \mathbf{W}_{(1)}^T$ is determined

$$\mathbf{W}_{(1)} \mathbf{W}_{(1)}^T = \beta_{(1)}^2 \cdot \begin{Bmatrix} a_{11} \mp s_{(1)} \\ a_{21} \\ a_{31} \\ \dots\dots\dots \\ a_{n1} \end{Bmatrix} \cdot \left\{ a_{11} \mp s_{(1)} \;\; a_{21} \; a_{31} \dots a_{n1} \right\}$$

$$= \frac{1}{s_{(1)}(s_{(1)} \mp a_{11})} \cdot \begin{Bmatrix} (a_{11} \mp s_{(1)})^2 & (a_{11} \mp s_{(1)}) \cdot a_{21} & \dots & (a_{11} \mp s_{(1)}) \cdot a_{n1} \\ a_{21}(a_{11} \mp s_{(1)}) & a_{21} \cdot a_{21} & \dots & a_{21} \cdot a_{n1} \\ \dots\dots & \dots\dots & \dots\dots & \dots\dots \\ a_{n1}(a_{11} \mp s_{(1)}) & a_{n1} \cdot a_{21} & \dots & a_{n1} \cdot a_{n1} \end{Bmatrix};$$

   - reflection matrix $\mathbf{H}_{(1)}$ is calculated by the formula $\mathbf{H}_{(1)} = \mathbf{E} - \mathbf{2W}_{(1)} \mathbf{W}_{(1)}^T$

$$\mathbf{H}_{(1)} = \begin{Bmatrix} 1-\dfrac{(a_{11} \mp s_{(1)})^2}{s_{(1)}(s_{(1)} \mp a_{11})} & \dfrac{(a_{11} \mp s_{(1)}) \cdot a_{21}}{s_{(1)}(s_{(1)} \mp a_{11})} & \cdots & \dfrac{(a_{11} \mp s_{(1)}) \cdot a_{n1}}{s_{(1)}(s_{(1)} \mp a_{11})} \\ \dfrac{a_{21}(a_{11} \mp s_{(1)})}{s_{(1)}(s_{(1)} \mp a_{11})} & 1-\dfrac{a_{21} \cdot a_{21}}{s_{(1)}(s_{(1)} \mp a_{11})} & \cdots & \dfrac{a_{21} \cdot a_{n1}}{s_{(1)}(s_{(1)} \mp a_{11})} \\ \cdots & \cdots & \cdots & \cdots \\ \dfrac{a_{n1}(a_{11} \mp s_{(1)})}{s_{(1)}(s_{(1)} \mp a_{11})} & \dfrac{a_{n1} \cdot a_{21}}{s_{(1)}(s_{(1)} \mp a_{11})} & \cdots & 1-\dfrac{a_{n1} \cdot a_{n1}}{s_{(1)}(s_{(1)} \mp a_{11})} \end{Bmatrix}; \tag{8.73}$$

The matrix $\mathbf{H}_{(1)}$ is orthogonal and this is so because the condition $\mathbf{H}_{(1)}\mathbf{H}_{(1)}^T = \mathbf{E}$ is true;

2. The new matrix $\mathbf{A}_1$ is established by the matrix multiplication $\mathbf{H}_{(1)}\mathbf{A}$, all the elements of which are located below the central diagonal in the first column and equal zero

$$\mathbf{A}_1 = \mathbf{H}_{(1)}\mathbf{A} = \begin{pmatrix} a_{11}^{(1)} & a_{12}^{(1)} & a_{13}^{(1)} & \cdots & a_{1,i}^{(1)} & \cdots & a_{1,n-2}^{(1)} & a_{1,n-1}^{(1)} & a_{1,n}^{(1)} \\ 0 & a_{22}^{(1)} & a_{23}^{(1)} & \cdots & a_{2,i}^{(1)} & \cdots & a_{2,n-2}^{(1)} & a_{2,n-1}^{(1)} & a_{2,n}^{(1)} \\ 0 & a_{32}^{(1)} & a_{33}^{(1)} & \cdots & a_{3,i}^{(1)} & \cdots & a_{3,n-2}^{(1)} & a_{3,n-1}^{(1)} & a_{3,n}^{(1)} \\ \cdots & \cdots & \cdots & \cdots & \cdots & \cdots & \cdots & \cdots & \cdots \\ 0 & a_{i,2}^{(1)} & a_{i,3}^{(1)} & \cdots & a_{i,i}^{(1)} & \cdots & a_{i,n-2}^{(1)} & a_{i,n-1}^{(1)} & a_{i,n}^{(1)} \\ \cdots & \cdots & \cdots & \cdots & \cdots & \cdots & \cdots & \cdots & \cdots \\ 0 & a_{n-2,2}^{(1)} & a_{n-2,3}^{(1)} & \cdots & a_{n-2,i}^{(1)} & \cdots & a_{n-2,n-2}^{(1)} & a_{n-2,n-1}^{(1)} & a_{n-2,n}^{(1)} \\ 0 & a_{n-1,2}^{(1)} & a_{n-1,3}^{(1)} & \cdots & a_{n-1,i}^{(1)} & \cdots & a_{n-1,n-2}^{(1)} & a_{n-1,n-1}^{(1)} & a_{n-1,n}^{(1)} \\ 0 & a_{n,2}^{(1)} & a_{n,3}^{(1)} & \cdots & a_{n,i}^{(1)} & \cdots & a_{n,n-2}^{(1)} & a_{n,n-1}^{(1)} & a_{n,n}^{(1)} \end{pmatrix}$$

3. At the following stage of the computational algorithm, the reflection matrix $\mathbf{H}_{(2)}$ is constructed whose multiplication by the matrix $\mathbf{A}_1$ allows cancelling all the matrix elements located below the central diagonal in the second column. When multiplying, the matrix $\mathbf{H}_{(2)}$ should not change the null values of the matrix elements in the first column. Let us designate the elements of the matrix $\mathbf{A}_1$ – $a_{ij}^{(1)}$. The reflection matrix $\mathbf{H}_{(2)}$ will be constructed by the elements $a_{ij}^{(1)}$ $(n \geq i,\ j \geq 2)$. Eventually, it looks as follows:

$$\mathbf{H}_{(2)} = \begin{Bmatrix} 1 & 0 & 0 & \dots & 0 \\ 0 & 1 - \dfrac{(a_{22}^{(1)} \mp s_{(2)})^2}{s_{(2)}(s_{(2)} \mp a_{22}^{(1)})} & \dfrac{(a_{22}^{(1)} \mp s_{(2)}) \cdot a_{32}^{(1)}}{s_{(2)}(s_{(2)} \mp a_{22}^{(1)})} & \dots & \dfrac{(a_{22}^{(1)} \mp s_{(2)}) \cdot a_{n2}^{(1)}}{s_{(2)}(s_{(2)} \mp a_{22}^{(1)})} \\ 0 & \dfrac{a_{32}^{(1)}(a_{22}^{(1)} \mp s_{(2)})}{s_{(2)}(s_{(2)} \mp a_{22}^{(1)})} & 1 - \dfrac{a_{32}^{(1)} \cdot a_{32}^{(1)}}{s_{(2)}(s_{(2)} \mp a_{22}^{(1)})} & \dots & \dfrac{a_{32}^{(1)} \cdot a_{n3}^{(1)}}{s_{(2)}(s_{(2)} \mp a_{22}^{(1)})} \\ \dots & \dots & \dots & \dots & \dots \\ 0 & \dfrac{a_{n2}^{(1)}(a_{22}^{(1)} \mp s_{(2)})}{s_{(2)}(s_{(2)} \mp a_{22}^{(1)})} & \dfrac{a_{n2}^{(1)} \cdot a_{32}^{(1)}}{s_{(2)}(s_{(2)} \mp a_{22}^{(1)})} & \dots & 1 - \dfrac{a_{n2}^{(1)} \cdot a_{n2}^{(1)}}{s_{(2)}(s_{(2)} \mp a_{22}^{(1)})} \end{Bmatrix}$$

4. Matrix $\mathbf{H}_{(2)}\mathbf{A}_1 = \mathbf{H}_{(2)}\mathbf{H}_{(1)}\mathbf{A}$ will contain zero elements in the first and second columns already (zero elements are located below the central diagonal). The presented procedure of the reflection matrix development should be repeated $(n-1)$ times. Eventually, the matrix $\mathbf{H}_{(n-1)} \dots \mathbf{H}_{(i)} \dots \mathbf{H}_{(2)}\mathbf{H}_{(1)}\mathbf{A}$ becomes the upper triangular matrix. Let us designate this matrix as $\mathbf{R}$.
5. Taking into account the aforementioned matrix transformations, the equation $\mathbf{AX} = \mathbf{B}$ will look as follows:

$$\mathbf{RX} = \mathbf{H}_{(n-1)} \dots \mathbf{H}_{(i)} \dots \mathbf{H}_{(2)}\mathbf{H}_{(1)}\mathbf{B} \tag{8.74}$$

Let us multiply both left sides of the latter equation by the matrix $\mathbf{Q} = \mathbf{H}_{(1)}^T\mathbf{H}_{(2)}^T \dots \mathbf{H}_{(i)}^T \dots \mathbf{H}_{(n-2)}^T\mathbf{H}_{(n-1)}^T$. Taking into account the orthogonality of the matrixes $\mathbf{H}_{(i)}$, such multiplication allows rewriting the Eq. (8.74) in the form which has been already discussed when presenting the rotation matrixes

$$\mathbf{QRX} = \mathbf{B} \tag{8.75}$$

In the same way as in the Eqs. (8.70), in the Eq. (8.75) $\mathbf{Q}$ – orthogonal matrix, and $\mathbf{R}$ – upper triangular matrix. The final stage of solving the system of linear equations does not differ from the aforementioned (Eqs. (8.71) and (8.72)).

8.3.3.3 Iteration Methods (Jacobi Methods)

The iteration methods of solving the system of nonlinear equations were considered in Chapter 4. It was mentioned that the methods can be also successfully applied for solving the systems of linear equations. Actually, the application of iteration methods can be rather effective when solving large systems of linear equations. When fulfilling the convergence conditions, the number of the necessary iterations required for the solution can be relatively small.

Let us give in matrix form some of the variants of the realization of iteration methods for solving the systems of linear algebraic equations of the type $\mathbf{AX} = \mathbf{B}$ which were not considered in Chapter 4:

- let us present the matrix $\mathbf{A}$ as the sum of three matrixes $\mathbf{A} = \mathbf{L} + \mathbf{D} + \mathbf{R}$, here $\mathbf{D}$, $\mathbf{L}$, $\mathbf{R}$ – diagonal matrix only filled with the diagonal elements a_{ii} of the matrix $\mathbf{A}$, and its non-diagonal parts – lower and upper triangular matrixes, respectively. The simplest iteration process is arranged by the following algorithm

$$\mathbf{X}^{(k+1)} = -\mathbf{D}^{-1}(\mathbf{L} + \mathbf{R})\mathbf{X}^{(k)} + \mathbf{D}^{-1}\mathbf{B} \tag{8.76}$$

  In accordance with the known theorems (for instance, [63]), the iteration process converges only if the diagonal dominance condition is fulfilled for the elements of the matrix $\mathbf{A}$

$$|a_{ii}| > \sum_{\substack{j=1 \\ j\neq i}} |a_{ij}|, \quad \forall\, i = 1,\ldots n; \tag{8.77}$$

- Seidel iteration algorithm (Gauss-Seidel) is developed by the matrix formula

$$\mathbf{X}^{(k+1)} = -(\mathbf{L} + \mathbf{D})^{-1}\mathbf{R}\,\mathbf{X}^{(k)} + (\mathbf{L} + \mathbf{D})^{-1}\mathbf{B} \tag{8.78}$$

  The convergence conditions in Seidel method are close to those providing the convergence in the ordinary iteration method;
- relaxation methods are arranged by the following algorithm

$$\mathbf{X}^{(k+1)} = (\mathbf{D} + \omega\mathbf{L})^{-1}((1-\omega)\mathbf{D} - \omega\mathbf{R})\mathbf{X}^{(k)} + \omega(\mathbf{D} + \omega\mathbf{L})^{-1}\mathbf{B} \tag{8.79}$$

In the formula (8.79) ω – relaxation coefficient whose value is taken in the interval $0 < \omega < 2$. The relaxation methods are called lower, if $\omega < 1$, and upper, if $\omega > 1$. The successful selection of the relaxation parameter can significantly increase the rate of the iteration process convergence, however, in any case, the condition (8.77) is not excluded to provide the convergence.

8.3.3.4 Sweep Methods

Sweep methods are spectacular and effective as applied to the systems of algebraic linear equations [63, 196]. These methods are applicable for significantly sparse matrixes in which the nonzero elements a_{ij} of the matrix **A** are placed on several diagonals of the matrix.

The simplest variant at which the sweep method is possible – the three-diagonal matrix **A**. In this matrix, the central diagonal and two ones adjoining it on the top and bottom are nonzero

$$\mathbf{A} = \begin{Bmatrix} a_{11} & a_{12} & 0 & \dots & 0 & 0 & 0 & \dots & 0 & 0 & 0 \\ a_{21} & a_{22} & a_{23} & \dots & 0 & 0 & 0 & \dots & 0 & 0 & 0 \\ 0 & a_{32} & a_{33} & \dots & 0 & 0 & 0 & \dots & 0 & 0 & 0 \\ \dots & \dots & \dots & \dots & \dots & \dots & \dots & \dots & \dots & \dots & \dots \\ 0 & 0 & 0 & \dots & a_{i-1,i-1} & a_{i-1,i} & 0 & \dots & 0 & 0 & 0 \\ 0 & 0 & 0 & \dots & a_{i,i-1} & a_{i,i} & a_{i,i+1} & \dots & 0 & 0 & 0 \\ 0 & 0 & 0 & \dots & 0 & a_{i+1,i} & a_{i+1,i+1} & \dots & 0 & 0 & 0 \\ \dots & \dots & \dots & \dots & \dots & \dots & \dots & \dots & \dots & \dots & \dots \\ 0 & 0 & 0 & \dots & 0 & 0 & 0 & \dots & a_{n-2,n-2} & a_{n-2,n-1} & 0 \\ 0 & 0 & 0 & \dots & 0 & 0 & 0 & \dots & a_{n-1,n-2} & a_{n-1,n-1} & a_{n-1,n} \\ 0 & 0 & 0 & \dots & 0 & 0 & 0 & \dots & 0 & a_{n,n-1} & a_{n,n} \end{Bmatrix}$$

The equation system $\mathbf{AX} = \mathbf{B}$ can be rewritten as follows:

$$a_{1,1}x_1 + a_{1,2}x_2 = b_1$$

$$a_{i-1,i}x_{i-1} + a_{i,i}x_i + a_{i+1,i}x_{i+1} = b_i, \quad i = 2, n-1$$

$$a_{n,n-1}x_{n-1} + a_{n,n}x_n = b_n \tag{8.80}$$

The sweep method – the method of excluding the unknown in which the assumption is used that it is possible to combine the unknown to be found x_{i-1}, x_i by the following recurrent correlation

$$x_{i-1} = L_{i-1}x_i + M_{i-1} \tag{8.81}$$

It can be easily seen that such assumption is true with the index value $i = 2$. For this case, the Eq. (8.81) is written down as follows:

$$x_1 = -\frac{a_{1,2}}{a_{1,1}}x_2 + \frac{b_1}{a_{1,1}} \tag{8.82}$$

If the correlation (8.81) is true, the following correlation is true as well

$$x_i = L_i x_{i+1} + M_i \tag{8.83}$$

Let us substitute the Eq. (8.81) into i equation of the system (8.80) and successively perform the following transformations

$$a_{i-1,i}(L_{i-1}x_i + M_{i-1}) + a_{i,i}x_i + a_{i+1,i}x_{i+1} = b_i, \quad i = 2, n-1;$$

$$x_i(a_{i-1,i}L_{i-1} + a_{i,i}) + a_{i+1,i}x_{i+1} = b_i - a_{i-1,i}M_{i-1}, \quad i = 2, n-1;$$

$$x_i = -\frac{a_{i+1,i}}{a_{i-1,i}L_{i-1} + a_{i,i}}x_{i+1} + \frac{b_i - a_{i-1,i}M_{i-1}}{a_{i-1,i}L_{i-1} + a_{i,i}}, \quad i = 2, n-1 \tag{8.84}$$

The Eq. (8.84) confirms the truthfulness of the assumption written down as (8.81). Comparing the Eqs. (8.83) and (8.84), we can write down the values of the coefficients L_i, M_i as the functions of the matrix **A** elements and values of the coefficients L_{i-1}, M_{i-1}

$$L_i = -\frac{a_{i+1,i}}{a_{i-1,i}L_{i-1} + a_{i,i}}; \quad M_i = \frac{b_i - a_{i-1,i}M_{i-1}}{a_{i-1,i}L_{i-1} + a_{i,i}}, \quad i = 2, n-1 \tag{8.85}$$

The coefficients L_i, M_i (these coefficients are called sweep) can be found for all the values $i = 1, n-1$:

- at $i = 1$ based on the Eq. (8.82) we have

$$L_1 = -\frac{a_{1,2}}{a_{1,1}}; \quad M_1 = \frac{b_1}{a_{1,1}}; \tag{8.86}$$

- for all the other values $i = 1, n-1$ sweep coefficients are found based on the Eqs. (8.85).

Now we find the unknown values of x_i by the known values of all the coefficients L_i, M_i. Let us consider two equations for the unknown x_{n-1}, x_n. The first equation is the equation from the system (8.80), and the second one – the Eq. (8.83) recorded for the index value $i = n-1$:

$$a_{n,n-1}x_{n-1} + a_{n,n}x_n = b_n;$$

$$x_{n-1} = L_{n-1}x_n + M_{n-1} \tag{8.87}$$

Solving the Eq. (8.87), we have the following solution for x_n

$$x_n = \frac{b_n - a_{n,n-1}M_{n-1}}{a_{n,n-1}L_{n-1} + a_{n,n}} \tag{8.88}$$

All the other unknown x_i are established in the cycle for $i = n-1$ to $i = 1$ by the formula (8.83).

In the method considered, the first stage at which the sweep coefficients L_i, M_i are calculated is called the direct sweep, and the second stage at which the unknown x_i are calculated is called the inverse sweep. In the same way as in Jacobi method, the sweep method is absolutely stable only if the following condition is fulfilled

$$|a_{i,i}| \geq |a_{i,i-1}| + |a_{i,i+1}| \tag{8.89}$$

Let us consider the sweep method application for the 5-diagonal matrix **A**. The equation system can be written down as follows:

$$c_0x_0 + d_0x_1 + e_0x_2 = f_0 \tag{8.90}$$

$$b_1x_0 + c_1x_1 + d_1x_2 + e_1x_3 = f_1 \tag{8.91}$$

$$a_2x_0 + b_2x_1 + c_2x_2 + d_2x_3 + e_2x_4 = f_2 \tag{8.92}$$

$$\ldots$$

$$a_i x_{i-2} + b_i x_{i-1} + c_i x_i + d_i x_{i+1} + e_i x_{i+2} = f_i, \quad i=3,\ N\text{-}3 \tag{8.93}$$

$$\ldots$$

$$a_{N-2} x_{N-4} + b_{N-2} x_{N-3} + c_{N-2} x_{N-2} + d_{N-2} x_{N-1} + e_{N-2} x_N = f_{N-2} \tag{8.94}$$

$$a_{N-1} x_{N-3} + b_{N-1} x_{N-2} + c_{N-1} x_{N-1} + d_{N-1} x_N = f_{N-1} \tag{8.95}$$

$$a_N x_{N-2} + b_N x_{N-1} + c_N x_N = f_N \tag{8.96}$$

To solve the Eq. (8.90)–(8.96), the following algorithm is applied [196]:

- it is assumed that the following correlations are true

$$x_i = \alpha_{i+1} x_{i+1} - \beta_{i+1} x_{i+2} + \gamma_{i+1}, \text{ for } \quad i=0,\ N-2; \tag{8.97}$$

$$x_{N-1} = \alpha_N x_N + \gamma_N, \text{ for } \quad i=N-1; \tag{8.98}$$

- in accordance with (8.97), the values for x_{i-1}, x_{i-2} are written down:

$$x_{i-1} = \alpha_i x_i - \beta_i x_{i+1} + \gamma_i; \tag{8.99}$$

$$x_{i-2} = \alpha_{i-1} x_{i-1} - \beta_{i-1} x_i + \gamma_{i-1}; \tag{8.100}$$

- in the Eq. (8.100) the value of x_{i-1} written down in the form (8.99) is substituted

$$x_{i-2} = \alpha_{i-1} x_{i-1} - \beta_{i-1} x_i + \gamma_{i-1} = \alpha_{i-1}(\alpha_i x_i - \beta_i x_{i+1} + \gamma_i) - \beta_{i-1} x_i + \gamma_{i-1};$$

$$x_{i-2} = (\alpha_{i-1}\alpha_i - \beta_{i-1}) x_i + \gamma_i) - \beta_i \alpha_{i-1} x_{i+1} + \alpha_{i-1}\gamma_i + \gamma_{i-1} \tag{8.101}$$

The Eq. (8.101) is true at *i=2, N–1;*

- expressions obtained for x_{i-2} (Eq. (8.101)), x_{i-1} (Eq. (8.99) and x_i (Eq. (8.97) are substituted into the Eq. (8.93) and this allows obtaining the bond of the unknown x_i, x_{i+1} and x_{i+2}

$$a_i((\alpha_{i-1}\alpha_i - \beta_{i-1}) x_i + \gamma_i) - \beta_i \alpha_{i-1} x_{i+1} + \alpha_{i-1}\gamma_i + \gamma_{i-1})$$
$$+ b_i(\alpha_i x_i - \beta_{i-1} x_{i+1} + \gamma_i) + c_i x_i + d_i x_{i+1} + e_i x_{i+2} = f_i$$

After the transformations, the equation can be rewritten as (8.97), at the same time, the coefficients α_{i+1}, β_{i+1}, γ_{i+1} can be recorded as (formulas are true for $2 \le i \le N-1$)

$$\alpha_{i+1} = \frac{1}{\Delta_i} \cdot (-d_i + \beta_i \cdot (a_i \cdot \alpha_{i-1} + b_i)),$$

$$\beta_{i+1} = \frac{e_i}{\Delta_i},$$

$$\gamma_{i+1} = \frac{1}{\Delta_i} \cdot (f_i - a_i \cdot \gamma_{i-1} - \gamma_i \cdot (a_i \cdot \alpha_{i-1} + b_i)),$$

$$\Delta_i = c_i - a_i \cdot \beta_{i-1} + \alpha_i \cdot (a_i \cdot \alpha_{i-1} + b_i); \qquad (8.102)$$

- Eqs. (8.102) are the equations for sweep coefficients whose values can be found by the direct sweep. When performing the direct sweep, the values of the coefficients $\alpha_1, \beta_1, \gamma_1$ and $\alpha_2, \beta_2, \gamma_2$ are found from the Eqs. (8.90), (8.91)

$$\alpha_1 = -\frac{d_0}{c_0}, \quad \beta_1 = \frac{e_0}{c_0}, \quad \gamma_1 = \frac{f_0}{c_0},$$

$$\alpha_2 = \frac{-d_1 + b_1 \cdot \beta_1}{c_1 + b_1 \cdot \alpha_1}, \quad \beta_2 = \frac{e_1}{c_1 + b_1 \cdot \alpha}, \quad \gamma_1 = \frac{f_1 - b_1 \cdot \gamma_1}{c_1 + b_1 \cdot \alpha}; \qquad (8.103)$$

- taking into account the found values of the coefficients $\alpha_i, \beta_i, \gamma_i$, the inverse sweep is performed following the correlations

$$x_N = \gamma_{N+1},$$

$$x_{N-1} = \alpha_N x_N + \gamma_N,$$

$$x_i = \alpha_{i+1} x_{i+1} - \beta_{i+1} x_{i+2} + \gamma_{i+1}, \ i = n-2, \ldots, 0 \qquad (8.104)$$

- aforementioned algorithm comprising the direct sweep by the Eqs. (8.103) and (8.102) and the inverse sweep performed by the Eqs. (8.104) is stable if the following conditions are fulfilled

$$|a_i| > 0 \text{ at } 2 \le i \le N,$$

$$|b_i| > 0 \text{ at } 1 \le i \le N,$$

$$|d_i| > 0 \text{ at } 0 \le i \le N-1,$$

$$|e_i| > 0 \text{ at } 0 \le i \le N-2,$$

$$|c_0| \geq |d_0| + |e_0|,$$

$$|c_1| \geq |b_1| + |d_1| + |e_1|,$$

$$|c_i| \geq |a_i| + |b_i| + |d_i| + |e_i|, \; 2 \leq i \leq N-2,$$

$$|c_{N-1}| \geq |a_{N-1}| + |b_{N-1}| + |d_{N-1}|,$$

$$|c_N| \geq |a_N| + |b_N|$$

It is proved in [196] that, at least, one of the foregoing inequalities needs to be strict. In this case, the computational algorithm is stable and the following correlation is additionally performed

$$|\alpha_i| + |\beta_i| \leq 1 \text{ at } 1 \leq i \leq N-1, \; |\alpha_N| \leq 1$$

Let us additionally mention that the monographs [167, 227] also contain other powerful computational procedures whose application is especially effective when solving the problems on computers with parallel architecture.

8.3.4 SOLUTION OF THE PROBLEM ON HEATING A SOLID MATERIAL AND ITS FURTHER COMBUSTION

Let us consider the problem on the process of the solid fuel ignition covered by an inert material, as an example. Let us accept that physical and chemical processes taking place during the solid fuel combustion correspond to the solid-phase combustion model [47, 98, 152, 161] and require the following factors to be taken into account:

- state of the gaseous phase (environment) to determine the convective, conductive and radiant components of the heat flow;
- state of the interface of the gaseous phase and solid fuel to evaluate power effects which can arise at phase transitions (melting and evaporation of the coating and solid fuel);
- thermophysical properties of the materials heated up (coating and solid fuel) which influence the heating up processes, etc.

The problem on heating up and combusting the solid fuel with the inert coating is solved at the following assumptions:

- solid fuel and inert coating are considered homogeneous media with the known thermophysical characteristics;
- heat transfer in the solid fuel is performed due to the heat conductivity mechanisms;
- one reaction of Arrhenius type can occur in the solid fuel;
- phase transitions can occur in the inert coating;
- solid fuel surface is heated up externally due to the convective heat transfer;
- optical transparency of the coating material and (or) fuel are not taken into account.

With the assumptions formulated, the equations of the solid fuel heating up in one-dimensional setting are written down as follows:

$$\rho c \frac{\partial T}{\partial t} = \frac{\partial}{\partial x} \lambda \frac{\partial T}{\partial x} + \rho Q \cdot \Phi(T, \beta)$$

$$\frac{\partial \beta_i}{\partial t} = \Phi(T, \beta) \tag{8.105}$$

The function $\Phi(T, \beta)$ in the Eq. (8.105) is different when calculating the coating (phase transitions are modeled)

$$\Phi(T, \beta) = k(1 - \beta) \quad \text{at } T \geq T_{nл} \tag{8.106}$$

and when calculating the solid fuel (solid-phase chemical reaction is modeled):

$$\Phi(T, \beta) = k \cdot \beta^{\alpha} \cdot (1 - \beta)^{\gamma} \cdot \exp(-\frac{E}{RT}) \tag{8.107}$$

In the aforementioned and further equations the following designations are used:

- t, x – process time and spatial coordinate (directed perpendicular to the fuel surface inside the heated layer);
- ρ – material density;
- c, λ – specific heat capacity and heat conductivity coefficient;
- T, β – temperature and depth of the chemical reaction or phase transition;
- Q, k, E/R, α, γ – coefficients in the laws for the chemical reaction and phase transition;

- T_s, T_* – temperature on the surface of the material heated up and ignition temperature (critical temperature);
- β_s, β_* – depth of the chemical reaction on the surface of the material heated up and its critical value;
- q_* – critical value of the heat flow from the gaseous phase into the material.

The Eqs. (8.105) can be used when calculating the solid fuel heating up, its ignition and further non-stationary combustion. In the Eqs. (8.105) the values of β_i change in the interval $1 > \beta_i > 0$.

Phase transitions linked with the melting of the solid inert material can be formally accepted as the endothermic reaction (reaction taking place with energy absorption) flowing by Arrhenius equation (Eq. (8.107), $\frac{E_i}{R_0 T} \approx 0$). If we accept the melting law as Arrenius law, the same equations, as when calculating the reaction in the solid fuel, can be applied when calculating the heating up of the inert material. The values of the chemical reaction energy Q in the correlation for the phase transition are taken as negative. When calculating the phase transitions by the correlation (8.106), the value Q is the specific melting heat whose value can be taken, for instance, in accordance with [212, 224].

Kinetic parameters in the Eq. (8.107) for the chemical reaction of Arrhenius type can be established, for example, by the techniques [43, 139]. For some fuel formulas, the values of kinetic parameters found experimentally can be obtained in special literature.

The Eqs. (8.105)–(8.107) to calculate the multilayer material heating up are solved when setting up the initial and boundary conditions. The setting up of the initial conditions for the time moment t_0 (beginning time of the equation integration) of the temperature field $T_0(x)$ and field for the depth of the chemical reactions (or depth of the phase transformation) $\beta_0(x)$. In particular, if the profiles of the temperature and depth of the chemical reactions (depth of the phase transformation) do not have peculiarities, when integrating the equations starting from the time moment $t_0 = 0$, the initial conditions are written down as follows:

$$t = 0 \rightarrow \quad T(x) = T_0; \quad \beta_i(x) = \beta_{i0} = 0,\ i = 1,2 \tag{8.108}$$

The boundary conditions for the material temperature $T(t,x)$ and chemical reaction depth $\beta_i(t,x)$ should be set up on the surface of the multilayer material ($x = 0$), on the boundary of different materials ($x = \Delta_i$) and in the not heated region of the solid fuel ($x \to \infty$). Based on the properties of the Eqs. (8.105)–(8.107), the boundary conditions will be as follows:

$$x = 0 \quad \frac{\partial T}{\partial x} = q_S\,;$$

$$x = \Delta_n \quad \lambda_n \frac{\partial T_n}{\partial x} = \lambda_m \frac{\partial T_m}{\partial x}\,; \tag{8.109}$$

$$x \to \infty; \quad \frac{\partial T}{\partial x} = 0; \quad \frac{\partial \beta_i}{\partial x} = 0, \ i = 1,2$$

The conditions on the material surface assert the coincidence of heat flows in the environment (in the gaseous phase) and surface layer of the heated up material. The conditions on the coating and fuel interface provide the heat flow continuity. The latter condition is equivalent to the one at which the fuel has neither heated up nor reacted at depth. This condition can be fulfilled automatically at correct designing of finite-difference algorithms.

The Eqs. (8.105) are integrated till the time point at which the temperature on the inert material surface will not reach the critical value (for instance, the known value of the melting or evaporation temperatures on the inert material surface)

$$T_S \geq T_*, \tag{8.110}$$

Let us accept that the critical temperature of the solid fuel is below the inert material temperature. In this case, the coating removal transits continuously into the solid fuel combustion.

After the coating removal starts, the Eqs. (8.105) are more convenient to solve in the moving coordinate system whose beginning coincides with the boundary of the gaseous phase and solid material being removed. Taking into account the Eqs. (7.73), (7.74), the Eqs. (8.105) are rewritten as follows:

$$\rho c\left(\frac{\partial T}{\partial t} - u\frac{\partial T}{\partial x}\right) = \frac{\partial}{\partial x}\lambda\frac{\partial T}{\partial x} + \rho Q \cdot \Phi(T,\beta)$$

$$\frac{\partial \beta_i}{\partial t} - u\frac{\partial \beta}{\partial x} = \Phi(T,\beta) \tag{8.111}$$

The Eqs. (8.111) are solved when setting up the initial and boundary conditions (they can be taken in the form (8.108), (8.109)), as well as when setting up the combustion conditions (Eq. (8.110)). When solving the problem considered, the dependence discussed in [98, 161, 204] is used as the combustion condition:

$$T_g = T_g(p) \tag{8.112}$$

When writing down the combustion condition, it is assumed that the liquid melt whose temperature coincides with the melting temperature T_g of the fuel (material) is formed on the surface of the burning material (or material being removed). The melt temperature is the function of the pressure p in the gaseous phase. For the materials, as a rule, the condition $T_* \geq T_g$ is fulfilled.

For the model (8.111) and (8.112) in computational algorithms, the peculiarities which can result in the stepwise change in the removal rate should be considered:

- temperature of the inert material removal is higher than the ignition temperature of the solid fuel placed under this inert material;
- after the beginning of the material removal (combustion) the value of the heat flow coming into the solid material can change. During some period of time, the heat flow value needs to change from the value q_s determined by the conditions in the gaseous phase to the value q_{s*} determined by the following correlation

$$q_{s*} = \lambda \frac{\partial T_s}{\partial x} = \rho c u (T_s - T_0) \tag{8.113}$$

The heat flow q_{s*} determined by the formula (8.113) corresponds to the stationary removal (combustion) of the solid material with the rate $u_{\mathrm{m}}(p)$. Taking into account (8.113), the total value of the heat flow entering the solid material is established by the formula [244]

$$q_\Sigma = q_s \cdot \delta + q_{s*} \cdot (1 - \delta);$$

$$\delta \approx (1 - \sqrt{\mathrm{Re}_x} \cdot \frac{\rho_m u_m}{\rho_g u_g})^{1,33}, \quad 1 \geq \delta \geq 0 \tag{8.114}$$

The Eqs. (8.111)–(8.114) will be solved by the finite-difference method according to the scheme (8.34). The computational area comprising the coating and solid fuel is split into I nodes ($1 \leq i \leq I$), the nodes can have different sizes.

The partial derivatives from the temperature T and chemical reaction depth β by time and spatial coordinate are approximated as follows:

$$\frac{\partial T(x_i,t_n)}{\partial t} \approx \frac{T_i^{n+1}-T_i^n}{t_{n+1}-t_n} = \frac{T_i^{n+1}-T_i^n}{\Delta t};$$

$$\frac{\partial T(x_i,t_n)}{\partial x} \approx \alpha \frac{T_{i+1}^n - T_{i-1}^n}{\Delta x_{i-1}+\Delta x_i} + (1-\alpha)\frac{T_{i+1}^{n+1}-T_{i-1}^{n+1}}{\Delta x_{i-1}+\Delta x_i}, \quad 1 \geq \alpha \geq 0;$$

$$\frac{\partial}{\partial x}\lambda(T_i^n)\frac{\partial T(x_i,t_n)}{\partial x} \approx \frac{\lambda_{i+1/2}}{\Delta x_i}(T_{i+1}^{n+1}-T_i^{n+1}) - \frac{\lambda_{i-1/2}}{\Delta x_i}(T_i^{n+1}-T_{i-1}^{n+1});$$

$$\frac{\partial \beta(x_i,t_n)}{\partial t} \approx \frac{\beta_i^{n+1}-\beta_i^n}{t_{n+1}-t_n} = \frac{\beta_i^{n+1}-\beta_i^n}{\Delta t};$$

$$\frac{\partial \beta}{\partial x} \approx \frac{\beta_{i+1}^{n+1}+\beta_{i+1}^n-\beta_i^{n+1}-\beta_i^n}{\Delta x_{i-1}+\Delta x_i} \tag{8.115}$$

During the approximation the following designations are accepted:

$$\Delta t = t_{n+1}-t_n; \quad \Delta x_{i-1} = x_i - x_{i-1}; \quad T_i^n = T(x_i,t_n); \quad \beta_i^n = \beta(x_i,t_n)$$

Taking into account the recorded approximations, the equations of heat conductivity and depth of decomposition (chemical reaction) with boundary conditions are as follows:

Heat conductivity equation

$$\frac{T_i^{n+1}-T_i^n}{\Delta t} - u(\alpha\frac{T_{i+1}^n-T_{i-1}^n}{(\Delta x_i+\Delta x_{i-1})} + (1-\alpha)\frac{T_{i+1}^{n+1}-T_{i-1}^{n+1}}{(\Delta x_i+\Delta x_{i-1})}) =$$

$$= \frac{2}{c\rho(\Delta x_i+\Delta x_{i-1})}(\frac{\lambda_{i+1/2}}{\Delta x_i}(T_{i+1}^{n+1}-T_i^{n+1})$$

$$-\frac{\lambda_{i-1/2}}{\Delta x_{i-1}}(T_i^{n+1}-T_{i-1}^{n+1})) + \Phi_T(T_i^n,\beta_i^n) \qquad i=2, I-1;$$

$$\lambda_1\frac{T_1^{n+1}-T_2^{n+1}}{\Delta x_1} = q_\Sigma;$$

$$\lambda_{I-1}\frac{T_{I-1}^{n+1}-T_I^{n+1}}{\Delta x_{I-1}}=0 \text{ or } T_I^{n+1}=T(0)$$

The equations are written down in the matrix form as follows:

$$\begin{pmatrix} a_{11} & a_{12} & 0 & \dots & 0 & 0 & 0 & \dots & 0 & 0 & 0 \\ a_{21} & a_{22} & a_{23} & \dots & 0 & 0 & 0 & \dots & 0 & 0 & 0 \\ 0 & a_{32} & a_{33} & \dots & 0 & 0 & 0 & \dots & 0 & 0 & 0 \\ \dots & \dots & \dots & \dots & \dots & \dots & \dots & \dots & \dots & \dots & \dots \\ 0 & 0 & 0 & \dots & a_{i-1,i-1} & a_{i-1,i} & 0 & \dots & 0 & 0 & 0 \\ 0 & 0 & 0 & \dots & a_{i,i-1} & a_{ii} & a_{i,i+1} & \dots & 0 & 0 & 0 \\ 0 & 0 & 0 & \dots & 0 & a_{i+1,i} & a_{i+1,i+1} & \dots & 0 & 0 & 0 \\ \dots & \dots & \dots & \dots & \dots & \dots & \dots & \dots & \dots & \dots & \dots \\ 0 & 0 & 0 & \dots & 0 & 0 & 0 & \dots & a_{I-2,I-2} & a_{I-2,I-1} & 0 \\ 0 & 0 & 0 & \dots & 0 & 0 & 0 & \dots & a_{I-1,I-2} & a_{I-1,I-1} & a_{I-1.I} \\ 0 & 0 & 0 & \dots & 0 & 0 & 0 & \dots & 0 & a_{I,I-1} & a_{I,I} \end{pmatrix} \times$$

$$\begin{pmatrix} T_1^{n+1} \\ T_2^{n+1} \\ T_3^{n+1} \\ \dots \\ T_{i-1}^{n+1} \\ T_i^{n+1} \\ T_{i+1}^{n+1} \\ \dots \\ T_{I-2}^{n+1} \\ T_{I-1}^{n+1} \\ T_I^{n+1} \end{pmatrix} = \begin{pmatrix} b_1 \\ b_2 \\ b_3 \\ \dots \\ b_{i-1} \\ b_i \\ b_{i+1} \\ \dots \\ b_{I-2} \\ b_{I-1} \\ b_I \end{pmatrix}$$

Here, the elements $a_{i-1,i}, a_{i,i}, a_{i+1,i}, b_i$ take the following values:

$$a_{11}=1; \quad a_{12}=-1; \quad b_1=\Delta x_1\cdot\frac{q_\Sigma}{\lambda};$$

$$a_{21} = -\begin{pmatrix} \frac{2}{c_2\rho_2} \cdot \frac{\lambda_{3/2}}{\Delta x_1} \\ -u \cdot (1-\alpha) \end{pmatrix} \cdot \Delta t; \quad a_{22} = -a_{21} - a_{23} + \left(\Delta x_1 + \Delta x_2\right);$$

$$a_{23} = -\begin{pmatrix} \frac{2}{c_2\rho_2} \cdot \frac{\lambda_{5/2}}{\Delta x_1} \\ +u \cdot (1-\alpha) \end{pmatrix} \cdot \Delta t; \quad b_2 = T_2^n \cdot \left(\Delta x_1 + \Delta x_2\right) + u \cdot \Delta t \cdot \alpha \cdot \left(T_3^n - T_1^n\right)$$

$$a_{i,i-1} = -\begin{pmatrix} \frac{2}{c_i\rho_i} \cdot \frac{\lambda_{i-1/2}}{\Delta x_{i-1}} \\ -u \cdot (1-\alpha) \end{pmatrix} \cdot \Delta t; \quad a_{ii} = -a_{i,i-1} - a_{i,i+1} + (\Delta x_{i-1} + \Delta x_i);$$

$$a_{i.i+1} = -\begin{pmatrix} \frac{2}{c_i\rho_i} \cdot \frac{\lambda_{i+1/2}}{\Delta x_i} \\ +u \cdot (1-\alpha) \end{pmatrix} \cdot \Delta t; \quad b_i = T_i^n \cdot (\Delta x_{i-1} + \Delta x_i) \qquad i=3,\ I-1;$$

$$+u \cdot \Delta t \cdot \alpha \cdot (T_{i+1}^n - T_{i-1}^n)$$

$$a_{I,I-1} = 1;\ a_{I,I} = -1;\ B_I = 0;\ ;$$

2. Equation for the depth of decomposition (chemical reaction)

$$\frac{\beta_i^{n+1} - \beta_i^n}{\Delta t} - u\frac{\beta_{i+1}^{n+1} + \beta_{i+1}^n - \beta_i^{n+1} - \beta_i^n}{\Delta x_{i-1} + \Delta x_i} = \Phi_\beta(T_i^n, \beta_i^n) \quad ; \quad i=1, I;$$

$$\beta_I^{n+1} = 0$$

During heating up, while the rate of the material decomposition (combustion) equals zero, the computational algorithm is developed as follows:

- finite-difference heat conductivity equation is solved at the first stage. At the same time, the value of the function $\Phi_T(T, \beta)$ is calculated by the values of the temperature T_i^n and depth of decomposition (chemical reaction) β_i^n calculated at the time moment t_n. The thermophysical properties of the materials heated up can change during their temperature growth. Therefore, the application of the inverse matrix method to solve the heat conductivity problem is inappropriate. To define a new field of temperatures at the time moment t_{n+1}, the sweep method should be used (formulas (8.80)–(8.87)). The condition of good conditionality (diagonal dominance – formula (8.89) is fulfilled for the system to be solved;

- based on the calculated temperature values, at the second stage, the field of the values of decomposition (chemical reaction) depth is calculated by the following algorithm

$$\beta_{\mathrm{I}}^{n+1} = 0;$$

$$M_i = u \cdot \frac{\Delta t}{\Delta x_{i-1} + \Delta x_i}; \quad R_i = 1 + M_i; \qquad i = I-1, 1; \tag{8.116}$$

$$\beta_i^{n+1} = \frac{1}{R_i} \cdot (M_i \cdot \beta_{i+1}^{n+1} + (M_i \cdot \beta_{i+1}^{n} + (1 - M_i) \cdot \beta_i^{n}) + \Phi_{\beta}(T_i^{n}, \beta_i^{n}))$$

By the recorded algorithm, the temperatures and decomposition (chemical reaction) depth are calculated till the moment when the temperature on the surface of the materials being heated reaches the temperature of melting (evaporation) beginning. This condition is written down by the formula (8.110). Further, the application of the recorded algorithms is possible when defining the value of the coating removal rate.

The removal rate is determined by the non-stationary law, at the same time, the removal rate is defined by the iteration algorithm. The iterations for the rate values are performed until the condition of the coating melting (or the condition by temperature on the burning solid fuel – condition (8.112)) when solving the heat conductivity equations is performed.

The iteration algorithm is developed as follows:

- temperature profile in the layer heated up is memorized $T_i^n, i = 1, I$;
- value of the removal rate $u_{\mathrm{m}} = u_{\mathrm{m}}^{(1)}$ is set up and the temperature value $T_g^{(1)} = T_1^{n+1}$ is set up on the coating surface by the solution of the heat conductivity equations. The temperature difference $T_g^{(1)}$ from the melting temperature $T_g(p) - \Delta T_s = T_g^{(1)} - T_g(p)$ is calculated on the surface;
- temperature profile $T_i^n, i = 1, I$ is recovered in the heated layer of the multilayer material, the new value of the removal rate is set up at the second iteration (and further) $u_{\mathrm{m}} = u_{\mathrm{m}}^{(2)}$ defined by the following formula

$$u_m^{(2)} = u_m^{(1)} - \frac{\partial u}{\partial T_g} \cdot \Delta T_s \tag{8.117}$$

and the temperature profile is recalculated by the heated layer re-solving heat conductivity equations;

- iterations by the formula (8.117) are performed till the value $\Delta T_s > \varepsilon$, where ε – set up accuracy of the calculations.

In the formula (8.117) the values of the partial derivative $\frac{\partial u}{\partial T_g}$ are established by the dependence $\frac{\partial u}{\partial T_g} = -\frac{u_{cm}}{(T_g - T_H)}$ which can be found from the equation of the heat balance on the melting material surface (balance of the heat coming from the gaseous phase and heat necessary for melting the heated up coating layer).

Below are the results obtained when calculating the heat-up and removal of the two-layer material with chemical reactions in the solid phase (endothermic for the coating and exothermic for the solid fuel). The results are obtained with the following main initial data:

- length of the cells used when forming the computational cell is selected in accordance with the geometric progression law; the first cell has the size $\Delta x_1 = 0{,}1 \cdot 10^{-6}$ m, and the rest increase with the progression coefficient $r = {\Delta x_{i+1}}/{\Delta x_i} = 1{,}05$. In the computational algorithm, the number of computational nodes by the spatial coordinate used when solving the problem is automatically selected – the new derivative can be added to the initially set up number of nodes in the calculations if the value of the derivative $\frac{\partial T}{\partial x}$ starts differing from zero at the boundary in the material depth;
- density of the inert material coating – $\rho_1 = 1200$ kg/m^3, and the solid fuel density – $\rho_2 = 1500$ kg/m^3;
- specific heat capacities of the materials – $c_1 = 1200$ K/(kg K); $c_2 = 1100$ J/(kg K);
- coefficients of heat conductivity of the materials – λ_1 = 0,6 W/(m K); $\lambda_2^{(1)} = 0{,}2$ W/(m K);
- initial temperature of the materials – $T_i = 300$ K;
- value of the heat flow coming into the surface layer of the solid material established by the formula – $q_\kappa = 1000 \cdot (3500 - T_1^n)$ W/m^2;
- thickness of the coating (inert material) – $h = 0{,}0003$ m.

The following values of the parameters characterizing the endothermic reaction are accepted in the calculations:

- temperature of the phase transition beginning – 500 K;

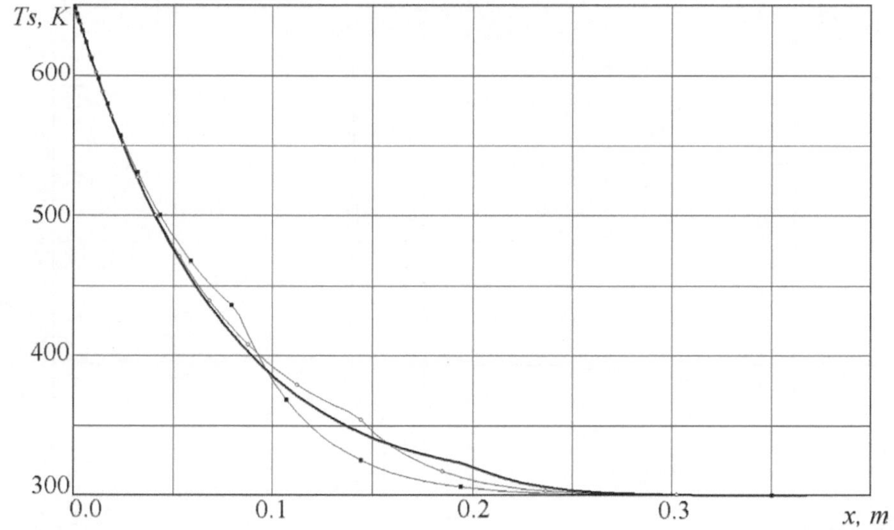

FIGURE 8.13 Temperature profile in the heated up layer of the two-layer material.

- reaction pre-exponential factor – $k = 1,0 \cdot 10^7$ 1/c;
- relation of the activation energy to the gas constant – $E / R = 0$;
- energy effect at the phase transition – $Q / c = -1,0 \cdot 10^{21}$ K.

In Figure 8.13 you can see the temperature profiles for the time moments 0.03 s (heavy line in the figure), 0.04 s (fine line marked with circumferences), 0.05 s (fine line marked with triangles). At the given curves, the curve breaking point *T(x)* corresponds to the interface of the inert material and solid fuel.

In Figure 8.14 you can see the results of the removal rate of the two-dimensional material. In the figure the solid heavy curve – non-stationary removal rate, fine line with marking – quasi-stationary values of the inert material removal rate and solid fuel combustion rate.

8.4 METHODS OF CONTROL VOLUMES

8.4.1 GENERAL IDEAS OF THE METHODS OF CONTROL VOLUMES

A large number of difference schemes applied to solve problems of physics can be referred to methods of control volumes. It should be mentioned

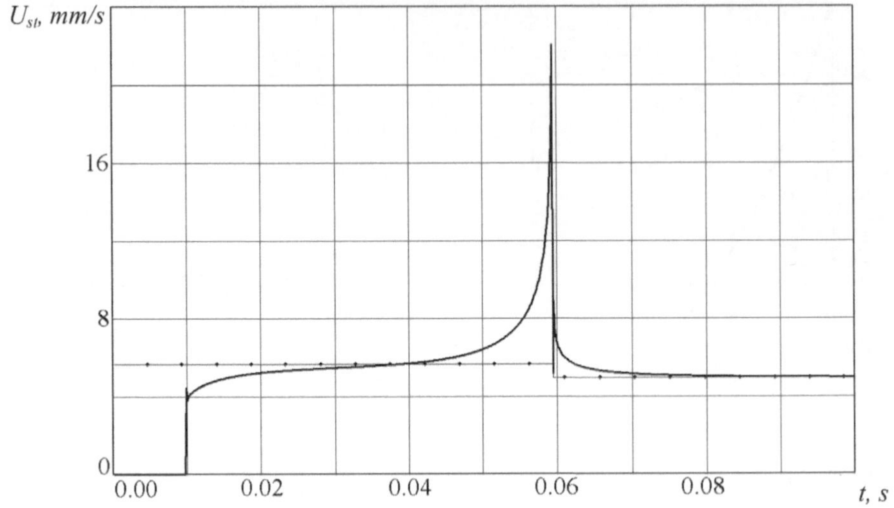

FIGURE 8.14 Dependence of the material removal rate on time

that such name of the method group in question is not common, nevertheless, the equation discretization ways in these methods are similar. In particular, the so-called integral-interpolatory methods developed in the works by the academician A. A. Samarsky and his scientific school can be referred to the methods of control volumes [190, 191, 197]. Patankar and Spolding developed the methods called finite-volumetric [170]. The conservative methods of flows (flow methods) are developed by the academician O. M. Belotserkovsky and his scientific school [27, 28]. At the same time, the methods by S. K. Godunov [241] and his scientific school widely spread in Russia and other countries can be also referred to conservative methods of flows, as well as its multiple modifications (for instance, [111, 125]).

Let us point out general moments found in various methods, which we refer to the methods of control volumes.

1. Application of staggered grids – the calculated derivatives are determined both in the centers of elementary volumes and on their boundaries. In Figure 8.15 you can see the elements of the computational grid (one-dimensional – Figure 8.15a, two-dimensional – Figure 8.15b). In the figures, the elementary volumes are marked with capital letters *I, J*, and the boundaries of elementary volumes – with small letters *i, j*.

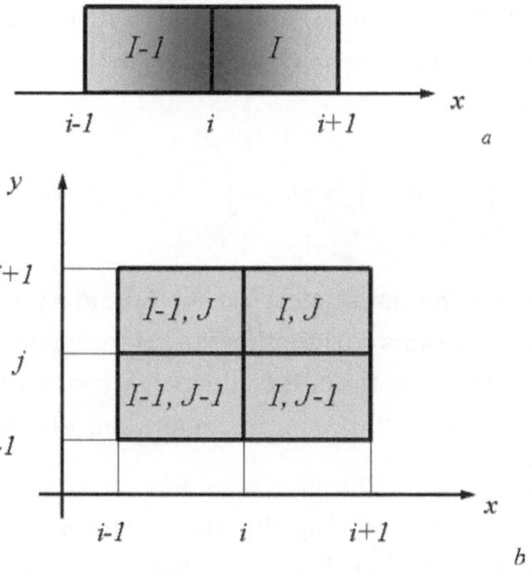

FIGURE 8.15 Control volumes in the computational area (a – one-dimensional variant; *b* – two-dimensional variant).

The values of the parameters referred to the centers of the segregated volumes are established for the segregated control volumes (for the equations of gas dynamics these are density, specific energy, velocity vector, etc.), to their boundaries (values of the same parameters, if required). Besides, the parameter values necessary at the approximation of differential equations recorded in the integral form are established. In particular, for the equations of gas dynamics these can be the control volume mass, its energy, values of the momentum vector, etc.;

2. The group of methods considered refers to conservative finite-difference schemes. The parameters referred to the control volume center are established based on the balance of flows entering the volume along all its boundaries and taking into account the sources contained in the volume. If we are talking about solving the problems of gas dynamics, it is necessary to provide the fulfillment of the conservation laws of mass, momentum and energy for all control volumes at each stage by time. When solving the heat conductivity equation, it is necessary to provide the heat balance for each volume at the arbitrary time point.
3. In the considered group of methods, when solving the equations, for instance, of the type

$$\frac{\partial f}{\partial t}+\frac{\partial g}{\partial q}=h(t,q,f,g)$$

the numerical integration by the spatial coordinate is applied instead of the differentiation procedure. Actually, the following equation is approximated

$$\int_{\Delta q} \frac{\partial f}{\partial t} dq + \int_{\Delta q} \frac{\partial g}{\partial q} dq = \int_{\Delta q} h(t,q,f,g) dq;$$

4. When defining the parameters on the boundary of control volumes, the known solutions of mathematical or physical problems are applied. At the same time, the linear interpolation of the parameter values calculated in the volume centers onto their boundaries is not used, as a rule.

Some of the methods satisfying the aforementioned principles are considered below. The information about other methods can be found, for example, in [14, 226].

8.4.2 METHOD BY S. PATANKAR AND D. SPOLDING

Application of integral-interpolatory approach to solve the stationary equation of diffusion

Let us consider the development of the difference scheme by the integral-interpolatory method on the example of the diffusion equation for the stationary one-dimensional case

$$\frac{\partial}{\partial x} a \frac{\partial f}{\partial x} + F(x) = 0 \tag{8.118}$$

Let us use the designations accepted in accordance with Figure 8.17. The numbers of control volumes are designated as $I-1, I$, and their volumes – with the indexes $i-1, i, i+1$.

One integration of the Eq. (8.118) by x allows obtaining the following correlation

$$(a\frac{\partial f}{\partial x})_I - (a\frac{\partial f}{\partial x})_{I-1} + \int_{I-1}^{I} F \cdot dx = 0$$

If we accept the profile of the function f in segregated volumes as linear, the latter equation in the finite-difference form can be rewritten as follows:

$$\frac{a_I(f_{i+1}-f_i)}{\Delta x_I}-\frac{a_{I-1}(f_i-f_{i-1})}{\Delta x_{I-1}}+\bar{F}\frac{\Delta x_{I-1}+\Delta x_I}{2}=0$$

$$\bar{F}=F_i+k_i\cdot f_i$$

or

$$b_i f_i = b_{i+1} f_{i+1} + b_{i-1} f_{i-1} + d_i \tag{8.119}$$

where

$$b_{i+1}=\frac{a_I}{\Delta x_I},\ b_{i-1}=\frac{a_{I-1}}{\Delta x_{I-1}},\ b_i=b_{i+1}+b_{i-1}-k_i\frac{\Delta x_{I-1}+\Delta x_I}{2},\ d_i=F\frac{\Delta x_{I-1}+\Delta x_I}{2}$$

Here F – value of the function $F(x)$ averaged by the volumes $(\Delta x_{I-1}+\Delta x_I)$.

For the Eq. (8.119), the rules which follow the physical assumptions (equation being solved is the heat conductivity or diffusion equation) can be formulated, including the conservation conditions of the discrete analog of the equation being solved:

- *rule 1* – correspondence of the flows at the control volume boundaries.

  The sense of the recorded rule is as follows. Since the problem (8.118) is stationary, the flow values on the right boundary of the control volume with the number *(I–1)* need to coincide with the flow values on the right boundary of the control volume with the number *I*. In the Eq. (8.118), the flow is the value $a\frac{\partial f}{\partial x}$. If the unsuccessful approximation of the flow value is selected, the formulated rule is not fulfilled. In this regard, the error which can result in the computational process instability will be accumulated;
- *rule 2* – correspondence of the coefficient signs b_{i-1}, b_i, b_{i+1} in the finite-difference formula (8.119) at the unknown values of the discrete unknown f_i.

  Let us assume that the function values in the points f_{i-1}, f_{i+1} increased by some value. From the physical ideas it follows that the function values in the point f_i should also increase. The similar effect should be found in the case when the values of the functions f_{i-1}, f_{i+1} decrease (f_i value needs to decrease). This can happen only if the signs at the coefficients b_{i-1}, b_i, b_{i+1} coincide. In particular, this can require the positive signs at these coefficients (further the coefficients b_{i-1}, b_i, b_{i+1} will be considered as positive);

- *rule 3* – negativity of the coefficient *k* when calculating the value of the source component $\bar{F}$.

  First of all, the requirements established by the rule 3 are based on the necessity of providing the positive value of the coefficient b_i. The unsuccessful approximation of the source element $F(x)$ can break this condition. Besides, the positive value of k_i is not essentially physical. Actually, if any disturbing factor resulting in the growth of the function f_i value arises, this disturbance will be enforced at the coefficient k_i positivity. From the physical ideas, everything should be vice versa for the system stability;
- *rule 4* – with the lack of the source component $F(x)$ the condition by the sum of the coefficients b_{i+1}, b_{i-1} needs to be fulfilled. The rule formulated can be substantiated by the following ideas.

Let the finite-difference approximation

$$b_i f_i = b_{i+1} f_{i+1} + b_{i-1} f_{i-1}$$

$$b_{i+1} = \frac{a_I}{\Delta x_I}, \quad b_{i-1} = \frac{a_{I-1}}{\Delta x_{I-1}}, \quad b_i = b_{i+1} + b_{i-1}$$

be true for the equation $\frac{\partial}{\partial x} a \frac{\partial f}{\partial x} = 0$ (Eq. (8.118) in which the component equals zero). Such approximation should be also true if the solution of the equation $\frac{\partial}{\partial x} a \frac{\partial (f + const)}{\partial x} = 0$ is sought. It is easy to make sure that the recorded condition will be fulfilled if at the finite-difference approximation the condition $b_i = b_{i+1} + b_{i-1}$ is fulfilled. One of the checks of the rule 4 fulfillment is the check of non-accumulation of computational errors if the value of the function $f(x) = C = const$.

Application of integral-interpolatory approach for solving the non-stationary equation of heat conductivity

Let us consider the application of the control volume method to record the non-stationary equation of heat conductivity different from (8.118) by the availability of the derivative from temperature by the process time

$$\rho c \frac{\partial T}{\partial t} = \frac{\partial}{\partial x} \lambda \frac{\partial T}{\partial x}$$

The discrete analysis for this equation is obtained by the integration of both parts of the equation by time (from time t till the time point $t+\Delta t$) and longitudinal coordinate (within the selected control volumes with the numbers *(I–1)* and *I*

$$\rho c \int_{x_{I-1}}^{x_I} \int_{t_n}^{t_n+\Delta t} \frac{\partial T}{\partial t} dt\, dx = \int_{t_n}^{t_n+\Delta t} \int_{x_{I-1}}^{x_I} \frac{\partial}{\partial x} \lambda \frac{\partial T}{\partial x} dx\, dt$$

Let us apply the following correlation when approximating the integral by time for the arbitrary function f

$$\int_{t_n}^{t_n+\Delta t} f\, dt = (\alpha f_i^{n+1} + (1-\alpha) f_i^{n})\Delta t$$

Here α – weight coefficient taking the values in the interval from 0 to 1 ($\alpha \in (0,1)$).

As applicable to the heat conductivity equation, the recorded approximation of the integral by time results in the finite-difference equation

$$\rho c \frac{\Delta x_{I-1} + \Delta x_I}{2\Delta t}(T_i^{n+1} - T_i^n) = \alpha\left(\frac{\lambda_I (T_{i+1}^{n+1} - T_i^{n+1})}{\Delta x_I} - \frac{\lambda_{I-1}(T_i^{n+1} - T_{i-1}^{n+1})}{\Delta x_{I-1}}\right)$$

$$+(1-\alpha)\left(\frac{\lambda_I (T_{i+1}^{n} - T_i^{n})}{\Delta x_I} - \frac{\lambda_{I-1}(T_i^{n} - T_{i-1}^{n})}{\Delta x_{I-1}}\right)$$

After the transformations, the equation can be brought to the following form (8.119)

$$b_i T_i^{n+1} = b_{i+1}(\alpha T_{i+1}^{n+1} + (1-\alpha)T_{i+1}^n) + b_{i-1}(\alpha T_{i-1}^{n+1} + (1-\alpha)T_{i-1}^n) + d_i T_i^n$$

$$b_{i+1} = \frac{\lambda_I}{\Delta x_I}, \quad b_{i-1} = \frac{\lambda_{I-1}}{\Delta x_{I-1}}, \quad d_i = \rho c \frac{\Delta x_{I-1} + \Delta x_I}{2\Delta t}, \quad b_i = d_i + \alpha(b_{i+1} + b_{i-1})$$

(8.120)

Taking the value $\alpha = 0$ in the latter equation, we obtain the explicit finite-difference scheme to determine the temperature values T_i^{n+1}. At $\alpha = 1$ we have the implicit finite-difference scheme. Crank-Nicolson scheme is obtained when it is assumed that $\alpha = 0.5$. Thus, the same formulas as with ordinary finite-difference schemes are obtained by integral-interpolatory method in the considered case (see Section 8.3.1).

Application of integral-interpolatory approach to solve the stationary equation with convective and diffusion components

Let us consider the stationary problem related to gas dynamics

$$\frac{d}{dx}\rho u f = \frac{d}{dx}\left(\lambda \frac{df}{dx}\right) \tag{8.121}$$

Here ρ, u, f, λ – gas density, its rate, substance transferred, heat conductivity coefficient, respectively.

If we accept that in this equation $\rho u = const, \lambda = const$ (the first condition is fulfilled identically, if the liquid or gas flow – $\frac{d}{dx}\rho u = 0$), this equation can be solved analytically. Let the equation solution be sought in the interval $0 \le x \le L$ and boundary values for the function f are known – $f(0) = f_0, f(L) = f_L$. In this case, the first integral of the Eq. (8.121) and solution for the function f are written down as follows:

$$\rho u f - \lambda \frac{df}{dx} = const \tag{8.122}$$

$$f = f_0 + (f_L - f_0)\frac{\exp(\frac{\rho u x}{\lambda}) - 1}{\exp(\frac{\rho u L}{\lambda}) - 1} \tag{8.123}$$

The value of the derivative $\frac{df}{dx}$ is established from the Eq. (8.123)

$$\frac{df}{dx} = \frac{\rho u}{\lambda} \frac{f_L - f_0}{\exp(\frac{\rho u L}{\lambda}) - 1} \exp(\frac{\rho u x}{\lambda}) \tag{8.124}$$

Taking into account (8.123), (8.124) the solution (8.122) is rewritten as follows:

$$\rho u(f + \frac{f_0 - f_L}{\exp\left(\frac{\rho u L}{\lambda}\right) - 1}) = const \tag{8.125}$$

In the similarity theory the value $Pe = \frac{\rho u L}{\lambda}$ – criteria correlation called Peclet number. Peclet number fixes the correlation of the convection and diffusion intensities. In particular, the value of the number $Pe = 0$ corresponds to the solution of the problem of pure diffusion, and value $Pe \to \pm\infty$ – solution of the problem of pure convection. Graphically, the solution of the Eq. (8.125) for the arbitrary value of Pe is given in Figure 8.16. The solutions corresponding to $Pe \to -\infty$, $Pe = -1$, $Pe = 0$, $Pe = 1$, $Pe \to +\infty$ are demonstrated successively, from the top to the bottom. The behavior of the solution of *f(x)* with large Peclet number values $(|Pe| >> 1)$ indicates that the finite-difference approximation of the Eq. (8.121) with central differences will be unsatisfactory for the convective component. This result was obtained before, when analyzing the stability of the scheme with central differences for the Eq. (8.10).

The application of the analytical solution obtained (8.125) for control volumes ($x \in (x_{I-1}, x_I)$, Figure 8.15) allows writing down the following finite-difference equation

$$(\rho u)_I (f_i + \frac{f_i - f_{i+1}}{\exp(\frac{\rho u \Delta x}{\lambda})_I - 1}) - (\rho u)_{I-1} (f_{i-1} + \frac{f_{i-1} - f_i}{\exp(\frac{\rho u \Delta x}{\lambda})_{I-1} - 1}) = 0$$

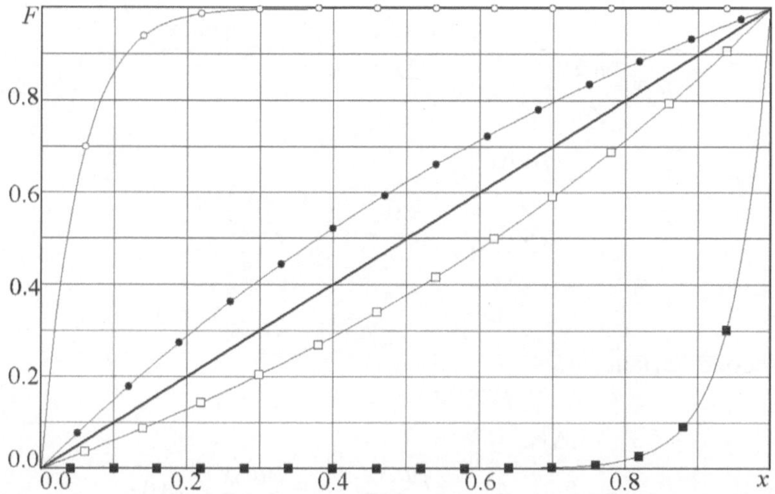

FIGURE 8.16 Solution of the differential equation of diffusion with different values of *Pe* number

This equation can be rewritten in the form (8.119)

$$b_i f_i = b_{i+1} f_{i+1} + b_{i-1} f_{i-1},$$

$$b_{i+1} = \frac{(\rho u)_I}{\exp(\frac{\rho u \Delta x}{\lambda})_I - 1}, \quad b_{i-1} = \frac{(\rho u)_{I-1} \exp(\frac{\rho u \Delta x}{\lambda})_{I-1}}{\exp(\frac{\rho u \Delta x}{\lambda})_{I-1} - 1},$$

$$b_i = b_{i+1} + b_{i-1} + (\rho u)_I - (\rho u)_{I-1}$$

When performing massive calculations on computers, the coefficients b_{i-1}, b_i, b_{i+1} are determined many times (thousand and million times). The availability of exponents for these coefficients significantly increases the computation time of these coefficients. Therefore, the following simplifications are proposed in [170]:

- first variant

$$\frac{\frac{\rho u \Delta x}{\lambda}}{\exp(\frac{\rho u \Delta x}{\lambda}) - 1} = -\frac{\rho u \Delta x}{\lambda}, \quad \text{if } \frac{\rho u \Delta x}{\lambda} < -2;$$

$$\frac{\frac{\rho u \Delta x}{\lambda}}{\exp(\frac{\rho u \Delta x}{\lambda}) - 1} = 1 - \frac{\rho u \Delta x}{2\lambda}, \quad \text{if } -2 \le \frac{\rho u \Delta x}{\lambda} \le 2;$$

$$\frac{\frac{\rho u \Delta x}{\lambda}}{\exp(\frac{\rho u \Delta x}{\lambda}) - 1} = 0, \quad \text{if } \frac{\rho u \Delta x}{\lambda} > 2;$$

- second variant

$$\frac{\frac{\rho u \Delta x}{\lambda}}{\exp(\frac{\rho u \Delta x}{\lambda}) - 1} = -\frac{\rho u \Delta x}{\lambda}, \quad \text{if } \frac{\rho u \Delta x}{\lambda} < -10;$$

$$\frac{\frac{\rho u \Delta x}{\lambda}}{\exp(\frac{\rho u \Delta x}{\lambda})-1}=(1+0{,}1\frac{\rho u \Delta x}{\lambda})^5-\frac{\rho u \Delta x}{\lambda}, \quad \text{if } -10 \le \frac{\rho u \Delta x}{\lambda}<0;$$

$$\frac{\frac{\rho u \Delta x}{\lambda}}{\exp(\frac{\rho u \Delta x}{\lambda})-1}=(1-0{,}1\frac{\rho u \Delta x}{\lambda})^5, \quad \text{if } 0 \le \frac{\rho u \Delta x}{\lambda} \le 10;$$

$$\frac{\frac{\rho u \Delta x}{\lambda}}{\exp(\frac{\rho u \Delta x}{\lambda})-1}=0, \quad \text{if } 10<\frac{\rho u \Delta x}{\lambda}$$

Application of integral-interpolatory approach to solve the system of stationary equations of gas dynamics (SIMPLE – method)

Let us consider the solution of the equations of gas dynamics by S. Patankar and D. Spolding method. To simplify the equation, let us write down in one-dimensional stationary setting:

$$\frac{\partial \rho u}{\partial x}=0$$
$$\frac{\partial \rho u u}{\partial x}+\frac{\partial p}{\partial x}=0 \tag{8.126}$$
$$\frac{\partial \rho u E}{\partial x}+\frac{\partial \rho u}{\partial x}=0$$

In the finite-difference form the second equation of the system (8.126) can be written down as follows:

$$b_i u_i = b_{i+1}u_{i+1}+b_{i-1}u_{i-1}+(p_I-p_{I-1})$$

$$b_{i+1}=\frac{(\rho u)_I}{\exp(\frac{\rho u \Delta x}{\lambda})_I-1}, \quad b_{i-1}=\frac{(\rho u)_{I-1}\exp(\frac{\rho u \Delta x}{\lambda})_{I-1}}{\exp(\frac{\rho u \Delta x}{\lambda})_{I-1}-1}$$

$$b_i=b_{i+1}+b_{i-1}+(\rho u)_I-(\rho u)_{I-1}$$

Let us accept that the following correlations are true

$$p=\overline{p}+p', \text{ where } p' << \overline{p}$$

$$u = \bar{u} + u' \text{, where } u' << \bar{u}$$

Then after substituting in the upper equation, we obtain the equation for corrections:

$$b_i u'_i = b_{i+1} u'_{i+1} + b_{i-1} u'_{i-1} + (p'_I - p'_{I-1})$$

If the values $\bar{p}, \bar{u}$ are close to their real values, the following is true

$$b_{i+1} u'_{i+1} + b_{i-1} u'_{i-1} \approx 0$$

Then $u'_i = \alpha_i \cdot (p'_I - p'_{I-1})$. Here $\alpha_i = 1/b_i$

Or $u_i = \bar{u}_i + \alpha_i \cdot (p'_I - p'_{I-1})$

Let us rewrite the continuity equation in the non-stationary form

$$\frac{\partial \rho}{\partial t} + \frac{\partial \rho u}{\partial x} = 0$$

The application of the method of control volumes for this equation allows obtaining the finite-difference dependence

$$\frac{\rho_I^{n+1} - \rho_I^n}{\Delta t} \Delta x + (\rho u)_{i+1} - (\rho u)_i = 0$$

$$\frac{\rho_I^{n+1} - \rho_I^n}{\Delta t} \Delta x + \rho_{i+1}(\bar{u}_{i+1} + \alpha_{i+1}(p'_{I+1} - p'_I)) - \rho_i(\bar{u}_i + \alpha_i(p'_I - p'_{I-1})) = 0$$

$$b_I p'_I = b_{I-1} p'_{I-1} + b_{I+1} p'_{I+1} + \varphi$$

Here $b_{I-1} = \rho_i \alpha_i$, $b_{I+1} = \rho_{i+1} \alpha_{i+1}$, $b_I = b_{I-1} + b_{I+1}$,

$$\varphi = \frac{\rho_I^n - \rho_i^{n+1}}{\Delta t} \cdot \Delta x + \rho_i \bar{u}_i - \rho_{i+1} \bar{u}_{i+1}$$

The condition $\varphi \to 0$ corresponds to the solution of stationary equations of gas dynamics.

The algorithm of solving equations of gas dynamics by the algorithm ***SIMPLE*** (Semi-Implicit Method for Pressure-Linked Equation) is as follows:

1. Pressure p field is set up;
2. Values of the velocities u are calculated;
3. Corrections for the pressures p' are calculated;
4. Calculation of the values of pressures and velocities taking into account the corrections calculated $p = \overline{p} + \kappa \cdot p'$, $u = \overline{u} + u'$. Here $\kappa \approx 0{,}8$ – relaxation coefficient.
5. Check of the function φ value. If its value considerably differs from zero, $\overline{p} = p$ is accepted and the procedure is repeated.

Application of the algorithm ***SIMPLE*** is especially effective when calculating the gas subsonic velocities (correction for the pressure is close to linear) and when calculating the incompressible (weakly compressible) liquids.

It should be pointed out that together with the algorithm ***SIMPLE***, other effective computational algorithms – its modifications are also used in practical applications. In particular, the algorithms ***SIMPLEC, SIMPLER***.

8.4.3 S.K. GODUNOV METHOD

The significant limitation of the method SIMPLE and its modifications SIMPLEC, SIMPLER developed by S. Patankar and D. Spolding is its low efficiency when solving the problems of transonic and supersonic gas dynamics. In particular, this is conditioned by the non-linearity of the equations of gas dynamics, and the non-linearity increases with the gas velocity growth due to compressibility properties demonstrated by gas. This is why it is impossible to apply the analytical expressions used when determining the gas parameters on the boundaries of control volumes and similar to (8.123) because of their low accuracy.

S.K. Godunov method is free from this shortcoming. In this method, the gas parameters on the boundaries of control volumes are found by the automodel solution of the linearized problem of gas dynamics (see Chapter 7) [241]. Different automodel solutions are used for different applications. Further, let us consider the variant of S.K. Godunov method as applied to the problems of ideal gas flow. The automodel solution of the problem on the decay of the arbitrary breakup in inert gas is used in these problems to define the parameters on the boundaries of control volumes [184].

Let us consider the problem whose computational scheme is given in Figure 8.17. The membrane M in the impact tube divides two semi-infinite areas *1, 2*, the values of gas-dynamic parameters (pressure, density, velocity,

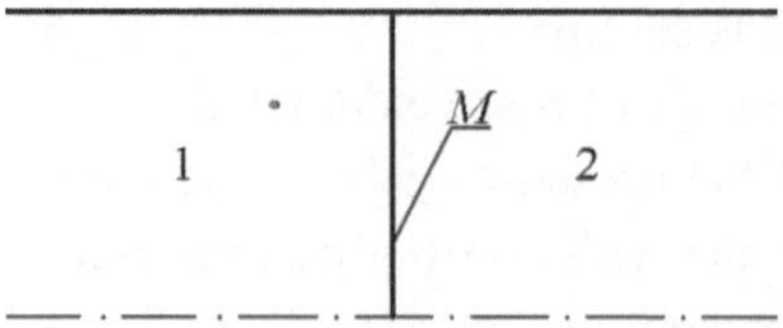

FIGURE 8.17 Scheme to the problem on the decay of the arbitrary breakup of gas-dynamic parameters.

energy) in which equal p_1, ρ_1, u_1, E_1 and p_2, ρ_2, u_2, E_2, respectively. To be definite, let the pressure in the area *1* be below the pressure in the area *2* ($p_1 \leq p_2$). In this case, when the membrane ***M*** breaks up, the process (decay of the parameter breakup) can progress following one of these schemes (Figure 8.18):

- shock waves d_1, d_2 divided by the contact breakup *k* spread in both directions (from the area *1* to the area *2*, and from the area *2* to the area *1*) (Figure 8.18a);
- shock wave *d* spreads from the right to the left (to area *1*), the contact breakup *k* follows it, and the depression wave *r* spreads to the higher pressure area (area *2*) (Figure 8.18b);
- depression waves r_1, r_2 divided by the contact breakup *k* spread in both directions (Figure 8.18c);
- depression waves r_1, r_2, divided by the area of zero pressure (vacuum) spread in both directions (Figure 8.18d).

The parameters of the breakup decay of gas-dynamic values after the membrane decomposition can be determined as follows:

1. We calculate the values of the spread velocity of the shock wave U_d, depression wave U_r, velocity value U_v in the computational area, corresponding to the possibility of the area arising with the pressure zero value (vacuum):

$$U_d = \frac{p_2 - p_1}{\sqrt{\rho_1 \cdot (\frac{k+1}{2} \cdot (p_2 + p_0) + \frac{k-1}{2} \cdot (p_1 + p_0))}}; \tag{8.127}$$

$$U_r = -\frac{2 \cdot c_2}{k-1} \cdot (1 - (\frac{p_1 + p_0}{p_2 + p_0})^{\frac{k-1}{2k}}); \tag{8.128}$$

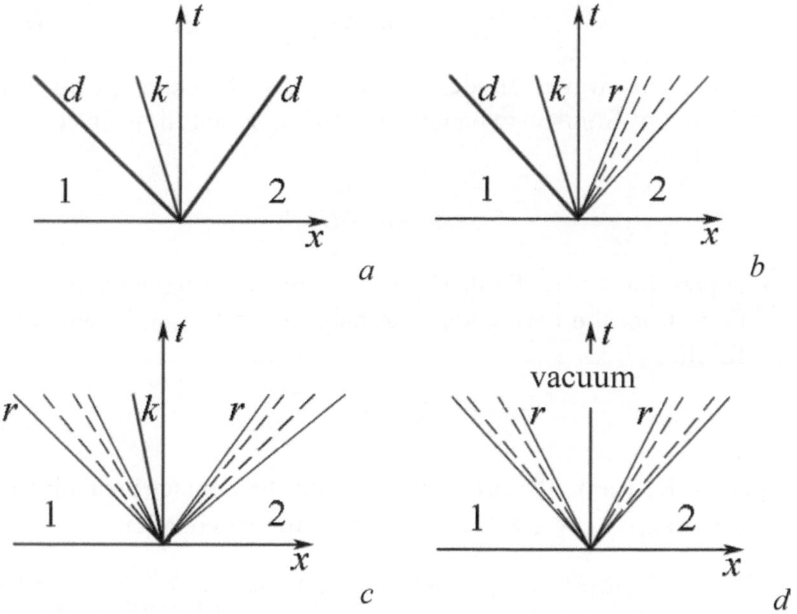

FIGURE 8.18 Decay schemes of the arbitrary discontinuity of parameters (a – decay with two shock waves; b – decay with a shock wave and depression wave; c – decay with two depression waves; d – decay with the vacuum area formation).

$$U_v = -\frac{2}{k-1} \cdot (c_1 + c_2) \tag{8.129}$$

Here the following is designated: $c_1 = \sqrt{k \cdot \frac{p_1 + p_0}{\rho_1}}$; $c_2 = \sqrt{k \cdot \frac{p_2 + p_0}{\rho_2}}$.

Besides, for the ideal gas we can accept that $p_0 = 0$;

2. Let us determine the variant of the decay of the parameter breakup after the membrane decomposition:
   - shock waves spread in both directions from the two media interface, if the following condition is fulfilled

$$u_1 - u_2 > U_d \tag{8.130}$$

   - shock wave spreads to the left from the two media interface, and depression wave – to the right, if the following conditions are fulfilled

$$U_r < u_1 - u_2 \leq U_d \tag{8.131}$$

- depression waves divided by the contact breakup spread in both directions from the two media interface, if the following conditions are fulfilled

$$U_v < u_1 - u_2 \leq U_r \tag{8.132}$$

- depression waves divided by the vacuum area spread in both directions from the two media interface, if the following condition is fulfilled

$$u_1 - u_2 \leq U_v; \tag{8.133}$$

3. Let us determine the pressure value on the contact breakup (inside the breakup decay area) solving the nonlinear equation

$$\Phi(P,\ p_1, \rho_1) + \Phi(P,\ p_2, \rho_2) = u_1 - u_2 \tag{8.134}$$

The following is designated in (8.134)

$$\Phi(P, p_l, \rho_l) = \begin{cases} \dfrac{P - p_l}{\rho_l c_l \sqrt{\dfrac{k+1}{2k}\dfrac{P + p_0}{p_l + p_0} + \dfrac{k-1}{2k}}}, if\ P \geq p_l \\ \dfrac{2}{k-1}\sqrt{k\dfrac{p_l + p_0}{\rho_l}}\left(\left(\dfrac{P + p_0}{p_l + p_0}\right)^{\frac{k-1}{2k}} - 1\right), if\ P < p_l \end{cases}$$

Here and further the index l takes the values $l=1,\ 2$. The Eq. (8.134) can be solved by any numerical method. However, when using this equation for the numerical solution of two- and three-dimensional equations of gas dynamics, the condition of the method effectiveness is important. The following can be accepted as the initial approximation for P when solving the Eq. (8.134)

$$P^0 = \frac{p_1\rho_2 c_2 + p_2\rho_1 c_1 + (u_1 - u_2)\rho_1 c_1 \rho_2 c_2}{\rho_1 c_1 + \rho_2 c_2}; \tag{8.135}$$

4. Let us determine the values of the mass velocities for the left (spreading from the right to the left) and for the right (spreading from the left to the right) waves

- for the shock wave

$$a_l = \sqrt{\rho_l(\frac{k+1}{2}(P+p_0)+\frac{k-1}{2}(p_l+p_0))} \quad (8.136)$$

- for the depression wave

$$a_l = \frac{k-1}{2k}\rho_l c_l \frac{1-\frac{P+p_0}{p_l+p_0}}{1-(\frac{P+p_0}{p_l+p_0})^{\frac{k-1}{2k}}}; \quad (8.137)$$

5. Let us determine the motion velocity of the contact breakup U

$$U = \frac{a_1 u_1 + a_2 u_2 + p_1 - p_2}{a_1 + a_2}; \quad (8.138)$$

6. Let us determine the velocities D of the left and right waves (for the depression wave – velocities D, D^* for ultimate characteristics) and gas density R on these waves
   - on the shock wave

$$D_l = u_l + (-1)^l \frac{a_l}{\rho_l} \quad (8.139)$$

$$R_l = \frac{\rho_l a_l}{a_l + (-1)^l \rho_l (u_l - U)} \quad (8.140)$$

   - on the depression wave

$$c_l^* = c_l - (-1)^l \frac{k-1}{2}(u_l - U) \quad (8.141)$$

$$D_l = u_l + (-1)^l c_l \quad (8.142)$$

$$D_l^* = U + (-1)^l c_l^* \quad (8.143)$$

$$R_l = \frac{k(P+p_0)}{(c_l^*)^2} \quad (8.144)$$

The considered algorithm of the calculation of gas parameter values on the contact breakup (Eqs. (8.127)–(8.144)) is the basis of S.K. Godunov method. Let us consider the solution of the equations of gas dynamics recorded in the non-stationary one-dimensional setting

$$\frac{\partial \rho}{\partial t}+\frac{\partial \rho u}{\partial x}=0$$

$$\frac{\partial \rho u}{\partial t}+\frac{\partial (p+\rho u^2)}{\partial x}=0$$

$$\frac{\partial \rho E}{\partial t}+\frac{\partial \rho u(E+{p}/{\rho})}{\partial x}=0 \qquad (8.145)$$

$$p=\rho(k-1)\left(E-\frac{1}{2}u^2\right)$$

The equation system (8.145) was solved on the grid demonstrated in Figure 8.14a. Let us consider the volumes with the numbers *I–1* and *I*. If we consider the processes on the boundary *i* of the volumes *I–1* and *I I–1* and *I* during the period of time corresponding to Courant-Friedrichs-Levy condition ($\Delta t=\min(\Delta t_{I-1},\Delta t_I)$; $\Delta t_{I-1}\le\frac{\Delta x_{I-1}}{|u_{I-1}|+c_{I-1}}$, $\Delta t_I\le\frac{\Delta x_I}{|u_I|+c_I}$), the interaction of gas flows located on both sides from the boundary *i* can be considered as the interaction of two flows located on both sides from the membrane M (Figure 8.16) in the problem on the decay of the arbitrary breakup of gas-dynamic parameters in the impact tube. The parameters on the boundary *i* of the volumes during the period of time from t_n to $t_{n+1}=t_n+\Delta t$ are defined as the parameters on the contact breakup and can be calculated by the algorithm (8.127)–(8.144)).

Taking the aforementioned into account, the explicit finite-difference scheme for calculating the parameters in the volume with the number *I* can be written down by S.K. Godunov method as follows:

$$\frac{\rho_I^{n+1}-\rho_I^n}{\Delta t}+\frac{(RU)_{i+1}^n-(RU)_i^n}{\Delta x_I}=0$$

$$\frac{(\rho u)_I^{n+1}-(\rho u)_I^n}{\Delta t}+\frac{(P+RU^2)_{i+1}^n-(P+RU^2)_i^n}{\Delta x_I}=0$$

$$\frac{(\rho E)_I^{n+1}-(\rho E)_I^n}{\Delta t}+\frac{(RUE+PU)_{i+1}^n-(RUE+PU)_i^n}{\Delta x_I}=0 \quad (8.146)$$

The solution algorithm of a problem in spatial setting with the help of S.K. Godunov method is similar to the one indicated before. At the first stage, the values of gas-dynamic parameters on all the boundaries of the control volume are calculated. At the same time, the variant of the breakup decay is established by the pressure values in the volumes adjacent to the boundary and values of the gas velocity components normal to the boundary. At the second stage, the equations of the type (8.146) are solved with the exception that these equations contain the derivatives by all the spatial coordinates.

8.4.4 LARGE-PARTICLE METHOD

The basis of the large-particle method is the method in the cells discussed in the works by F. Harlow [53]. The basics of the large-particle method are developed by the academician O. M. Belotserkovsky and his scientific school [27, 28].

The large-particle method should be referred to the methods of physical separation. Let us consider the main stages of the algorithm for the equation system of gas flow in the channel with the variable area F written down as follows:

$$\frac{\partial}{\partial t}\rho F+\frac{\partial}{\partial x}\rho Fu=f_\rho$$

$$\frac{\partial}{\partial t}\rho Fu+\frac{\partial}{\partial x}\rho Fu^2+F\frac{\partial p}{\partial x}=f_u$$

$$\frac{\partial}{\partial t}\rho FE+\frac{\partial}{\partial x}\rho FuE+\frac{\partial}{\partial x}Fpu=f_E$$

$$p=\rho(k-1)\left(E-\frac{1}{2}u^2\right) \quad (8.147)$$

Here f_ρ, f_u, f_E – source components, F – cross-section area of the channel ($F=F(t,x)$).

In accordance with the theory of large-particle method, the Eq. (8.147) can be split as follows:

Euler stage

The mixture flow is assumed as frozen. Only the pressure field influences the changes in the large particle parameters (inside the control volume). Besides, it is assumed that the geometric dimensions of the large particle change only at the final stage. Then the equations of the stage are written down as follows:

$$\rho F = const$$

$$\rho F \frac{\partial}{\partial t} u + F \frac{\partial p}{\partial x} = 0$$

$$\rho F \frac{\partial}{\partial t} E + \frac{\partial}{\partial x} F p u = 0$$

$$p = \rho(k-1)\left(E - \frac{1}{2}u^2\right) \tag{8.148}$$

Lagrangian stage

At this stage, the transfer effects through the boundaries of large particles are calculated, at the same time, the values of the overflow velocities on the boundaries of each particle and values of the mass flows, momentum and energy through the boundaries are calculated. The specific recording of these parameters is performed at the finite-difference stage formalization with the application of the formulas of the first and second order of accuracy;

Final stage

In accordance with the conservativeness conditions, the balance of masses, momentum and energy is brought for each large particle. Further, the new values of density, velocity and total energy are calculated in the centers of computational nodes. At this stage, the calculation is performed in accordance with the following equation system

$$\frac{\partial}{\partial t} \rho F + \frac{\partial}{\partial x} \rho F u = f_\rho$$

$$\frac{\partial}{\partial t} \rho F u + \frac{\partial}{\partial x} \rho F u^2 = f_u$$

$$\frac{\partial}{\partial t} \rho F E + \frac{\partial}{\partial x} \rho F u E = f_E$$

$$p = \rho(k-1)\left(E - \frac{1}{2}u^2\right) \tag{8.149}$$

The main finite-difference realization of the stages of the large-particle method for the Eq. (8.147) is written down in the explicit form as follows:

Euler Stage

$$u_I^{n+1} = u_I^n - \frac{\Delta t}{\Delta x_I + 0,5(\Delta x_{I-1} + \Delta x_{I+1})} \frac{p_{I+1}^n - p_{I-1}^n}{\rho_I^n},$$

$$\tilde{E}_I^{n+1} = E_I^n - \frac{\Delta t}{\Delta x_I + 0,5(\Delta x_{I-1} + \Delta x_{I+1})} \frac{(Fpu)_{I+1}^n - (Fpu)_{I-1}^n}{(F\rho)_I^n}; \tag{8.150}$$

Lagrangian stage

$$\Delta M_i^n = [\rho_i^n]\,\tilde{u}_i^{n+1} F_i^n \Delta t$$

$$\Delta I_i^n = \Delta M_i^n\,[\tilde{u}_i^{n+1}]$$

$$\Delta \mathrm{E}_i^n = \Delta M_i^n\,[\tilde{E}_i^{n+1}]$$

$$\tilde{u}_i^{n+1} = 0,5(\tilde{u}_I^{n+1} + \tilde{u}_{I-1}^{n+1})$$

$$\left[a_{i+\frac{1}{2}}\right] = \begin{cases} a_i, \tilde{u}_{i+\frac{1}{2}}^{n+1} \geq 0 \\ a_{i+1}, \tilde{u}_{i+\frac{1}{2}}^{n+1} < 0 \end{cases}; \quad a = \{\rho, u, E\}; \tag{8.151}$$

Final stage

$$M_I^{n+1} = M_I^n + \Delta M_i^n - \Delta M_{i+1}^n$$

$$I_I^{n+1} = I_I^n + \Delta I_i^n - \Delta I_{i+1}^n$$

$$\mathrm{E}_I^{n+1} = \mathrm{E}_I^n + \Delta \mathrm{E}_i^n - \Delta \mathrm{E}_{i+1}^n$$

$$\rho_I^{n+1} = \frac{M_I^{n+1}}{W_I^{n+1}},\ u_I^{n+1} = \frac{I_I^{n+1}}{M_I^{n+1}},\ E_I^{n+1} = \frac{\mathrm{E}_I^{n+1}}{M_I^{n+1}} \tag{8.152}$$

The shortcoming of the main scheme of the large-particle method [28] is its relatively low stability especially, when calculating the areas with low subsonic velocities (Mach numbers M<0.3). Below you can see the algorithms of the large-particle method described in [244] which allow improving the stability of the numerical calculation to Courant numbers $Ku \approx 1$. It should be pointed out that the schemes [244] are successfully used when solving the problems of internal ballistics in one-dimensional non-stationary setting, for the multi-component gas, as well.

The main modifications of the large-particle method applied in the schemes [244] consist in the following.

1. The analysis indicates that the relatively low stability of the scheme (8.150)–(8.152) is conditioned by the unsuccessful approximation of differential equations solved at Euler stage of the method. The application of implicit schemes at Euler stage is the solution of the problem of stability improvement. However, such approach is cumbersome. In [244], the Eqs. (8.150) are solved after solving the equation

$$\frac{\partial p}{\partial t} = -(k-1)p\frac{1}{F}\frac{\partial Fu}{\partial x} \tag{8.153}$$

The Eq. (8.153) can be obtained from the Eqs. (8.147). It can be approximated as follows:

$$\frac{\tilde{p}_I^{n+1} - p_I^n}{\Delta t} = -(k-1)\frac{p_I^n}{F_I^n}\frac{(Fu)_{I+1}^n - (Fu)_{I-1}^n}{\Delta x_I + 0{,}5(\Delta x_{I-1} + \Delta x_{I+1})} \tag{8.154}$$

$$\frac{\tilde{p}_I^{n+1} - p_I^n}{\Delta t} = -(k-1)\frac{\tilde{p}_I^{n+1}}{F_I^n}\frac{(Fu)_{I+1}^n - (Fu)_{I-1}^n}{\Delta x_I + 0{,}5(\Delta x_{I-1} + \Delta x_{I+1})} \tag{8.155}$$

The schemes (8.154) and (8.155) are explicit and the values $\tilde{p}_I^{n+1}$ calculated by these equations need to be used instead of the values p_I^n in the equations of Euler stage (8.150). The recorded modifications allow improving the stability of the explicit scheme of the large-particle method to the values $Ku \approx 0.5$.

The permitted value of Courant number $Ku \approx 0.5$ can be also provided if rewriting the Eq. (8.153) as follows:

$$\frac{\partial \ln p}{\partial t} = -(k-1)\frac{1}{F}\frac{\partial Fu}{\partial x} \tag{8.156}$$

The finite-difference approximation of the Eq. (8.156) allows writing down the equation for $\tilde{p}_I^{n+1}$ as follows:

$$\tilde{p}_I^{n+1} = p_I^n \exp(-(k-1)\frac{\Delta t}{F_I^n}\frac{(Fu)_{I+1}^n - (Fu)_{I-1}^n}{\Delta x_I + 0,5(\Delta x_{I-1} + \Delta x_{I+1})}) \tag{8.157}$$

2. The application of implicit approximation schemes of the Eq. (8.153) for pressure allows increasing the permitted Courant number up to the values $Ku \approx 1.0$. Preliminarily, it is necessary to perform the following equation transformations.

In accordance with the approximation theory of finite-difference schemes (see Section 8.3.2) we can write down the following

$$\frac{\partial p}{\partial t} \approx \frac{\tilde{p}_I^{n+1} - p_I^n}{\Delta t} - \frac{\Delta t}{2}\frac{\partial^2 p}{\partial t^2}$$

At the same time, the partial derivative $\frac{\partial^2 p}{\partial t^2}$ can be obtained from the Eq. (8.153)

$$\frac{\partial^2 p}{\partial t^2} = (k-1)\frac{p}{F}(\frac{\partial}{\partial x}\frac{F}{\rho}\frac{\partial p}{\partial x} + (k-1)\frac{1}{F}(\frac{\partial Fu}{\partial x})^2)$$

Taking into account the latter two equations, (8.153) can be rewritten as follows:

$$\frac{\tilde{p}_I^{n+1} - p_I^n}{\Delta t} \approx -(k-1)\frac{p}{F}\frac{\partial Fu}{\partial x} + \frac{\Delta t}{2}(k-1)\frac{p}{F}(\frac{\partial}{\partial x}\frac{F}{\rho}\frac{\partial p}{\partial x} + (k-1)\frac{1}{F}(\frac{\partial Fu}{\partial x})^2)$$

The implicit finite-difference approximation of the latter equation is written down as the system of linear three-diagonal algebraic equations

$$-a_I\tilde{p}_{I-1}^{n+1} + c_I\tilde{p}_I^{n+1} - b_I\tilde{p}_{I+1}^{n+1} = f_I, \quad I = 1, I_{\max} \tag{8.158}$$

The following is designated:

$$a_I = (k-1)\left(\frac{p}{F}\right)_{I-1}^n \frac{(\Delta t)^2}{\Delta x_I(\Delta x_{I-1} + \Delta x_I)}\left(\frac{F}{\rho}\right)_i^n$$

$$b_I = (k-1)\left(\frac{p}{F}\right)_{I+1}^n \frac{(\Delta t)^2}{\Delta x_I(\Delta x_{I+1} + \Delta x_I)}\left(\frac{F}{\rho}\right)_{i+1}^n$$

$$c_I = 1 + a_I + b_I$$

$$f_I = p_I^n\left\{1 + (k-1)\frac{\Delta t}{\Delta x_I}\frac{(Fu)_{i+1}^n - (Fu)_i^n}{F_I^n}\left[\frac{(k-1)}{2}\frac{\Delta t}{\Delta x_I}\frac{(Fu)_{i+1}^n - (Fu)_i^n}{F_I^n} - 1\right]\right\}$$

The equation system (8.158) is solved with the known boundary conditions for $\tilde{p}_I^{n+1}$ (with $I = 1$ and $I = I_{\max}$). The test calculations demonstrate that the following conditions can be accepted as the boundary ones

$$\tilde{p}_1^{n+1} = p_1^n, \quad \tilde{p}_{I_{\max}}^{n+1} = p_{I_{\max}}^n$$

The scheme (8.158) provides the high stability, it can be applied when solving spatial problems (using the method of alternate directions when solving the equations for pressure). However, this scheme has shortcomings. In particular, when solving spatial problems, it is difficult to apply the implicit scheme of the type (8.158) to solve problems in the regions of complex area when applying the irregular difference grids, etc. Besides, when calculating the stagnant areas with minimal subsonic velocities (with Mach numbers $M<0.1$), we observe the great calculation error demonstrated in the form of gas and velocity pulses.

3. The finite-difference approximation with the calculation of intermediary pressure values on the boundaries of large particles (control volumes) is the significant reserve for improving the computation accuracy and stability at Euler stage. This fact is explained in [170] and consists in the following. If the pressure field has the parameter distribution with some pressure pulsation $\pm\Delta p$ between the neighboring particles, the scheme of Euler stage (8.150) does not react at the pulsation availability, since $p_{i+1} - p_{i-1} \approx 0$.

The best results on the application of the large-particle method in the problems of internal ballistics, and this is confirmed by the multiple test calculations, are provided by the following finite-difference realization of Euler stage

$$u_I^{n+1} = u_I^n - \frac{\Delta t}{\Delta x_I} \frac{\tilde{p}_{i+1}^{n+1} - \tilde{p}_i^{n+1}}{\rho_I^n}$$

$$\tilde{E}_I^{n+1} = E_I^n - \frac{\Delta t}{\Delta x_I} \frac{(F\tilde{p}\tilde{u})_{i+1}^{n+1} - (F\tilde{p}\tilde{u})_i^{n+1}}{(F\rho)_I^n} \quad (8.159)$$

In the Eqs. (8.159) the values of the pressures $\tilde{p}_{i+1}^{n+1}$, $\tilde{p}_i^{n+1}$ and velocities $\tilde{u}_{i+1}^{n+1}, \tilde{u}_i^{n+1}$ on the boundaries of large particles (control volumes) are established by the results of solving the problem on the decay of weak breakup of gas-dynamic parameters. For the boundary with the number *i* (the boundary between large particles with the numbers *I–1* and *I*) the calculated correlations for $\tilde{p}_i^{n+1}$ and $\tilde{u}_i^{n+1}$ are written down as follows:

$$\tilde{p}_i^{n+1} = 0,5(p_{I-1}^n + p_I^n + a_i(u_{I-1}^n - u_I^n))$$

$$\tilde{u}_i^{n+1} = 0,5(u_{I-1}^n + u_I^n + \frac{p_{I-1}^n - p_I^n}{a_i})$$

$$a_i = \sqrt{\frac{(k_{I-1} + k_I)(p_{I-1}^n + p_I^n)(\rho_{I-1}^n + \rho_I^n)}{8}} \quad (8.160)$$

The computational algorithms of Euler stage (8.159) and (8.160) are used when solving spatial problems of internal ballistics, on orthogonal grids, as well [48, 244].

Let us consider the realization peculiarities of the large-particle method algorithms when solving the problems of internal ballistics in the areas with the complex shape. The approach connected with dividing the internal volume of the engine chamber into several computational areas linked with each other gas-dynamically is developed in [244]. The settings of the gas-dynamic problem different by the number of spatial derivatives considered can be used in different areas. When cross-linking separate areas with each other, the laws of mass conservation, momentum and energy are provided. The ideology of such approach is close to the ideology of the method of finite elements realized in the software package ANSYS [109]. Such ideology is convenient to solve practical problems and it is implemented in the software package of solving the problems of internal ballistics [10].

Let us point out the requirements to the models, computational algorithms and software package realized when developing the complex of programs [10]:

- application of computational algorithms of solving problems of gas dynamics needs to be equally appropriate when calculating the areas of different dimensionality ("zero-dimensional" areas, channels, areas with the characteristic two-dimensional or spatial modes of flow);
- computational algorithms need to give the possibility to calculate the velocity module and its direction in one-, two- or three-dimensional variant;
- elementary computational cell (large particle, control volume) can have the arbitrary number of faces, the concrete number of which is set up in the initial data. The cell faces can connect the neighboring computational cells, they can be the impermeable boundary, perforated boundary, surface of the heated up or burning fuel. It is necessary to provide the possibility of calculating the flows along the cell faces using all the algorithms in the software package implemented (for instance, the boundary of the area with the environment – critical section of the nozzle block, etc.);
- computational algorithm needs to allow dividing the computational process along the cell faces and volume, which facilitates the application of multiprocessor computer facilities and allows applying the methods of control volumes.

The equations of gas-dynamics in the following form [11] can be used in the inter-ballistic calculations in the arbitrary area (ignition system, volume, channel, etc.):

$$H_1H_2H_3\frac{\partial \rho}{\partial t}+\frac{\partial}{\partial q_1}\rho v_1H_2H_3+\frac{\partial}{\partial q_2}\rho v_2H_1H_3+\frac{\partial}{\partial q_3}\rho v_3H_1H_2=0$$

$$H_1H_2H_3\frac{\partial \rho\alpha_i}{\partial t}+\frac{\partial}{\partial q_1}\rho\alpha_i v_1H_2H_3$$

$$+\frac{\partial}{\partial q_2}\rho\alpha_i v_2H_1H_3+\frac{\partial}{\partial q_3}\rho\alpha_i v_3H_1H_2=0,\ i=1,\ N,$$

$$H_1H_2H_3\frac{\partial}{\partial t}\rho v_1+\frac{\partial}{\partial q_1}\rho v_1^2H_2H_3+\frac{\partial}{\partial q_2}\rho v_1v_2H_1H_3+\frac{\partial}{\partial q_3}\rho v_1v_3H_1H_2$$

$$=-H_2H_3\cdot\frac{\partial p}{\partial q_1}-\rho v_2H_3d_{12}-\rho v_3H_2d_{13}$$

$$H_1H_2H_3\frac{\partial}{\partial t}\rho v_2+\frac{\partial}{\partial q_1}\rho v_1\, v_2H_2H_3+\frac{\partial}{\partial q_2}\rho v_2^2\, H_1H_3+\frac{\partial}{\partial q_3}\rho v_2v_3H_1H_2=$$
$$=-H_1H_3\cdot\frac{\partial p}{\partial q_2}-\rho v_1H_3d_{21}-\rho v_3H_1d_{23}$$

(8.161)

$$H_1H_2H_3\frac{\partial}{\partial t}\rho v_3+\frac{\partial}{\partial q_1}\rho v_3\, v_1\, H_2H_3+\frac{\partial}{\partial q_2}\rho v_3v_2H_1H_3+\frac{\partial}{\partial q_3}\rho v_3^2H_1H_2$$
$$=-H_1H_2\cdot\frac{\partial p}{\partial q_3}-\rho v_1H_2d_{31}-\rho v_2H_1d_{32}$$

$$H_1H_2H_3\frac{\partial}{\partial t}\rho E+\frac{\partial}{\partial q_1}\rho v_1(E+{}^{p}\!/\!_{\rho})H_2H_3$$
$$+\frac{\partial}{\partial q_2}\rho v_2(E+{}^{p}\!/\!_{\rho})H_1H_3+\frac{\partial}{\partial q_3}\rho v_3(E+{}^{p}\!/\!_{\rho})H_1H_2=0$$

In the Eq. (8.161) the continuity equations for the total density of combustion products, continuity equations for the i density component (there are only N components in the combustion products), equations of momentum conservation (three equations – for the velocity components v_1, v_2, v_3), equation of energy conservation E (sum of the internal and kinetic energies) are successively written down. Here H_1, H_2, H_3 – Lame coefficients with different values in different coordinate systems. The coefficients d_{12}, d_{13}, d_{21}, d_{23}, d_{31}, d_{32} also depend on the coordinate system selected. In particular, the following correlations are true for the cylindrical coordinate system:

$$q_1=x,\quad q_2=r,\quad q_3=\vartheta,\quad H_1=1,\quad H_2=1,\quad H_3=r;$$

$$d_{11}=0,\quad d_{13}=0,\quad d_{21}=0,\quad d_{23}=-v_3,\quad d_{31}=0,\quad d_{32}=v_3$$

The aforementioned equation system is true for the arbitrary coordinate system whose elementary volume is demonstrated in Figure 7.3 (in the figure the elementary volume faces are designated as ds_1, ds_2, ds_3, respectively).

The equation system can be represented in the recorded form during the numerical solution, however, the system simplifications are also possible (for instance, if all the cells available in the computational area border with the impermeable grid along two parallel faces). Let us point out that the consideration of cells with additional faces (number of faces can exceed six) is possible in the computational algorithms of process calculation. In this case, it is necessary to consider the operation procedures with fractional particles similar to the ones discussed in [28]. In this case, to provide the computational accuracy, it is preferable to apply the conservation equations recorded in the integral form (integrated along the coordinates q_1, q_2, q_3).

Let us formulate the elementary volume model in the computational area inside the engine chamber:

- location of the elementary cell is established in the cylindrical coordinate system bound with the engine central axis (axis x);
- elementary cell can have several faces (boundaries), the part of which connects the cell with the adjacent cells (these faces are completely open), part of faces (boundaries) are partially perforated, part of faces (boundaries) are impermeable solid boundaries, the fuel combustion products come to the part of faces (boundaries);
- elementary cell can contain the sources of mass and energy, for instance, the granulated ignition composition (in particular, it is necessary when modeling the elementary cell as the igniter body);
- open elementary cells are not necessarily perpendicular (perpendicularity of the faces or their orthogonality are important when modeling the gas-dynamic processes in the assumptions of the viscous gas flow);
- heat losses can occur through the impermeable boundaries of the elementary cell (due to the heat transfer and heat conductivity processes) and momentum (due to friction) whose models are established additionally.

The foregoing formulated conditions allow writing down the computational algorithms of gas-dynamic processes for elementary cells. Let us consider the main correlations based on the application of the aforementioned algorithms of the large-particle method.

At the first (Euler) stage, the divergent preservation equation components are not taken into account. The pressure values are calculated on the cell boundaries (Figure 8.19, boundaries *AB, CD*). At the second (Lagrangian) stage, the flows on the boundaries of the adjacent cells are calculated. At the final stage, the balance of masses, energies and momentum is calculated in each computational cell. Let us point out that at such explanation of the computational method the algorithm logics during the calculation of spatial flows is the same as at the calculation of one-dimensional flows.

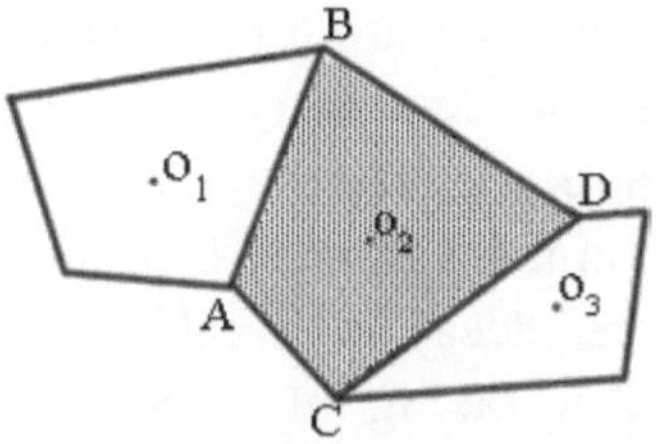

FIGURE 8.19 View of non-orthogonal computational cells.

For the additional simplification of the computation logics the cycles are arranged as follows:

- at the first stage, the pressure values on the boundaries are calculated for all the boundaries;
- at the second stage, the intermediary values of flow velocity and energy, as well as the normal velocity components on the boundaries (taking into account the calculated pressure values) are calculated for all the boundaries;
- at the third stage, the flows of mass, momentum and energy are calculated for each computational cell, the internal sources of mass, momentum and energy are calculated as well;
- at the fourth stage, the step by time is selected and masses, energy and momentum in the cell are calculated;
- at the final stage, the values of gas dynamic magnitudes are calculated in each cell.

The first two (or three) stages of the method can be combined in the computational algorithm.

Let us consider the realization of the aforementioned stages in more detail.

1. The pressure values are determined on the boundaries of the computational cells (these are the boundaries AB and *CD* for the cells with the center in O2), if the values of gas-dynamic values (ρ, p, u, v, w, E, etc.) in the cells with the centers in *O1, O2, O3* are known. In the same way as in S.K. Godunov method [241], we assume that two semi-infinite flows interact on the cell boundaries. If the interaction time of these flows is limited (computation step by time is coordinated with the step along the spatial coordinate by Courant-Friedrichs-Levy condition), the pressure computation error on the boundary of adjacent cells will be acceptable. Two elementary solutions for the pressure value on the boundary can be written down in the same way

as in the problem on the decay of arbitrary breakup of the parameters. For definiteness, when solving the elementary problem on the breakup decay, it is accepted that the pressure, for instance, on the left side from the boundary exceeds the pressure on the right side from the boundary. Besides, it is assumed that the parameter decay is "weak" [184]. Further, the pressure dependence on the boundary, taking into account the process development variant, is written down. In the first variant, the shock waves spread in both directions from the boundary. In the second variant, the shock wave spreads in one direction from the boundary, and the depression wave – in the opposite direction.

In the algorithms of the large-particle method not only the pressure should be known on the boundaries of the adjacent cells but also the pressure product by the value of the flow velocity normal component. The algorithms of the breakup decay allow also obtaining the value of the flow velocity normal component. The analysis indicates that when calculating the processes in engine chamber, the application of this algorithm is quite satisfactory for the computational accuracy applicable in practice.

2. In accordance with the main idea of the large-particle method, it is assumed that the gas density in all the computational volume is unchanged when performing the first stage of the method (Euler stage). This allows excluding the mass conservation equation from the calculations at this stage. The remaining equations are additionally simplified. In particular, the initial equation system (1) can be rewritten as follows:

$$H_1H_2H_3\,\rho\frac{\partial}{\partial t}v_1 = -H_2H_3\cdot\frac{\partial p}{\partial q_1} - \rho v_2 H_3 d_{12} - \rho v_3 H_2 d_{13}$$

$$H_1H_2H_3\,\rho\frac{\partial}{\partial t}v_2 = -H_1H_3\cdot\frac{\partial p}{\partial q_2} - \rho v_1 H_3 d_{21} - \rho v_3 H_1 d_{23}$$

$$H_1H_2H_3\,\rho\frac{\partial}{\partial t}v_3 = -H_1H_2\cdot\frac{\partial p}{\partial q_3} - \rho v_1 H_2 d_{31} - \rho v_2 H_1 d_{32}$$

$$H_1H_2H_3\,\rho\frac{\partial}{\partial t}E + \frac{\partial}{\partial q_1}v_1 p H_2H_3 + \frac{\partial}{\partial q_2}v_2 p H_1H_3 + \frac{\partial}{\partial q_3}v_3 p H_1H_2 = 0 \tag{8.162}$$

To record the equations in the finite-difference form, the equation system should be written down as follows:

$$\rho\frac{\partial}{\partial t}v_1 = -\frac{\partial p}{H_1\partial q_1} - \frac{\rho v_2 d_{12}}{H_1H_2} - \frac{\rho v_3 d_{13}}{H_1H_3}$$

$$\rho\frac{\partial}{\partial t}v_2 = -\frac{\partial p}{H_2\partial q_2} - \frac{\rho v_1 d_{21}}{H_1H_2} - \frac{\rho v_3 d_{23}}{H_2H_3}$$

$$\rho\frac{\partial}{\partial t}v_3 = -\frac{\partial p}{H_3\partial q_3} - \frac{\rho v_1 d_{31}}{H_1H_3} - \frac{\rho v_2 d_{32}}{H_2H_3}$$

$$\rho\frac{\partial}{\partial t}E = -\frac{1}{H_1H_2H_3}\cdot\left(\frac{\partial}{\partial q_1}v_1 pH_2H_3 + \frac{\partial}{\partial q_2}v_2 pH_1H_3 + \frac{\partial}{\partial q_3}v_3 pH_1H_2\right) \tag{8.163}$$

In this form, when writing down the Eqs. (8.163) for the elementary volume (at the finite-difference solution of equations) with the edges $H_1\Delta_{q1}$, $H_2\Delta_{q2}$, $H_3\Delta_{q3}$, the values of the areas of the elementary volume faces perpendicular to the axes q_3, q_1, q_2 respectively, can be applied instead of the expressions H_1H_2, H_2H_3, H_1H_3, and the elementary volume value – instead of the product $H_1H_2H_3$. The partial derivatives $\frac{\partial p}{\partial q_i}$ (or $\frac{\partial pv_j}{\partial q_i}$) can be calculated using the pressure gradient by the normal $\frac{\partial p}{\partial n}$:

$$\frac{\partial p}{\partial n}\mathbf{n} = \frac{\partial p}{H_1\partial q_1}\mathbf{i} + \frac{\partial p}{H_2\partial q_2}\mathbf{j} + \frac{\partial p}{H_3\partial q_3}\mathbf{k}$$

Thus, the computation algorithm at the second stage is as follows:

- gradient values $\frac{\partial p}{\partial n}, \frac{\partial pv_1}{\partial n}, \frac{\partial pv_2}{\partial n}, \frac{\partial pv_3}{\partial n}$ are calculated for each face (the calculation is performed by the known pressure and velocity values in the cell center and on the face considered, as well as by the known value of the distance from the cell center to the face considered);
- values of the derivatives $\frac{\partial p}{H_i\partial q_i}, \frac{\partial pv_j}{H_i\partial q_i}$ are determined by the known values of $\frac{\partial p}{\partial n}, \frac{\partial pv_1}{\partial n}, \frac{\partial pv_2}{\partial n}, \frac{\partial pv_3}{\partial n}$ and directing cosines $(\mathbf{n}\cdot\mathbf{i})$, $(\mathbf{n}\cdot\mathbf{j})$, $(\mathbf{n}\cdot\mathbf{k})$;

- we find the intermediary values of velocities and energy from the Eqs. (8.163) with the selected step Δt by time, summing up the results by each face;
- normal velocity components are found by interpolation on each boundary in the cell centers by the known velocity values.

Let us point out that the expressions $\rho v_2 H_3 d_{12}$, $\rho v_3 H_2 d_{13}$, $\rho v_1 H_3 d_{21}$, $\rho v_3 H_1 d_{23}$, $\rho v_1 H_2 d_{31}$, $\rho v_2 H_1 d_{32}$ in the right sides of the Eqs. (8.163) can be taken into account either at this stage or the final stage of the method.

3. At the next stages of the computational algorithm, the following equations are solved

$$H_1 H_2 H_3 \frac{\partial \rho}{\partial t} = -\left(\frac{\partial}{\partial q_1} \rho v_1 H_2 H_3 + \frac{\partial}{\partial q_2} \rho v_2 H_1 H_3 + \frac{\partial}{\partial q_3} \rho v_3 H_1 H_2\right)$$

$$H_1 H_2 H_3 \frac{\partial \rho \alpha_i}{\partial t} = -\begin{pmatrix} \dfrac{\partial}{\partial q_1} \rho \alpha_i v_1 H_2 H_3 + \dfrac{\partial}{\partial q_2} \rho \alpha_i v_2 H_1 H_3 + \\ \dfrac{\partial}{\partial q_3} \rho \alpha_i v_3 H_1 H_2 \end{pmatrix}, \qquad i=1, N,$$

$$H_1 H_2 H_3 \frac{\partial}{\partial t} \rho v_1 = -\left(\frac{\partial}{\partial q_1} \rho v_1^2 H_2 H_3 + \frac{\partial}{\partial q_2} \rho v_1 v_2 H_1 H_3 + \frac{\partial}{\partial q_3} \rho v_1 v_3 H_1 H_2\right)$$

$$H_1 H_2 H_3 \frac{\partial}{\partial t} \rho v_2 = -\left(\frac{\partial}{\partial q_1} \rho v_1 v_2 H_2 H_3 + \frac{\partial}{\partial q_2} \rho v_2^2 H_1 H_3 + \frac{\partial}{\partial q_3} \rho v_2 v_3 H_1 H_2\right)$$

$$H_1 H_2 H_3 \frac{\partial}{\partial t} \rho v_3 = -\left(\frac{\partial}{\partial q_1} \rho v_3 v_1 H_2 H_3 + \frac{\partial}{\partial q_2} \rho v_3 v_2 H_1 H_3 + \frac{\partial}{\partial q_3} \rho v_3^2 H_1 H_2\right)$$

$$H_1 H_2 H_3 \frac{\partial}{\partial t} \rho E = -\left(\frac{\partial}{\partial q_1} \rho v_1 E H_2 H_3 + \frac{\partial}{\partial q_2} \rho v_2 E H_1 H_3 + \frac{\partial}{\partial q_3} \rho v_3 E H_1 H_2\right)$$

(8.164)

As above, when solving the Eqs. (8.164) for the elementary volume (at finite-difference solution of the equations) with edges $H_1 \Delta q_1$, $H_2 \Delta q_2$, $H_3 \Delta q_3$, we can use the values of the areas of the elementary volume faces perpendicular to the axes q_3, q_1, q_2, respectively, instead of the expressions $H_1 H_2$, $H_2 H_3$, $H_1 H_3$, and the elementary volume value – instead of the product $H_1 H_2 H_3$.

Let the step by time equal Δt. We integrate both parts by the spatial derivatives and time. The first equation of the system (8.164) in the finite-difference form is rewritten as follows:

$$H_1\Delta q_1\, H_2\Delta q_2\, H_3\Delta q_3\, (\rho_{i,j,k}^{n+1} - \rho_{i,j,k}^{n}) = -\begin{pmatrix} (\rho v_1)_{i+1,j,k}^{n} - \\ (\rho v_1)_{i,j,k}^{n} \end{pmatrix} H_2\Delta q_2\, H_3\Delta q_3$$
$$+ ((\rho v_2)_{i,j+,k}^{n} - (\rho v_2)_{i,j,k}^{n})\, H_1\Delta q_1\, H_3\Delta q_3$$
$$+ ((\rho v_3)_{i,j,k+1}^{n} - (\rho v_3)_{i,j,k}^{n})\, H_1\Delta q_1\, H_2\Delta q_2)\Delta t$$

In the finite-difference equation the flows are summed up by the six faces of the elementary volume. Similar equations can be written down, if the number of faces in the elementary cell differs from six and faces are not perpendicular to the axial coordinates. For this case, the equation of mass conservation is written down as follows:

$$\Delta W \cdot (\rho^{n+1} - \rho^{n}) = \sum_{N} \langle \rho^{n} \rangle u_n \Delta S \cdot \Delta t \tag{8.165}$$

Here ΔW – elementary volume value, ΔS – area of boundaries through which the combustion products overflow, N – number of faces, u_n – flow velocity component normal to the face, $\langle \rho^n \rangle$ – gas density defined taking into account the gas flow direction through the cell boundary. Similarly, all the other system Eqs. (8.164) are written down. If source components are available in the right parts of the Eqs. (8.161) (e.g., the ignition composition is located in the volume), they can be taken into account at this stage (additional components are available in the right side of the Eqs. (8.165)).

4. Calculation of the flows of mass, momentum and energy through the elementary cell allows establishing the new values of $\rho^{n+1}\Delta W$, $\rho^{n+1}\mathbf{v}^{n+1}\Delta W$, $\rho^{n+1}E^{n+1}\Delta W$. At the final stage, ρ^{n+1}, $\mathbf{v}^{n+1}$, E^{n+1} are calculated by the ordinary algebraic computations.

The application of S.K. Godunov method, logically, looks even simpler to solve the considered problem. The algorithm for calculating the Eqs. (8.161) with this method is realized as follows:

- at the first stage, the pressure values are established on the boundaries between the neighboring cells. The pressure values can be determined in the assumption that the parameter breakup in the adjacent cells is

"weak." However, the operation of such programs demonstrates that the pressure determination using this assumption (assumption on "weak" breakup) when calculating the engine output to the quasi-stationary mode results in the computational algorithm instability. The computational algorithm stability is much higher, if the pressure on the boundary of the adjacent cells is determined by the algorithm for the arbitrary breakup of gas-dynamic parameters;
- at the second stage, the values of all gas-dynamic parameters are established on the boundaries of adjacent cells. The flows of mass, momentum and energy through the boundaries are additionally found;
- at the third stage, the step by time is selected and mass, energy and momentum in the computational area cells are calculated;
- at the final stage, the values of gas-dynamic values are calculated in each cell.

It should be pointed out that computation accuracy with the help of large-particle and S.K. Godunov methods are practically the same (both methods are of the first order of accuracy). The continuous software package exploitation demonstrates that the large-particle method is somewhat more reliable than S.K. Godunov method for solving the problems of internal ballistics in heat engines (as a rule, the calculations by large-particle method provide the stability with Courant numbers $Ku \leq 0.95$, and by S.K. Godunov Method – with Courant numbers $Ku \leq 0.90$). Nevertheless, it is advisable to have the possibility of performing calculations by different methods in the software package applied.

As an example, in Figures 8.20–8.28 you can see the results of the solid fuel ignition calculations in the axisymmetric solid-fuel engine (Figure 5.4). In all the calculations, the mass consumption of combustion products from

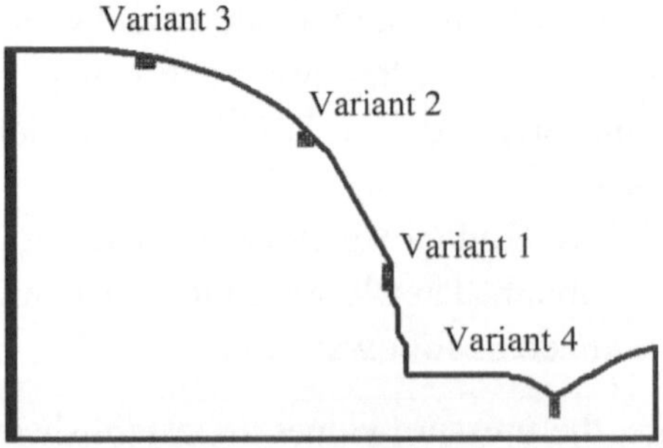

FIGURE 8.20 Computational scheme of the combustion chamber with different igniter locations.

the igniter body is taken as the same. The problem is solved with the help of large-particle and S.K. Godunov methods by the aforementioned algorithms.

In accordance with the assumptions, it is accepted in the developed models that the fuel ignites after the surface layer reaches some critical temperature. However, depending on the igniter location and remoteness, the field of heat flows is heterogeneous in different sections of the combustion chamber. In this regard, not all the charge surface ignites at the initial time point but only some, its most heated part. It is demonstrated in [244] that for different designs of solid-fuel engines the character of flame spread along the solid fuel charge is mainly determined by the igniter location. At the same time, in particular, it is shown that the values of the time points of the fuel involvement into combustion, flame spreading velocities along the fuel surface, etc. are different for different asymmetry variants of the fuel charge placement into the combustion chamber. Taking into account the results [244], it can be assumed that the igniter location is also principal in the gas generator scheme considered.

To investigate the influence of the igniter location on the fuel ignition character, it is advisable to consider the temperature field changes (or heat flow field changes) in time and space. Four different igniter installation schemes in the chamber are considered in the calculations (Figure 8.20): variant 1 – corresponds to the igniter location in the area of nozzle bottom and gas duct connection; variant 2 – igniter is located in the middle of the nozzle bottom; variant 3 – igniter is located in the area of the nozzle bottom and fuel shell; variant 4 – igniter is located in the vicinity of the minimal section of the nozzle block. In the latter case, the plug is placed in the minimal section of the block preventing the output of the combustion products into the nozzle part to the pressure values in the combustion chamber ~ 1.2 MPa.

For variant 1 of the igniter body location, in Figure 8.21 you can see the pressure computation results in the computational area for different stages of the initial region of the engine unit operation: igniter operation stage (t=0.005 s); fuel ignition stage (t=0.010 s); joint combustion of the igniting composition and fuel stage (t=0.015 s).

In Figure 8.21, the fuel area involved into the combustion is marked with the black line (left boundary), the igniter location is marked with black on the nozzle bottom. When only the igniter combustion products enter the free volume, the pressure rises throughout the whole volume of the computational area (Figure 8.21a). At the moment of the fuel surface heat-up to the ignition temperature, it ignites, at the same time, in the place of the gas duct

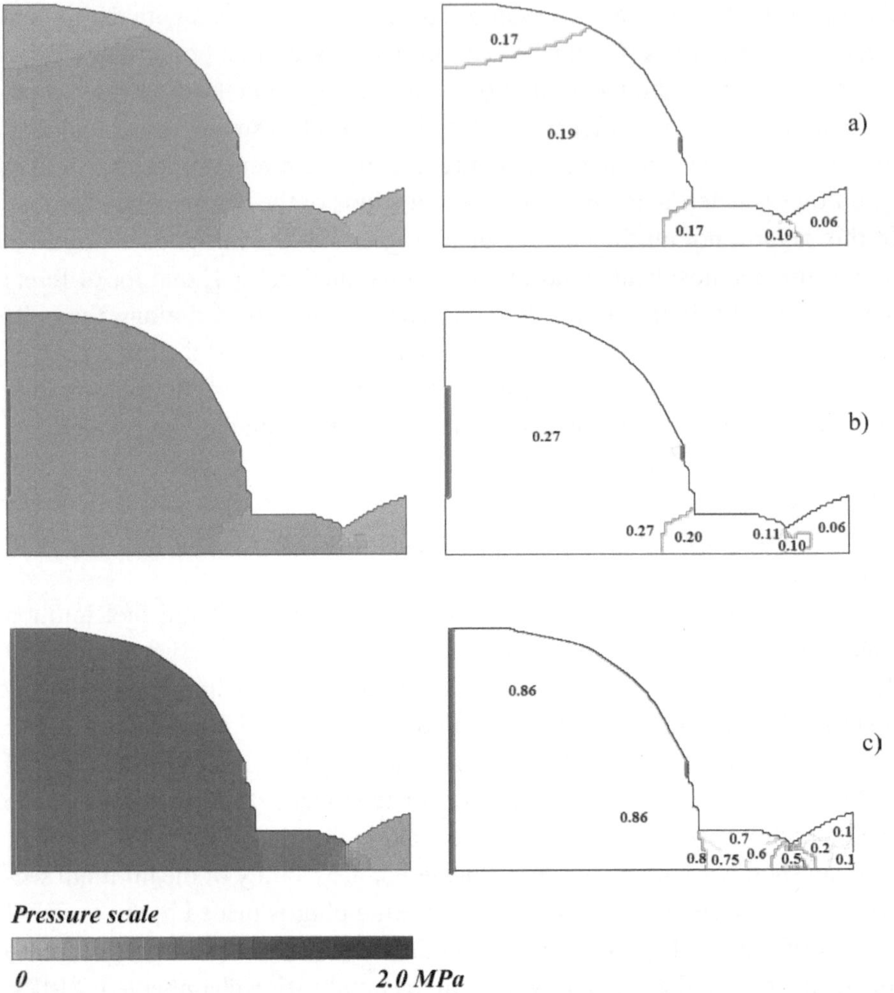

FIGURE 8.21 Pressure field in the free volume of the engine chamber (a – 0.0050 s; b – 0.0100 s; c – 0.0150 s).

location the expected pressure decrease is observed caused by the section area diminishing (Figure 8.21b). After the fuel ignites throughout the whole surface, the further pressure increase is observed in the free volume of the combustion chamber. The pressure field is given in Figure 8.21c, at the same time, the isolines do not change with further development of the processes, only the pressure values are changing.

For the first computational case, the temperature fields of the combustion products in the engine free volume during the fuel heat-up, ignition and

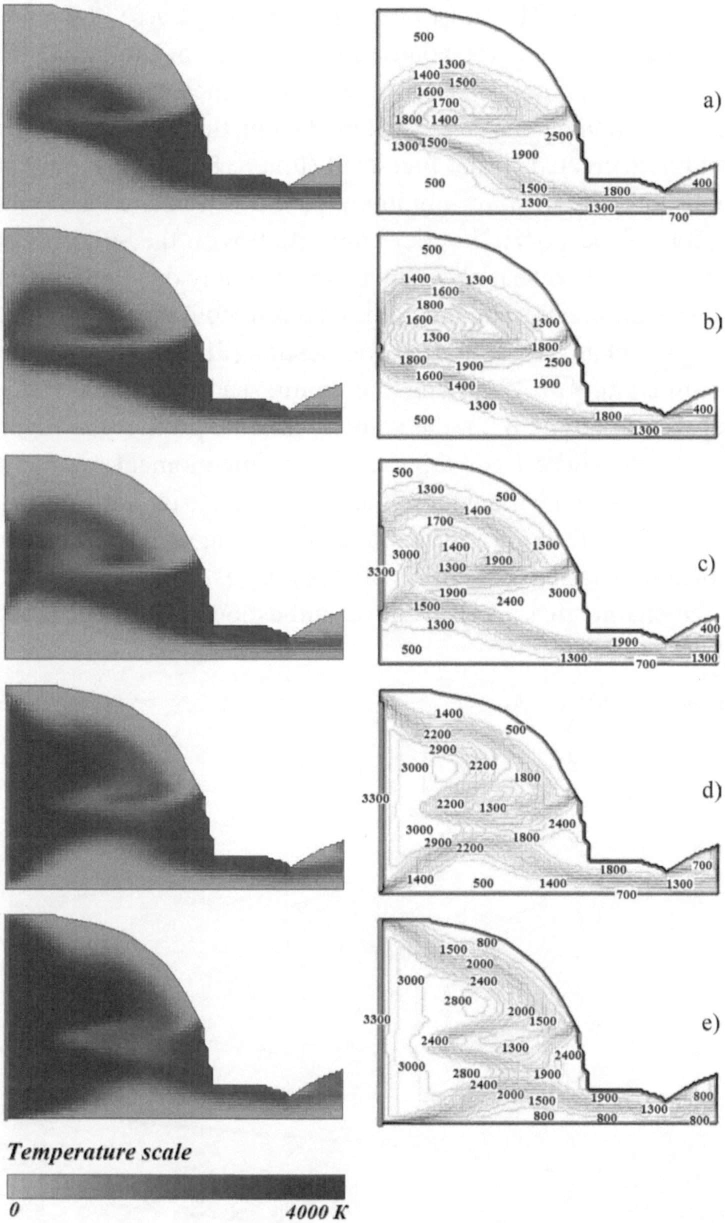

FIGURE 8.22 Temperature field in the free volume of the engine unit combustion chamber (computational case 1) (a – 0.0050 s; b – 0.0882 s; c – 0.0099 s; d – 0.0120 s; e – 0.0135 s).

combustion are demonstrated in Figure 8.22. During the igniter operation, the medium is heated up in the direction of the solid-fuel charge and nozzle (Figure 8.22a). The most heat-stressed charge area is the area closest to the initiator, thus corresponding to the point of the initial fuel ignition (Figure 8.22b). After 30% of the charge surface are involved into the combustion, the flame is further distributed along the charge end-face due to the thermal field created by the fuel itself (Figure 8.22c). It should be taken into account that the medium minimal velocities (under 5 m/s) take place in the region of the nozzle bottom connection with the solid-fuel charge, thus indicating the fuel heat-up in the given area only due to the radiant heat flow (Figure 8.22d). After the fuel ignition throughout the whole surface, the igniter only maintains the solid fuel combustion (Figure 8.22e).

The temperature profile along the charge end-face for different time points is given in Figure 8.23, the velocity module profile – in Figure 8.24. In both cases, the curve 1 corresponds to the time moment prior to the fuel ignition, curve 7 – to the moment of the fuel complete ignition, and the intermediary curves correspond to the intermediary time values. Figure 8.23 is an additional illustration of the foregoing processes (Figure 8.22) and allows defining the charge area involved into combustion at different time points.

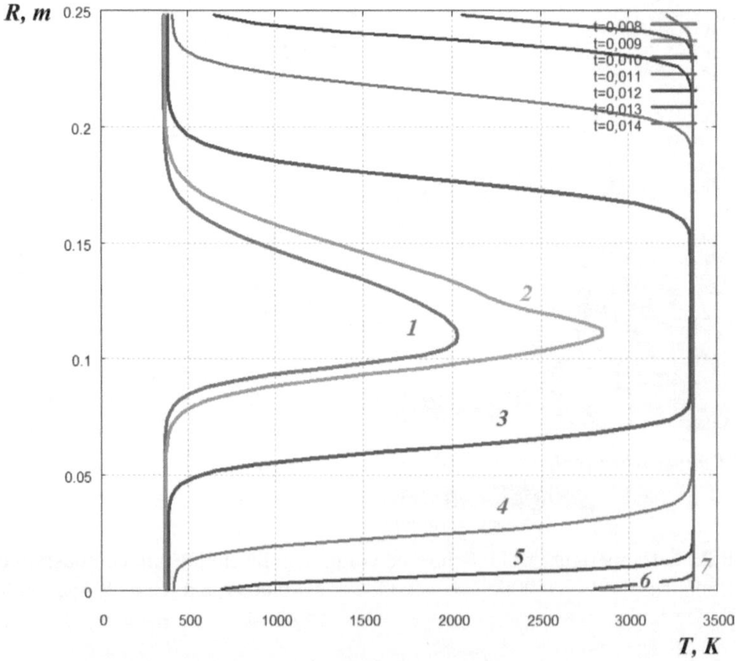

FIGURE 8.23 Temperature distribution along the solid fuel end-face (*1* – 0.008 s, *2* – 0.009 s, *3* – 0.010 s, *4* – 0.011 s, *5* – 0.012 s, *6* – 0.013 s, *7* – 0.014 s).

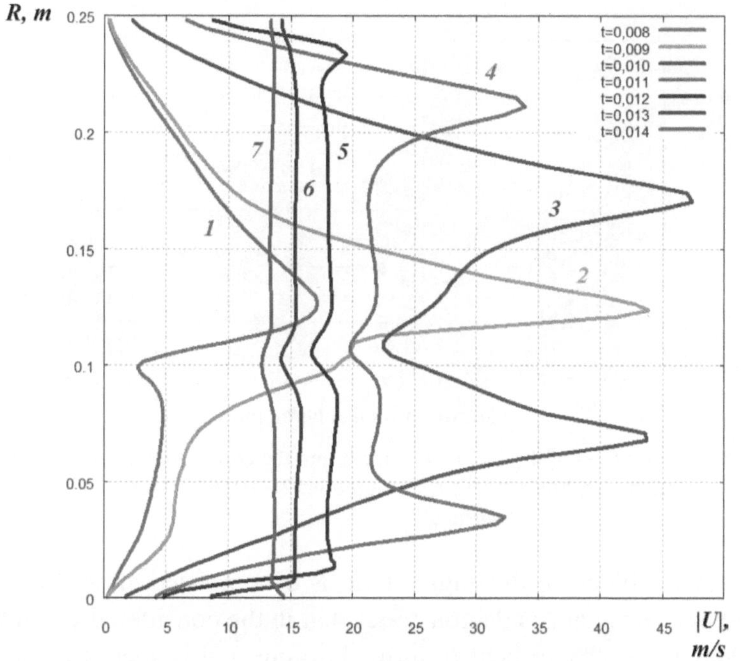

FIGURE 8.24 Velocity module distribution along the solid fuel end-face (*1* – 0.008 s, *2* – 0.009 s, *3* – 0.010 s, *4* – 0.011 s, *5* – 0.012 s, *6* – 0.013 s, *7* – 0.014 s).

In particular, when 90% of the fuel charge surface are combusted, the velocity values along the charge end-face are established and have the values in the range 13–16 m/s (curves 6, 7, Figure 8.24).

In Figure 8.25 you can see the diagram illustrating the duration of the fuel involvement into combustion at different variants of the igniter location. The analysis of the results demonstrates that the longest is the fuel heat-up when the igniter is located on the nozzle bottom of the engine unit. Less time is required when the igniter is located on the side surface of the combustion chamber (variant 3), and the least time corresponds to the igniter location in the minimal section of the nozzle block (if the plug is available). For the objective interpretation of the information given in Figure 8.24, the character of the temperature field change in the combustion chamber is considered below when the igniter is located in accordance with the variants 2, 3 and 4 (Figure 8.20).

The application of the variant 2 of the igniter location is characterized by the longest fuel involvement into the combustion (Figure 8.26). It was assumed by the conditions of the problem being solved that the plug, placed

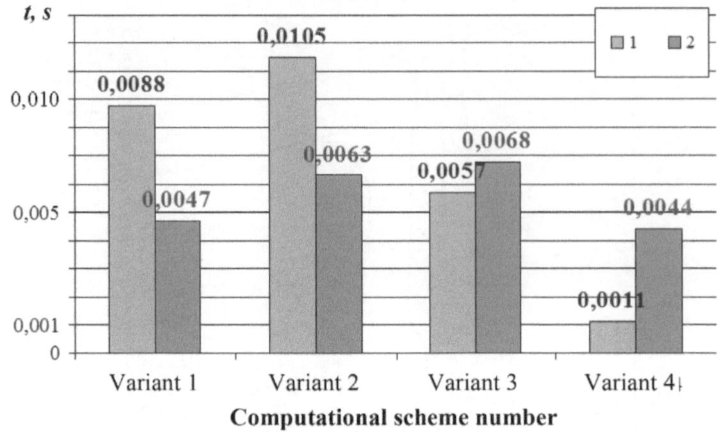

FIGURE 8.25 Duration of the fuel involvement into the combustion (*1* – time period from the igniter operation start till the fuel surface ignition; *2* – time period required for the flame distribution along the whole fuel surface).

into the nozzle block of the engine unit, is decomposed after the igniter is switched on. The analysis demonstrates that in the considered variant of the igniter location ~50% of heat from the burning fuel is spent to heat up the chamber free volume, and is further lost when the combustion products outflow through the free boundary (Figure 8.26a and 8.26b). This explains the relatively long fuel heat-up till its combustion (t=0.0105 c). The most heat-stressed section is located in the upper part of the charge, this is the region of the fuel initial initiation (Figures 8.26b and 8.26c). It should be pointed out that the uniform temperature field distribution throughout the combustion chamber is observed in the considered variant, thus giving the possibility to minimize the volume of stagnant areas in the nozzle bottom region (Figure 8.26d and 8.26e).

The variant 3 of the igniter location (Figure 8.27) is characterized by the longest time interval spent for the flame distribution along the charge surface (t=0.0068 s). However, the fuel combustion in this scheme starts within t=0.0057 s, which is the least from all the schemes of the igniter location on the nozzle bottom.

The fast fuel ignition in the case corresponding to the variant 3 is provided by the lack of thermal energy losses of the igniter combustion products through the nozzle and heat accumulation in the immediate vicinity from the fuel surface (Figures 8.27a and 8.27b). Besides, the igniter location in the place of nozzle bottom connection with the gas generator body allows excluding the long heat-up of the stagnant area. After the fuel ignition and

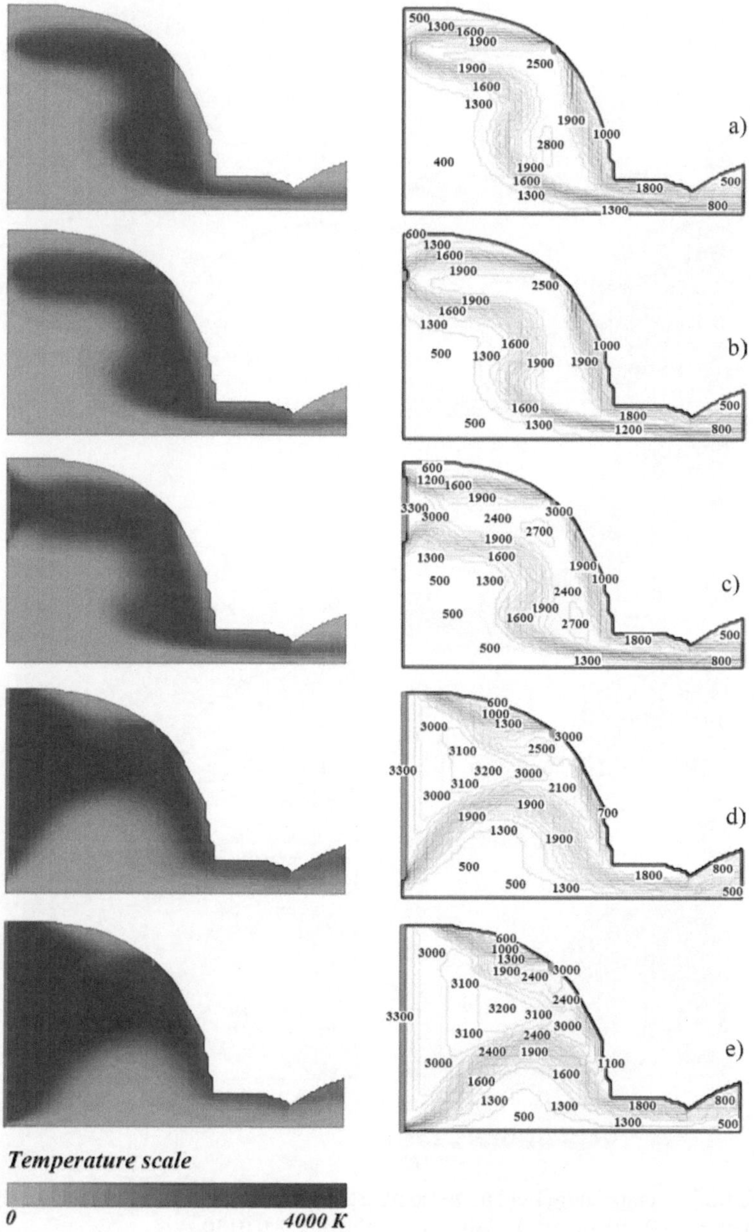

FIGURE 8.26 Temperature field in the free volume of the combustion chamber (computational case 2) (a – 0.0075 s; b – 0.0105 s; c – 0.0120 s; d – 0.0150 s; e – 0.0170 s).

flame spreading along 30% of its surface, the further charge combustion is maintained by the own combustion products (Figure 8.27c). When ~90%

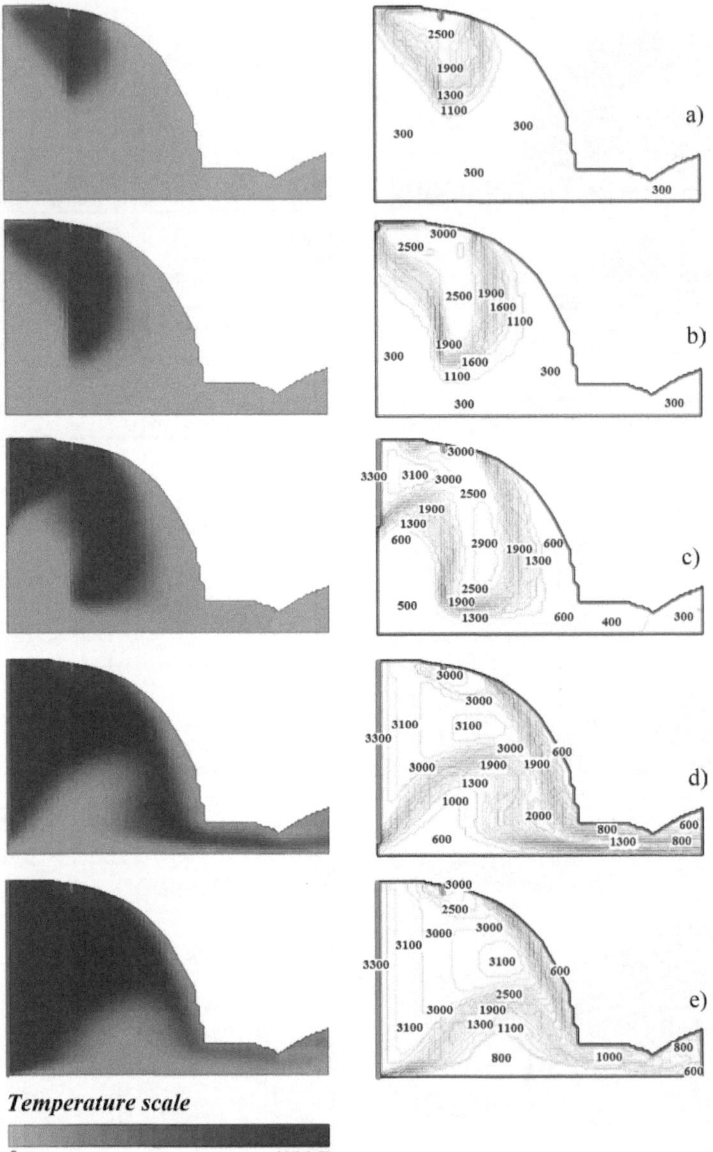

FIGURE 8.27 Temperature field in the combustion chamber volume (computational case 3) (a – 0.0030 s; b – 0.0060 s; c – 0.0080 s; d – 0.0110 s; e – 0.0130 s).

of the charge surface are involved into the combustion, the relatively cold gas is completely pushed to the gas generator nozzle bottom (Figure 8.27d). Further, after the ignition of the whole surface, the cold gas is pushed into the gas duct and nozzle block (Figure 8.27e).

The variant 4 of the igniter location (Figure 8.28) is the case of the solid fuel combustion in the closed vessel. In this regard, the high velocity increase of pressure and temperature of the combustion products in the combustion

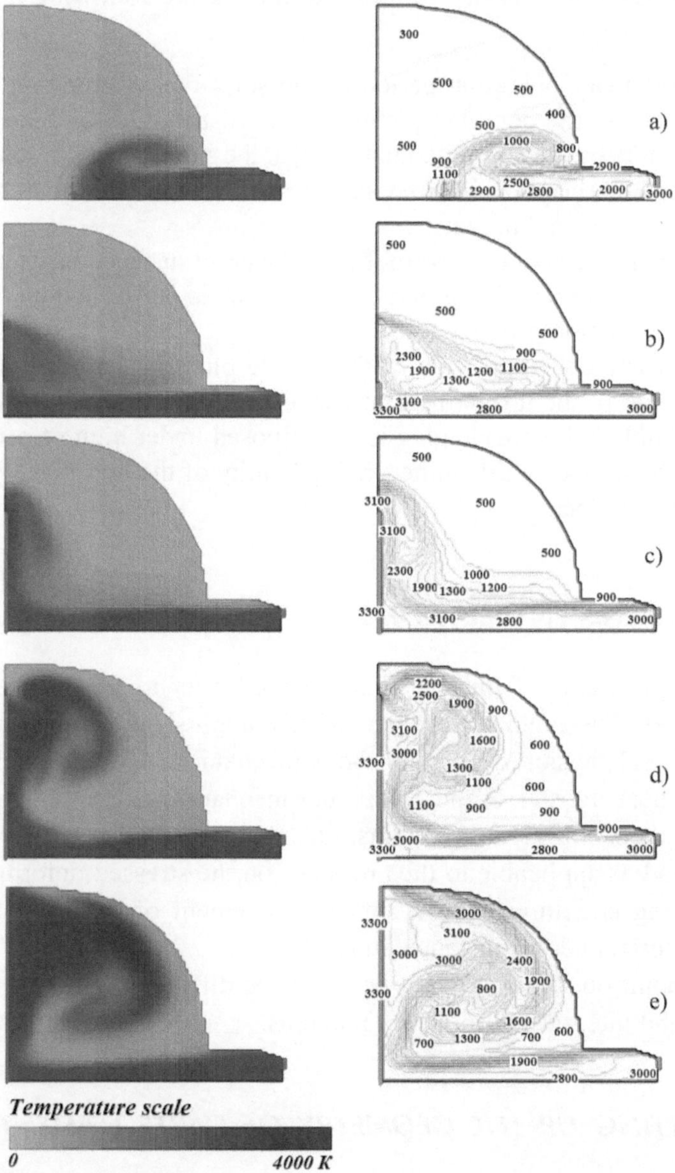

FIGURE 8.28 Temperature field in the combustion chamber volume (computational case 4) (a – 0.0005 s; b – 0.0012 s; c – 0.0015 s; d – 0.0030 s; e – 0.0060 s).

chamber volume is characteristic for this variant. Besides, the front of gaseous combustion products spreads from the igniter along the solid fuel charge surface (Figures 8.28b, 8.28c and 8.28e). Such character of the combustion product movement contributes to the rapid ignition of the fuel charge.

The investigations performed allowed making the following important practical conclusions:

- non-stationary operation period of the solid-fuel engine unit is characterized by the complex picture of the distribution of gas-dynamic parameters by the computational area, at the same time, the correct and objective description is possible only with the help of two- and (or) three-dimensional models;
- after the ignition of the part of the solid charge surface, the heat-up and ignition of the remaining part of the surface is performed by the solid fuel combustion products;
- at the early decomposition of the nozzle plug the igniter location in the region of the nozzle bottom connection with the engine unit shell is optimal. If the nozzle plug is decomposed under high pressure, it is advisable to locate the igniter in the vicinity of the minimal section of the nozzle block.

8.5 FINITE-ELEMENT METHODS

The first applications of finite-element method refer to the middle of XX century [109]. The methods became widely applied in various problems of mathematical physics (deformable body mechanics, incompressible liquid mechanics, heat transfer in the continuum mechanics, etc.), constructional mechanics, theory of electric circuits, etc. Below are only some elements of the method as applicable to the problems on the stressed-deformed state of engineering structures. The systematic statement of the finite-element method can be found, for instance, in [162].

Let us point out the following peculiarities differing it from the finite-difference and integral-interpolatory methods.

8.5.1 SETTING UP THE GEOMETRY OF FINITE ELEMENTS

The computational algorithms are developed for the finite elements whose shapes can be linear, flat or spatial.

When solving the problems of stressed-deformed state of solids, the elastic elements, like the elastic spring, rod, beam can be taken as linear ones. When solving the problems of electrical engineering – elements, like the resistor, capacity, inductivity, etc.

As a rule, flat finite elements have a triangular or (and) rectangular shape (Figure 8.4a). Flat elements can have different degrees of freedom (linear, quadratic, cubic and other elements, Figure 8.29).

The spatial finite elements are selected in the form of tetrahedrons, pentahedrons or hexahedrons (Figure 8.4b). The application of finite elements of different shapes and dimensionality in the problem being solved is acceptable.

Depending on the problem being solved, it is necessary to set up the finite element physical properties and bonds apart from setting up its geometry and orientation in space.

8.5.2 FUNCTIONAL DEPENDENCIES INSIDE THE FINITE ELEMENT

The unknown function being the solution of the set up problem is the polynomial of one or another degree inside each finite element, depending on the finite element type selected.

The models of linear interpolation can be used for linear finite elements. Thus, for the function *f* changing in the interval from x_1 to x_2, the dependence *f(x)* can be written down as follows:

$$f(x) = L_1(x) f(x_1) + L_2(x) f(x_2) \qquad (8.166)$$

Here $L_1(x) = \dfrac{x_2 - x}{x_2 - x_1}$, $L_2(x) = \dfrac{x - x_1}{x_2 - x_1}$ (functions $L_1(x), L_2(x)$ are called the shape coefficients). The equality $L_1(x) + L_2(x) = 1$ is an important property of shape coefficients. Besides, it can be pointed out that the Eq. (8.166) is very similar to the Eq. (2.27) in Chapter 2 obtained using the interpolation of Lagrangian polynomials. Lagrangian polynomials of different degrees are often used when interpolating in the methods of finite elements. At the same time, the shape coefficients in the node points (points in which the function values are known) take the values 0 or 1. Hermitian polynomials are often used together with Lagrangian polynomials.

Let us consider how the function values are interpolated inside the finite element with linear triangular finite elements (Figure 8.30). In Figure 8.30a

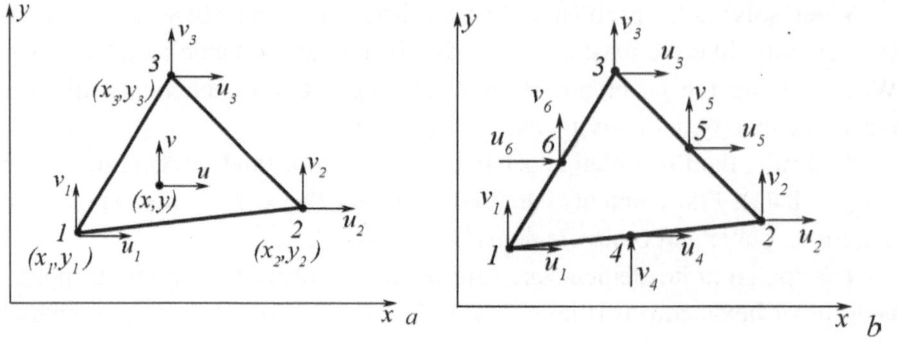

FIGURE 8.29 Examples of flat finite elements (a – linear triangular element; *b* – quadratic triangular element).

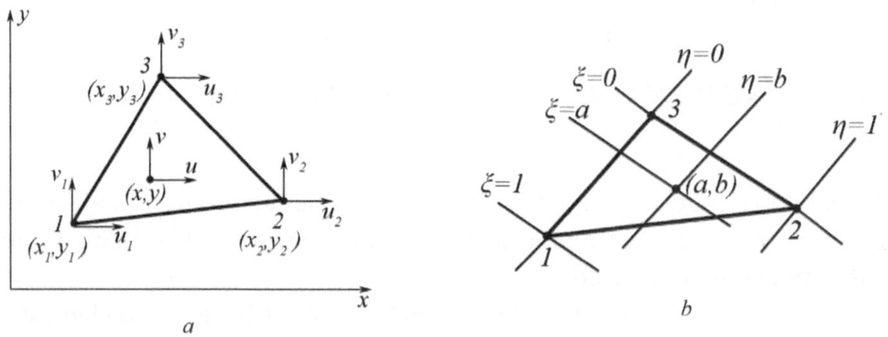

FIGURE 8.30 Interpolation with the triangular element.

the element is placed in the global coordinate system *xy*, in Figure 8.30b the local coordinate system $\xi\eta$ is demonstrated linked with the apexes of the triangular element *1, 2, 3*. The correlations that allow linking the coordinates of the triangular apexes in the global and local coordinate system are relatively simple and are not given. At the same time, the recording of the interpolation dependence for the arbitrary function *f* changing in the triangular element is very simple in the local coordinate system.

The following interpolation dependence for the function $f(\xi,\eta)$ can be written down based on Figure 8.30b

$$f(\xi,\eta) = L_1(\xi,\eta)f(\xi_1,\eta_1) + L_2(\xi,\eta)f(\xi_2,\eta_2) + L_3(\xi,\eta)f(\xi_3,\eta_3) \tag{8.167}$$

Here $L_1(\xi,\eta)=\xi$, $L_2(\xi,\eta)=\eta$, $L_3(\xi,\eta)=1-\xi-\eta$ – shape coefficient.

The indicated parameters are sufficient to record the general interpolation dependence for other variants of finite elements:

- quadratic triangular element (Figure 8.31)

$$f(\xi,\eta)=\sum_{i=1}^{6}L_i(\xi,\eta)f(\xi_i,\eta_i) \tag{8.168}$$

Here $L_1(\xi,\eta)=\xi(2\xi-1)$, $L_2(\xi,\eta)=\eta(2\eta-1)$, $L_3(\xi,\eta)=(1-\xi-\eta)(1-2\xi-2\eta)$, $L_4(\xi,\eta)=4\xi\eta$, $L_5(\xi,\eta)=4\eta(1-\xi-\eta)$, $L_6(\xi,\eta)=4(1-\xi-\eta)\xi$. The condition $\sum_{i=1}^{6}L_i(\xi,\eta)=1$ is fulfilled;

- linear quadrangular element (Figure 8.32)

$$f(\xi,\eta)=\sum_{i=1}^{4}L_i(\xi,\eta)f(\xi_i,\eta_i) \tag{8.169}$$

Here $L_1(\xi,\eta)=\frac{1}{4}(1-\xi)(1-\eta)$, $L_2(\xi,\eta)=\frac{1}{4}(1+\xi)(1-\eta)$, $L_3(\xi,\eta)=\frac{1}{4}(1+\xi)(1+\eta)$, $L_4(\xi,\eta)=\frac{1}{4}(1-\xi)(1+\eta)$. The condition $\sum_{i=1}^{4}L_i(\xi,\eta)=1$ is fulfilled;

- quadratic quadrangular element (Figure 8.33)

$$f(\xi,\eta)=\sum_{i=1}^{8}L_i(\xi,\eta)f(\xi_i,\eta_i) \tag{8.170}$$

Here $L_1(\xi,\eta)=\frac{1}{4}(1-\xi)(\eta-1)(1+\xi+\eta)$, $L_2(\xi,\eta)=\frac{1}{4}(1+\xi)(\eta-1)(1-\xi+\eta)$, $L_3(\xi,\eta)=\frac{1}{4}(1+\xi)(\eta+1)(\xi+\eta-1)$, $L_4(\xi,\eta)=\frac{1}{4}(\xi-1)(\eta+1)(\xi-\eta+1)$, $L_5(\xi,\eta)=\frac{1}{2}(1-\xi^2)(1-\eta)$, $L_6(\xi,\eta)=\frac{1}{2}(1+\xi)(1-\eta^2)$, $L_7(\xi,\eta)=\frac{1}{2}(1-\xi^2)(1+\eta)$, $L_8(\xi,\eta)=\frac{1}{2}(1-\xi)(1-\eta^2)$. The condition $\sum_{i=1}^{8}L_i(\xi,\eta)=1$ is fulfilled.

The variants of the function f value interpolation using other types of finite elements are considered in [162].

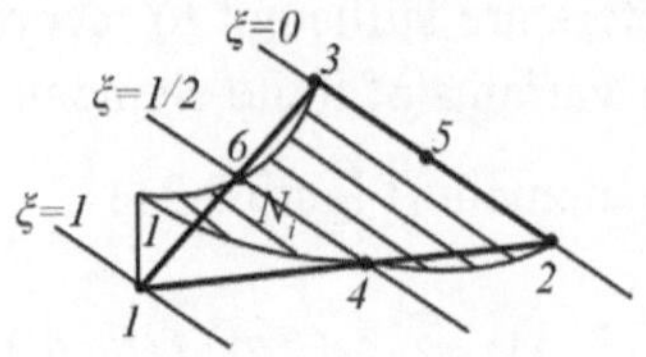

FIGURE 8.31 Interpolation with the quadratic triangular element.

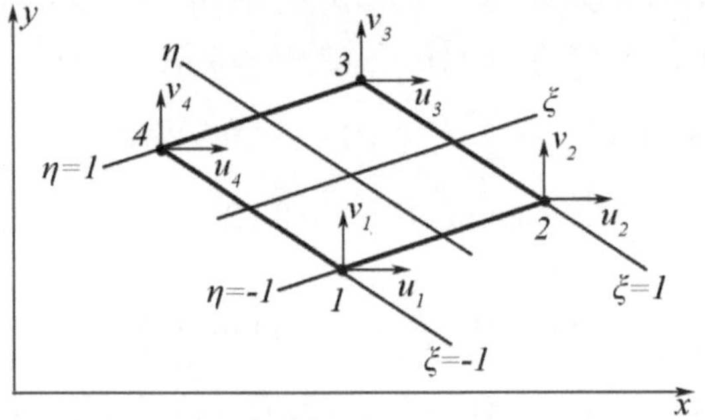

FIGURE 8.32 Interpolation with the linear quadrangular element.

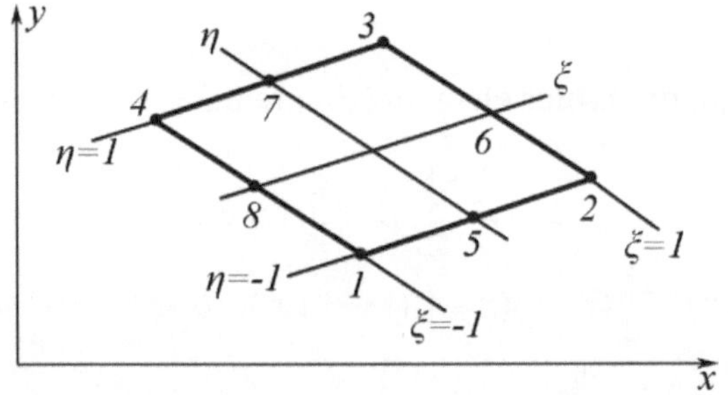

FIGURE 8.33 Interpolation with the quadratic quadrangular element.

8.5.3 INITIAL PROBLEM DISCRETIZATION

Problem discretization in the finite-element method of using the polynomial interpolation can be performed for the differential equation or the equivalent variant formulated in the variation setting. Thus, for instance, the stationary heat conductivity equation $\frac{\partial^2 T}{\partial x^2} + \frac{\partial^2 T}{\partial y^2} = 0$ set up on the plane *xy* with boundary Dirichlet conditions (temperature values are set up on the boundaries) and Neumann conditions (values of heat flows are set up on the boundaries) is equivalent to the problem on the functional minimization $\Phi = \frac{1}{2}\iint [(\frac{\partial \hat{T}}{\partial x})^2 + (\frac{\partial \hat{T}}{\partial y})^2] dxdy$. Here $\hat{T}(x, y)$ – functions from many allowable testing functions set up in the plane xy. Testing functions need to be continuous, have piecewise-continuous first derivatives and satisfy the boundary conditions of the initial problem [162].

Both approaches are equally often used in different versions of the finite element method, however, it can be asserted that the discretization of initial differential equations has been more preferable lately.

8.5.4 APPLIED COMPUTATIONAL METHODS

The discretization of differential equations or variation problem for the population of finite elements brings the initial problem to the solution of the system of linear algebraic equations of the type $\mathbf{AX} = \mathbf{B}$. As a rule, the matrix of the coefficients $\mathbf{A}$ in this equation system has a diagonal form or is very disperse. The methods of solving such equation systems are considered in Section 8.3.3.

Let us consider the simplest examples of solving some problems on stressed-deformed state of solids with the help of the finite-element method taken from [109]. Other applications of the finite-element method in problems on stressed-deformed state can be found in [48, 85, 109, 162, 239, 240, 243], etc.

It was mentioned before that the simplest finite elements – linear elements, the unknown function for these elements is the function of one spatial variable. In problems on stressed-deformed state of a solid, the elastic element like elastic spring (Figure 8.34a), rod (Figure 8.34b), beam (Figure 8.34c) can be taken as linear.

The element of elastic spring type is characterized by the rigidity k connecting the change in the linear size $u = \Delta x$ of the spring with the force value $P - P = k\,u$ applied to the spring.

For the element demonstrated in Figure 8.34a, the forces P_i, P_j are applied in the nodes i and j leading to the movement of these node points onto the values u_1, u_2, respectively. The bond of forces and movements in the nodes *1* and *2* can be written down as follows:

$$P_i = k(u_i - u_j)$$

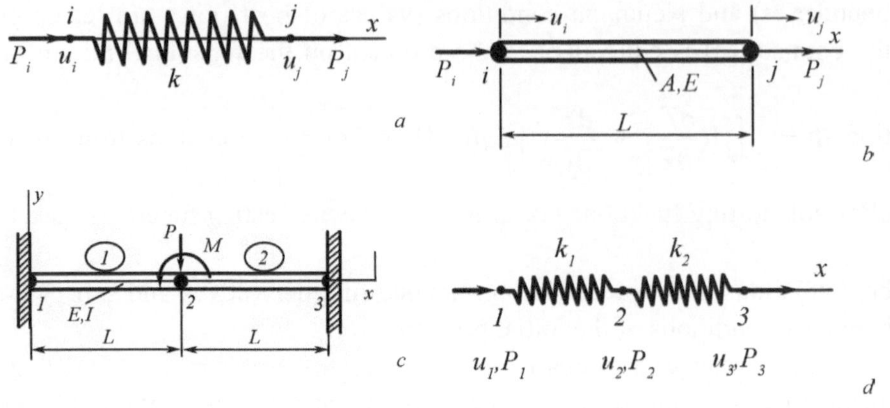

FIGURE 8.34 Types of finite elements (a – elastic spring, *b* – elastic rod, *c* – elastic beam, *d* – two springs).

$$P_j = k(u_j - u_i)$$

The system of linear algebraic equations for the considered simplest linear element in the matrix form is the following

$$\begin{pmatrix} k & -k \\ -k & k \end{pmatrix} \begin{pmatrix} u_i \\ u_j \end{pmatrix} = \begin{pmatrix} P_i \\ P_j \end{pmatrix}$$

In Figure 8.34*d* you can see the system scheme consisting of two linear elastic elements. The rigidity of these elements differs. By analogy with the aforementioned, the following matrix equations can be written down for the first and second elements

$$\begin{pmatrix} k_1 & -k_1 \\ -k_1 & k_1 \end{pmatrix} \begin{pmatrix} u_1 \\ u_2 \end{pmatrix} = \begin{pmatrix} P_1^{(1)} \\ P_2^{(1)} \end{pmatrix}$$

$$\begin{pmatrix} k_2 & -k_2 \\ -k_2 & k_2 \end{pmatrix} \begin{pmatrix} u_2 \\ u_3 \end{pmatrix} = \begin{pmatrix} P_2^{(2)} \\ P_3^{(2)} \end{pmatrix}$$

In the linear problems, in the considered one as well, the application of superposition principle is possible, giving the possibility to sum up the lower and upper matrix equations

$$\left(\begin{pmatrix} k_1 & -k_1 & 0 \\ -k_1 & k_1 & 0 \\ 0 & 0 & 0 \end{pmatrix} + \begin{pmatrix} 0 & 0 & 0 \\ 0 & k_2 & -k_2 \\ 0 & -k_2 & k_2 \end{pmatrix}\right)\begin{pmatrix} u_1 \\ u_2 \\ u_3 \end{pmatrix} = \begin{pmatrix} P_1^{(1)} \\ P_2^{(1)} + P_2^{(2)} \\ P_3^{(2)} \end{pmatrix}$$

After designating $P_1 = P_1^{(1)}$, $P_2 = P_2^{(1)} + P_2^{(2)}$, $P_3 = P_3^{(2)}$, the latter matrix equation is eventually as follows:

$$\begin{pmatrix} k_1 & -k_1 & 0 \\ -k_1 & k_1 + k_2 & -k_2 \\ 0 & -k_2 & k_2 \end{pmatrix}\begin{pmatrix} u_1 \\ u_2 \\ u_3 \end{pmatrix} = \begin{pmatrix} P_1 \\ P_2 \\ P_3 \end{pmatrix}$$

Let us consider the following example. The rod (Figure 8.34b) is characterized by the length L, cross-section area F and elasticity modulus E. The elasticity modulus binds the axial motion $u(x)$ and relative deformation $\varepsilon(x) = \frac{du(x)}{dx}$ with the strains $\sigma(x)$, occurring in the rod, by the formula

$$\sigma(x) = E\varepsilon(x)$$

From [207] it is known that the rod rigidity can be established by the dependence $k = \frac{EF}{L}$. Thus, the considered problem comes down to the problem on elastic element deformation discussed above. In this problem, the matrix equation connecting the movements u and force loads f are as follows:

$$\frac{EF}{L}\begin{pmatrix} 1 & -1 \\ -1 & 1 \end{pmatrix}\begin{pmatrix} u_1 \\ u_2 \end{pmatrix} = \begin{pmatrix} P_1 \\ P_2 \end{pmatrix}$$

The next example of the finite element is "beam" (Figure 8.34c) characterized by the length L, moment of the cross-section area inertia I and elasticity modulus E. The beam has two degrees of freedom, it is characterized by the linear deflection Δy and rotation angle $\theta = \frac{dy}{dx}$. The cutting forces P_1, P_2 and bending moments M_1, M_2 acting in the plane xy act on the left and right boundaries of the beam. The change in the rotation angle for them is established by the correlation $-EI\frac{d\theta(x)}{dx} = M(x)$.

The matrix equation binding the cutting forces P_1, P_2 and bending moments M_1, M_2 with the deflections y_1, y_2 and angular movements θ_1, θ_2 in the matrix form is written down as follows [109]

$$\frac{2EI}{L^3}\begin{pmatrix} 6 & -6 & 3L & 3L \\ -6 & 6 & -3L & -3L \\ 3L & -3L & 2L^2 & L^2 \\ 3L & -3L & L^2 & 2L^2 \end{pmatrix}\begin{pmatrix} y_1 \\ y_2 \\ \theta_1 \\ \theta_2 \end{pmatrix} = \begin{pmatrix} P_1 \\ P_2 \\ M_1 \\ M_2 \end{pmatrix}$$

In practice, the combination of the elements "rod+beam" is also considered. Such combination is considered as linear (the superposition principle is permitted) and allows considering not only the shear loads and bend, but also the force loads along the element (axial loads).

For flat finite elements, we can develop the matrix equation binding the movement $u(x)$ with the relative deformation $\varepsilon(x) = \frac{du(x)}{dx}$ or (and) strains $\sigma(x) = E\varepsilon(x)$. Taking into account the coefficients of the form L_i, the movements in the flat finite element can be demonstrated as $u(x) = \sum_{i=1}^{N} L_i(x)u(x_i)$ or $u(\xi) = \sum_{i=1}^{N} L_i(\xi)u(\xi_i)$.

The relative deformation can be established as follows:

$$\varepsilon(x) = \frac{du(x)}{dx} = \frac{d}{dx}\left(\sum_{i=1}^{N} L_i(x)u(x_i)\right) = \frac{d}{d\xi}\left(\sum_{i=1}^{N} L_i(\xi)u(\xi_i)\right)\frac{d\xi}{dx}$$

In the matrix form the latter equation can be written down as follows:

$$\varepsilon(x) = \begin{pmatrix} B_1 & \dots & B_i & \dots & B_N \end{pmatrix} \cdot \begin{pmatrix} u_1 \\ \dots \\ u_2 \\ \dots \\ u_N \end{pmatrix} = [B]\{u\}$$

Here $[B]$ – the matrix of the differential movements

$$[B] = \frac{d}{d\xi}\left(L_1(\xi), \dots, L_i(\xi), \dots, L_N(\xi)\right)\frac{d\xi}{dx}$$

The strains $\sigma(x) = E\varepsilon(x)$ will be recorded in the matrix form

$$\sigma = E[B]\{u\}$$

In [109] the further transformation of the latter matrix equation is given, which is brought to the following form

$$[k]\{u\} = \{P\}$$

In the latter matrix equation $[k]$ – element rigidity matrix, $\{P\}$ – force action matrix. The rigidity matrix is established by the integration by the finite element volume

$$[k] = \int_W ([B]^T E[B])\,dW$$

In the problems on the stressed-deformed state for flat finite elements, its thickness is principal. Depending on the ratio of the element thickness to its characteristic size, the equations from the theory of thin or thick plates are used as computational correlations. Taking into account the equations for the rigidity matrix [*k*], the possible change in the thickness along the finite element should be also considered.

The applications of the finite-element method when solving heat-conductivity problems in composite solid fuels (metallized and non-metallized) are considered in [13].

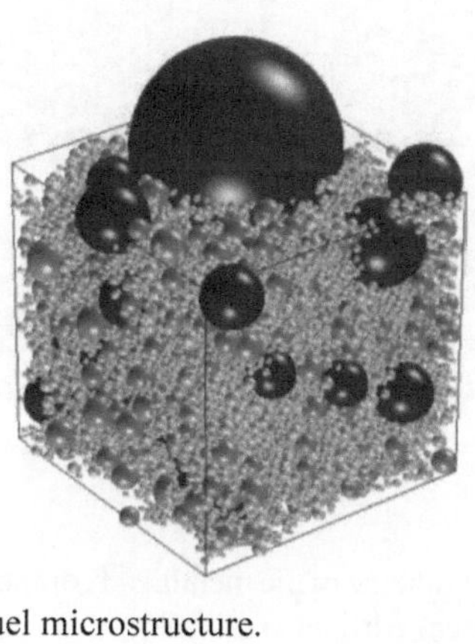

FIGURE 8.35 Composite fuel microstructure.

The necessity and practicability of applying finite elements in the problems on solid fuel heat-up are illustrated by Figure 8.35. The composite fuel contains the combustible (e.g., natural rubber), which is also the binder of all components, oxidizer (e.g., ammonium perchlorate), flammable metals (e.g., aluminum) [135]. The oxidizer particle sizes can be in the range 10–100 mcm, aluminum particles – 1–10 mcm. The thickness of the solid fuel heated up layer in the engine unit combustion chamber does not exceed 300–500 mcm. Such structure of the composite solid fuel makes the solid fuel model impracticable as the homogeneous material applied when calculating the heat-up of ballistite fuels [18, 204]. In fact, the thermophysical characteristics of basic materials contained by the composite fuel can differ in over thousand times.

In Figure 8.36 you can see the computational schemes used when analyzing the process of the solid fuel heat-up. Due to the limited computational resources, the non-stationary process is analyzed in two-dimensional setting for the fuel element limited by the sizes $L \times H$. The computational area is formed by the set of finite elements of triangular type. The number of finite elements is taken in the range from 5000 up to 50,000. For comparison, in Figure 8.37 you can see the computational results for the composite metallized fuel heat-up (Figure 8.37a) and non-metallized fuel (Figure 8.37b).

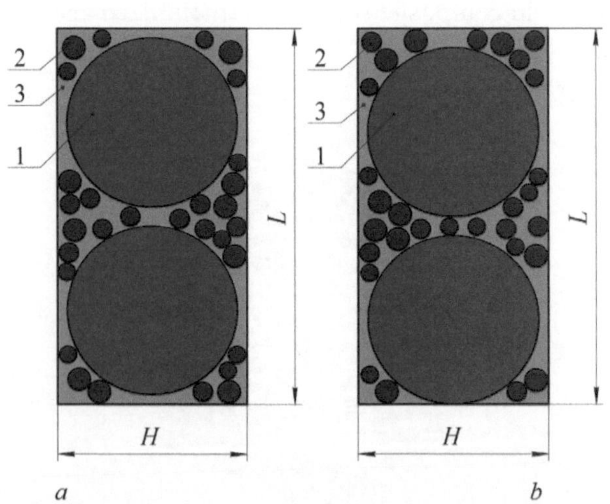

FIGURE 8.36 Computational scheme of the metallized composite solid fuel (*1* – ammonium perchlorate particle; *2* – aluminum particle; *3* – binder).

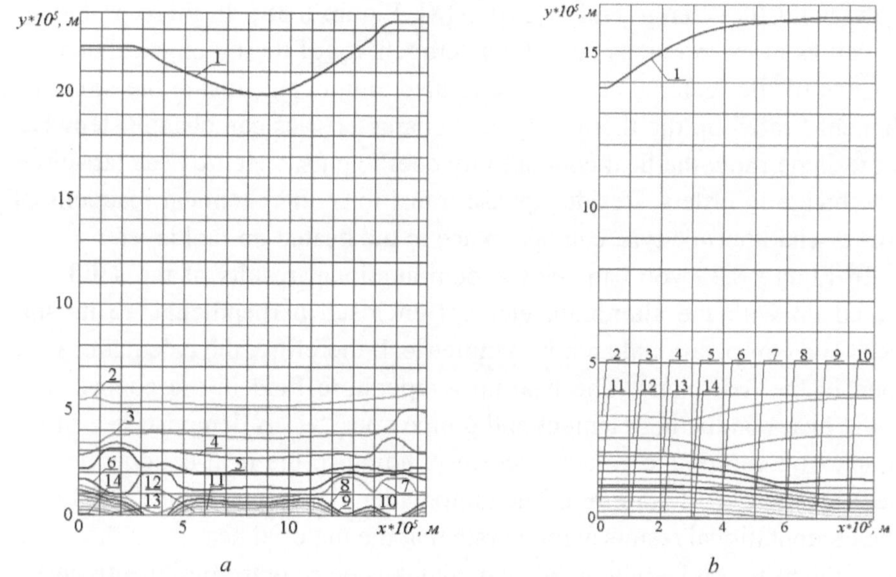

FIGURE 8.37 Temperature field in the heated up area of the composite solid fuel (a – metalized fuel, b – non-metalized fuel) (Isotherms: *1* – 293 K; *2* – 313 K; *3* – 333 K; *4* – 353 K; *5* – 373 K; *6* – 393 K; *7* – 413 K; *8* – 433 K; *9* – 453 K; *10* – 473 K; *11* – 493 K; *12* – 513 K; *13* – 533 K; *14* – 570 K).

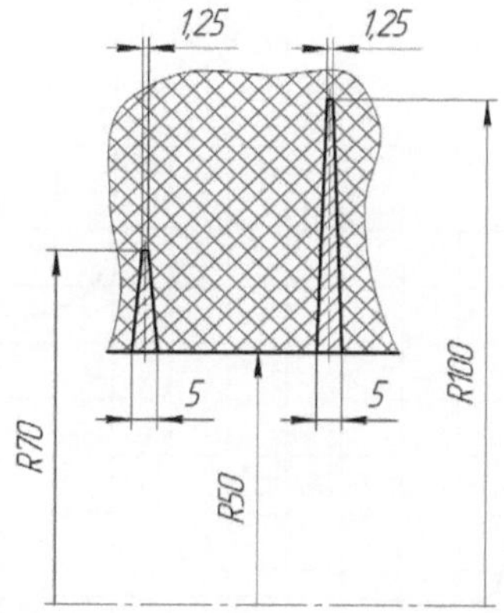

FIGURE 8.38 Scheme of the non-recoverable heat-conducting element placed in the solid fuel charge.

Another problem connected with the heated up state of the solid fuel is the problem on regulating the ignition processes and fuel combustion rate using thermal knives or non-recoverable heat-conducting elements located in the solid fuel charge body ([210, 218], Figure 8.38). In these problems, as well as in the previous one, the heterogeneity of thermophysical characteristics of the materials heated up is also significant. Inside the structure element heated up the thermophysical characteristics can change stepwise, at the same time, the heat conductivity coefficients, specific heat capacities can change in orders. Besides, phase transitions and chemical reactions of exo- or endothermic type can take place in the heated up fuel layer.

In Figure 8.39 you can see the computational results of the solid fuel heated up with the aluminum element enclosed perpendicular to its surface. The computational area is symmetrical, therefore, the calculations are done in the half-plane. The heat-up temperature field in the contact area of the heat-conducting element and gunpowder "H" is demonstrated in the figure. The coordinate values *z*, corresponding to the heat-conducting element location – $z \leq 2.50$ mm. The gunpowder "H" is placed at $z>2.50$ mm. The computational results demonstrate that the material heat-up in the vicinity of the heat-conducting metal element takes place with significantly lower velocity than the gunpowder heat-up. Finally, the gunpowder placed in the distance from the heat-conducting element ignites earlier than the gunpowder contacting the metal.

The aforementioned regularities are important when designing new elements contained by solid fuel charges.

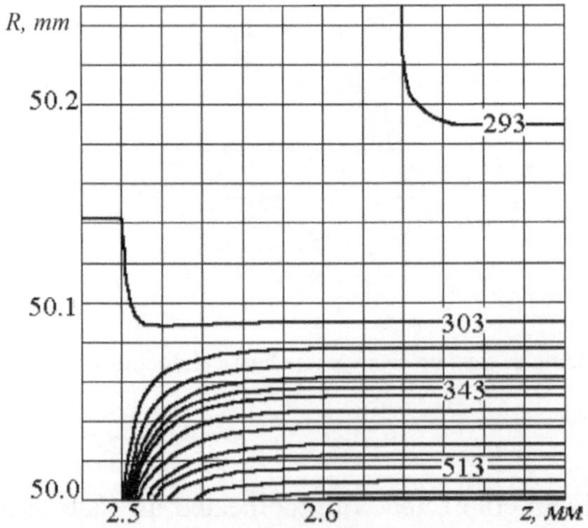

FIGURE 8.39 Temperature field in the contact area of heat-conducting element and gunpowder.

Other applications of finite-element method can be found in numerous domestic and foreign literature. Besides, it should be taken into account that the main algorithms of the finite-element method are implemented in the software commercial and non-commercial products applied in practice (for instance, ANSYS, Solid Works, Pro Engineer, etc.).

8.6 BOUNDARY-ELEMENT METHODS

The aforementioned numerical methods of solving the problems of mathematical physics are based on the function approximation notions inside the computational area. These methods are highly efficient and are widely spread when solving modeling problems in engineering. Starting from 1970-s, another group of methods based on the approximation application on the computational area boundaries has been intensively developed. After the approximation, the unknown function behavior is calculated inside the computational area by analytical correlations. This group of methods is called the boundary-element methods [69, 119]. These methods are sometimes called panel, potential methods [29, 77, 187, 239]. The boundary-element methods are successfully applied when solving different problems of mathematical physics – problems on hydraulics and subsonic gas dynamics, diffusion and heat conductivity, problems on stressed-deformed state of solids, etc. The main peculiarity of problems, solving which the application of boundary-element methods is possible, is the availability of potentials (the unknown function needs to be potential). Another method peculiarity, which is especially important when applying the complex boundary-element method, is the possibility of solving problems formulated in two-dimensional spatial setting.

Let us consider individual cases of problems solved by the boundary-element method.

8.6.1 VORTEX-FREE FLOWS OF THE IDEAL LIQUID

Let us consider Cauchy-Riemann equations [132] recorded in Cartesian coordinate system xy for the function ψ and potential φ

$$\frac{\partial \varphi}{\partial x} = \frac{\partial \psi}{\partial y}$$

$$\frac{\partial \varphi}{\partial y} = -\frac{\partial \psi}{\partial x} \tag{8.171}$$

Cauchy-Riemann equations are true for vortex-free flow of incompressible liquid [32, 42], and the velocity vector components are easily found by the known potential value φ

$$u = \frac{\partial \varphi}{\partial x}, \quad v = \frac{\partial \varphi}{\partial y}$$

The Eqs. (8.171) can be transformed into the form of Laplace equations

$$\nabla^2 \psi = \frac{\partial^2 \psi}{\partial x^2} + \frac{\partial^2 \psi}{\partial y^2} = 0 \tag{8.172}$$

$$\nabla^2 \varphi = \frac{\partial^2 \varphi}{\partial x^2} + \frac{\partial^2 \varphi}{\partial y^2} = 0 \tag{8.173}$$

The form of equations for the function of current ψ and potential φ allows ascertaining that the equipotential lines are orthogonal to the current lines. This gives the possibility to introduce the complex function ω depending on the complex variable $z = x + iy$ and being the following

$$\omega(z) = \varphi(x, y) + i\psi(x, y) \tag{8.174}$$

Let the function $\omega(z)$ be represented in the analytical form in some singly-connected area D with the boundary Γ, and the point z_0 is inside the area D (Figure 8.40). Then Cauchy integral formula is written down as follows:

$$\omega(z_0) = \frac{1}{2\pi i} \int_{\Gamma} \frac{\omega(\xi)}{\xi - z_0} d\xi \tag{8.175}$$

The formula (8.175) is the basis of the complex method of integral correlations [69]. If the value $\omega(\xi)$ is known on the boundary of the computational area Γ, the Eq. (8.175) allows calculating the value of the function $\omega(z)$ in any point $z = z_0$ inside the area D. In particular, let $\omega(\xi)$ be the polynomial $p_m(\xi)$ of m degree

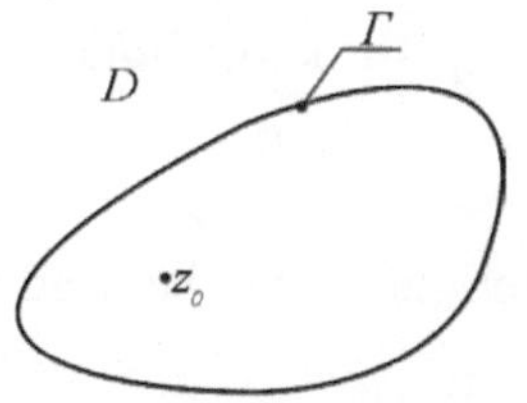

FIGURE 8.40 Computational area in the boundary-element method.

$$p_m(\xi) = C_0 + C_1\xi + C_2\xi^2 + \ldots + C_m\xi^m$$

in which coefficients C_i – complex numbers ($C_i = \alpha_i + i\beta_i$), $i = 1, m$. α_i, β_i – actual coefficients. In this case, the following is true

$$\omega(z_0) = \frac{1}{2\pi i} \int_{\tilde{A}} \frac{p_m(\xi)}{\xi - z_0} d\xi = p_m(z_0)$$

The point $z = z_0$ is taken arbitrarily, therefore, the following more general expression is true

$$\omega(z) = p_m(z) \quad \forall z \in D \cup \tilde{A} \tag{8.176}$$

For practical applications the following theorem on the derivatives of the function $\omega(z_0)$ is also important

$$\omega^{(n)}(z_0) = \frac{n!}{2\pi i} \int_{\Gamma} \frac{\omega(\xi)}{(\xi - z_0)^{n+1}} d\xi \tag{8.177}$$

In all the aforementioned integrals by the contour Γ, when integrating, the contour is bypassed in the positive direction (counterclockwise).

8.6.2 FLOW IN POROUS MEDIA

The flow in porous media is subject to Darcy law which states that the liquid flow velocity along the direction dl is directly proportional to the pressure drop (discharge) $d\varphi$, medium permeability k, liquid specific weight in the medium ρg, area of the flow cross-section F and inversely proportional to the medium viscosity μ

$$W = -\frac{k\rho g}{\mu} F \frac{d\varphi}{dl}$$

If Darcy law is true, the potential φ can be used to describe the liquid flow in the porous medium

$$\varphi = \frac{p}{\rho g} + z$$

Here p – hydrostatic pressure, z – vertical distance from some zero level.

The potential φ (total discharge energy) satisfies Laplace equation – $\nabla^2\varphi = \frac{\partial^2\varphi}{\partial x^2} + \frac{\partial^2\varphi}{\partial y^2} = 0$. For the considered flow we can introduce the function ψ characterizing the current lines of the liquid particles. Poisson equation is true for the current function – $\nabla^2\psi = \frac{\partial^2\psi}{\partial x^2} + \frac{\partial^2\psi}{\partial y^2} = 0$.

Further, the problem is brought to the type solved by the method of complex boundary conditions similarly to the aforementioned problem on the vortex-free liquid flow (equations 8.172, 8.173).

8.6.3 DIFFUSION IN THE LIQUID SOLUTION

In accordance with Fick equation of diffusion, the mass (admixture) flow $\mathbf{m} = (m_x, m_y, m_z)$ of the dissolved substance for the time unit through the area unit is proportional to the concentration gradient C of the dissolved substance in this direction – $\mathbf{m} = -D\nabla C$. Here D – diffusion coefficient. Taking into account the correlation for the mass flow, the continuity equation is written down as follows:

$$\frac{\partial C}{\partial t} + \nabla\mathbf{m} = 0$$

The equation $\nabla\mathbf{m} = 0$ is true for the established liquid flow or

$$\nabla^2 C = \frac{\partial^2 C}{\partial x^2} + \frac{\partial^2 C}{\partial y^2} = 0$$

Poisson equation $\nabla^2\psi = \frac{\partial^2\psi}{\partial x^2} + \frac{\partial^2\psi}{\partial y^2} = 0$ is true for the equations of the current line ψ of the dissolved substance (admixtures), thus bringing the problem on the dissolved substance (admixture) to the aforementioned ones.

The individual case of the diffusion equation is the heat conductivity equation which for the isotropic material with the constant value of the temperature conductivity in the stationary two-dimensional case is written down as follows:

$$\nabla^2 T = \frac{\partial^2 T}{\partial x^2} + \frac{\partial^2 T}{\partial y^2} = 0$$

The heat conductivity equation is brought to the standard problem on the complex method of boundary elements introducing the function ψ for the heat flow lines, which satisfies Laplace equation in the stationary case.

8.6.4 PROBLEM ON DEVELOPING THE ORTHOGONAL CURVILINEAR FINITE-DIFFERENCE GRID

Cauchy-Riemann equations for the current function ψ and potential φ recorded in the form (8.172) and (8.173) can be applied to develop the orthogonal curvilinear grids in the areas with curvilinear side boundaries of singly-connected channel areas (for instance, the areas given in Figure 8.41). In fact, it was mentioned before that the equipotential and current lines are inter-orthogonal and this allows accepting the method of complex boundary elements when developing the finite-difference grid.

In practice, when developing the finite-difference grid, its congelation or depression in some regions of the computational area are necessary. The

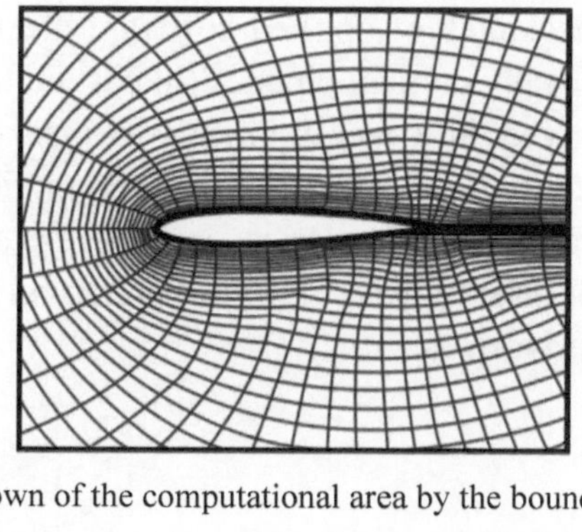

FIGURE 8.41 Breakdown of the computational area by the boundary-element method.

quality management of the finite-difference grid being developed can be provided solving Poisson equation

$$\frac{\partial^2\psi}{\partial x^2}+\frac{\partial^2\psi}{\partial y^2}=f_1(x,y)$$

$$\frac{\partial^2\varphi}{\partial x^2}+\frac{\partial^2\varphi}{\partial y^2}=f_2(x,y)$$

The functions $f_1(x,y)$, $f_2(x,y)$ should be selected in these equations. The application of the boundary-element method in such problem becomes labor-consuming. The acceptable algorithm of solving this problem is discussed, for instance, in [14].

Let us point out that [69] also contains other problems found in engineering (e.g., problems on stressed-deformed state of a solid, problems on elasticity theory), for the solution of which the algorithms of boundary-element method can be applied.

8.7 SOLUTION OF THE SYSTEMS OF DIFFERENTIAL EQUATIONS IN PARTIAL DERIVATIVES USING COMPUTER ENGINEERING

Differential equations (systems of differential equations) in partial derivatives are presented as the most complicated equations of mathematical physics. There are many methods of their solution, however, this only highlights their complexity. There is no universal approach which could allow equally precisely solving the heat conductivity equation and equations of gas dynamics. There is no method which would allow equally precisely solving the equations of gas dynamics at subsonic and supersonic gas velocities, etc. At present, the packages of applied software are developed in which different problems of mathematical physics (in particular, ANSYS package) are attempted to be solved, however, it is early to speak about the imperfection of algorithms used in these packages. In mathematical packages widely applied (MathCad, MatLab, Mathematica, etc.) there is also the function and software which can be used when solving the relatively simple differential equations in partial derivatives.

TABLE 8.1 Examples of the *MathCad* Functions Providing the Solution of Differential Equations in Partial Derivatives

No	Function name	Function attribution	Description of function arguments
1	***pdesolve*** (*u, x, xrange, t, trange, xpts, tpts*); ***numol*** (*xrange, xpts, trange, tpts, Ndpe, Nae, rhs, init, bc*); ***multigrid*** (*M, ncycle*); ***relax*** (*M1, M2, M3, M4, M5, F, U, r*)	***pdesolve*** and ***numol*** provide the solution of the equation or equation system of parabolic or hyperbolic type at the set up boundary (*xrange*) and initial (*trange*) conditions; ***multigrid*** and ***relax*** provide Poisson equation in the rectangular (square) area with zero boundary conditions;	*u* – vector of the integrated function; *x* – spatial coordinate; xrange – vector of the spatial coordinate values for boundary conditions; *t* – time (or other argument); *trange* – vector of the time values for boundary conditions; *xpts, tpts* – number of spatial and time discretization points of the computational area; *Ndpe* – number of differential equations; *Nae* – number of additional algebraic equations; *rhs* – vector of the right sides of the equation; *init* – initial conditions for the vector of functions integrated; *bc* – functional matrix of boundary conditions; *M* – matrix containing the values of the right sides of Poisson equation; *ncycle* – number of cycles on each iteration level; *M1, M2, M3, M4, M5* – matrixes of the coefficients in Poisson equation; *F* – vector of the right sides of differential equations; *U* – boundary and initial conditions; *r* – relaxation coefficient used in the iteration process

TABLE 8.2 Examples of the Subprograms of the Library *IMSL* Providing the Solution of Differential Equations in Partial Derivatives

No	Appeal to the subprogram or function	Attribution of the subprogram or function	Description of arguments of the subprogram or function
1	*call* ***molch*** *(ido, fcnut, fcnbc, npdes, t, tend, nx, xbreak, tol, hinit, y, Ldy)*;	Solution of the system of differential equations in partial derivatives with convective and diffusion components	*ido* – characteristic indicating the calculation state; *fcnut* – name of the user subprogram calculating the right sides of differential equations (values of the derivatives by time); *fcnbc* – name of the user subprogram calculating the boundary conditions; *npdes* – number of differential equations; *t, tend* – initial and final values of the argument by which the integration is performed; *nx* – number of the points of finite-difference grid; *xbreak* – array preserving the conjugating points of Hermite cubic splines; *tol* – error value allowed during the integration; *hinit* – initial step of the integration by time; *y* – vector of the variables integrated; *Ldy*=npdes
2	*call* ***fps2h*** *(prhs, brhs, coefu, nx, ny, ax, bx, ay, by, ibcty, iorder, u, Ldu)*	Solution of Poisson or Helmholtz equations in two-dimensional rectangular area	*prhs* – user function calculating the equation right side; *brhs* – user function calculating the boundary conditions; *coefu* – coefficient at the function u; *nx, ny* – number of computational nodes by the coordinates; *ax, bx* – left and right boundaries of the integration area; *ay, by* – upper and lower boundaries of the integration area; *ibcty* – characteristic defining the type of boundary conditions; *iorder* – accuracy order of the finite-difference approximation; *u* – function integrated; *Ldu=nx*

Some of the functions comprised by the mathematical package *MathCad* [86, 141] are given in Table 8.1. In Table 8.2 you can see separate subprograms in the library *IMSL* of the compiler *Visual Fortran* [22].

In *MathCad* the functions ***pdesolve*** and ***numol*** provide the solution of differential equations and systems of differential equations in partial derivatives of parabolic and hyperbolic types. The first of these functions inside the computational block cannot be used many times. The function ***numol*** is free from this shortcoming.

The functions ***multigrid*** and ***relax*** provide the solution of Poisson equation. The function ***relax*** is more universal. The application of the function ***multigrid*** assumes the setting up of zero boundary conditions.

The software in the library *IMSL* of the compiler *Visual Fortran* provides the solution of more complicated problems than in the package *MathCad*. In particular, in the Table 8.2 you can see the description of two programs – ***molch*** and ***fps2h***.

The first software (***molch***) provides the solution of hyperbolic or parabolic equations containing the partial derivative of the first order by time and partial derivatives of the first and (or) second order by the spatial coordinate. The problem solution is shown in the form of Hermite cubic splines.

The second software (***fps2h***) allows solving Poisson or Helmholtz equations in two-dimensional rectangular area. The boundary conditions can be set up in Dirichlet, Neumann forms or accepted periodically [22, 63]. The finite-difference schemes of the second or fourth orders of accuracy by spatial coordinates can be used to solve differential equations.

Apart from the software presented in Table 8.2, there are others in the library *IMSL*. For instance, the software ***fps3h*** allows solving Poisson or Helmholtz equations in three-dimensional area (inside the parallelepiped). The software ***pde_1d_mg*** allows solving the system of differential equations in partial derivatives of hyperbolic type (transfer equations) with complex nonlinear boundary conditions.

It should be pointed out that the list of functions applied in *MathCad* and software in the library *IMSL* of the compiler *Visual Fortran* is continuously appended and this provides a user with new richer possibilities.

KEYWORDS

- **Cartesian coordinate system**
- **continuum mechanics**
- **cylindrical coordinate system**
- **Dufort-Frankel scheme**
- **integral-interpolatory approach**
- **large-particle method**
- **Lax theorem**
- **Neumann method**
- **transfer equation**

BIBLIOGRAPHY

1. Abgaryan K. A., Rapoport I. M. Rocket dynamics. M.: Mechanical Engineering, 1969, 379 p.
2. Abugov D. I., Bobylev V. M. Theory and calculation of solid fuel rocket engines. M.: Mechanical Engineering, 1987, 272 p.
3. Averson A. E., Barzykin V. V., Merzhanov A. G. Application of mathematical apparatus of the transient heat conduction in the ignition theory. Heat and mass transfer in physical-chemical transformations. V.2, Minsk: Science and Technology, 1968, 464 p.
4. Aivazyan S. A., Eniukov I. G., Meshakin L. D. Fundamentals of data modeling and preprocessing. M.: Finance and Statistics, 1983, 472 p.
5. Akulich I. L. Mathematical programming examples and problems. M.: Higher School, 1986, 319 p.
6. Alabuzhev P. M. Theory of similarity and dimensions. Modeling, Novosibirsk, 1968.
7. Alexeev B. V., Grishin A. M. Physical gas dynamics of reactants. M.: Higher School, 1985, 464 p.
8. Alemasov V. E., Dregalin A. F., Tishin A. P. Theory of rocket engines. M.: Mechanical Engineering, 1989, 464 p.
9. Aliev A. V. Mathematical modeling in power engineering. Part 1. Construction of mathematical models: Textbook for students, Izhevsk: Publishing House of ISTU, 2001, 164 p.
10. Aliev A. V. Software package "Solid fuel engine". Catalog of innovation design of Izhevsk State Technical University, Izhevsk: ISTU, 2001, 24 p.
11. Aliev A. V., Andreev V. V. Development of parallel algorithms for the calculation of gas dynamics by the method of large particles. Intelligent systems in manufacturing, Izhevsk: ISTU, № 1 (7), 2006, 4–17 pp.
12. Aliev A. V., Saushin P. N. Airbags: Issues of ballistic design. Automotive Industry, № 5, 2008, 32–35 pp.
13. Aliev A. V., Suvorov Z. P. Modeling of heat conduction processes in the medium with essentially heterogeneous properties. Bulletin of ISTU, № 4 (44), 2009, 182–186 pp.
14. Anderson D., Tannehill J., Pletcher R. Computational fluid mechanics and heat transfer. 2 vols. M.: Mir, 1990, 726 p.
15. Arsenin V. Ya. Methods of mathematical physics and special functions, M. Nauka, 1974, 432 p.
16. Arfken G. Mathematical methods in physics M.: Atomizdat, 1970, 712 p.
17. Arkhangelsky A. Ya. Programming in Delphi 7. M.: Binom-Press, 2003, 659 p.
18. Assovsky I. G. Combustion physics and internal ballistics. M.: Nauka, 2005, 357 p.
19. Afanasiev V. A., Afanasieva N. U., Kazakov V. S. Theory of automatic control. 2 parts, Izhevsk: ISTU, 2007, 388 p. + 192 p.
20. Bartenyev O. V. Graphics OpenGL: Programming in Fortran. M.: Dialog MIFI 2000, 368 p.
21. Bartenyev O. V. Modern Fortran. M.: Dialog MIFI, 1998, 397 p.

22. Bartenyev O. V. Fortran for professionals. The mathematical library IMSL. 3 books. M.: Dialog MIFI, 2000–2001, 448 p. + 320 p. + 368 p.
23. Bakhvalov N. S., Zhidkov N. P., Kobelkov V. P. Numerical methods. M.: Nauka, 1987, 600 p.
24. Belov I. A., Isaev S. A., Korobkov V. A. Problems and methods of calculating separated flows of an incompressible fluid, L.: Sudostroenie, 1989, 256 p.
25. Belotserkovsky O. M. Computational mechanics. Modern problems and results. M.: Nauka, 1991, 183 p.
26. Belotserkovsky O. M. Direct numerical simulation of free developed turbulence. Mg. Computing mathematic and math. physics, 1985, Vol. 25, № 12. 1856–1882 pp.
27. Belotserkovsky O. M. Numerical modeling in continuum mechanics. M.: Nauka, 1984, 520 p.
28. Belotserkovsky O. M., Davydov Yu. M. Method of large particles. Computer experiment. M.: Nauka, 1982, 392 p.
29. Belotserkovsky S. M., Nisht M. I. Separated and unseparated overflow around thin wings by an ideal fluid. M.: Nauka, 1978.
30. Belotserkovsky S. M., Oparin A. M., Chechetkin V. M. Turbulence: New approaches. M.: Nauka, 2002, 286 p.
31. Besekersky V. A., Popov E. P. The theory of automatic control systems. M.: Nauka, 1972, 768 p.
32. Betchelor J. Introduction to fluid dynamics. M.: Mir, 1973, 760 p.
33. Billing V. A. Visual C ++. Book for programmers. M.: Publishing Department "Russian Edition" LLP "Chanel Trading Ltd," 1996, 352 p.
34. Blackwell D., Girshik M. I. The theory of games and statistical decisions. M.: IL, 1958.
35. Bobrovsky S. Pentagon technology at the Russian programmers service. Software Engineering, St. Petersburg: Peter, 2003, 222 p.
36. Bobylev V. M. Solid fuel rocket engine as a means to control the movement of missiles. M.: Mechanical Engineering, 1992, 160 p.
37. Bogomolov K. L., Tishkin V. F. Dirichlet cells in the metric of the shortest path. Mathematical modeling, 2006. V.18. № 8, 37–48 pp.
38. Bogoryad I. B. Fluctuations of viscous liquid in the cavity of a solid body, Tomsk: TSU, 1999, 133p.
39. Borevich Z. I. Determinants and matrices. M.: Nauka, 1970, 200 p.
40. Brown S. Visual Basic 6: Training Course, St. Petersburg: Peter, 2002, 576 p.
41. Bulgakov V. K., Lipanov A. M. Theory of erosive burning of solid propellants. M.: Nauka, 2001, 138 p.
42. Wallander S. V. Lectures on fluid mechanics, L.: Publishing House of Leningrad University, 1978, 296 p.
43. Varnatts Y., Maas W., Dibble R. Combustion. Physical and chemical aspects, modeling, experiments, the formation of contaminants. M.: Fizmatlit, 2003, 352 p.
44. Vasilenko V. A. Spline functions: Theory, algorithms and programs. M.: Nauka, 1983, 215 p.
45. Vasilyev F. P. Methods of optimization. M.: Factor press, 2002, 442 p.
46. Verzhbitcky V. M. Fundamentals of numerical methods: Textbook for universities. M.: Higher School, 2009, 840 p.
47. Vilyunov V. N. Theory of ignition of condensed substances, Novosibirsk: Nauka, 1984, 192 p.

48. Internal Ballistics of SPREs. A. V. Aliev, G. N. Amarantov, V. F. Akhmadeev, et al.; ed. By A. M. Lipanova, Yu. M. Milekhin. M.: Mechanical Engineering, 2007, 504 p.
49. Intrachamber processes and energy conversion in space power systems. B. T. Erokhin, V. M. Bytskevich, V. N. Durnev, et al. M.: VINITI RAN, 2001, 480 p.
50. Volkov K. N., Emelyanov V. N. Large-eddy simulation in the calculations of turbulent flows. M.: Fizmatlit 2008, 368 p.
51. Volkov K. N., Emelyanov V. N. Flow and heat transfer in channels and rotating cavities. M.: Fizmatlit 2010, 488 p.
52. Volosevych P. P., Levanov E. I. Self-similar solutions of gas dynamics and heat transfer. M.: Publishing House MFTI, 1997, 240 p.
53. Computational methods in fluid dynamics. Ed. by S. S. Grigorian, Yu. D. Shmyglevsky M.: Mir, 1967, 384 p.
54. Computational methods in mathematical physics. P. N. Vabishchevich, V. M. Goloviznin, G. G. Elenin, et al. Gen. ed. A. A. Samarsky M.: Publishing house MSU, 1986, 150 p.
55. Gantmakher F. R. Theory of matrices. M.: Nauka, 1966, 576 p.
56. Gelman V. Ya. Solving mathematical problems with Excel, St. Petersburg, Peter, 2007, 237 p.
57. Genetic algorithms, artificial neural networks and the problems of virtual reality. G. K. Voronovsky, K. V. Makhotilo, S. N. Petrashev, S. A. Sergeev, Kharkov.: OSNOVA, 1997, 112 p.
58. Gill F., Murray W., Wright M. Practical Optimization. M.: Mir, 1985, 509 p.
59. Ginzburg I. P. Aerogasdynamics. M.: Higher School, 1966, 404 p.
60. Ginzburg I. P. Applied fluid dynamics, L.: ed. by Leningrad University, 1958, 338 p.
61. Ginzburg I. P. Friction and heat transfer during the gas mixture moving, L.: ed. LSU, 1975, 302 p.
62. Godunov S. K. Equations of mathematical physics. M.: Nauka, 1971, 416 p.
63. Godunov S. K., Ryabenky V. S. Difference schemes. M.: Nauka, 1973, 400 p.
64. Goloskokov D. P. Equations of mathematical physics. Solving problems in the MAPLE system. Textbook for HEIs, St. Petersburg: Peter, 2004, 539 p.
65. Grebyonkin V. I., Kuznetsov N. P., Cherepov V. I. Power characteristics of solid propulsion engines and propulsion systems for special purposes, Izhevsk: Ed. ISTU, 2003, 352 p.
66. Grenander U., Freiberger W. Short course of computational probability and statistics. M.: Nauka, 1978, 192 p.
67. Grigorenko N. L. Mathematical methods for the management of multiple dynamic processes. M.: Publishing House MSU, 1990, 201 p.
68. Grishin V. K. Statistical methods for analysis and design of experiments, M. Ed. MSU, 1975, 128 p.
69. Gromadka II.T., Lei C. Complex boundary element method. M.: Mir, 1990, 303 p.
70. Grossman K., Kaplan A. A. Nonlinear programming based on unconstrained optimization. M.: Nauka, 1981, 183 p.
71. Gukhman A. A. Application of similarity theory to study the processes of heat and mass transfer. M.: Higher School, 1974, 328 p.
72. Davydov Yu. M. Tables for calculating the external transonic gas flows. M.: VTs AN SSSR, 1983, 361 p.
73. Darhvelidze P. G., Markov E. P. Delphi – visual programming environment, St. Petersburg: BHV – Saint Petersburg, 1996, 352 p.

74. Special purpose impulse type solid fuel engines. Basics of design, construction and working out experience. I. M. Gladkof, Yu. P. Ermakov, B. Ya. Malkin, et al. M.: CRI information, 1990, 116 p.
75. Demidovich B. P., Maron I. A. Fundamentals of computational mathematics. M.: Nauka, 1966, 664 p.
76. Demidovich B. P., Maron I. A., Shuvalova E. Z. Numerical methods of analysis. M.: Nauka, 1967, 368 p.
77. Jeffries G., Swirls B. Methods of mathematical physics. 3 vols. M.: Mir, 1970, 424 p.; 352 p.; 344 p.
78. Joseph D. Stability of fluid motion. M.: Mir, 1981, 638 p.
79. George A., Liu J. Numerical solution of large sparse systems of equations. M.: Mir, 1984, 333 p.
80. Ditman A. O., Savchuk V. D., Yakubov I. R. Methods of analogies in the aircraft aerodynamics. M.: Mechanical Engineering, 1987, 150 p.
81. Dmitrievsky A. A., Lysenko L. N. External ballistics. M.: Mechanical Engineering, 2005, 608 p.
82. Daubechies I. Ten lectures on wavelets, Izhevsk: Research Center "Regular and Chaotic Dynamics," 2001, 464 p.
83. Dow R. B. Fundamentals of the modern shells theory. M.: Nauka, 1964, 567 p.
84. Drakin I. I. Basics of unmanned aircraft designing based on economic efficiency. M.: Mechanicaml Engineering, 1973, 224 p.
85. Dulnev G. N., Parfenov V. G., Sigalov A. V. The use of computers to solve heat transfer problems. M.: Higher School, 1990, 207 p.
86. Dyakonov V. P. Computer mathematics. M.: Knowledge, 2001, 1296 p.
87. Dyachenko V. F. Basic concepts of computational mathematics. M.: Nauka, 1977, 128 p.
88. Daniel K. Application of statistics in industrial experiment. M.: Mir, 1979.
89. Dyunze M. F., Zhimolokhin V. G. Solid fuel rocket engines for space systems. M.: Mechanical Engineering, 1982.
90. Emelyanov V. N. Introduction to the theory of difference schemes, St. Petersburg: Ed. BSTU, 2006, 192 p.
91. Erokhin B. T. Theory of intrachamber processes and designing of solid fuel rocket engines. M.: Mechanical Engineering, 1991, 560 p.
92. Erokhin B. T. Theory, analysis and design of rocket engines. M.: MGAPI, 2004, 864 p.
93. Erokhin B. T., Bogoslovsky V. N. Theory of heat and mass transfer processes and design of solid fuel rocket launch. M.: Leader-M, 2008, 382 p.
94. Erokhin B. T., Lipanov A. M. Unsteady and quasi-stationary modes of solid fuel rocket engines. M.: Mechanical Engineering, 1977, 200 p.
95. Zavyalov Yu. S., Kvasov B. I., Miroshnichenko V. L. Methods of spline functions. M.: Nauka, 1980, 352 p.
96. Zach D. Visual Basic.net. Self-teaching guide, St. Petersburg: Peter, 2003, 558 p.
97. Zarubin V. S. Mathematical modeling in engineering: Textbook for HEIs. Ed. by V. S. Zarubin, A. P. Krishchenko M.: ed. Bauman MSTU, 2003, 496 p.
98. Zeldovich Ya. B., Leipunsky O. I., Librovich V. B. Theory of unsteady combustion of gunpowder. M.: Nauka, 1975, 131 p.
99. Zubov V. I. Dynamics of the controlled systems. M.: Higher School, 1982, 200 p.
100. Ievlev V. M. Turbulent motion of high temperature continuous media. M.: Nauka, 1975, 256 p.
101. Ievlev V. M. Numerical simulation of turbulent flows. M.: Nauka, 1990, 216 p.

102. Ikramov Kh. D. Numerical solution of matrix equations. M.: Nauka, 1984, 192 p.
103. Irtegov D. V. Introduction to operating systems, St. Petersburg: BHV – Saint Petersburg, 2002, 624 p.
104. Investigation of solid fuel rocket engine. Transl. from English. Ed. By M. Summerfield. M.: Publishing House of foreign lit., 1963, 440 p.
105. Joss J., Joseph D. The elementary theory of stability and bifurcations. M.: Mir, 1983, 301 p.
106. Kavtaradze R. Z. Local heat transfer in reciprocating engines. M.: ed. Bauman MSTU, 2001, 592 p.
107. Kalinin V. V., Kovalev Yu. N., Lipanov A. M. Unsteady processes and design methods of units of SFREs. M.: Mechanical Engineering, 1986, 216 p.
108. Kamke E. Reference on ordinary differential equations. M.: Nauka, 1976, 576 p.
109. Kaplun A. B., Morozov E. M., Olferyeva M.A. ANSYS in the hands of an engineer: A practical guide, M: Editorial URSS, 2004, 272 p.
110. Karmanov V. G. Mathematical programming. M.: Fizmatlit 2004, 264 p.
111. Kireev V. I., Vaynovsky A. S. Numerical simulation of gas-dynamic flows. M.: ed. MAI, 1991, 254 p.
112. Kiryanov D. V., Kiryanova E. N. Computational physics. M.: ed. Polibuk Multimedia, 2006, 352 p.
113. Kovenya V. M., Yanenko N. N. Splitting method in problems of gas dynamics, Novosibirsk: Nauka CO, 1981, 304 p.
114. Computers and nonlinear phenomena: Informatics and modern science. Edited and compiled by A. A. Samarsky. M.: Nauka, 1988, 192 p.
115. Computers, models, computational experiment. Introduction to computer science from the standpoint of mathematical modeling. Edited and compiled by A. A. Samarsky. M.: Nauka, 1988, 172 p.
116. Konovalov A. A. Theory of technical systems. Marketing aspect, Yekaterinburg: UIF "Nauka," 1993, 312 p.
117. Korn G., Korn T., Mathematical handbook for scientists and engineers. M.: Nauka, 1974, 720 p.
118. Korolev L. N. The structure of computers and mathematical software. M.: Nauka, 1978, 352 p.
119. Crouch S., Starfield A. Boundary element methods in solid mechanics. M.: Mir, 1987, 328 p.
120. Krutov V. I. Automatic regulation and control of internal combustion engines. M.: Mechanical Engineering, 1989, 416 p.
121. Krutko P. D. Inverse problems of the dynamics of automatic control theory. Series of lectures. M.: Mechanical Engineering, 2004, 576 p.
122. Kuznetsov A. A. Optimization of parameters of ballistic missiles by efficiency. M.: Mechanical Engineering, 1986, 160 p.
123. Kuzmin M. P. Electrical simulation of unsteady heat transfer processes. M.: Energiya, 1974, 416 p.
124. Kuzmin M. P., Lagun I. M. Unsteady thermal conditions of structural elements of aircraft engines. M.: Mechanical Engineering, 1980, 240 p.
125. Kulikovsky A. G., Pogorelov N. V., Semenov A. Yu. Mathematical problems of the numerical solution of hyperbolic systems. M.: FIZMATLIT 2001, 608 p.
126. Kutateladze S. S. Analysis of similarity in thermal physics, Novosibirsk: Nauka, 1982, 280 p.

127. Kutateladze S. S., Leontiev A. I. Heat and mass transfer and friction in a turbulent boundary layer. M.: Énergoizdat, 1972, 342 p.
128. Kutergin V. A. Artificial objects and constructive processes, Izhevsk: ed. IPM RAS, 2007, 551 p.
129. Kuhling H. Reference-book on physics M.: Mir, 1982.
130. Kert B. E., Kozlov V. I., Makarovets N. A. Mathematical modeling and experimental development of separation systems of rocket projectiles, Tula: ed. FSUE "SSPE Splav," 2006, 652 p.
131. Lavrentiev M. A., Shabat B. V. Methods of complex variable theory. M.: Nauka, 1973, 736 p.
132. Lavrentiev M. A., Shabat B. V. Problems of hydrodynamics and their mathematical models. M.: Nauka, 1973, 416 p.
133. Landau L. D., Lifshitz E. M. Theoretical physics. 10 vols. M.: Nauka, 1986.
134. Lipanov A.M. Theoretical hydromechanics of Newtonian fluids. M.: Nauka, 2011, 551 p.
135. Lipanov A. M. Physical-chemical and mathematical models of the combustion of mixed fuels. Preprint, Izhevsk: IPM RAS, 2007, 112 p.
136. Lipanov A. M., Aliev A. V. Designing solid fuel rocket engines. M.: Mechanical Engineering, 1995, 400 p.
137. Lipanov A. M., Kisarov Yu. F., Kluchnikov I. G. Numerical experiment in classical fluid mechanics of turbulent flows, Ekaterinburg: Ural Branch of RAS, 2001, 164 p.
138. Loytsansky L. G. Fluid and gas mechanics. M.: Nauka, 1970, 904.
139. Losev S. A., Umansky S. Ya., Yakubov I. T. Physical-chemical processes in gas dynamics. Reference-book. 2 vols. M.: ed. Research Institute of Mechanics, MSU, 1995, 350 p. + 368 p.
140. Lykov A. V. Heat and mass transfer (reference), M. Energiya, 1972, 560 p.
141. Makarov E. G. Engineering calculations in MathCad 14, St. Petersburg: Peter, 2007, 592 p.
142. McCracken D., Dorn W. Numerical methods and programming in FORTRAN. M.: Mir, 1977, 590 p.
143. Maksimei I. V. Simulation modeling on computer simulation. M.: Radio and Communications, 1988, 232 p.
144. Marchuk G. I. Methods of computational mathematics. M.: Nauka, 1977, 456 p.
145. Marchuk G. I. Theoretical model of weather forecast. Report of the Academy of Sciences USSR, 1964. Vol. 155, № 5.
146. Marchuk G. I., Shaidurov V. V. Improving the accuracy of solutions of difference schemes. M.: Nauka, 1979, 320 p.
147. The mathematical theory of combustion and explosion. Ya. B. Zeldovich, G. I. Barenblatt, V. B. Librovich, G. M. Makhviladze M.: Nauka, 1980, 478 p.
148. Mathematical modeling of high-temperature processes in power plants. V. E. Alemasov, A. F. Dregalin, V. G. Kryukov, V. I. Naumov M.: Nauka, 1989, 256 p.
149. Methods for calculating turbulent flows: Trans. from English. Ed. by V. Coleman. M.: Mir, 1984, 464 p.
150. Mityukov N. V. Simulation modeling in military history. M.: ed. LCI, 2007, 280 p.
151. Modeling of separated flows on computer. O. M. Belotserkovsky, S. M. Belotserkovsky, Yu. M. Davydov, M. I. Nisht. M.: Scientific Council on the complex problem "Cybernetics" USSR Academy of Sciences, 1984.

152. Modeling of combustion processes of solid fuels. L. K. Gusachenko, V. E. Zarko, V. Ya. Ziryanov, V. P. Bobrishev; ed. by G. V. Sakovich, Novosibirsk: Nauka, 1985, 182 p.
153. Models for calculating the natural frequencies of acoustic oscillations in the SFRE chamber. A. V. Aliev, O. V. Mishchenkova, V. I. Sarabiev, V. I. Babin. Rocket engines and the problems of space exploration. M.: Torus Press, 2005, pp. 295–303.
154. Moiseev N. N. Mathematical problems of system analysis, Moscow Book House "LIBROKOM," 2011, 488 p.
155. Moiseev N. N., Ivanilov Yu. P., Stolyarov E. M. Methods of optimization. -M.: Nauka, Main edition of physical and mathematical literature, 1978, 352 p.
156. Moiseeva N. K. Functional and value analysis in mechanical engineering. M.: Mechanical Engineering, 1987, 320 p.
157. Molchanov A. A. Modeling and design of complex systems, Kiev: Higher School, 1988.
158. Nonlinear effect of formation of self-sustaining high temperature gas layer in nonstationary processes of magnetic hydrodynamics. Tikhonov A. N., Samarsky A. A., Zaklyazminsky L. A., et al. Report of the Academy of Sciences USSR, 1967, Vol.173, № 4. 808–811 pp.
159. Nemnyugin M. A., Stesik O. L. Modern Fortran. Teach yourself, St. Petersburg: BHV – Saint Petersburg, 2004, 496 p.
160. Novikov F. A., Yatsenko A. D. Microsoft Office as a whole. St. Petersburg: BHV – Saint Petersburg, 1995, 336 p.
161. Novozhylov B. V. Unsteady combustion of solid propellants. M.: Nauka, 1973, 176 p.
162. Norrie D., de Vries J. Introduction to finite element method. M.: Mir, 1981, 304 p.
163. Ovsyannikov L. V. Lectures on the basics of gas dynamics. M.: Nauka, 1981, 368 p.
164. Oran E., Boris J. Numerical simulation of reacting flows. M.: Mir, 1990, 660 p.
165. Controls thrust vectoring missiles: Design schemes, calculation, experiment. R. V. Antonov, V. I. Grebyonkin, N. P. Kuznetsov, et al., Moscow – Izhevsk: Research Center "Regular and Chaotic Dynamics," 2006, 550 p.
166. Orlovsky S. A. Decision-making under the information ambiguity. M.: Nauka, 1981.
167. Ortega J. Introduction to parallel and vector methods for solving linear systems. M.: Mir, 1991, 367 p.
168. Fundamentals of the theory of automatic control of missile propulsion systems. A. I. Babkin, S. V. Belov, N. B. Rutkovsky, E. V. Soloviev M.: Mechanical Engineering, 1986, 453 p.
169. Application packages. Functional content; Ed. By A. A. Samarsky. M.: Nauka, 1986, 140 p.
170. Patankar S. Numerical methods for solving the problems of heat transfer and fluid dynamics. M.: Energoatomizdat, 1984, 152 p.
171. Pashkov L. T. Fundamentals of the combustion theory. Textbook. M.: ed. MEI 2002, 136 p.
172. Peregudov F. I., Tarasenko F. P. Introduction to system analysis. M.: Higher School, 1989.
173. Petrovsky I. G. Lectures on the theory of ordinary differential equations. M.: Nauka, 1970, 279 p.
174. Polovinkin A. A. Fundamentals of engineering creativity. M.: Mechanical Engineering, 1988.
175. Pravdin V. M., Shanin A. P. Ballistics of unguided aircraft, Snezhinsk: ed. VNIITF, 1999, 496 p.

176. Prisnyakov V. F. Dynamics of the solid fuel rocket engines. Textbook for universities. M.: Mechanical Engineering, 1984, 248 p.
177. Problems of corrected and guided aerial bombs creation; Ed. By E. S. Shahidzhanova. M.: Engineer, 2003, 528 p.
178. Pugachev V. S. Probability theory and mathematical statistics. M.: 1979, 496 p.
179. Razorenov G. N., Bakhramov E. A., Titov Yu. F. Aircraft control systems (ballistic missiles and their warheads). M.: Mechanical Engineering, 2003, 584 p.
180. Solid fuel rocket engines with adjustable traction module. V. I. Petrenko, V. L. Popov, A. M. Rusak, V. I. Feofilaktov – Miass: Publishing House SRC "Acad. V. P. Makeev Design Bureau," 1994, 246 p.
181. Adjustable solid fuel propulsion systems: Methods for calculating workflow experimental studies. V. G. Zezin, V. I. Petrenko, V. L. Popov, et al.; Ed. by V. I. Petrenko – Ufa: Publishing House "Dauria," 1996, 296 p.
182. Adjustable power plants for solid propellants. A. A. Kimyaev, V. I. Petrenko, V. L. Popov, S. G. Yarushin – Perm: ed. PGTU, 1999, 168 p.
183. Solution of one-dimensional gas dynamics problems in mobile grids. G. B. Alalykin, S. K. Godunov, N. L. Kireeva, L. A. Pliner M.: Nauka, 1970, 112 p.
184. Rozhdestvensky B. L., Yanenko N. N. System of quasilinear equations. M.: Nauka, 1978, 688 p.
185. Roach P. Computational fluid dynamics. M.: Mir, 1980, 616 p.
186. Rizhikov Yu. I. Modern Fortran: Textbook, St. Petersburg.: CORONA print, 2007, 288 p.
187. Ryabenky V. S. The method of difference potentials for some problems of continuum mechanics. M.: Nauka, 1987, 320 p.
188. Savelyev Yu. P. Lectures on the flight dynamics equations and external ballistics. 2 books, St. Petersburg, 2003, 352 p. + 575 p.
189. Samarsky A. A. Mathematical modeling and computational experiment. Bulletin of the Academy of Sciences of the USSR, 1979. № 5, 38–49 pp.
190. Samarsky A. A. Theory of difference schemes. M.: Nauka, 1977, 656 p.
191. Samarsky A. A., Andreev V. B. Difference methods for elliptic equations. M.: Nauka, 1976, 352 p.
192. Samarsky A. A., Gulin A. V. Stability of difference schemes. M.: Nauka, 1973, 416 p.
193. Samarsky A. A., Gulin A. V. Numerical methods. M.: Nauka, 1989, 432 p.
194. Samarsky V. A., Dorodnitsyn V. A., Kurdumov S. P., Popov Yu. P. Formation of T-layers during braking plasma by magnetic field. Report of Academy of Sciences of USSR, 1967. Vol. 216, № 6, 1254–1257 pp.
195. Samarsky A. A., Mikhailov A. P. Mathematical modeling: Ideas. Methods. Examples. M.: Nauka. Fizmatlit, 1997, 320 p.
196. Samarsky A. A., Nikolaev E. S. Methods for solving grid equations. M.: Nauka. Fizmatlit, 1978, 591 p.
197. Samarsky A. A., Popov Yu. P. Difference schemes of gas dynamics. M.: Nauka. Fizmatlit, 1975, 350 p.
198. Collection of scientific Fortran programs. In 2 volumes, Trans. from English. M.: Statistics, 1974.
199. Sebisi T., Bradshaw P. Convective heat transfer. M.: Mir, 1987, 591 p.
200. Sedov L. I. Similarity and dimensional methods in mechanics. M.: Nauka, 1967.
201. Sedov L. I. Continuum mechanics. 2 vols. M.: Nauka, 1970.
202. Smirnov V. I. Course of higher mathematics. 4 vols. M.: Nauka, 1974.

203. Solovey E. Ya. Dynamics of targeting systems of guided air bombs; ed. by E. S. Shahidzhanova. M.: Mechanical Engineering, 2006, 328 p.
204. Sorkin R. E. Gas dynamics of solid fuel rocket engine. M.: Nauka, 1967, 368 p.
205. Sorkin R. E. Theory of intrachamber processes in solid fuels missile systems. M.: Nauka, 1983, 288 p.
206. Spolding D. B. Combustion and mass transfer. M.: Mechanical Engineering, 1985, 238 p.
207. Reference-book for engineers and students: Higher mathematics. Physics. Theoretical mechanics. Strength of materials. Polyanin A. D., Polyanin V. D., Popov V. A., et al. M.: International Education Program, 1996, 432 p.
208. Stechkin S. B., Subbotin Yu. N. Splines in computational mathematics. M.: Nauka, 1976, 248 p.
209. Tanenbaum E. Computer architecture, St. Petersburg: Peter, 2003, 704p.
210. Solid fuel propulsion units. Yu. S. Solomonov, A. M. Lipanov, A. V. Aliev, et al. M.: Mechanical Engineering, 2011, 416 p.
211. Tenenev V. A., Yakimovich B. A. Genetic algorithms in system modeling, Izhevsk: Publishing House of ISTU, 2010, 308 p.
212. Thermal decomposition and combustion of explosives and powders. G. B. Manelis, G. M. Nazin, Yu. I. Rubtsov, V. A. Strunin. M.: Nauka, 1996, 223 p.
213. Technology of system modeling. G. A. Avramchuk, A. A. Vavilov, S. V. Emelyanov et al. Ed. by S. V. Emelyanov, et al. M.: Mechanical Engineering; Berlin: Tekhnik, 1988, 520 p.
214. Thompson J. M. T. Instability and catastrophes in science and technology. M.: Mir, 1985, 254 p.
215. Tyurin Yu. N. Statistical methods of analysis of expert assessments. M.: Nauka, 1977.
216. Wild D. Optimal design. M.: Mir, 1981, 272 p.
217. Umnyashkin V. A., Sazonov V. V., Filkin N. M. Operational characteristics of the car. Textbook, Izhevsk: Publishing House of ISTU, 2002, 180 p.
218. Controlled power plants for solid propellants. V. I. Petrenko, M. I. Sokolovsky, G. A. Zykov, et al. M.: Mechanical Engineering, 2003, 464 p.
219. Urmaev A. S. Fundamentals of modeling on analog computers. M.: Nauka, 1974, 320 p.
220. Stability of engines workflow in aircraft. M. A. Ilchenko, V. V. Kryutchenko, Yu. S. Mnatsakanian, et al. M.: Mechanical Engineering, 1995, 320 p.
221. Fedorenko R. P. Introduction to computational physics. M.: MIPT, 1994, 528 p.
222. Fedorov V. A., Milman O. O. Thermal-hydraulic oscillations and instability in heat exchange systems with two-phase flow. M.: ed. MEI, 1998, 244 p.
223. Feynman R., Leighton R., Sands M. Feynman lectures on physics. 3, 4 vols. M.: Mir, 1976, 496 p.
224. Physical quantities: Reference-book. A. N. Babichev, N. A. Babushkina, A. M. Bratkovsky, et al.; Ed. by I. S. Grigoriev, E. Z. Meylihova. M.: Energoatomizdat, 1991, 1232 p.
225. Fishburn B. Utility theory for decision-making. M.: Nauka, 1978.
226. Fletcher K. Computational methods in fluid dynamics. In 2 vols. M.: Mir, 1991, 502 p. + 552 p.
227. Forsythe J., Malcolm M., Mouler K. Machine methods of mathematical calculations. M.: Mir, 1980, 280 p.
228. Forsythe J., Mouler K. Numerical solution of systems of linear algebraic equations. M.: Mir, 1969, 168 p.
229. Huntley G. Dimensional analysis. M.: Mir, 1970.
230. Hemming R. V. Numerical methods. M.: Mir, 1972, 400 p.
231. Hill P. Science and art of designing. M.: Mir, 1973, 263 p.

232. Chemical thermodynamics. A. M. Kutepov, A. D. Polyanin, Z. D. Zapryanov, et al. M.: Bureau Quantum, 1996, 336 p.
233. Himmelblau D. Applied nonlinear programming: Trans. from English, Bykhovskaya I. M., Vavilov B.T. Ed. by M. L. Byhovsky M.: Mir, 1975, 536 p.
234. Hitrik M. S., Fedorov S. M. Dynamics of control systems of missiles with onboard digital computers. M.: Mechanical Engineering, 1972.
235. Hockney R., Eastwood J. Numerical simulation by the particle method. M.: Mir, 1987, 640 p.
236. Tsisar I. F., Neiman V. G. Computer modeling in economics. M.: Dialog-MIFI 2002, 304 p.
237. Chernorutsky I. G. Methods of optimization in control theory, St. Petersburg: Peter, 2004, 256 p.
238. Chesnokova O. V. Delphi 2007. Algorithms and programs. Ed. by Alekseev E. R. M.: NT Press, 2008, 368 p.
239. Numerical study of modern problems in gas dynamics; Ed. by O. M. Belotserkovsky M.: Nauka, 1974, 398 p.
240. Numerical modeling in aerohydrodynamics; Ed. by G. G. Cherny M.: Nauka, 1986, 263 p.
241. Numerical solution of multidimensional problems of gas dynamics. S. K. Godunov, A. V. Zabrodin, M. Ya. Ivanov, et al. M.: Nauka, 1976, 400 p.
242. Numerical methods for the dynamics of the aircraft in conditions of aerodynamic interference. N. A. Baranov, A. S. Belotserkovsky, M. I. Kanevsky, L. I. Turchak. M.: Nauka, 2001, 207 p.
243. Numerical methods in fluid mechanics: Trans. from English. Ed. by O. M. Belotserkovsky. M.: Mir, 1973, 304 p.
244. Numerical experiment in the theory of solid fuel rocket engines. A. M. Lipanov, V. P. Bobryshev, A. V. Aliev, et al., Yekaterinburg: UIF "Nauka," 1994, 303 p.
245. Shenk H. Theory of engineering experiment. M.: Mir, 1972, 383 p.
246. Shishkov A. A. Gas dynamics of powder rocket engines. M.: Mechanical Engineering, 1974, 156 p.
247. Shishkov A. A., Rumyantsev B. V. Gasifies of missile systems. M.: Mechanical Engineering, 1981, 152 p.
248. Shishkov A. A., Panin S. D., Rumyantsev V. V. Workflows in solid fuel rocket engines. M.: Mechanical Engineering, 1989, 239 p.
249. Schlichting H. Boundary layer theory. M.: Nauka, 1974, 711 p.
250. Shup T. Solving of engineering problems on a computer. M.: Mir, 1982, 238 p.
251. Shchigolev B. M. Mathematical processing of observations. M.: Nauka, 1969, 344 p.
252. Experimental methods for determining the parameters of special purpose engines. I. M. Gladkof, V. S. Mukhammedov, E. L. Valuev V. I. Cherepov. M.: STC "Informtechnica," 1993, 300 p.
253. Althea J. Coombes M. Expert systems: Concepts and examples. M.: Finance and Statistics, 1987, 191 p.
254. Emmons H. Fundamentals of gas dynamics. M.: ed. for. lit., 1963, 703 p.
255. Condensed energy system. Short encyclopedic dictionary. Ed. by B. P. Zhukov Ed. 2nd, revisions. M.: Yanus-K, 2000, 596 p.
256. Yanenko N. N., Karnachuk V. I., Konovalov A. N. Problems of mathematical technology. Computational methods of continuum mechanics. 1977, № 8, 129–157 p.

EPILOGUE

The last pages of the book are finished, but it is impossible to say that the authors solved all the issues mentioned in the preface. We would like to return to one or another chapter and add to it. First of all, this is connected with the violent development of such subject as "mathematical modeling." Thus, the sections linked with the description of specialized software packages created, for instance, to solve the problems of experiment processing, to solve the problems of continuum mechanics, etc. are provoked. Problems of choice (Chapter 4) are extremely important in engineering. The free hold of approaches which allow solving the selection problems requires the more detailed acquaintance with the methods of mathematical programming. The number of practical examples in this chapter can be increased. From the point of the subject "mathematical modeling" the dynamic models described in Chapter 5 can become the subject of a separate book. It can be predicted that using the dynamic models, the possibility of developing the evolutionary theory of Universe development combining the micro-world and macro-world together will appear in the nearest future.

Let us point out one important thing. Multi-processor computers are the architecture of the modern and future computer engineering. For such equipment, the necessity in applying the so-called "parallel algorithms" which would allow increasing the efficiency of using the multi-processor equipment increases. This aspect was touched upon when writing the book, however, it requires more attention.

In our opinion, the book advantages are as follows:

- mathematics – "queen of sciences," and its possibilities in the application to solve different practical problems need to be promoted, and we did it;
- mathematics – beautiful science, however, mathematicians, as a rule, do not understand engineers, and engineers do not understand mathematicians. The conversation between a mathematician and engineer can occur in the presence of "an interpreter." Our aim is to be "interpreters;"
- mathematics and mathematical modeling will live and develop since they provide the solution of not only quantitative but also qualitative

problems. It is difficult to overestimate the role of these subjects in XXI century, when the processors are tried to be built in all engineering objects (actually, computers with the applied software), which control the behavior of this equipment maintaining their normal operation, under the action of various accidental disturbances, as well.

Our work on developing the issues described in this book will continue by all means!

INDEX

G

H

U

V

W

Z